10th Anniversary of Cells—Advances in Plant, Algae and Fungi Cell Biology

10th Anniversary of Cells—Advances in Plant, Algae and Fungi Cell Biology

Editor

Suleyman Allakhverdiev

MDPI

Basel • Beijing • Wuhan • Barcelona • Belgrade • Novi Sad • Cluj • Manchester

Editor
Suleyman Allakhverdiev
Laboratory of Controlled
Photobiosynthesis
Institute of Plant Physiology,
RAS
Moscow
Russia

Editorial Office
MDPI
St. Alban-Anlage 66
4052 Basel, Switzerland

This is a reprint of articles from the Special Issue published online in the open access journal *Cells* (ISSN 2073-4409) (available at: www.mdpi.com/journal/cells/special_issues/10th_Anniversary_Cells_Advances_in_Plant_Algae_Fungi_Cell_Biology).

For citation purposes, cite each article independently as indicated on the article page online and as indicated below:

Lastname, A.A.; Lastname, B.B. Article Title. *Journal Name* **Year**, *Volume Number*, Page Range.

ISBN 978-3-0365-9145-2 (Hbk)
ISBN 978-3-0365-9144-5 (PDF)
doi.org/10.3390/books978-3-0365-9144-5

Contents

About the Editor

Suleyman Allakhverdiev

Suleyman I. Allakhverdiev is the Head of the Controlled Photobiosynthesis Laboratory at the K.A. Timiryazev Institute of Plant Physiology of the Russian Academy of Sciences (RAS). He obtained his BS and MS in Physics from the Department of Physics, Azerbaijan State University, Baku. He obtained his Dr. Sci. Degree in Plant Physiology and Biochemistry from the Institute of Plant Physiology, RAS (2002, Moscow). In 2018, he was recognized as being first in the ranking of the most highly cited scientists in Russia. In 2019, he was the winner of the K.A. Timiryazev Prize of the Russian Academy of Sciences. In 2021, he became the laureate of "Global Energy Prize" in the "Non-conventional energy" category for his outstanding contribution to the development of alternative energy, particularly his design for a system of artificial photosynthesis and for his cycle of scientific works on bioenergy and hydrogen energy. In 2022, The Honored Scientist of the Russian Federation Title was given to him as an assessment of his outstanding contribution to plant biology development and the study of the fundamental principles of photosynthesis. In 2022, he was elected as a corresponding member of the Russian Academy of Sciences. In 2018–2023, he was recognized by Thomson Reuters-WoS (Clarivate Analytics) as one of the Highly Cited world-class researchers selected for their exceptional research performance, and entered the list of the most highly cited scientists of the world (1% of the total number of researchers).

Preface

Welcome to this reprint commemorating the "10th Anniversary of Cells—Advances in Plant, Algae, and Fungi Cell Biology." This compilation is a testament to a decade of groundbreaking research and scientific excellence in the realms of plant, algae, and fungi cell biology. Within these pages, we invite you to embark on a journey through the remarkable discoveries that have shaped our understanding of these diverse and vital biological domains. This reprint represents the transformation of a remarkable Special Issue into a cohesive volume, offering an expanded exploration of the subject matter. It serves as both a reflection of the progress made over the past ten years and a glimpse into the promising future of cellular biology. The articles within this reprint encompass a rich tapestry of topics, from genomics to cellular defense mechanisms, and from the intricacies of mycorrhizal fungi to the physiology of algae thriving in extreme environments. Each article is a testament to the tireless dedication of researchers who have contributed their expertise to advance our knowledge in these areas. As we celebrate this milestone, we extend our heartfelt gratitude to the authors whose commitment to scientific inquiry has made this reprint possible. Their rigorous work and dedication to the field are evident in the depth and quality of their contributions. We also acknowledge the indispensable role played by our reviewers, whose expertise and insights have shaped the content within these pages. This reprint is intended to be a valuable resource for scientists, researchers, educators, and anyone passionate about the fascinating world of plant, algae, and fungi cell biology. We hope that the insights shared within will inspire future generations of scientists and contribute to further advancements in this dynamic field. In closing, we invite you to immerse yourself in the pages of this reprint and explore the wealth of knowledge it contains. It is a celebration of the remarkable progress made over the past decade and a testament to the limitless potential of scientific exploration. Thank you for joining us in commemorating the "10th Anniversary of Cells—Advances in Plant, Algae, and Fungi Cell Biology."

Suleyman Allakhverdiev
Editor

cells

MDPI

Editorial

10th Anniversary of Cells—Advances in Plant, Algae and Fungi Cell Biology

Suleyman I. Allakhverdiev

K.A. Timiryazev Institute of Plant Physiology, Russian Academy of Sciences, Botanicheskaya Street 35, Moscow 127276, Russia; suleyman.allakhverdiev@gmail.com

check for updates

Citation: Allakhverdiev, S.I. 10th Anniversary of Cells—Advances in Plant, Algae and Fungi Cell Biology. *Cells* **2022**, *11*, 3759. https://doi.org/10.3390/cells11233759

Received: 17 November 2022
Accepted: 21 November 2022
Published: 24 November 2022

Publisher's Note: MDPI stays neutral with regard to jurisdictional claims in published maps and institutional affiliations.

In 2021, the 10th anniversary of the publication of *Cells* occurred. To date, the journal has published more than 4500 papers, and the journal website attracts more than 60,000 monthly page views. To mark this important milestone, we organized a Special Issue entitled "10th Anniversary of Cells—Advances in Plant, Algae and Fungi Cell Biology". For this Special Issue, we collected high-quality research articles in relevant research fields.

Cyanobacteria from the genus *Arthrospira/Limnospira* are considered haloalkalo-tolerant organisms with optimal growth temperatures of approximately 35 °C. They are most abundant in soda lakes in tropical and subtropical regions [1]. Indeed, the paper by Misztak et al. is devoted to the comprehensive genome-based characterization and physiological investigation of the new strain O9.13F, which was isolated from the winter freezing Solenoye Lake in Western Siberia [1]. Comparative genomics showed that no unique genes were found for the Siberian strain related to its tolerance to low temperatures and high salinity, while the fatty acid composition was specific and unique to this strain. The authors showed that the optimal cultivation temperature for strain O9.13F (20 °C) is lower than that for *Arthrospira/Limnospira* strains (35 °C) [1].

Coccoids are plant sap-sucking hemipterans that are considered serious pests in agriculture, horticulture, and forestry [2]. Szklarzewicz and coauthors [2] investigated the symbiotic systems of eight Polish species of scale insects of the Coccidae family: *Parthenolecanium corni*, *Parthenolecanium fletcheri*, *Parthenolecanium pomeranicum*, *Psilococcus ruber*, *Sphaerolecanium prunasti*, *Eriopeltis festucae*, *Lecanopsis formicarum* and *Eulecanium tiliae*. The histological, ultrastructural and molecular analyses showed that all these species host fungal symbionts in fat body cells. Analyses of ITS2 and beta-tubulin gene sequences, as well as fluorescence in situ hybridization, confirmed that they should all be classified into the genus *Ophiocordyceps*. *Ophiocordyceps* fungi are commonly known as virulent, specialized entomopathogens. The essential role of the fungal symbionts observed in the biology of the soft scale insects examined was confirmed by their transovarial transmission between generations [2].

Another thing to note is that the arbuscular mycorrhizal fungus *Glomus Viscosum* proves tolerant to *Verticillium* wilt in artichoke by modulating antioxidant defense systems [3]. The artichoke (*Cynara scolymus* L.) is a horticultural species of relevant economic interest belonging to the Asteraceae family that is widely cultivated in the Mediterranean basin and widespread throughout the world [3]. Several studies have demonstrated that even some nonfood byproducts of artichokes are widely used as hepatoprotective, antioxidant, anticarcinogenic, hypoglycemic, and hypocholesterolemic agents (see [3]). The health-promoting properties and important nutritional values of artichokes have been extensively related to the high content of some bioactive phenolic compounds, such as caffeoylquinic acid derivatives and flavonoids, showing strong scavenging activity against reactive oxygen species (ROS) and free radicals [3]. On the other hand, Verticillium wilt, caused by the fungal pathogen *Verticillium dahliae*, is the most severe disease threatening artichoke (*Cynara scolymus* L.) plants. Villani et al. evaluated the effect of the AMF *Glomus viscosum Nicolson* in enhancing plant tolerance towards the pathogen *V. dahliae*. The role of the antioxidant systems involved in the complex network of the pathogen–fungi–plant interaction was investigated. The results obtained showed that the AMF-inoculated plants exhibited significant increases in the activities of antioxidant enzymes, a higher content of ascorbate (ASC) and glutathione (GSH), and a decrease in the levels of lipid peroxidation and hydrogen peroxide (H_2O_2). Hence, *G. viscosum* may represent an effective strategy for mitigating *V. dahliae* pathogenicity in artichokes, improving nutritional value and benefiting human health [3].

The chlorococcal green alga *Parachlorella kessleri* is a favorable organism, as it can produce both starch and neutral lipids, and *P. kessleri* commonly divides into more than two daughter cells by means of a specific mechanism—multiple fission. Zachleder et al. used synchronized cultures of algae to study the effects of supra-optimal temperatures [4]. Synchronized cultures were grown at optimal (30 °C) and supra-optimal (40 °C) temperatures and incident light intensities of 110 and 500 μmol photons $m^{-2}s^{-1}$. The time course of cell reproduction (*DNA* replication, cellular division), growth (total *RNA*, protein, cell dry matter, cell size), and synthesis of energy reserves (net starch, neutral lipid) was studied [4]. At 40 °C, cell reproduction was arrested, but the growth and accumulation of energy reserves continued; this led to the production of giant cells enriched in protein, starch, and neutral lipids [4].

The plant hormone cytokinin (CK) is central to plant life, regulating many processes. Pizarro et al. investigated the effects of CK on cellular trafficking and on the cytoskeleton in plant cells [5]. They found that in addition to the xylanase receptor-like protein (RLP) LeEIX2, CK affects the distribution of the flagellin receptor-like kinase (RLK) flagellin-sensing 2 (FLS2), increasing its presence in the plasma membrane. FLS2, first characterized in Arabidopsis, acts as the PRR for the bacterial PAMP (pathogen-associated molecular pattern) flagellin in several plant species (see [5]). Examining cellular trafficking compartments and the cytoskeleton in general, the authors showed that CK affects endosome distribution and increases the number of endomembrane compartments. CK also caused disorganization and reduction in actin filaments, but not tubulin. The results are in agreement with those previously reported for fungi, suggesting a fundamental role for CK in regulating cellular integrity and trafficking as a mechanism for controlling and executing CK-mediated processes [5].

The identification of putative virulence genes by means of DNA methylation studies in the cereal pathogen *Fusarium graminearum* was described by Tini et al. [6]. The *DNA* isolated from SC50 and SC50 × 3 was subjected to a methylation content-sensitive enzyme and a double-digest, restriction-site-associated *DNA* technique (ddRAD-MCSeEd). *DNA* methylation analysis indicated 1024 genes whose methylation levels changed in response to inoculation on a healthy host after subculturing. These results demonstrate that the physiological shifts following subculturing have an impact on genomic *DNA* methylation levels and suggest that the ddRAD-MCSeEd approach can be an important tool for detecting genes potentially related to fungal virulence [6].

Ivanov et al. examined the potential for pilot-scale starch production in *C. reinhardtii* using a supraoptimal temperature, a method that has already been proven to cause a rapid 2-fold increase in starch yields under laboratory conditions [7]. The experiments described in this paper are the first attempt to employ a supraoptimal temperature in the production of starch in microalgae at the pilot scale and showed that exposure to a supraoptimal temperature (39 °C) causes a complete block of nuclear and cellular division accompanied by an increased accumulation of starch. The method was successfully applied and resulted in an almost 3-fold increase in the starch content of *C. reinhardtii* dry matter. Nevertheless, technical challenges, such as bioreactor design and light availability within the culture, still need to be addressed (see [7]).

It has been challenging to simultaneously improve photosynthesis and stress tolerance in plants. Liu et al. hypothesized that the ectopic expression of a CAM-specific phosphoenolpyruvate carboxylase (PEPC) would enhance both photosynthesis and abiotic stress tolerance [8]. Their experiments showed a higher photosynthetic rate and biomass production under normal conditions, along with significant carbon metabolism changes in malate accumulation, the carbon isotope ratio δ13C, and the expression of multiple orthologues of CAM-related genes. Furthermore, AaPEPC1 overexpression enhanced proline biosynthesis and improved salt and drought tolerance in the transgenic plants [8].

Zinc is implicated in numerous cellular processes, including cell division and DNA and protein synthesis, as well as in the catalytic activity of many enzymes. Two major membrane protein families facilitate zinc homeostasis in the animal kingdom, i.e., Zrt/Irt-like proteins and Zn transporters (ZnTs), which essentially conduct zinc flux in opposite directions. Human ZIPs (hZIPs) regulate the importing of extracellular zinc to the cytosol, being critical in preventing the overaccumulation of this potentially toxic metal and crucial for diverse physiological and pathological processes, including the development of neurodegenerative disorders and several cancers [9]. Becares et al. reported that yeast is able to produce four full-length hZIP members belonging to three different subfamilies [9]. One target (hZIP1) was purified in the high quantity and homogeneity required for downstream biochemical analysis.

The primary purpose of the review by Barre et al. is to give an overview of the types of glycans present in the glycan shield of different pathogenic enveloped viruses and how legume lectins with different specificities can act as carbohydrate-binding agents (CBAs) for these viruses. In addition, the biomedical and therapeutic potential of plant lectins as antiviral drugs has been discussed [10].

Plants regularly encounter a wide range of abiotic and biotic stresses in nature. Abiotic stress includes drought, salinity, extreme temperatures, radiation, floods, and heavy metals, whereas biotic stressors include insects, animal herbivores, and microbial pathogens [11]. Plants have evolved various strategies to detect herbivores and mount an effective defense system against them. Interestingly, ion channels or transporters are the first responders against herbivores. A review by Gandhi et al. described a comprehensive overview of the role of ion channels in early defense signaling against herbivorous insects [11].

A review by Hameed et al. was dedicated to the effects of salt stress on chloroplasts, their structures, and various biochemical reactions occurring in them [12]. The authors showed that the presence of salt in plant cells disrupts basic metabolic processes, contributing to severe negative effects on plant development and growth. For example, changes in chloroplast size, number, lamellar organization, lipid and starch accumulation interfere with cross-membrane transportation. Research has shown that the maintenance of normal chloroplast physiology is necessary for the survival of the entire plant [12].

The core abscisic acid (ABA) signaling pathway consists of receptors, phosphatases, kinases and transcription factors, among them ABA INSENSITIVE 5 (ABI5) and ABRE BINDING FACTORs/ABRE-BINDING PROTEINs (ABFs/AREBs), which belong to the BASIC LEUCINE ZIPPER (bZIP) family and control the expression of stress-responsive genes. ABI5 is mostly active in seeds and prevents germination and postgerminative growth under unfavorable conditions. The review by Collin et al. focuses on recent reports regarding ABI5 and ABF/AREB functions during abiotic stress [13]. They discuss the increased stress tolerance of transgenic plants overexpressing genes encoding ABA-dependent bZIPs. The authors suggest that ABI5 and ABFs/AREBs are crucial ABA-dependent transcription factors regulating processes essential for plant adaptation to stress at different developmental stages [13].

In the review paper by Sharkey [14], the pentose phosphate pathway (PPP) is divided into an oxidative branch that makes pentose phosphates and a nonoxidative branch that consumes pentose phosphates, although the nonoxidative branch is considered reversible. The reaction sequence in the Calvin–Benson cycle is the opposite of the typical direction of the nonoxidative PPP. This can be considered an anabolic version of the nonoxidative PPP. In addition to the strong association between the nonoxidative PPP and photosynthesis metabolism, there is recent evidence that oxidative PPP reactions form a shunt around the nonoxidative PPP section of the Calvin–Benson cycle. A constitutive operation of this shunt occurs in the cytosol and gives rise to an unusual labelling pattern of photosynthetic metabolites, while an inducible shunt in the stroma may occur in response to stress (see [14]).

We would like to express our sincerest thanks to our readers, innumerable authors, anonymous peer reviewers, editors, and all the people working in some way for the journal who have made substantial contributions for years. Without your support, we would never have made it.

For the details of the *Cells* 2021 Best Paper Awards for Anniversary Special Issues please click here https://www.mdpi.com/journal/cells/awards.pdf/0/pdf_32_2021_2_award.pdf (accessed on 28 February 2022).

Funding: This work was supported by the Russian Science Foundation (RNF) grants (19-14-00118 and 22-44-08001) and the state contract of the Ministry of Science and Higher Education of the Russian Federation (Project No. 122050400128-1).

Institutional Review Board Statement: Not applicable.

Informed Consent Statement: Not applicable.

Data Availability Statement: The datasets generated during and/or analyzed during the current study are available from the corresponding author on reasonable request.

Conflicts of Interest: The author declares no conflict of interest.

References

1. Misztak, A.E.; Waleron, M.; Furmaniak, M.; Waleron, M.M.; Bazhenova, O.; Daroch, M. Waleron KF. Comparative genomics and physiological investigation of a new *Arthrospira/Limnospira* strain O9.13F isolated from an alkaline, winter freezing, Siberian Lake. *Cells* **2021**, *10*, 3411. [CrossRef] [PubMed]
2. Szklarzewicz, T.; Michalik, K.; Grzywacz, B.; Kalandyk-Kołodziejczyk, M.; Michalik, A. Fungal associates of soft scale insects (*Coccomorpha: Coccidae*). *Cells* **2021**, *10*, 1922. [CrossRef] [PubMed]
3. Villani, A.; Tommasi, F.; Paciolla, C. The arbuscular mycorrhizal fungus *Glomus viscosum* improves the tolerance to verticillium wilt in artichoke by modulating the antioxidant defense systems. *Cells* **2021**, *10*, 1944. [CrossRef] [PubMed]
4. Zachleder, V.; Kselíková, V.; Ivanov, I.N.; Bialevich, V.; Vítová, M.; Ota, S.; Takeshita, T.; Kawano, S.; Bišová, K. Supra-optimal temperature: An efficient approach for overaccumulation of starch in the green alga *Parachlorella kessleri*. *Cells* **2021**, *10*, 1806. [CrossRef] [PubMed]
5. Pizarro, L.; Munoz, D.; Marash, I.; Gupta, R.; Anand, G.; Leibman-Markus, M.; Bar, M. Cytokinin Modulates Cellular Trafficking and the Cytoskeleton, Enhancing Defense Responses. *Cells* **2021**, *10*, 1634. [CrossRef] [PubMed]
6. Tini, F.; Beccari, G.; Marconi, G.; Porceddu, A.; Sulyok, M.; Gardiner, D.M.; Albertini, E.; Covarelli, L. Identification of putative virulence genes by DNA methylation studies in the cereal pathogen *Fusarium graminearum*. *Cells* **2021**, *10*, 1192. [CrossRef] [PubMed]
7. Ivanov, I.N.; Zachleder, V.; Vítová, M.; Barbosa, M.J.; Bišová, K. Starch production in *Chlamydomonas reinhardtii* through supraoptimal temperature in a pilot-scale photobioreactor. *Cells* **2021**, *10*, 1084. [CrossRef] [PubMed]
8. Liu, D.; Hu, R.; Zhang, J.; Guo, H.B.; Cheng, H.; Li, L.; Borland, A.M.; Qin, H.; Chen, J.G.; Muchero, W.; et al. Overexpression of an agave phosphoenolpyruvate carboxylase improves plant growth and stress tolerance. *Cells* **2021**, *10*, 582. [CrossRef] [PubMed]
9. Becares, E.R.; Pedersen, P.A.; Gourdon, P.; Gotfryd, K. Overproduction of human Zip (SLC39) zinc transporters in *Saccharomyces cerevisiae* for biophysical characterization. *Cells* **2021**, *10*, 213. [CrossRef] [PubMed]
10. Barre, A.; Van Damme, E.J.M.; Klonjkowski, B.; Simplicien, M.; Sudor, J.; Benoist, H.; Rougé, P. Legume lectins with different specificities as potential glycan probes for pathogenic enveloped viruses. *Cells* **2022**, *11*, 339. [CrossRef] [PubMed]
11. Gandhi, A.; Kariyat, R.; Harikishore, A.; Ayati, M.; Bhunia, A.; Sahoo, N. Deciphering the role of ion channels in early defense signaling against herbivorous insects. *Cells* **2021**, *10*, 2219. [CrossRef] [PubMed]
12. Hameed, A.; Ahmed, M.Z.; Hussain, T.; Aziz, I.; Ahmad, N.; Gul, B.; Nielsen, B.L. Effects of salinity stress on chloroplast structure and function. *Cells* **2021**, *10*, 2023. [CrossRef] [PubMed]
13. Collin, A.; Daszkowska-Golec, A.; Szarejko, I. Updates on the role of ABSCISIC ACID INSENSITIVE 5 (ABI5) and ABSCISIC ACID-RESPONSIVE ELEMENT BINDING FACTORs (ABFs) in ABA signaling in different developmental stages in plants. *Cells* **2021**, *10*, 1996. [CrossRef] [PubMed]
14. Sharkey, T.D. Pentose phosphate pathway reactions in photosynthesizing cells. *Cells* **2021**, *10*, 1547. [CrossRef] [PubMed]

cells MDPI

Article

Comparative Genomics and Physiological Investigation of a New *Arthrospira/Limnospira* Strain O9.13F Isolated from an Alkaline, Winter Freezing, Siberian Lake

Agnieszka E. Misztak [1,†,‡], Malgorzata Waleron [1,*,†], Magda Furmaniak [2], Michal M. Waleron [2], Olga Bazhenova [3], Maurycy Daroch [4] and Krzysztof F. Waleron [2,*]

1 Laboratory of Plant Protection and Biotechnology, Intercollegiate Faculty of Biotechnology UG and MUG, University of Gdansk, 80-307 Gdansk, Poland; amisztak@uliege.be
2 Department of Pharmaceutical Microbiology, Faculty of Pharmacy, Medical University of Gdansk, 80-416 Gdansk, Poland; magda.furmaniak@acteryon.com (M.F.); michal.waleron@gumed.edu.pl (M.M.W.)
3 Department of Ecology, Nature Management and Biology, Omsk State Agrarian University Named after P.A. Stolypin, 644008 Omsk, Russia; olga52@bk.ru
4 School of Environment and Energy, Peking University Shenzhen Graduate School, Shenzhen 518055, China; m.daroch@pkusz.edu.cn
* Correspondence: malgorzata.waleron@biotech.ug.edu.pl (M.W.); krzysztof.waleron@gumed.edu.pl (K.F.W.)
† Equal contribution.
‡ Current address: Unit of Animal Genomics, GIGA Institute, University of Liège, 4031 Liège, Belgium.

Citation: Misztak, A.E.; Waleron, M.; Furmaniak, M.; Waleron, M.M.; Bazhenova, O.; Daroch, M.; Waleron, K.F. Comparative Genomics and Physiological Investigation of a New *Arthrospira/Limnospira* Strain O9.13F Isolated from an Alkaline, Winter Freezing, Siberian Lake. *Cells* **2021**, *10*, 3411. https://doi.org/10.3390/cells10123411

Academic Editor: Suleyman Allakhverdiev

Received: 15 October 2021
Accepted: 1 December 2021
Published: 3 December 2021

Abstract: Cyanobacteria from the genus *Arthrospira/Limnospira* are considered haloalkalotolerant organisms with optimal growth temperatures around 35 °C. They are most abundant in soda lakes in tropical and subtropical regions. Here, we report the comprehensive genome-based characterisation and physiological investigation of the new strain O9.13F that was isolated in a temperate climate zone from the winter freezing Solenoye Lake in Western Siberia. Based on genomic analyses, the Siberian strain belongs to the *Arthrospira/Limnospira* genus. The described strain O9.13F showed the highest relative growth index upon cultivation at 20 °C, lower than the temperature 35 °C reported as optimal for the *Arthrospira/Limnospira* strains. We assessed the composition of fatty acids, proteins and photosynthetic pigments in the biomass of strain O9.13F grown at different temperatures, showing its potential suitability for cultivation in a temperate climate zone. We observed a decrease of gamma-linolenic acid favouring palmitic acid in the case of strain O9.13F compared to tropical strains. Comparative genomics showed no unique genes had been found for the Siberian strain related to its tolerance to low temperatures. In addition, this strain does not possess a different set of genes associated with the salinity stress response from those typically found in tropical strains. We confirmed the absence of plasmids and functional prophage sequences. The genome consists of a 4.94 Mbp with a GC% of 44.47% and 5355 encoded proteins. The *Arthrospira/Limnospira* strain O9.13F presented in this work is the first representative of a new clade III based on the 16S rRNA gene, for which a genomic sequence is available in public databases (PKGD00000000).

Keywords: *Arthrospira*; haloalkalotolerant cyanobacteria; metagenomics; phylogenomics; fatty acid

1. Introduction

Arthrospira Stizenberger ex Gomont 1892 [1]/*Limnospira* sensu Nowicka-Krawczyk [2] that was formerly called *Spirulina* [3] is a non-nitrogen fixing, filamentous cyanobacterium representing edible, haloalkalotolerant organisms. The genus *Arthrospira* has been observed to occur worldwide in a varied range of haloalkaline habitats, from freshwater alkaline conditions to hypersaline environments [1,4,5]. In tropical and subtropical regions, it forms exceptionally high blooms turning soda lakes into one of the world's most productive ecosystems [6]. The occurrence of *Arthrospira* has been reported all around the globe (apart from polar regions) from the natural water bodies characterised by a high average temperature during summer and not entirely freezing during winter. Such environments are found throughout the world in Africa's Rift Valley, in Central Asia, in the North American Desert, on the Andean Plateau, in the rain-shadowed regions of California and Nevada, on the Central Mexican Plateau, on a high-altitude wetland in the Chilean Altiplano, in Manitoba (Canada), in Wadi al Natrun (Egypt), on the Decan Plateau (India), as well as in salty puddles (in Baranda, Europe), in a tropical crater lake located on Mayotte Island (Comoros Archipelago, Western Indian Ocean), in Western Siberia (Russia) and in Eastern Australia [5–11]. In contrast, the commercial biomass of *Arthrospira* is cultivated only in areas with high yearly average temperatures due to economic reasons. The comparative metagenomic analyses of water, sediments and mats from Canadian, Siberian and American soda lakes revealed a common microbiome characteristic for this haloalkaline environment [8,10,12,13]. Interestingly, the genus *Arthrospira/Limnospira* is dominant in tropical soda lakes in Africa. In contrast, it is absent in the Southern Siberian soda lakes (Kulunda Steppe, Altai) [14] and has not been observed in this area since 1931 when N.N. Voronihin described *Spirulina/Athrospira fusiformis*. However, *Arthrospira* was detected in Solonoye Lake in Western Siberia [7].

The nutritional value of the *Arthrospira* biomass is commonly known and well documented. Besides its nutritional properties, *Arthrospira* evinces therapeutic and industrial potential. Biological activity of several *Arthrospira* compounds, namely, C-phycocyanin, γ-linolenic acid (GLA), calcium spirulan and immulina have all been documented in the fields of cancer-fighting, immunomodulation, cardiovascular diseases prevention and others [15]. In the era of alternative energy resources, the *Arthrospira* biomass is also a promising biofuel, producing mostly hydrogen [16–19]. *Arthrospira* sp. PCC 8005 is also a part of the fourth compartment of the Micro-Ecological Life Support System Alternative (MELiSSA), aiming to elaborate artificial ecosystems for space expeditions [20,21]. The biomass of *Arthrospira* is commercially available under the name 'Spirulina' due to a taxonomic revision proposed by Geitler in 1932 [3] that was later shown to be wrong [22]. Since then, it has been shown that many strains listed as *Spirulina* belong to *Arthrospira* [23]. However, this commercial name is not likely to be replaced by the correct one in the foreseeable future due to its practical and technological meaning [24]. The taxonomy of *Arthrospira* was primarily based on morphology, although ecology and the geographic origin of the organisms was also partly considered to distinguish many species [23,25]. However, the distinction of the *Arthrospira* species based on morphology is questionable and can be misleading, since, e.g., the helix geometry and gas vacuole presence can be influenced by physical and chemical conditions [26–28].

So far, for the genetic characterisation and taxonomy of *Arthrospira* only three genetic markers: the 16S rRNA gene, 16S-23S rRNA Internally Transcribed Spacer (ITS) and phycocyanin operon fragment *cpcBA* have been used. The phylogenetic studies of *Arthrospira* based on the 16S rRNA gene divided the genus into three clads (Comte et al. 2013). Meanwhile, only two clusters (I and II) and five subclusters (I.A, I.B, II.A, II.B, II.A/II.B) have been discriminated based on the sequence polymorphism of the ITS [29,30]. The intrageneric relatedness has also been assessed by a sequence analysis of a region of the phycocyanin operon, consisting of the intergenic spacer (IGS) and parts of the flanking *cpcB* and *cpcA* genes [5,31–33]. According to Papapanagiotou and Gkelis [34], the combined analysis of molecular (16S rRNA, ITS and *cpcBA*-IGS sequences) and phenotypic (13 morphological and morphometric) characters proposed and described by [35] data were consistent and divided the *Arthrospira* strains into three clusters/taxa. Unfortunately, it is difficult to assess whether the genetic relationship within the genus *Arthrospira* is the same for all three markers, as the phylogenies based on each of them were carried out on different sets of strains. It should be emphasised that the 16S rRNA analysis [2] was not supported by other markers that are typically used in the phylogenetic analysis of cyanobacteria, such as the ITS or phycocyanin operon fragment *cpcBA*. Moreover, since there are no isolates of *Arthrospira jenneri* strains observed in an environmental sample by [2] or deposited in culture collections, further analyses of the biochemical properties and genetic traits that would allow the comparison of both the *Athrospira/Limnospira* genera are not possible. Likewise, no phylogenomic analyses can be performed. Due to the lack of verified isolated strains and/or original genetic material of *A. jenneri*, which can be used for comparative analysis, it is debatable whether this reclassification is accurate. The modern taxonomy of prokaryotes that can be cultivated in vitro should not be based solely on the study of only one or a few genes nor exclusively based on their morphological features, which, in the case of *Arthrospira*, are highly variable depending on numerous factors, which have been accurately documented in the literature [36]. Therefore, in this work, in which we use genomic data for the classification of a new Siberian isolate, we retain the taxonomic name *Arthrospira*.

Up to now (7 October 2021), 17 *Arthrospira* genomes have been sequenced: *Arthrospira* sp. PCC 8005 [37], *A. platensis* NIES-39 [38], *A. platensis* C1 [39], *A. maxima* CS-328 [40], *A. platensis* Paraca P0 [41], *A. platensis* YZ [42], *Arthrospira* sp. TJSD091 [43], *Arthrospira* sp. PLM2.Bin9 [13,44], *A. platensis* NIES-46 [19], *A. platensis* FACHB-835 [45], *A. fusiformis* SAG 85.79 [31], *Arthrospira* sp. BM01 [46], *A. fusiformis* KN (JACGXW000000000), *A. platensis* FACHB-971 [45], *A. platensis* FACHB-439 [45], *Arthrospira* sp. PCC 9108 (CP066886), *Arthrospira* sp. SH-MAG29 [9] and *Arthrospira* sp. TJSD092. This publicly available genomic data can be used for comparative genomics and phylogenomic analyses allowing a reliable identification of the newly isolated Siberian strain O9.13F.

This study aimed to perform a comprehensive genetic and physiological investigation of the new *Arthrospira/Limnospira* sp. strain O9.13F isolated from an alkaline, winter freezing Siberian lake. The study aimed (1) to classify and determine the systematic position of the new strain, (2) to characterise physiologically the strain capable of surviving the winter period in the Siberian lake, (3) to perform a comparative genomic analysis of Siberian and tropical *Arthrospira/Limnospira* strains and (4) to verify whether the O9.13F strain is suitable for cultivation in the temperate climate zone.

2. Materials and Methods

2.1. Strain Isolation and Growth Conditions

On 11 September 2013, the thick cyanobacteria bloom from Solenoye Lake (Supplementary Figure S1) was collected into sterile, capped containers and transported to the laboratory. Then the biomass was resuspended in a liquid Zarrouk medium [47] and treated with trimethoprim (60 µg/mL), cycloheximide (200 µg/mL) and amphotericin B (5 µg/mL) to remove fungi. Next, a series of repassages on the solid Zarrouk medium was performed to separate cyanobacteria present in the bloom from each other and co-occurring bacteria. The axenity of the *Arthrospira* cultures was examined according to the procedure recommended by Rippka [48].

Moreover, to be certain that the new strains were unicyanobacterial, eight single filaments with different morphologies were isolated from the fresh biomass from an alkaline pond. From each of the filaments, pure cyanobacterial cultures (named O9.13A-H) were obtained. All new isolates were identified by sequencing of the 16S rRNA gene and screened by PCR with primers specific for the ITS subclusters [29,30] and the ITS_CL_III primer designed in this work. As the newly isolated strains were morphologically and genetically identical, only the O9.13 F strain was selected for further analyses and remained a part of the *Arthrospira* strains collection at the Intercollegiate Faculty of Biotechnology, University of Gdansk, Poland.

To determine the optimal growing conditions, the *Arthrospira* culture was grown with and without shaking at 150 rpm in the Zarrouk liquid medium pH~10–11 under a 16 h light/8 h dark photoperiod under illumination of 30–40 µmol photons m^{-2} s^{-1} provided by fluorescent lamps at four temperatures: 15 °C, 20 °C, 28 °C and 35 °C and in three salinities, NaCl concentration of 25 g/L, 50 g/L and 100 g/L, in triplicates, starting with the culture of similar optical density (3 McF in the case of strain O9.13F). Two other *Arthrospira* reference strains, PCC 8005 and PCC 7345, were grown in the same conditions, except temperature 15 °C at which these strains die off. The growth index was expressed as a ratio of the initial culture optical density to the final measured value of a particular sample after 8 days of incubation (averaged from all replicates). To assess the properties of the biomass, the fatty acids (FA), biosynthetic pigments and proteins were extracted from the above-mentioned cultures.

2.2. *Phenotypic Characteristics of Arthrospira sp. O9.13F*

2.2.1. Acclimation of *Arthrospira* Strains to Long-Term Stress of Low Temperature

Arthrospira, O9.13F and four strains PCC 8005, PCC 7345, SAG 49.88 and CCALA 023 were cultivated without shaking under cold room conditions in the Zarrouk medium under a constant lighting of 10 µmol photons m^{-2} s^{-1} with a daily amplitude of 5–8 °C for 10 months. After 3 and 10 months of cultivation in the cold room, all samples were inoculated onto the solid Zarrouk medium and grown at 20 °C under a light intensity of 20 µmol photons m^{-2} s^{-1} for two months. The photographic documentation was made using a Nikon 80i light microscope.

2.2.2. The Survivability of *Arthrospira* Strains in Different Stress Conditions

The *Arthrospira* strains O9.13F, PCC 8005 and PCC 7345 were subjected to stress conditions caused by the addition of different substances, cultivation in water acquired from industrial ponds or changes in medium salinity. The stress factors were selected based on several criteria: (i) They can appear in the natural pond next to the farmlands. (ii) The used stressor would be useful for the removal of an excess of associated microorganisms from the biomass before various procedures (e.g., genome sequencing). (iii) They would be beneficial in growing in numerous industrial and natural ponds. The tested stressors and conditions included: working solution of fungicide 0.35% (Curzate Cu 49.5 containing cymoxanil and copper oxychloride), working solution of herbicide 0.75 mg/L (glyphosate), working solution of insecticide 0.04% (Clothianidin), hydrogen peroxide 50, 100 and 200 mM, silver nanoparticles (AgNPs) 3 µg/mL, industrial water rich in calcium, potassium and sodium carbonate with pH = 13 (Poland); undiluted and industrial water mixed 1:1 with a Zarrouk medium, water from Gdańsk Bay (estimated 7–8‰ of salinity) and the addition of NaCl to total salinity 150 and 200 g/L. The biomass was cultivated at 28 °C under stress for one week. Afterwards, the biomass was washed with distilled water and transferred to a solid Zarrouk medium. Due to the filamentous character of the *Arthrospira* biomass, as well as the tendency of the biomass to create mats, a standard means of measuring bacterial growth (optical density) is not applicable for assessing the growth kinetics of this organism. Therefore, a method based on the condition, duration of the lag phase and visible increase in the number of viable filaments was developed. Here, the recovery rate was defined as to whether the trichomes survived the stress conditions and, if so, after what time the active growth was resumed upon reintroduction to physiological conditions. This process was monitored using the following criteria: 0—degradation of the filaments, 1—whole, nonfragmented filaments present, no growth observed, 2—some growth observed after two weeks of a lag phase, 3—weak growth after one week of a lag phase, 4—stable growth after a short lag phase and 5—abundant growth without a lag phase.

2.2.3. Fatty Acids Methyl Ester Analysis (FAME)

The total lipids were extracted in triplicates according to the Bligh and Dyer protocol [49] by homogenising the biomass in chloroform/methanol under ultrasonication. Following evaporation, the lipids were derivatised to FAMEs with sulphuric acid (2%) in methanol. The resulting FAMEs were extracted to hexane and analysed using a Shimadzu GC-2010 equipped with a flame ionization detector and a 60 m × 0.25 mm BPX70 column (SGE Analytical Science). Standard mixtures of fatty acids methyl esters (Supelco® 37 Component FAME Mix, 1890, cis/trans) were used to identify the major fatty acids based on their retention time. Individual FAs were expressed as percentages of the total FAs present in the chromatogram.

2.2.4. Proteins Content Analysis

The total amounts of proteins in the extracts obtained by ultrasonication of the biomass were measured following the protocol of Bradford [50]. The 1 mL samples (triplicates) of the culture of *Arthrospira* sp. O9.13F cultivated in various conditions were carefully washed with distilled water three times and centrifuged for 20 min at 13,200× *g* rpm. The cell lysis was performed with 1 mL of 0.1 M NaOH, and the cells were incubated for 30 min at 95 °C with shaking. Afterwards, the tubes were left to cool down, acidified with a 5 M solution of HCl and centrifuged for 10 min under 13,200 rpm. To each well in the 96-well plate, 250 μL of the Bradford reagent (Sigma Aldrich #B6916) and 10 μL of the samples (in triplicates) were added. The spectrophotometric measurements were done on an Envision 2105 Multimode Plate Reader (Perkin Elmer) after 30 min of incubation at room temperature without shaking in the wavelength 595 nm. The protein concentration was determined by comparison to a standard curve prepared based on a BSA solution in the range between 0.1–1.4 mg/mL.

2.2.5. The Measurements of Photosynthetic Dyes Content

The biomass obtained from 1 mL of the tested culture in triplicates was centrifuged for 20 min at 13,200× *g* rpm, the supernatant was removed, and 1 mL of 90% methanol was added. The biomass was further lysed by ultrasonication for 20 min and centrifuged again for 15 min at 13,200× *g* rpm. The absorbance was measured by the spectrophotometer at 470 nm, 665 nm and 720nm. The concentration of chlorophyll *a* and total carotenoids was established according to equations by [51,52]:

$$\text{Chl}\alpha\ [\mu\text{g/mL}] = 12.9447 \times (A_{665} - A_{720})$$

$$\text{Carotenoids}\ [\mu\text{g/mL}] = 1000 \times (A_{470} - A_{720}) - 2.86\ \text{Chl}\alpha\ [\mu\text{g/mL}]/221$$

2.2.6. Statistical Analysis

The analysis of differences between the groups was performed using the Shapiro–Wilks test followed by Kruskal–Wallis test using the criterium of Fisher's least significant difference (LSD) in R [53].

2.3. *Microscopic Observations*

2.3.1. Light Microscopy

The microscopic observations of *Arthrospira* sp. O9.13F filaments were performed, and the photographs were taken with a Nikon 80i microscope with a Nomarski contrast at ×400 and ×1000 magnification with the NIS Elements version D software from Nikon.

2.3.2. Transmission Electron Microscopy (TEM)

Preparation for the ultrathin sectioning and preparation for TEM were carried out as described previously [23,54,55]. Samples were fixed in 2.5% formaldehyde and 2.5% glutaraldehyde in a 0.05 M cacodylate buffer (pH 7.0) for 4 h at room temperature. Then, the material was rinsed in the same buffer and postfixed in 1% osmium tetroxide in the cacodylate buffer at 4 °C overnight. Specimens were treated with 1% uranyl acetate in distilled water for 1 h, dehydrated in an acetone series and embedded in Spurr's resin. Serial ultrathin (60–100 nm) sections were cut with a diamond knife on a Sorvall MT2B microtome. The material on the grids was poststained with a saturated solution of uranyl acetate in 50% ethanol and 0.04% lead citrate. Observations were made using a Philips CM 100 transmission electron microscope operating at 80 kV. For the scanning electron microscope, samples from fresh cultures were fixed in 2.5% formaldehyde and 2.5% glutaraldehyde in 0.05 M Na-cacodylate buffer (pH = 7.0) for 4 h at room temperature, dehydrated in an ethanol series (30, 50, 75, 96, 100%) and critical point dried using liquid CO_2 and a K850 critical point dryer (EMITECH, Ashford, England). The dried specimens were mounted on stubs with SPI carbon conductive double-sided adhesive discs, gold coated (350 Å) (SPI-Module Sputter Coater, Structure Probe, Inc., Chester, PA, USA) and examined in a Phillips XL-30 SEM operating at an accelerating voltage of 15 kV.

2.4. Molecular Characteristics

2.4.1. DNA Extraction

For DNA extraction, 10 mL of the dense *Arthrospira* culture cultivated at 20 °C was centrifuged, and the collected biomass was carefully washed 3 times with sterile distilled water prior to nucleic acid extraction. Afterwards, the cell lysis and nucleic acids extraction were carried out according to the protocol proposed by the Joint Genome Institute for bacterial DNA isolation using CTAB [56] followed by RNA digestion using Turbo RNase (Ambion). The DNA quantity and quality were assessed first using a NanoDrop Spectrophotometer and later with agarose gel electrophoresis.

2.4.2. PCR Amplification Sequencing and Phylogenetic Analysis

The sequences of the *cpcBA*, ITS and 16S rRNA gene of the new Siberian isolates were determined after cloning as previously described [57]. To confirm this assignment, one clone for each ribosomal operon variant was sequenced with primers 359F, 16S979F and 23S30R [58–60]. For amplification and sequencing of phycocyanin operon, fragment primers described by [32] were used. Sequencing was carried out using an ABI PRISM DNA Sequencer (Perkin-Elmer) according to the manufacturer's manual. Both strands were sequenced using the forward and reverse PCR primers. Moreover, the clones were assigned to the different ITS clusters as described by [30] with an additional reverse primer ITS_CL_III (TTGACTATCAGAGGAAATCCG) designed by us specific for cluster III based on the sequence from *Arthrospira* PCC9901 (MF509176) and which was used with the primer 16S3′F [29].

Further, the obtained sequences were subjected to a BLAST sequence similarity search analysis to identify the nearest taxa. The obtained sequences of the 16S rRNA gene, *cpcBA* operon, ITS and sequences of the closest representative taxa that belong to cyanobacteria were aligned using the MAFFT algorithm in Geneious v9.1.8. [61]. The phylogenetic analysis was performed with the MEGA v. X software, (www.megasoftware.net (accessed on 23 October 2020)), and trees were constructed using the Neighbour Joining algorithm with a Tamura 3-parameter and Maximum Likelihood algorithm, and the Hasegawa–Kishino–Yano models selected on the model test module were implemented in MEGA. A bootstrap analysis with 1000 replications was performed to assess the robustness of the clusters.

The nucleotide sequence data reported in this paper are available under the following accession numbers: 16S rDNA, MF509165, ITS; MF509166–MF509169, *cpcBA* and MF509119. As the 16S rRNA gene sequences obtained for all eight clones were identical, only one was deposited in the GenBank database. A similar situation occurred with the *cpcBA* operon sequence. In the case of the ITS, the sequences of two different genetic variants obtained for the two clones O9.13F and O913H were submitted.

2.4.3. Whole-Genome Sequencing and Annotation

Genome sequencing was performed on an Illumina HiSeq 2000 platform in the Institute of Biochemistry and Biophysics of the Polish Academy of Sciences, Warsaw, Poland. The genome assembly was performed using computational resources provided by the grid computer "durandal" operated by InBioS-PhytoSYSTEMS at the University of Liège (Belgium).

The raw reads were first filtered using Kraken v1.0 software [62] followed by quality and adapter trimming performed using Trimmomatic v0.32 [63]. The Kmer length and distribution were analysed using kmergenie v1.6949 [64]. Processed reads were de novo assembled using Spades v3.10.1 [65], Velvet v1.2.10 [66], MIRA v1.1.1 [67], Geneious v9.1.8 [61] and Ray v2.3.1 [68]. The resulting sets of contigs were integrated, and the final assembly was corrected by CISA v1.3 (Lin and Liao, 2013) software and finally scaffolded with SSPACE v3.0 [69]. The assembly was evaluated using Quast v4.5 software [70].

Annotation was performed by the NCBI Prokaryotic Genome Annotation Pipeline [71,72]. Functional annotations were carried out by the KEGG Automatic Annotation Server [73] and AutoFACT [74], for which the annotation was based on the COG, Smart and Pfam databases. Signal peptides and transmembrane helices were predicted with a Phobius server [75].

R-M systems were identified using the REBASE® database [76], Clustered Regularly Interspaced Short Palindromic Repeats (CRISPR) sequences were predicted using the CRISPRFinder tool [77]. CRISPR-associated genes used in comparative genomics were obtained from the MaGe platform [78]. The identification of prophage regions was performed using the PHASTER tool [79]. Plasmid sequences were detected using the PlasmidFinder server [80].

The screening for the secondary metabolite biosynthesis gene clusters in the *Arthrospira* sp. strain O9.13F genome sequence was assessed with AntiSMASH [81,82].

2.4.4. Phylogenomic Analyses

The phylogenomic analysis is based on the sequences of 363 of the most conserved proteins comparison using the PhyloPhlAn computational pipeline (v3) [83]. The calculation of the average nucleotide identity (ANI) was performed using the PyAni module (v0.2.10) [84] with MUMmer (v3.23) [85] as an alignment method. In silico DNA–DNA hybridisation (*is*DDH) values were calculated using the Genome-to-Genome Distance Calculator (v2.1) [86] with BLAST+ [87] as a local alignment tool.

3. Results

3.1. *Phenotypic Characteristics of* Arthrospira *sp. O9.13F*

3.1.1. Morphology of *Arthrospira* sp. O9.13F

Upon introduction into the laboratory conditions, trichomes of *Arthrospira* sp. O9.13F were in the shape of tight left-handed open helices (Figure 1). With time and without the influence of natural UV light, the trichomes became looser. Macroscopically, the biomass of strain O9.13F either created a layer just under the surface of the medium or remained on the bottom of the bottle with a tendency to create mats. The trichomes were motile and covered with a layer consisting of exopolysaccharides. The morphology and ultrastructures of the *Arthrospira* sp. O9.13F were typical for representatives of this genus. The TEM (Figures 1A and 2) and light microscopy show the cell dimensions and morphology of the filaments grown on a solid (Figure 1B) and liquid medium (Figure 1C).

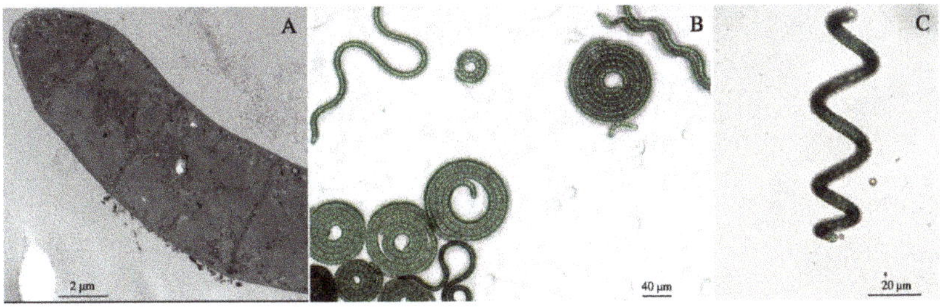

Figure 1. Images of *Arthrospira* sp. O9.13F. Photographs were taken by TEM (**A**) and light microscopy of *Arthrospira* culture grown on solid (**B**) and liquid Zarrouk medium (**C**).

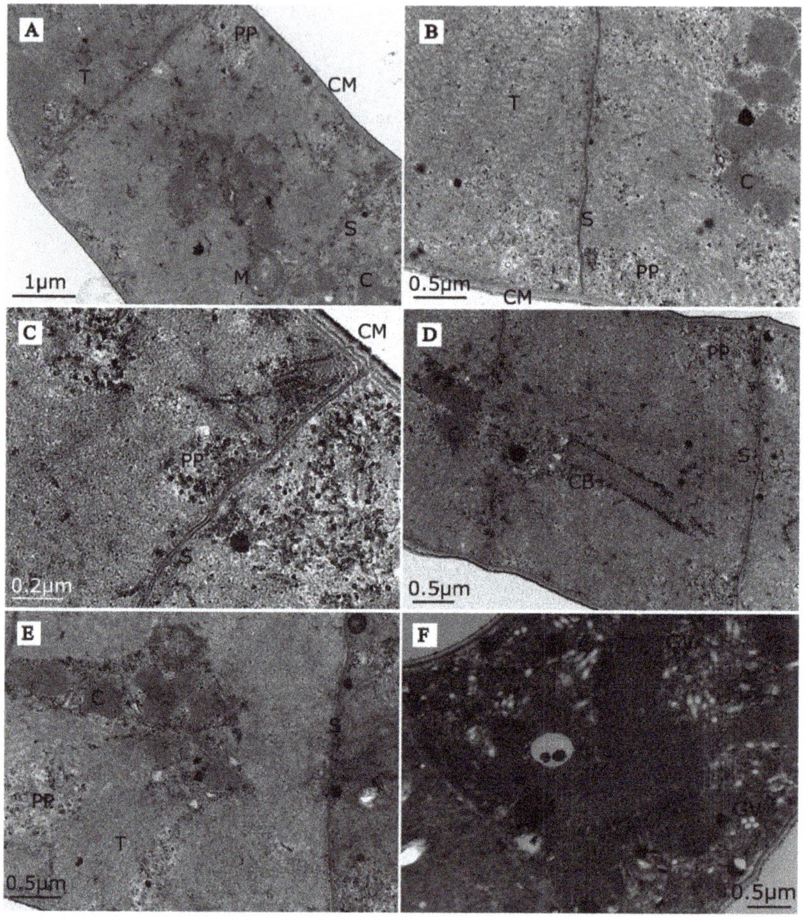

Figure 2. TEM photography of cellular ultrastructures of *Arthrospira* sp. O9.13F. Panels (**A–F**) represent different cross-sections of cells illustrating ultrastructures recognised. Abbreviations as follows: cell membrane (CM), septum (S), gas vesicles (GV), thylakoids with bead-like ribosomes (T), mesosome (M), carboxysomes (C), cylindrical bodies (CB), polyphosphate granule (PP).

We identified all the ultrastructures recognised in other strains of *Arthrospira* investigated by Van Eykelenburg [54] and Tomaselli [23]—thylakoids, cylindrical bodies, mesosomes, gas vesicles, carboxysomes, cyanophycin granules, polyhydroxyalkanoate inclusions, cell membranes and septa (Figure 2A–E). Abundant gas vesicles with a diameter of 45–60 nm and a length up to 600 nm scattered throughout the cells were visible in the EM pictures of strain O9.13F (Figure 2F). The density is comparable to other strains as visible in the TEM microscopic images by [23,54].

3.1.2. Growth Conditions

Cell growth occurred at pH 7.0–10.0, and it tolerated from 1 up to 150 (g/L) of NaCl. The growth temperature of the strain O9.13F ranged between 15 and 35 °C, with optimum growth at 20 °C (Supplementary Figure S2). The optimal growth index for O9.13F of 20 °C was lower than the optimal temperature of the tropical strains *Arthrospira* sp. PCC8005 (India) [30] and *A. platensis* PCC7345 (saline marsh Del Mar Slough, CA, USA) for which the largest increase of biomass production was observed at 28 °C (Figure 3). Two reference strains, PCC8005 and PCC7345, were not grown at a temperature of 15 °C because in this condition, both strains stopped their metabolism, and filaments started to fall apart; therefore, the measurements of proteins and fatty acids were below the sensitivity limit of the method. Interestingly, for the strain O9.13F grown at 15 °C, its growth index was higher in the medium with higher salinity. It was around 4 McF in the medium supplemented with 150–200 g/L of NaCl, while at lower concentrations of 25–50 g/L, it reached a maximum of 2.5. The highest growth was observed in the medium with a salinity of 150 g/L (Supplementary Figure S2).

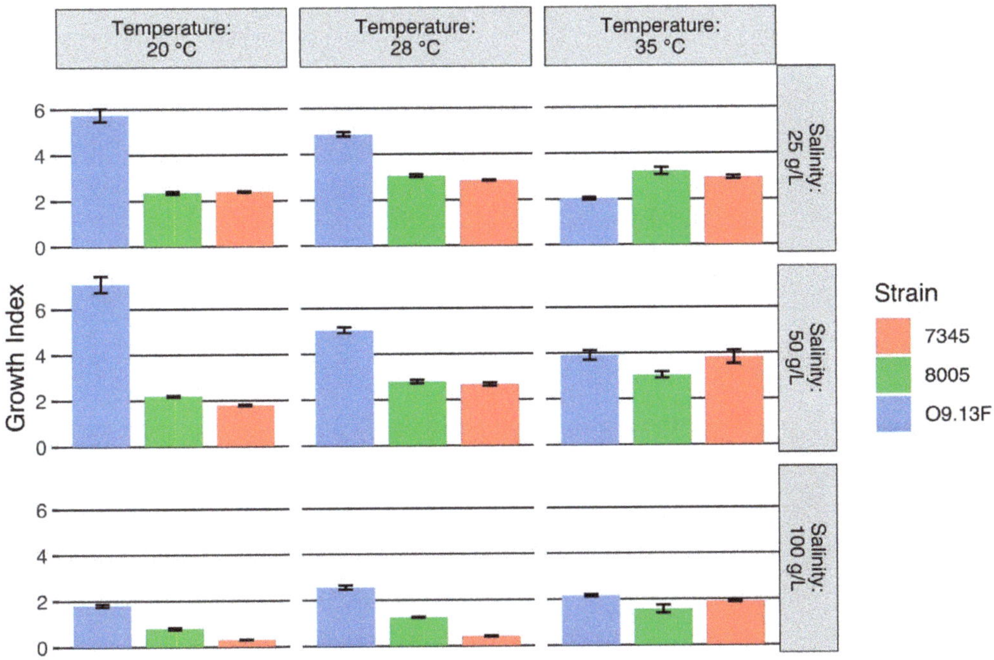

Figure 3. Effect of salinity (Zarrouk medium supplemented with NaCl to total salinity up to 25 g/L, 50 g/L, 100 g/L) on the growth of *Arthrospira* sp. strains, O9.13F, PCC7345 and PCC8005. Strains were grown for 8 days at different temperatures (20 °C, 28 °C and 35 °C). The optical density measurements were performed in triplicates for each strain in each of the conditions. The standard deviations are shown. The growth index is the quotient of the optical density [McF] of the culture after 8 days incubation and the optical density of the culture at the start point of the experiment.

Accordingly, the highest amount of proteins (1.01 mg $\pm$ 0.22 /mL) and photosynthetic dyes: chlorophyll (6.26 $\pm$ 1.16 mg/mL) and total carotenoids (1.59 $\pm$ 0.13 mg/mL) extracted from the biomass of O9.13F, were obtained from the culture cultivated at 20 °C. Under the conditions of the highest salinity, 100 g/L, all three strains had a significantly lower biomass increase than in other salinity values, regardless of the temperature at which the cultures were grown. However, the biomass grown at 35 °C in the 100 g/L salinity medium was the lowest (Figure 4).

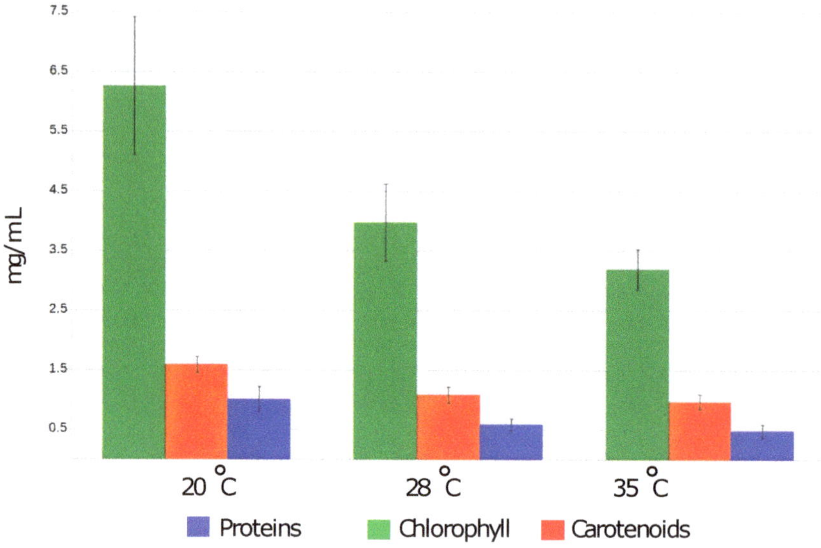

Figure 4. Quantity of chlorophyll, total carotenoids and proteins [mg/mL] extracted from biomass of *Arthrospira* sp. O9.13F cultivated at three different temperatures, 20 °C, 28 °C and 35 °C in Zarrouk medium with a total salinity 25 g/L. Standard deviations based on three replicate analyses are shown.

3.1.3. Acclimation to Prolonged Cold Stress

As the strain was isolated from a Siberian lake that is covered by ice in the winter, we tested the tolerance in laboratory conditions at a low temperature and compared it with other strains that originated from warmer regions: PCC8005 (India), PCC7345 (USA), *A. maxima* SAG 49.88 (Italy) and *A. fusiformis* CCALA23 (Kenya). The experiment was performed in a cold room, in low light 10 µmol photons m^{-2} s^{-1} conditions, for 10 months with subculturing after three months.

The general observation was that strains lost the intensity of pigmentation, and some filaments started to decay after this period. One of the strains, *A. maxima* SAG 49.88, changed the morphology and began to coil filaments. After the inoculation of tested strains filaments on a solid Zarrouk medium and transfer to the higher temperature 20 °C with a change of light intensity 20 µmol photons m^{-2} s^{-1}, the strains recovered to the normal pigmentation, and growth was visible after a week of incubation; with exception of the strain *A. fusiformis* CCALA 23 that started to grow as a visible mat of filaments on the plate after six weeks (Figure S3).

3.1.4. The Sensitivity of *Arthrospira* sp. O9.13F to Various Stressors

Based on the survival rate of the *Arthrospira* strains, the tested stress factors were divided into three levels of severity: low, intermediate and high. The best growth (low-stress severity) with no noticeable lag phase after the passage to the fresh medium was observed after the addition of hydrogen peroxide (in all three concentrations). The biomass of all tested strains was also showing minimal inhibition of the growth upon exposition to the working concentration of insecticide, water from Gdańsk Bay (lower salinity) and water from an industrial pond (pH > 13) diluted with a Zarrouk medium 1:1. The most differential and strain-dependent effect (intermediate stress severity) was observed in the case of exposition to undiluted industrial water (reach in calcium, potassium and sodium carbonate and pH > 13), silver nanoparticles (15 µg/mL) and herbicide and salinity of 100 g/L. Among the most harmful factors (high-stress severity) were high salinity, 150 and 200 g/L of total salt and a working concentration of 0.04% of fungicide. From the compilation of the results of the sensitivity of the strains to the applied stresses, we concluded that the most sensitive strain was *Arthrospira* sp. O9.13F (Figure 5).

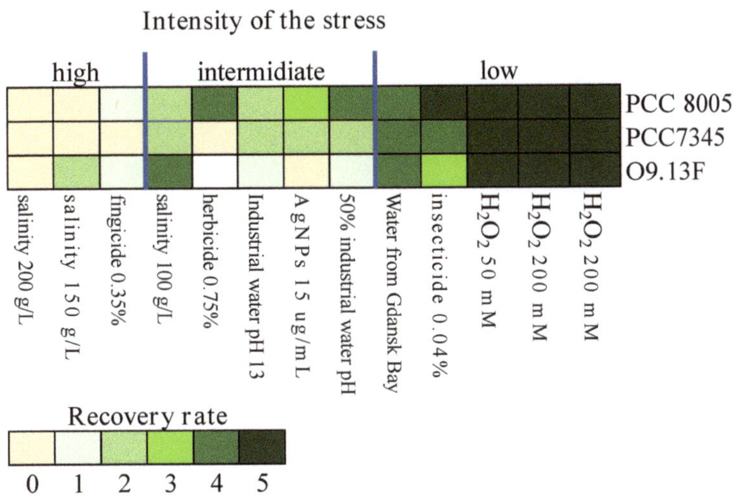

Figure 5. The heatmap presents the recovery rates of eight tested *Arthrospira* strains, O9.13F, PCC 7345 and PCC 8005 cultivated in stress conditions. Above the heatmap, three levels of the stress severity (low, intermediate, high) are grouped according to the overall similarity of the different strains' responses. The colour scale given in the heatmap key corresponds to the recovery rate. The survivability of the biomass was assessed based on the condition, duration of the lag phase and visible increase in the number of viable filaments and recorded by the following system: 0—degradation of the filaments, 1—whole, nonfragmented filaments present, no growth observed, 2—growth observed after two weeks of a lag phase, 3—very weak growth recorded after one week of a lag phase, 4—stable growth after a short lag phase, 5—very abundant growth without a lag phase.

3.1.5. Fatty Acid Composition

The major fatty acids detected in the *Arthrospira* strains were palmitic acid (16:0), gamma-linolenic acid (18:3n6), linoleic acid (18:2n6c) and palmitoleic acid (16:1). However, the fatty acid content of the *Arthrospira* sp. strain O9.13F was not the same as in the strains PCC8005 and PCC7345. The most significant difference in fatty acid content in the *Arthrospira* biomasses was observed in the percentage of palmitic and gamma-linolenic acid (Table 1). The observed decrease of gamma-linolenic acid in favour of palmitic acid in strain O9.13F could be due to the stressful growth conditions and the fatty acid content. All tree strains were cultivated at 28 °C, which is not optimal for the growth of the Siberian strain O9.13F (Table 1). Indeed, the same observation was recorded when the fatty acid content was established for the *Arthrospira* sp. strain O9.13F grown at 20 °C, 28 °C and 35 °C (Supplementary Figure S4).

Table 1. Total FA composition (%) of *Arthrospira* strains O9.13F, PCC 7345 and PCC 8005 cultivated at 28 °C.

Fatty Acid	O9.13F	PCC 8005	PCC 7345
16:0	36.71 ± 3.93	30.81 ± 1.2	31.64 ± 3.87
16:1	8.80 ± 1.1	10.29 ± 0.65	9.97 ± 1.13
18:0	1.12 ± 0.33	0.87 ± 0.16	1.62 ± 0.68
18:1n9c	2.40 ± 0.97	2.59 ± 0.32	1.54 ± 0.66
18:1n11c	1.67 ± 0.3	1.29 ± 0.04	1.32 ± 0.35
18:2n6c	18.86 ± 0.34	18.78 ± 0.9	19.66 ± 2.43
18:3n6	30.45 ± 4.45	35.37 ± 1.98	34.26 ± 6.16

n = 3; 16:0—Palmitic acid; 16:1—Palmitoleic acid; 18:0—Stearic acid; 18:1n9c—Oleic acid; 18:1n11c—Vaccenic acid; 18:2n6c—Linoleic acid; 18:3n6—γ-Linolenic acid. The data are means ± standard deviations based on three replicate analyses.

3.2. Taxonomy of Arthrospira sp. Strain O9.13F

The new Siberian isolate of *Arthrospira* was identified based on the 16S rRNA gene, ITS and the phycocyanin operon fragment *cpcBA* sequencing and three different genomic analyses: ANI, *is*DDH and core proteome-based phylogeny.

The cloning and sequencing of the amplified 16S and *cpcBA* revealed that all eight isolates O9.13A-H (each originated from a single filament) have identical (100%) sequences. As a result of the ITS cloning and sequencing, we obtained two different genetic variants representing the ITS clusters I and III according to the numbering proposed by [30,88], respectively. These observations are in agreement with the results of the detection of the ITS variants by PCR with subcluster-specific primers described by Baurain et al. [30] and the reverse primer ITS_CLIII designed by us.

As the newly isolated strains were morphologically and genetically identical, only strain O9.13 F was used for further analyses. The phylogenetic analyses based on the 16S rRNA gene, ITS and phycocyanin operon fragment *cpcBA* were performed on the same dataset.

The obtained 16S rRNA gene sequence of the isolate O9.13F was 99.91% similar to *A. erdosensis* 'Inner Mongolia' (JN831261), *Arthrospira* sp. PCC 9901 (MG777134) and 99.72% similar to the other *Arthrospira* sp. strains TJSD092 (CP028914) and PCC 8005 (FO818640).

The phylogenetic analysis based on the small ribosomal subunit sequence showed that the Siberian strain groups with the strain PCC 9901 belonging to the new III clade described by Comte et al. [88] (Supplementary Figure S5). However, it is worth noting that the sequence 16S rRNA gene of the BM01 strain is only 92.5% similar to the sequences of other *Arthospira* strains, and it was excluded from the analysis.

Likewise, in 16S rRNA-based phylogeny, the analysis of the ITS region showed only two phylogroups within the *Arthrospira* genus called ITS clusters and five different genotypes named subclusters according to Baurain et al. [30]. ITS cluster I gathers strains belonging to clades I and III described based on the 16S rRNA gene, while ITS cluster II consists of all strains from the 16S clade II. Strains belonging to 16S clades I and III formed clearly separated subclusters within ITS cluster I. The results of the cloning and sequencing of ITS from strain O9.13F showed that two different genetic variants (genotypes) had been detected within the chromosome. The ITS sequence of clone I is grouped together with strains from 16S clade I, while the ITS clone III is grouped together with strains from 16S clade III (Supplementary Figure S6).

Based on the sequence of the phycocyanin operon fragment *cpcBA*, the newly described strain O9.13F is grouped with strains that belong to clade I (based on the 16S rRNA gene). However, unlike the results of ITS-based phylogeny, strains carrying the 16S gene sequences belonging to clade III appear to be more closely related to strains with the ITS cluster II.B. Moreover, strains belonging to the ITS subcluster II.A described by Baurain et al. [30] are more related with strains cluster I (Supplementary Figure S7).

The phylogenomic analysis is based on 363 of the most conserved proteins retrieved from 17 *Arthrospira* strains for which the genomic sequences are available in databases. The use of the most evolutionarily conserved proteins whose genes occur in a single copy in the cell and an analysis based on genomic sequences reduce the impact of intragene or operons recombination on the results of phylogenomic analyses. The topology of the ML tree based on core protein showed that the *Arthrospira* genus consists of only two groups, and the O9.13F strain clusters together with the *Arthrospira* strains PCC8005, TJSD091, TJSD092 and CS-328, all of which are representatives of 16S clade I and ITS cluster I (Figure 6). The presence of two phylogenetic groups revealed by our phylogenomic analyses is consistent with the analyses described above and with the previously described analyses of ITS sequences by Baurain et al. [30]. Phylogenesis based on the analysis of core proteins enabled the precise classification of all strains for which the genomes are known. Also, for strains for which the identification relying on single locus, 16S rRNA gene, was doubtful, e.g., *Arthrospira* sp. BM01. The same goes for strains *A. platensis* Paraca and *Arthrospira* sp. PLM2.Bin9 that were ungrouped when the fragment of the *cpcBA* phycocyanin operon was analysed [31].

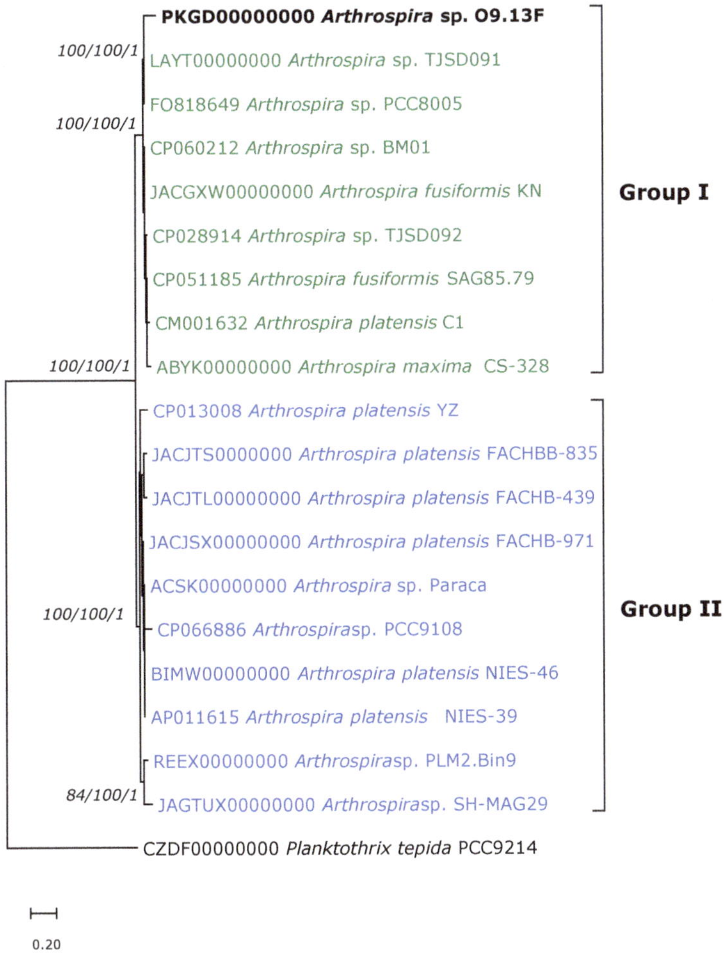

Figure 6. Maximum Likelihood tree indicating the current relatedness of *Arthrospira* sp. O9.13F.

The phylogenetic tree establishing *Arthrospira* sp. O9.13F's phylogenetic position is built based on the sequences of 363 of the most conserved proteins extracted from

Arthrospira proteomes that are available in the Genbank database. The Maximum Likelihood tree was constructed using the PhyloPhlAn computational pipeline (Segata et al. 2013). The distance tree was inferred using Geneious software v 9.1.8. The gene sequences of *Planktothrix tepida* PCC9214 (CZDF00000000) were used as an outgroup. The bootstrap consensus was inferred from 1000 replicates. Bootstrapping values <50% were cut off. Numbers above branches indicate the bootstrap value for ML, the NJ method and the posterior probabilities for the Bayesian inference method, respectively. The sequence of strain O9.13F is indicated in bold font. The colours correspond to the 16S clades: green -clade 1, blue—clade II (according to [30]) and red—clade III (according to [88]).

To compare the relatedness of the *Arthrospira* sp. O9.13F to other available genomes from the *Arthrospira* genus, we analysed the Average Nucleotide Identity (Figure 7) and in silico DNA–DNA hybridisation (Figure 8). Both methods confirmed the assignment of the strain to the *Arthrospira* genus, presenting similarity to the available genomes between 92.9% and 99.4% based on ANI and between 48.1% and 86.1% based on *is*DDH (Supplementary Table S1). Moreover, the presented results for all genomes were consistent with *Arthrospira* division into two main clusters based on core protein phylogeny and ITS sequence [30].

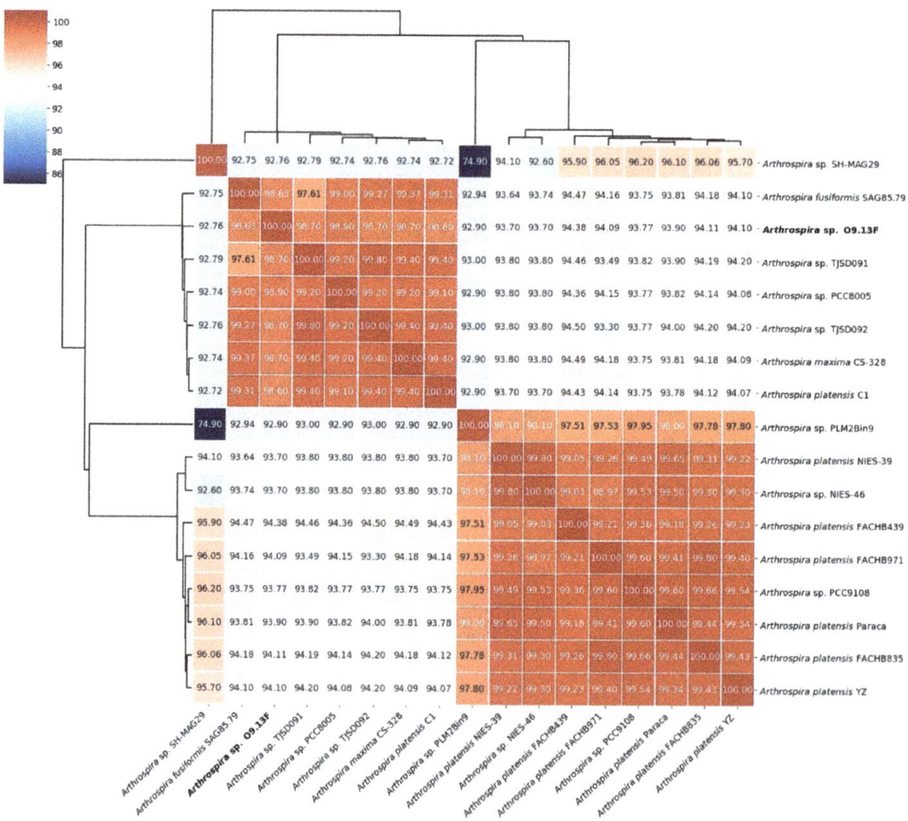

Figure 7. Average Nucleotide Identity (ANI) of 17 *Arthrospira* genomes available in the GenBank database as calculated using PyAni v0.2.10 [84], with MUMer v3.23 [85] as the alignment method. Depicted linkage is calculated as UPGMA, distance is estimated using the Euclidean method.

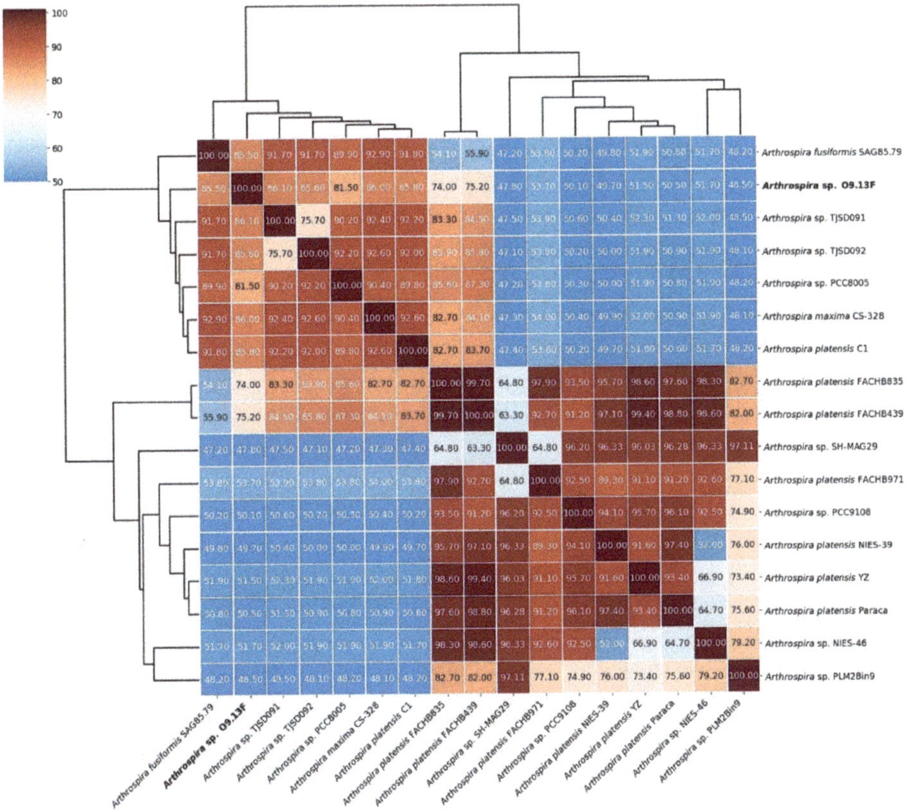

Figure 8. In silico DNA–DNA hybridization of 17 *Arthrospira* genomes available in the GenBank calculated with GGDC 2.1 [86] with BLAST+ as a local alignment tool.

3.3. General Features of the Arthrospira sp. O9.13F Genome

The genome of *Arthrospira* sp. O9.13F is deposited in the GenBank database under accession number PKGD00000000, bioproject PRJNA384118. The draft genome of *Arthrospira* sp. O9.13F consisted of 928 scaffolds of a total length of 4 945 448 bp and a G + C content of 44.47%. A total of 5355 genes were annotated of which 4688 (87.5%) were protein-coding genes, 625 (11.67%) were pseudogenes and 42 (0.78%) genes were encoding RNA and three CRISPR clusters (Table 2). A total of 3278 (69.92%) of predicted genes were assigned to COGs. Most genes were assigned to protein families responsible for replication, recombination and repair as well as signal transduction. The distribution of genes into COGs is presented in Table 3. The additional annotation of the genome with the KEGG Automatic Annotation Server (https://www.genome.jp/kegg/ (accessed on 3 January 2019)) ascribed 1287 genes into 189 pathways, the highest numbers of genes were mapped to the following processes: ABC transporters (55), porphyrin and chlorophyll metabolism (41), photosynthesis (40), purine metabolism (39), two-component system (33), pyrimidine metabolism (32) and oxidative phosphorylation (31).

Table 2. Genome statistics of *Arthrospira* sp. O9.13F.

Attribute	Value	% of Total
Genome size (bp)	4,945,448	100
DNA coding (bp)	3,759,108	76.01
DNA G + C (bp)	2,198,162	44.5
DNA scaffolds	928	-
Total genes	5355	100
Protein coding genes	4688	87.5
RNA genes	42	0.78
Pseudogenes	625	11.67
Genes in internal clusters	-	-
Genes with function prediction	3363	62.80
Genes assigned to COGs	3278	69.92
Genes with Pfam domains	3237	69.05
Genes with signal peptides	400	8.53
Genes with transmembrane helices	862	18.39
CRISPR repeats	3	0.06

Table 3. The number of genes associated with general COG functional categories.

Code	Value	% of Total *	Description
J	146	4.45	Translation, ribosomal structure and biogenesis
A	0	0	RNA processing and modification
K	145	4.42	Transcription
L	311	9.49	Replication, recombination and repair
B	2	0.06	Chromatin structure and dynamics
D	25	0.76	Cell cycle control, cell division, chromosome partitioning
V	84	2.56	Defence mechanisms
T	308	9.4	Signal transduction mechanisms
M	209	6.38	Cell wall/membrane biogenesis
N	39	1.19	Cell motility
U	44	1.34	Intracellular trafficking and secretion
O	156	4.76	Posttranslational modification, protein turnover, chaperones
C	174	5.31	Energy production and conversion
G	140	4.27	Carbohydrate transport and metabolism
E	195	5.95	Amino acid transport and metabolism
F	58	1.77	Nucleotide transport and metabolism
H	115	3.51	Coenzyme transport and metabolism
I	73	2.23	Lipid transport and metabolism
P	138	4.21	Inorganic ion transport and metabolism
Q	95	2.90	Secondary metabolites biosynthesis, transport and catabolism
R	499	15.22	General function prediction only
S	322	9.82	Function unknown
-	1811	33.82	Not in COGs

* The % of total is based on the total number of protein coding genes in the genome.

3.4. Comparative Genome Analyses

A comparative analysis of the *Arthrospira* sp. O9.13F genome with other *Arthrospira* genomes revealed the absence of genetic determinants related to the synthesis of cyanobacterial toxins, such as microcystins, nodularin or cylindrospermopsin [89]. The secondary metabolites analysis of the *Arthrospira* sp. O9.13F genome revealed the presence of five different secondary metabolite gene clusters that were present in all of the studied *Arthrospira* genomes: four cyanobactin/bacteriocin gene clusters and one phytoene synthase. The presence of phytoene is obvious because there are different carotenoids in the *Arthrospira* cells, and a complete pathway for carotenoids synthesis was present within the genome of this cyanobacterium. The secondary metabolites analysis did not indicate the presence of siderophore biosynthetic gene clusters, which agreed with the negative results of the CAS plate assay [90].

The sequences of genes involved in gliding motility present in *A. platensis* C1 and *A. platensis* NIES-39 were compared with the *Arthrospira* sp. O9.13F genome. Genes encoding the S-layer gliding motility protein (oscillin) [91], as well as genes encoding type IV pilus-related proteins involved in twitching motility (*pilA*, *pilC*-like, *pilD* and *pilT2* genes) are present in the *Arthrospira* sp. O9.13F genome.

The buoyancy of *Arthrospira* sp. O9.13F is regulated by gas vesicles (Figure 2F) encoded by certain identified genes: *gvpA*, *gvpC*, *gvpK*, *gvpFL*, *gvpV*, *gvpW*, *gvpN* and *gvpJ*. GvpA proteins form the vesicle core, while GvpC proteins provide structural support and influence the shape of the gas vesicles [92]. The proteins GvpK, GvpFL and GvpN possibly stabilise the structure of the gas vesicle. The GvpJ protein is considered to determine the shape of the gas vesicle. The functions of the GvpV and GvpW proteins remain unknown [92,93].

No plasmid sequences were observed during the DNA extraction and analysis of the genomic data for the *Arthrospira* sp. O9.13F isolate.

One incomplete prophage region was identified in the *Arthrospira* genome of 10.6 kb, with 10 total proteins, including seven phage hits and three hypothetical proteins. All 10 proteins are common in other *Arthrospira* genomes. The phage with the highest protein similarity rate to the identified region was *Citrobacter* phage Margaery (NC 028755.1).

We checked the existence of loci coding RM and CRISPR/Cas systems in the genome of the analysed cyanobacterium. The genome sequence of the *Arthrospira* sp. O9.13F strain contains three sets of the type I R-M systems (hsdM, hsdR and hsdS genes), and isoschizomers of Hin1I, Eco88I, Mph1103I, BshNI and Pfl23II complete the type II R-M systems. Genes encoding two isoschizomers of the type II solitary methyltransferases, namely, BspPI and Bsp143I, were also identified.

A total of 26 different CRISPR/Cas system direct repeats (DR) sequences were found in the *Arthrospira* sp. O9.13F genome, with six confirmed and 20 questionable. It also contains CRISPR/Cas essential *cas1* and *cas2* genes as well as the *cas6* gene and CRISPR-associated protein *csx3* gene. Repeat Associated Mysterious Protein (RAMP) genes, involved in CRISPR/Cas defence processes, were also identified, namely, *cmr2*, *cmr4*, *cmr6*, *csm1*, *csm2*, *csm3*, *csm4* and *csm5*. The organisation of the CRISPR/Cas operons is shown in Supplementary Figure S8.

3.4.1. Genes Responsible for the Stress Adaptation

Cyanobacteria of the *Arthrospira* genus are known to be resistant to temperatures up to 40 °C, alkaline pH and high salt concentration. Numerous genes responsible for their adaptation capacity have been detected in the genomes of these cyanobacteria. So far, the genes encoding Na^+/H^+ antiporters, CO_2 uptake (NDH^{-1}) as well as genes responsible for the biosynthesis of three compatible solutes (sucrose, trehalose, glucosylglycerol or glucosylglycerate) have been described [15,94,95]. To verify whether the Siberian strain O9.13F differs from tropical ones, the presence of 68 genes related to the stress response was checked in its genome (Supplementary Table S3). We confirmed the presence of 60 genes. The sequences of these genes are almost identical or differ only in single polymorphisms between O9.13F and other *Arthrospira* strains. It is worth emphasising that the sequences of the Siberian strain that persists during the winter period in a freezing lake are the same as for the tropical strains. What is more, it should be noted that the absence of eight genes in the genome of the O.M13F strain does not necessarily mean that some pathways are incomplete because the sequence of its genome is not closed, and its size of approximately 4.95 Mbp is smaller than in the case of most *Arthrospira* genomes (from 5.75 to 6.78 Mbp) that are available in the GenBank database (Supplementary Table S2).

3.4.2. *Arthrospira* sp. O9.13F Unique Genes

A comparative genomic analysis of *Arthrospira* sp. O9.13F and other *Arthrospira* genomes revealed 24 genes that are unique for this strain (Table 4). Among them, we found 14 genes encoding hypothetical proteins, RNA-binding S4 domain-containing protein as well as two genes encoding four helix bundle proteins. Moreover, unique CDs encode proteins involved in the immunity system, N-6 DNA methylase, which is most similar with Methylase AflIII, the DUF433 domain-containing protein that might be a part of the putative toxin/antitoxin system ([96]) as well as the WYL domain-containing protein, which is a regulator of the Type VI-D CRISPR-Cas system [97] and thus might regulate the response to environmental stresses [98] (Table 4).

Table 4. *Arthrospira* sp. O9.13F-specific CDs.

	ORF	Description	The Most Similar Sequence in Genbank	Identity
1	B9S53_26655	hypothetical protein	-	-
2	B9S53_26345	hypothetical protein	AI-2E family transporter [*Natronoflexus* pectinivorans] WP_165921875.1	61%
3	B9S53_26040	IS91 family transposase	transposase [*Lentisphaerae* bacterium] MBM4144559.1	73%
4	B9S53_26035	hypothetical protein	WYL domain-containing protein [*Lentisphaerae* bacterium] MBM4165248.1	83%
5	B9S53_25925	hypothetical protein	-	-
6	B9S53_25920	RNA-binding protein	RNA-binding S4 domain-containing protein [*Flavobacteriales* bacterium] TVQ76998.1	78%
7	B9S53_25790	hypothetical protein	-	-
8	B9S53_25595	four helix bundle protein	four helix bundle protein [*Chloroflexi* bacterium] HFC09004.1	67%
9	B9S53_25590	hypothetical protein	four helix bundle protein [*Chloroflexi* bacterium] HFC09004.1	53%
10	B9S53_25580	four helix bundle protein	-	
11	B9S53_24895	hypothetical protein	N-6 DNA methylase [*Sphaerospermopsis* sp. FACHB-1094] WP_190648063.1	59%
12	B9S53_23130	hypothetical protein	ROK family protein [*Pseudanabaena* sp. FACHB-1277] WP_190353199.1	43%
13	B9S53_22765	hypothetical protein	-	-
14	B9S53_21915	hypothetical protein	-	-
15	B9S53_21010	IS4 family transposase	transposase [*Phormidium* sp. FACHB-1136] WP_199324687.1	75%
16	B9S53_20985	hypothetical protein	-	-
17	B9S53_20435	hypothetical protein	hypothetical protein [*Cylindrospermopsis* raciborskii] WP_071249329.1	67%
18	B9S53_19635	hypothetical protein	-	-
19	B9S53_12695	hypothetical protein	-	-
20	B9S53_12545	hypothetical protein	-	-
21	B9S53_11425	hypothetical protein	aldolase [*Chloroflexi* bacterium] MBT7081734.1	50%
22	B9S53_11330	IS4 family transposase	transposase, partial [*Leptolyngbya* sp. PCC 6406] WP_155834780.1	75%
23	B9S53_06535	DUF2281 domain-containing protein	DUF2281 domain-containing protein [*Geitlerinema* sp. P-1104] NMG59147.1	94%
24	B9S53_01945	hypothetical protein	hypothetical protein [*Nodularia* sp. LEGE 06071] WP_193996843.1	32%

4. Discussion

The temperature of the water in Solenoye Lake varies between 0 and 30 °C during vegetation season. The water salinity level varies from mesohaline during spring to polyhaline in the remaining seasons, and seasonal pH variation is in the range of 8.94–9.94 [7]. The lake is located on salt clay, and it freezes only on the surface during the winter. The ice cover remains for three months, from December to the beginning of March. The surface water temperature in these months ranges between −3 °C and −15 °C, with an average of approximately −10 °C. In winter, brine, which has a large heat accumulation capacity, may keep a higher temperature on the lower water layers. In addition, the salinity of the water lowers the temperature of the ice cover formation. The ice formation on the lake's surface increases salinity in the lower water layers and allows the medium to remain liquid [99]. The cyanobacteria blooms in Solenoye Lake are observed annually. The composition of the lake's phytoplankton has been investigated since 2007 [7]. During the entire ice-free period, cyanobacteria abundantly develop in the lake as does the species of the *Arthrospira* genus. *Arthrospira* forms the bloom in the warmer season, although its vegetation is noted in the lake even in the winter under the ice. The presence of *Arthrospira* in Solenoye Lake was noted even a few decades ago when there was a resort where the healing properties of the microflora in the lake were used to treat a variety of human diseases [7].

This study is the first comprehensive, physiological and genomic description of the *Arthrospira* strain O9.13F isolated from Solenoye Lake in September 2013. It is the first well-characterised isolate of *Arthrospira* found in Siberia. The described strain O9.13F is cultivable in laboratory conditions and showed the highest relative growth index and highest accumulation of proteins and photosynthetic pigments in the biomass upon cultivation at 20 °C. The observed optimal growth temperature is significantly lower than the temperature of 35 °C reported in the literature as optimal for most *Arthrospira* strains that typically originate from tropical areas. The earlier study of Kumar et al. 2011 showed that tropical strains of *Arthrospira* could not effectively grow and accumulate biopigments at 20 °C. We obtained similar observations during our research. Two reference strains, PCC 8005 and PCC 7345, grew less efficiently at 20 °C than the Siberian strain (Figure 4). While at the temperature of 15 °C, both strains stopped their metabolism, and filaments started to fall apart; therefore, the measurements of proteins and fatty acids were below the sensitivity limit of the method.

Nevertheless, the O9.13F strain was able to grow at 15 °C. Interestingly, we observed the highest growth index in the Zarrouk medium at this temperature with the most elevated tested salinity of 200 g/L. Furthermore, we showed that *Arthrospira* O9.13F could persist in temperatures around 9 °C for several months. They were unable, however to survive freezing at −20 °C in laboratory conditions. The waters of Lake Solenoje do not freeze to the bottom. Considering the properties of the O9.13F strain and the fact that in the ice-covered lake, the water in the deeper layers is more saline and warmer than those closer to the surface [99], we can hypothesise that the Siberian *Arthrospira* strain can survive the winter periods on the bottom of the lake and cause annual blooms observed in the summer.

The fatty acid content of the biomass of the Siberian strain does not deviate much from the biomass of two other analysed strains originating from tropical climates when each was cultivated in their optimal temperatures: 20 °C for O9.13F and 28 °C for PCC 8005 and PCC 7345. The same trend of a decrease of gamma-linolenic acid in favour of palmitic acid caused by a change of the temperature was observed for the Siberian strain as for the tropical ones. A drop in temperature from 30 to 20 °C led to a decrease in palmitic acid accompanied by an increase in linoleic, and the gamma-linolenic acid content was observed earlier by [100], suggesting an increase in membrane fluidity in response to the temperature change. Colla et al. [101] observed that elevation of the growth temperature from 30 °C to 35 °C increased the palmitoleic and linoleic acid concentration but decreased gamma-linolenic acid. Thus, the observed reduced amount of gamma-linolenic acid in the biomass of the Siberian strain after the elevation of the growth temperature from 20 °C to 28 °C confirm its acclimation to cooler than tropical climate.

Additionally, we concluded that it is possible to recover the *Arthrospira* strains after long incubation periods at a low temperature (9 °C) and in low light conditions (10 μmol photons m^{-2} s^{-1}). This observation is valuable in case of storage problems with this species in culture collections as the *Arthrospira* strains are known to not survive the cryopreservation and lyophilisation procedures [102].

Based on this study's detailed genetic characteristics, the newly isolated Siberian strain O9.13F belongs to the *Arthrospira/Limnospira* genus. On the genomic level, the Siberian strain O9.13F is highly similar to 17 other *Arthrospira/Limnospira* genomes that are available in the GenBank. Undoubtedly it is not a new species, and based on genomic analyses it belongs to the group gathering strains *A. maxima* CS-328, *A. platensis* C1, *Arthrospira* sp. PCC 8005, TJSD091, TJSD092, BM1 and *Arthrospira fusiformis* KN and SAG 85.79. The second group consists of the strains *A. platensis* Paraca, NIES-39, NIES-46, YZ, FACHBB-835, FACHBB-439, FACHBB-971 and *Arthrospira* sp. PCC9108, PLM2.Bin9 and SH-MAG29. Presently published *Arthrospira* genomes belong to the I.A, I.B and II.A ITS subclusters distinguished by Baurain et al. The strain *Arthrospira* sp. O9.13F presented in this work is not only the first strain isolated from a climate other than a tropical or subtropical one (Siberia, Russia) but is also a representative of a new clade III based on the 16S rRNA gene (according to Comte et al. 2013), for which the genomic sequence is available in public databases (PKGD00000000).

What is more, here we report that the Siberian strain O9.13F has two different ITS variants in contrast to most *Arthrospira* strains that possess only one variant of the ITS within the ribosomal operon, except for strain PCC 7345 that was shown by [30].

The discrepancy between the phylogenies based on ITS and *cpcBA* has already been observed in the works of [5,32], in which the possibility of recombination has been postulated. Our results of the *cpcBA* sequence analysis do not agree with the data presented by [31,34]. These authors only used the noncoding IGS fragment between the *cpcBA* genes. It could be caused by the fact that we used, apart from the IGS region, also fragments of the flanking *cpcB* and *cpcA* genes, similar to Manen and Falquet 2002 and Dadheech et al., 2010 earlier [5,32].

The results of the phylogenetic analyses based on the 16S rRNA gene, ITS and phycocyanin operon fragment *cpcBA* are not consistent, indicating that these three genetic markers, which are typically used in the phylogenetic analysis of cyanobacteria, are not appropriate for the classification of the new isolates of *Arthrospira*. They do not allow for establishing a taxonomy of this genus as their results are not congruent. Furthermore, the position of the PCC9901 strain, which represents the 16S clade III, is undetermined.

Nevertheless, the results obtained in this study reveal the presence of only two groups within the genus *Arthrospira*. Finally, the strain O9.13F constantly merges with the same group of strains from the first clad 16S and the first ITS cluster. And these observations correlate with the results of genomic analyses, ANI, *is*DDH and core protein-based phylogeny.

The genomic comparisons showed that the O9.13F strain, like other members of this genus, possesses extensive R-M and CRISPR/Cas systems that enhance their protection system against exogenic invasive genetic elements, such as plasmids and viruses as well as intragenomic rearrangements. In addition, the gas vesicle genes operon of strain O9.13F is similar to other *Arthrospira* strains. The virtual analysis agreed with the TEM analysis, and both confirmed that the Siberian strain produces gas vesicles to regulate buoyancy depending on the light intensity.

Comparative genomics revealed no apparent characteristic genes that may be directly responsible for the adaptation of the Siberian strain to low temperatures of its habitat during the winter period. However, among unique CDs, we found those which encode proteins that might regulate the response to environmental stresses, such as methylase, the putative toxin/antitoxin system or a regulator of the CRISPR-Cas system [96–98]. What is more, strain O9.13F did not possess a different set of genes associated with the stress response than tropical strains.

The observed discrepancy between different physiological characteristics (observed acclimatisation) with the simultaneous lack of significant genetic differences (no signs of adaptation in the form of unique genes or different genes conditioning the survival of the low-temperature period) remains an intriguing observation that requires further research. Furthermore, in further studies aimed at explaining the mechanisms that allow the Siberian strain to survive the winter period, the influence of other microorganisms that coexist in the waters of Lake Solonye should be considered.

Considering all the data we collected; we can conclude that the genus *Arthrospira* has a much broader acclimation capacity than previously reported. The observed physiological properties of the Siberian strain indicate their usefulness in terms of production in temperate climate zones with the use of local water resources, such as water from the Baltic Sea.

Supplementary Materials: The following are available online at https://www.mdpi.com/article/10.3390/cells10123411/s1, Figure S1: The cyanobacterial bloom, Lake Solenoye, Figure S2: The growth index quantity of chlorophyll, total carotenoids extracted from biomass of *Arthrospira* sp. O9.13F cultivated at different temperatures and salt concentrations, Figure S3: Acclimation of *Arthrospira* strains to long-term cold stress, Figure S4: The percentage of the total fatty acids in the biomass of *Arthrospira* sp. O9.13F cultivated at 20, 28 and 37 °C, Figure S5: Phylogenetic tree of *Arthrospira* strains based on 16S rRNA partial sequences, Figure S6: Phylogenetic tree of *Arthrospira* strains based on ITS rRNA partial sequences, Figure S7: Phylogenetic tree of *Arthrospira* strains based on *cpcBA* sequences, Figure S8: Organisation of CRISPR/Cas operons in *Arthrospira* sp. O9.13F genome, Table S1. ANI and *is*DDH values between *Arthospira sp.* O9.13F and other 16 *Arthrospira* strains for which genomes are available in the GenBank, Table S2: Summary of statistics of *Arthrospira* genomes available on NCBI. Table S3: Presence of genes responsible for the stress adaptation in *Arthrospira* genomes. References [103,104] are refered to in Supplementary Materials.

Author Contributions: Conceptualization, M.W., K.F.W. and O.B.; methodology M.W., K.F.W., O.B., A.E.M., M.F., M.M.W. and M.D.; validation M.W., K.F.W., O.B., A.E.M., M.F., M.M.W. and M.D.; formal analysis, M.W., K.F.W., A.E.M. and M.M.W.; investigation, M.W., K.F.W., O.B., A.E.M., M.F., M.M.W. and M.D.; resources, M.W. and K.F.W.; data curation, M.W., K.F.W. and M.M.W.; writing—original draft preparation, M.W., K.F.W., M.M.W., A.E.M. and M.D.; visualization, M.W., K.F.W., A.E.M., M.F. and M.M.W.; supervision, M.W., K.F.W. and O.B.; project administration M.W. and K.F.W.; funding acquisition, M.W., K.F.W. and O.B. All authors have read and agreed to the published version of the manuscript.

Funding: This work was supported by the University of Gdansk project 531-N107-D801-21. Sample collection was funded by the "European Union—European Social Fund" (UDA-POKL.04.01.01-00-017/10) (2012–2013), University of Gdansk, Gdansk, Poland. Genome sequencing was supported by the Ministry of Science and Higher Education Republic of Poland from the quality-promoting subsidy under the Leading National Research Centre (KNOW) program 2012–2017 Faculty of Pharmacy– [dec. MNiSW-DS-6002-4693-23/WA/12] and the National Natural Science Foundation of China (32071480).

Institutional Review Board Statement: Not applicable.

Informed Consent Statement: Not applicable.

Data Availability Statement: The data that support the findings of this study are available from the corresponding authors (K.F.W. and M.W.), upon request. The draft genome sequence of *Arthrospira* sp. O9.13F has been deposited in NCBI/Genbank under accession number PKGD00000000. The nucleotide sequence data reported in this paper are available under the following accession numbers: 16S rDNA; MF509165, ITS; MF509166–MF509169, *cpcBA* and MF509119.

Conflicts of Interest: The authors declare no conflict of interest.

References

1. Castenholz, R.W.; Rippka, R.; Herdman, M.; Wilmotte, A. Form- Arthrospira. In *Bergey's Manual of Systematic Bacteriology*; Staley, J.T., Bryant, M.P., Pfenning, N., Holt, J.G., Eds.; William & Wilkins: Baltimore, MD, USA, 1989; Volume 3, pp. 1774–1775. ISBN 978-1-118-96060-8.
2. Nowicka-Krawczyk, P.; Muhlsteinova, R.; Hauer, T. Detailed Characterization of the Arthrospira Type Species Separating Commercially Grown Taxa into the New Genus Limnospira (Cyanobacteria). *Sci. Rep.* **2019**, *9*, 694. [CrossRef]

3. Geitler, L. *Dr. L. Rabenhorst's Kryptogamen-Flora von Deutschland, und der Schweiz. Zweite, vollständig neu bearbeitete Auflage; Vierzehnter Band. Die Algen herausgegeben von Prof. Dr. R. Kolkwitz, Berlin, Cyanophyceae von Europa unter Berücksichtigung der anderen Kontinente/von Dr. Lothar Geitler; Kryptogamen-Flora von Deutschland, und der Schweiz. Zweite, vollständig neu bearbeitete Auflage*, 2nd ed.; Akademische Verlagsgesellschaft m. b. H.: Leipzig, Germany, 1930; Volume 14.
4. Ciferri, O. Spirulina, the Edible Microorganism. *Microbiol. Rev.* **1984**, *47*, 551–578. [CrossRef]
5. Dadheech, P.K.; Ballot, A.; Casper, P.; Kotut, K.; Novelo, E.; Lemma, B.; Pröschold, T.; Krienitz, L. Phylogenetic Relationship and Divergence among Planktonic Strains of Arthrospira (Oscillatoriales, Cyanobacteria) of African, Asian and American Origin Deduced by 16S–23S ITS and Phycocyanin Operon Sequences. *Phycologia* **2010**, *49*, 361–372. [CrossRef]
6. Linhart, C.; Schagerl, M. Seasonal Succession of the Travertine-Forming Desmid Oocardium Stratum. *J. Phycol.* **2015**, *51*, 1055–1065. [CrossRef] [PubMed]
7. Bazhenova, O.; Konovalova, O. Phytoplankton of Lake Solenoye (Omsk) as a Promising Source of Bioresources. *Contemp. Probl. Ecol.* **2012**, *5*, 275–280. [CrossRef]
8. Leboulanger, C.; Agogué, H.; Bernard, C.; Bouvy, M.; Carré, C.; Cellamare, M.; Duval, C.; Fouilland, E.; Got, P.; Intertaglia, L.; et al. Microbial Diversity and Cyanobacterial Production in Dziani Dzaha Crater Lake, a Unique Tropical Thalassohaline Environment. *PLoS ONE* **2017**, *12*, e0168879. [CrossRef] [PubMed]
9. Castro-Severyn, J.; Pardo-Esté, C.; Mendez, K.; Fortt, J.; Marquez, S.; Molina, F.; Castro-Nallar, E.; Remonsellez, F.; Saavedra, C. Living to the High Extreme: Unraveling the Composition, Structure, and Functional Insights of Bacterial Communities Thriving in the Arsenic-Rich Salar de Huasco Altiplanic Ecosystem. *Microbiol. Spectr.* **2021**, *9*, e0044421. [CrossRef] [PubMed]
10. Sorokin, D.Y.; Berben, T.; Melton, E.D.; Overmars, L.; Vavourakis, C.D.; Muyzer, G. Microbial Diversity and Biogeochemical Cycling in Soda Lakes. *Extremophiles* **2014**, *18*, 791–809. [CrossRef]
11. Fužinato, S.; Fodora, A.; Subakov-Simić, G. Arthrospira Fusiformis (Voronichin) Komarek et Lund (Cyanoprokaryota) A New Species for Europe. *Algol. Stud.* **2010**, *134*, 17–24. [CrossRef]
12. Vavourakis, C.; Mehrshad, M.; Balkema, C.; van Hall, R.; Andrei, A.-S.; Ghai, R.; Sorokin, D.; Muyzer, G. Metagenomes and Metatranscriptomes Shed New Light on the Microbial-Mediated Sulfur Cycle in a Siberian Soda Lake. *BMC Biol.* **2019**, *17*, 69. [CrossRef] [PubMed]
13. Zorz, J.; Sharp, C.; Kleiner, M.; Gordon, P.; Pon, R.; Dong, X.; Strous, M. A Shared Core Microbiome in Soda Lakes Separated by Large Distances. *Nat. Commun.* **2019**, *10*, 4230. [CrossRef] [PubMed]
14. Grant, W.D.; Sorokin, D.Y. Distribution and Diversity of Soda Lake Alkaliphiles. In *Extremophiles Handbook*; Horikoshi, K., Ed.; Springer: Tokyo, Japan, 2011; pp. 27–54. ISBN 978-4-431-53898-1.
15. Furmaniak, M.; Misztak, A.; Franczuk, M.; Wilmotte, A.; Waleron, M.; Waleron, K. Edible Cyanobacterial Genus Arthrospira: Actual State of the Art in Cultivation Methods, Genetics, and Application in Medicine. *Front. Microbiol.* **2017**, *8*, 2541. [CrossRef]
16. Ananyev, G.; Carrieri, D.; Dismukes, G. Optimization of Metabolic Capacity and Flux through Environmental Cues to Maximize Hydrogen Production by the Cyanobacterium "Arthrospira (Spirulina) Maxima". *Appl. Environ. Microbiol.* **2008**, *74*, 6102–6113. [CrossRef] [PubMed]
17. Ainas, M.; Hasnaoui, S.; Bouarab, R.; Abdi, N.; Drouiche, N.; Mameri, N. Hydrogen Production with the Cyanobacterium Spirulina Platensis. *Int. J. Hydrogen Energy* **2017**, *42*, 4902–4907. [CrossRef]
18. Krishnan, A.; Qian, X.; Ananyev, G.; Lun, D.; Dismukes, G. Rewiring of Cyanobacterial Metabolism for Hydrogen Production: Synthetic Biology Approaches and Challenges. *Adv. Exp. Med. Biol.* **2018**, *1080*, 171–213. [CrossRef] [PubMed]
19. Suzuki, S.; Yamaguchi, H.; Kawachi, M. The Draft Genome of a Hydrogen-Producing Cyanobacterium, Arthrospira Platensis NIES-46. *J. Genom.* **2019**, *7*, 56–59. [CrossRef]
20. Badri, H.; Monsieurs, P.; Coninx, I.; Ruddy, W.; Leys, N. Molecular Investigation of the Radiation Resistance of Edible Cyanobacterium Arthrospira Sp. PCC 8005. *MicrobiologyOpen* **2015**, *4*, 187–207. [CrossRef] [PubMed]
21. Hendrickx, L.; De Wever, H.; Hermans, V.; Mastroleo, F.; Morin, N.; Wilmotte, A.; Janssen, P.; Mergeay, M. Microbial Ecology of the Closed Artificial Ecosystem MELiSSA (Micro-Ecological Life Support System Alternative): Reinventing and Compartmentalizing the Earth's Food and Oxygen Regeneration System for Long-Haul Space Exploration Missions. *Res. Microbiol.* **2006**, *157*, 77–86. [CrossRef] [PubMed]
22. Gugliemi, G.; Rippka, R.; Tandeau De Marsac, N. Main Properties That Justify the Different Taxonomic Position of *Spirulina* spp. and *Arthrospira* spp. among Cyanobacteria. *Bull. Inst. Oceanogr. Monaco* **1993**, *12*, 13–23.
23. Tomaselli, L. Morphology, Ultrastructure and Taxonomy of Arthrospira (Spirulina Maxima and Arthrospira (Spirulina) Platensis. In *Spirulina platensis (Arthrospira): Physiology, Cell-Biology and Biotechnology*; Vonshak, A., Ed.; Taylor & Francis Ltd.: London, UK, 1997; pp. 1–15. ISBN 978-0-429-07994-8.
24. Komárek, J.; Lund, J.W.G. What Is «Spirulina Platensis» in Fact? *Algol. Stud. Hydrobiol. Suppl. Vol.* **1990**, *58*, 1–13.
25. Komárek, J.; Anagnostidis, K. *Süßwasserflora von Mitteleuropa, Bd. 19/2: Cyanoprokaryota: Bd. 2/Part 2: Oscillatoriales*; Elsveire GmbH: Munchen, Germany, 2005; ISBN 3-8274-0919-5.
26. Bai, N.; Seshadri, C.V. On Coiling and Uncoiling of Trichomes in the Genus Spirulina. *Algol. Stud.* **1980**, *26*, 32–47.
27. Bai, N. Competitive Exclusion or Morphological Transformation? A Case Study with Spirulina Fusiformis. *Arch. Hydrobiol.* **1985**, *71*, 191–199.
28. Mühling, M.; Harris, N.; Belay, A.; Whitton, B. Reversal of Helix Orientation in the Cyanobacterium Arthrospira. *J. Phycol.* **2003**, *39*, 360–367. [CrossRef]

29. Scheldeman, P.; Baurain, D.; Bouhy, R.; Scott, M.; Mühling, M.; Whitton, B.; Belay, A.; Wilmotte, A. Arthrospira ('Spirulina') Strains from Four Continents Are Resolved into Only Two Clusters, Based on Amplified Ribosomal DNA Restriction Analysis of the Internally Transcribed Spacer. *FEMS Microbiol. Lett.* **1999**, *172*, 213–222. [CrossRef] [PubMed]
30. Baurain, D.; Renquin, L.; Grubisic, S.; Scheldeman, P.; Belay, A.; Wilmotte, A. Remarkable Conservation of Internally Transcribed Spacer Sequences of Arthrospira ("Spirulina") (Cyanophyceae, Cyanobacteria) Strains from Four Continents and of Recent and 30-Year-Old Dried Samples from Africa. *J. Phycol.* **2002**, *38*, 384–393. [CrossRef]
31. Hicks, M.; Tran-Dao, T.-K.; Mulroney, L.; Bernick, D. De-Novo Assembly of Limnospira Fusiformis Using Ultra-Long Reads. *Front. Microbiol.* **2021**, *12*, 657995. [CrossRef] [PubMed]
32. Manen, J.-F.; Falquet, J. The CpcB-CpcA Locus as a Tool for the Genetic Characterization of the Genus Arthrospira (Cyanobacteria): Evidence for Horizontal Transfer. *Int. J. Syst. Evol. Microbiol.* **2002**, *52*, 861–867. [CrossRef]
33. Ballot, A.; Dadheech, P.; Krienitz, L. Phylogenetic Relationship of Arthrospira, Phormidium and Spirulina Strains from Kenyan and Indian Waterbodies. *Algol. Stud.* **2004**, *113*, 37–56. [CrossRef]
34. Papapanagiotou, G.; Gkelis, S. Taxonomic Revision of Commercially Used Arthrospira (Cyanobacteria) Strains: A Polyphasic Approach. *Eur. J. Phycol.* **2019**, *54*, 595–608. [CrossRef]
35. Mühling, M.; Somerfield, P.; Harris, N.; Belay, A.; Whitton, B. Phenotypic Analysis of Arthrospira (Spirulina) Strains (Cyanobacteria). *Phycologia* **2006**, *45*, 148–157. [CrossRef]
36. Ogato, T.; Kifle, D.; Fetahi, T.; Sitotaw, B. Evaluation of Growth and Biomass Production of Arthrospira (Spirulina) Fusiformis in Laboratory Cultures Using Waters from the Ethiopian Soda Lakes Chitu and Shala. *J. Appl. Phycol.* **2014**, *26*, 2273–2282. [CrossRef]
37. Janssen, P.; Morin, N.; Mergeay, M.; Leroy, B.; Ruddy, W.; Vallaeys, T.; Waleron, K.; Waleron, M.; Wilmotte, A.; Quillardet, P.; et al. Genome Sequence of the Edible Cyanobacterium Arthrospira Sp. PCC 8005. *J. Bacteriol.* **2010**, *192*, 2465–2466. [CrossRef] [PubMed]
38. Fujisawa, T.; Narikawa, R.; Okamoto, S.; Ehira, S.; Yoshimura, H.; Suzuki, I.; Masuda, T.; Mochimaru, M.; Takaichi, S.; Awai, K.; et al. Genomic Structure of an Economically Important Cyanobacterium, Arthrospira (Spirulina) Platensis NIES-39. *DNA Res. Int. J. Rapid Publ. Rep. Genes Genomes* **2010**, *17*, 85–103. [CrossRef] [PubMed]
39. Cheevadhanarak, S.; Paithoonrangsarid, K.; Prommeenate, P.; Kaewngam, W.; Musigkain, A.; Tragoonrung, S.; Tabata, S.; Kaneko, T.; Chaijaruwanich, J.; Sangsrakru, D.; et al. Draft Genome Sequence of Arthrospira Platensis C1 (PCC9438). *Stand. Genom. Sci.* **2012**, *6*, 43–53. [CrossRef] [PubMed]
40. Carrieri, D.; Ananyev, G.; Lenz, O.; Bryant, D.; Dismukes, G. Contribution of a Sodium Ion Gradient to Energy Conservation during Fermentation in the Cyanobacterium Arthrospira (Spirulina) Maxima CS-328. *Appl. Environ. Microbiol.* **2011**, *77*, 7185–7194. [CrossRef] [PubMed]
41. Lefort, F.; Calmin, G.; Crovadore, J.; Falquet, J.; Hurni, J.-P.; Østerås, M.; Haldemann, F.; Farinelli, L. Whole-Genome Shotgun Sequence of Arthrospira Platensis Strain Paraca, a Cultivated and Edible Cyanobacterium. *Genome Announc.* **2014**, *2*, e00751-14. [CrossRef] [PubMed]
42. Xu, T.; Qin, S.; Hu, Y.; Song, Z.; Ying, J.; Li, P.; Dong, W.; Zhao, F.; Yang, H.; Bao, Q. Whole Genomic DNA Sequencing and Comparative Genomic Analysis of Arthrospira Platensis: High Genome Plasticity and Genetic Diversity. *DNA Res. Int. J. Rapid Publ. Rep. Genes Genomes* **2016**, *23*, 325–338. [CrossRef]
43. Dong, S.; Chen, J.; Wang, S.; Wu, Y.; Hou, H.; Li, M.; Yan, C. Draft Genome Sequence of Cyanobacteria Arthrospira Sp. TJSD091 Isolated from Seaside Wetland. *Mar. Genom.* **2015**, *24*, 197–198. [CrossRef] [PubMed]
44. Sharp, C.; Urschel, S.; Dong, X.; Brady, A.; Slater, G.; Strous, M. Robust, High-Productivity Phototrophic Carbon Capture at High PH and Alkalinity Using Natural Microbial Communities. *Biotechnol. Biofuels* **2017**, *10*, 84. [CrossRef] [PubMed]
45. Chen, M.-Y.; Teng, W.; Zhao, L.; Hu, C.-X.; Zhou, Y.; Han, B.-P.; Song, L.-R.; Shu, W.-S. Comparative Genomics Reveals Insights into Cyanobacterial Evolution and Habitat Adaptation. *ISME J.* **2020**, *15*, 211–227. [CrossRef] [PubMed]
46. Maghembe, R.; Angelina, M.; Harish, A.; Nyandoro, S.; Lyantagaye, S.; Hati-Kaul, R. Draft Genome Sequence of Limnospira Sp. Strain BM01, Isolated from a Hypersaline Lake of the Momela Ecosystem in Tanzania. *Microbiol. Resour. Announc.* **2021**, *10*, e00132-21. [CrossRef] [PubMed]
47. Zarrouk, C.; Zarrouk, C. *Contribution à l 'Étude d 'une Cyanophycée Influence de Divers Facteurs Physiques et Chimiques sur la Croissance et la Photosynthése de Spirulina Maxima Geitler*; Academie de Paris: Paris, France, 1966.
48. Rippka, R. Isolation and Purification of Cyanobacteria. *Methods Enzymol.* **1988**, *167*, 3–27. [CrossRef] [PubMed]
49. Bligh, E.L.G.; Dyer, W.J.A. A Rapid Method of Total Lipid Extraction And Purification. *Can. J. Biochem. Physiol.* **1959**, *37*, 911–917. [CrossRef] [PubMed]
50. Bradford, M.M. A Rapid and Sensitive Method for the Quantitation of Microgram Quantities of Protein Utilizing the Principle of Protein-Dye Binding. *Anal. Biochem.* **1976**, *72*, 248–254. [CrossRef]
51. Ritchie, R.J. Consistent Sets of Spectrophotometric Chlorophyll Equations for Acetone, Methanol and Ethanol Solvents. *Photosynth. Res.* **2006**, *89*, 27–41. [CrossRef] [PubMed]
52. Wellburn, A. The Spectral Determination of Chlorophylls a and b, as Well as Total Carotenoids, Using Various Solvents with Spectrophotometers of Different Resolution. *J. Plant Physiol.* **1994**, *144*, 307–313. [CrossRef]
53. Pinheiro, J.; Bates, D.; DebRoy, S.; Sarkar, D.; R Core Team. nlme: Linear and Nonlinear Mixed Effects Models. R package version 3.1-153, 2021. Available online: https://CRAN.R-project.org/package=nlme (accessed on 1 October 2020).

54. Eykelenburg, C. The Ultrastructure of Spirulina Platensis in Relation to Temperature and Light Intensity. *Antonie Leeuwenhoek* **1979**, *45*, 369–390. [CrossRef] [PubMed]
55. Hayat, M.A. *Principles and Techniques of Electron Microscopy: Biological Applications*; Edward Arnold: London, UK, 1981; ISBN 978-0-7131-2830-7.
56. William, S.; Feil, H.; Copeland, A. Bacterial Genomic DNA Isolation Using CTAB 2012. Available online: http://1ofdmq2n8tc36m6i46scovo2e.wpengine.netdna-cdn.com/wp-content/uploads/2014/02/JGI-Bacterial-DNA-isolation-CTAB-Protocol-2012.pdf (accessed on 1 October 2019).
57. Waleron, M.; Waleron, K.; Vincent, W.F.; Wilmotte, A. Allochthonous Inputs of Riverine Picocyanobacteria to Coastal Waters in the Arctic Ocean. *FEMS Microbiol. Ecol.* **2007**, *59*, 356–365. [CrossRef] [PubMed]
58. Hrouzek, P.; Ventura, S.; Lukesova, A.; Mugnai, M.; Turicchia, S.; Komárek, J. Diversity of Soil Nostoc Strains: Phylogenetic and Phenotypic Variability. *Algol. Stud.* **2005**, *117*, 251–264. [CrossRef]
59. Taton, A.; Grubisic, S.; Brambilla, E.; De Wit, R.; Wilmotte, A. Cyanobacterial Diversity in Natural and Artificial Microbial Mats of Lake Fryxell (McMurdo Dry Valleys, Antarctica): A Morphological and Molecular Approach. *Appl. Environ. Microbiol.* **2003**, *69*, 5157–5169. [CrossRef] [PubMed]
60. Wilmotte, A.; Van der Auwera, G.; Wachter, R. Structure of the 16 S Ribosomal RNA of the Thermophilic Cyanobacterium Chlorogloeopsis HTF ('Mastigocladus Laminosus HTF') Strain PCC7518, and Phylogenetic Analysis. *FEBS Lett.* **1993**, *317*, 96–100. [CrossRef]
61. Kearse, M.; Moir, R.; Wilson, A.; Stones-Havas, S.; Cheung, M.; Sturrock, S.; Buxton, S.; Cooper, A.; Markowitz, S.; Duran, C.; et al. Geneious Basic: An Integrated and Extendable Desktop Software Platform for the Organization and Analysis of Sequence Data. *Bioinform. Oxf. Engl.* **2012**, *28*, 1647–1649. [CrossRef] [PubMed]
62. Wood, D.; Salzberg, S.; Wood, D.E.; Salzberg, S.L. Kraken: Ultrafast Metagenomic Sequence Classification Using Exact Alignment. *Genome Biol.* **2014**, *15*, R46. [CrossRef] [PubMed]
63. Bolger, A.; Lohse, M.; Usadel, B. Trimmomatic: A Flexible Trimmer for Illumina Sequence Data. *Bioinform. Oxf. Engl.* **2014**, *30*, 2114–2120. [CrossRef] [PubMed]
64. Chikhi, R.; Medvedev, P. Informed and Automated K-Mer Size Selection for Genome Assembly. *Bioinform. Oxf. Engl.* **2013**, *30*, 31–37. [CrossRef] [PubMed]
65. Bankevich, A.; Nurk, S.; Antipov, D.; Gurevich, A.; Dvorkin, M.; Kulikov, A.; Lesin, V.; Nikolenko, S.; Pham, S.; Prjibelski, A.; et al. SPAdes: A New Genome Assembly Algorithm and Its Applications to Single-Cell Sequencing. *J. Comput. Biol. J. Comput. Mol. Cell Biol.* **2012**, *19*, 455–477. [CrossRef] [PubMed]
66. Zerbino, D.; Birney, E. Velvet: Algorithms for De Novo Short Read Assembly Using De Bruijn Graphs. *Genome Res.* **2008**, *18*, 821–829. [CrossRef]
67. Chevreux, B.; Wetter, T.; Suhai, S. Genome Sequence Assembly Using Trace Signals and Additional Sequence Information. In Proceedings of the Computer Science and Biology: Proceedings of the German Conference on Bioinformatics (GCB'99), Braunschweig, Germany, 7–10 October 1999; Volume 99, p. 56.
68. Boisvert, S.; Laviolette, F.; Corbeil, J. Ray: Simultaneous Assembly of Reads from a Mix of High-Throughput Sequencing Technologies. *J. Comput. Biol. J. Comput. Mol. Cell Biol.* **2010**, *17*, 1519–1533. [CrossRef] [PubMed]
69. Boetzer, M.; Henkel, C.; Jansen, H.; Butler, D.; Pirovano, W. Scaffolding Pre-Assembled Contigs Using SSPACE. *Bioinform. Oxf. Engl.* **2011**, *27*, 578–579. [CrossRef] [PubMed]
70. Gurevich, A.; Saveliev, V.; Vyahhi, N.; Tesler, G. QUAST: Quality Assessment Tool for Genome Assemblies. *Bioinform. Oxf. Engl.* **2013**, *29*, 1072–1075. [CrossRef] [PubMed]
71. Tatusova, T.; DiCuccio, M.; Badretdin, A.; Chetvernin, V.; Nawrocki, E.; Zaslavsky, L.; Lomsadze, A.; Pruitt, K.; Borodovsky, M.; Ostell, J. NCBI Prokaryotic Genome Annotation Pipeline. *Nucleic Acids Res.* **2016**, *44*, gkw569. [CrossRef] [PubMed]
72. Haft, D.; DiCuccio, M.; Badretdin, A.; Brover, V.; Chetvernin, V.; O'Neill, K.; Li, W.; Chitsaz, F.; Derbyshire, M.; Gonzales, N.; et al. RefSeq: An Update on Prokaryotic Genome Annotation and Curation. *Nucleic Acids Res.* **2017**, *46*, D851–D860. [CrossRef] [PubMed]
73. Moriya, Y.; Itoh, M.; Okuda, S.; Yoshizawa, A.; Kanehisa, M.; Moriya, Y.; Itoh, M.; Okuda, S.; Yoshizawa, A.C.; Kanehisa, M. KAAS: An Automatic Genome Annotation and Pathway Reconstruction Server. *Nucleic Acids Res.* **2007**, *35*, W182–W185. [CrossRef] [PubMed]
74. Koski, L.; Gray, M.; Lang, B.; Burger, G. AutoFACT: An Automatic Functional Annotation and Classification Tool. *BMC Bioinform.* **2005**, *6*, 151. [CrossRef] [PubMed]
75. Käll, L.; Krogh, A.; Sonnhammer, E. Advantages of Combined Transmembrane Topology and Signal Peptide Prediction—The Phobius Web Server. *Nucleic Acids Res.* **2007**, *35*, W429–W432. [CrossRef] [PubMed]
76. Roberts, R.; Vincze, T.; Posfai, J.; Macelis, D. REBASE-A Database for DNA Restriction and Modification: Enzymes, Genes and Genomes. *Nucleic Acids Res.* **2015**, *43*, D298–D299. [CrossRef] [PubMed]
77. Grissa, I.; Vergnaud, G.; Pourcel, C. CRISPRFinder: A Web Tool to Identify Clustered Regularly Interspaced Short Palindromic Repeats. *Nucleic Acids Res.* **2007**, *35*, W52–W57. [CrossRef] [PubMed]
78. Vallenet, D.; Calteau, A.; Dubois, M.; Amours, P.; Bazin, A.; Beuvin, M.; Burlot, L.; Bussell, X.; Fouteau, S.; Gautreau, G.; et al. MicroScope: An Integrated Platform for the Annotation and Exploration of Microbial Gene Functions through Genomic, Pangenomic and Metabolic Comparative Analysis. *Nucleic Acids Res.* **2019**, *48*, D579–D589. [CrossRef] [PubMed]

79. Arndt, D.; Grant, J.; Marcu, A.; Sajed, T.; Pon, A.; Liang, Y.; Wishart, D. PHASTER: A Better, Faster Version of the PHAST Phage Search Tool. *Nucleic Acids Res.* **2016**, *44*, gkw387. [CrossRef] [PubMed]
80. Carattoli, A.; Zankari, E.; Garcia-Fernandez, A.; Larsen, M.; Lund, O.; Villa, L.; Aarestrup, F.; Hasman, H. In Silico Detection and Typing of Plasmids Using PlasmidFinder and Plasmid Multilocus Sequence Typing. *Antimicrob. Agents Chemother.* **2014**, *58*, 3895–3903. [CrossRef] [PubMed]
81. Weber, T.; Blin, K.; Duddela, S.; Krug, D.; Kim, T.Y.; Bruccoleri, R.; Lee, S.Y.; Fischbach, M.; Müller, R.; Wohlleben, W.; et al. AntiSMASH 3.0-A Comprehensive Resource for the Genome Mining of Biosynthetic Gene Clusters. *Nucleic Acids Res.* **2015**, *43*, W237–W243. [CrossRef] [PubMed]
82. Blin, K.; Medema, M.; Kottmann, R.; Lee, S.Y.; Weber, T. The AntiSMASH Database, a Comprehensive Database of Microbial Secondary Metabolite Biosynthetic Gene Clusters. *Nucleic Acids Res.* **2016**, *45*, gkw960. [CrossRef] [PubMed]
83. Segata, N.; Börnigen, D.; Morgan, X.; Huttenhower, C. PhyloPhlAn Is a New Method for Improved Phylogenetic and Taxonomic Placement of Microbes. *Nat. Commun.* **2013**, *4*, 2304. [CrossRef] [PubMed]
84. Pritchard, L.; Glover, R.; Humphris, S.; Elphinstone, J.; Toth, I. Genomics and Taxonomy in Diagnostics for Food Security: Soft-Rotting Enterobacterial Plant Pathogens. *Anal. Methods* **2015**, *8*, 12–24. [CrossRef]
85. Kurtz, S.; Phillippy, A.; Delcher, A.; Smoot, M.; Shumway, M.; Antonescu, C.; Salzberg, S. Versatile and Open Software for Comparing Large Genomes. *Genome Biol.* **2004**, *5*, R12. [CrossRef] [PubMed]
86. Meier-Kolthoff, J.; Auch, A.; Klenk, H.-P.; Göker, M. Genome Sequence-Based Species Delimitation with Confidence Intervals and Improved Distance Functions. *BMC Bioinform.* **2013**, *14*, 60. [CrossRef] [PubMed]
87. Camacho, C.; Coulouris, G.; Avagyan, V.; Ma, N.; Papadopoulos, J.; Bealer, K.; Madden, T.L. BLAST+: Architecture and Applications. *BMC Bioinform.* **2009**, *10*, 421. [CrossRef] [PubMed]
88. Comte, K.; Coursin, T.; Carre-Mlouka, A. A New Genotype in the Genus Arthrospira (Oscillatoriales, Cyanobacteria) Revealed by a Mosaic-like Structure of the 16S-23SrRNA Intergenic Spacer Region in Strain PCC 9901. *Phycologia* **2013**, *52*, 333–337. [CrossRef]
89. Burns, B.; Saker, M.; Moffitt, M.; Neilan, B. Molecular Detection of Genes Responsible for Cyanobacterial Toxin Production in the Genera Microcystis, Nodularia, and Cylindrospermopsis. *Methods Mol. Biol.* **2004**, *268*, 213–222. [CrossRef] [PubMed]
90. Waleron, M.; Misztak, A.; Waleron, M.; Furmaniak, M.; Mrozik, A.; Waleron, K. Arthrospiribacter Ruber Gen. Nov., Sp. Nov., a Novel Bacterium Isolated from Arthrospira Cultures. *Syst. Appl. Microbiol.* **2020**, *43*, 126072. [CrossRef] [PubMed]
91. Hoiczyk, E.; Baumeister, W. Oscillin, an Extracellular, Ca2+-Binding Glycoprotein Essential for the Gliding Motility of Cyanobacteria. *Mol. Microbiol.* **1997**, *26*, 699–708. [CrossRef]
92. Miklaszewska, M.; Waleron, M.; Morin, N.; Calusinska, M.; Wilmotte, A.; Marsac, N.; Rippka, R.; Waleron, K. Elucidation of the Gas Vesicle Gene Clusters in Cyanobacteria of the Genus Arthrospira (Oscillatoriales, Cyanophyta) and Correlation with ITS Phylogeny. *Eur. J. Phycol.* **2012**, *47*, 233–244. [CrossRef]
93. Carre-Mlouka, A.; Comte, K.; Castets, A.-M.; Bouchier, C.; Marsac, N. The Gas Vesicle Gene Cluster from Microcystis Aeruginosa and DNA Rearrangements That Lead to Loss of Cell Buoyancy. *J. Bacteriol.* **2004**, *186*, 2355–2365. [CrossRef] [PubMed]
94. Hagemann, M. Molecular Biology of Cyanobacterial Salt Acclimation. *FEMS Microbiol. Rev.* **2011**, *35*, 87–123. [CrossRef] [PubMed]
95. Hagemann, M. Chapter Two—Genomics of Salt Acclimation: Synthesis of Compatible Solutes among Cyanobacteria. In *Advances in Botanical Research*; Chauvat, F., Cassier-Chauvat, C., Eds.; Genomics of Cyanobacteria; Academic Press: Cambridge, MA, USA, 2013; Volume 65, pp. 27–55.
96. Matelska, D.; Ginalski, K. Comprehensive Classification of the PIN Domain-like Superfamily. *Nucleic Acids Res.* **2017**, *45*, 6995–7020. [CrossRef] [PubMed]
97. Zhang, H.; Dong, C.; Li, L.; Wasney, G.; Min, J. Structural Insights into the Modulatory Role of the Accessory Protein WYL1 in the Type VI-D CRISPR-Cas System. *Nucleic Acids Res.* **2019**, *47*, 5420–5428. [CrossRef] [PubMed]
98. Koonin, E.; Zhang, F. Coupling Immunity and Programmed Cell Suicide in Prokaryotes: Life-or-Death Choices. *BioEssays* **2016**, *39*, 1–9. [CrossRef] [PubMed]
99. Stappen, G.; Litvinenko, L.; Litvinenko, A.; Boyko, E.; Marden, B.; Sorgeloos, P. A Survey of Artemia Resources of Southwest Siberia (Russian Federation). *Rev. Fish. Sci.* **2009**, *17*, 117–148. [CrossRef]
100. Mühling, M.; Belay, A.; Whitton, B. Variation in Fatty Acid Composition of Arthrospira (Spirulina) Strains. *J. Appl. Phycol.* **2005**, *17*, 137–146. [CrossRef]
101. Colla, L.M.; Bertolin, T.E.; Costa, J.A.V. Fatty Acids Profile of Spirulina Platensis Grown Under Different Temperatures and Nitrogen Concentrations. *Z. Nat. C* **2004**, *59*, 55–59. [CrossRef] [PubMed]
102. Shiraishi, H. Cryopreservation of the Edible Alkalophilic Cyanobacterium Arthrospira Platensis. *Biosci. Biotechnol. Biochem.* **2016**, *80*, 2051–2057. [CrossRef] [PubMed]
103. Goris, J.; Konstantinidis, K.T.; Klappenbach, J.A.; Coenye, T.; Vandamme, P.; Tiedje, J.M. DNA-DNA hybridization values and their relationship to whole-genome sequence similarities. *Int. J. Syst. Evol. Microbiol.* **2007**, *57*, 81–91. [CrossRef] [PubMed]
104. Richter, M.; Rosselló-Móra, R. Shifting the genomic gold standard for the prokaryotic species definition. *Proc. Natl. Acad. Sci. USA* **2009**, *106*, 19126–19131. [CrossRef]

cells

MDPI

Article

The Arbuscular Mycorrhizal Fungus *Glomus viscosum* Improves the Tolerance to Verticillium Wilt in Artichoke by Modulating the Antioxidant Defense Systems

Alessandra Villani, Franca Tommasi and Costantino Paciolla *

Department of Biology, University of Bari "Aldo Moro", Via E. Orabona 4, 70125 Bari, Italy; alessandra.villani@uniba.it (A.V.); franca.tommasi@uniba.it (F.T.)
* Correspondence: costantino.paciolla@uniba.it; Tel.: +39-080-544-3557

Abstract: Verticillium wilt, caused by the fungal pathogen *Verticillium dahliae*, is the most severe disease that threatens artichoke (*Cynara scolymus* L.) plants. Arbuscular mycorrhizal fungi (AMF) may represent a useful biological control strategy against this pathogen attack, replacing chemical compounds that, up to now, have been not very effective. In this study, we evaluated the effect of the AMF *Glomus viscosum* Nicolson in enhancing the plant tolerance towards the pathogen *V. dahliae*. The role of the ascorbate-glutathione (ASC-GSH) cycle and other antioxidant systems involved in the complex network of the pathogen-fungi-plant interaction have been investigated. The results obtained showed that the AMF *G. viscosum* is able to enhance the defense antioxidant systems in artichoke plants affected by *V. dahliae*, alleviating the oxidative stress symptoms. AMF-inoculated plants exhibited significant increases in ascorbate peroxidase (APX), monodehydroascorbate reductase (MDHAR), and superoxide dismutase (SOD) activities, a higher content of ascorbate (ASC) and glutathione (GSH), and a decrease in the levels of lipid peroxidation and hydrogen peroxide (H_2O_2). Hence, *G. viscosum* may represent an effective strategy for mitigating *V. dahliae* pathogenicity in artichokes, enhancing the plant defense systems, and improving the nutritional values and benefit to human health.

Keywords: Verticillium wilt; *Glomus viscosum* Nicolson; arbuscular mycorrhizal fungi; oxidative stress; antioxidant systems; defense ability

Citation: Villani, A.; Tommasi, F.; Paciolla, C. The Arbuscular Mycorrhizal Fungus *Glomus viscosum* Improves the Tolerance to Verticillium Wilt in Artichoke by Modulating the Antioxidant Defense Systems. *Cells* **2021**, *10*, 1944. https://doi.org/10.3390/cells10081944

Academic Editor: Suleyman Allakhverdiev

Received: 24 June 2021
Accepted: 28 July 2021
Published: 30 July 2021

1. Introduction

The artichoke (*Cynara scolymus* L.) is a horticultural species of relevant economic interest belonging to the Asteraceae family, widely cultivated in the Mediterranean basin and widespread throughout the world [1,2]. This perennial crop is well known for the antioxidative, antimicrobial, and probiotic properties of its edible parts, including the inner fleshy leaves (bracts) and the receptacle [1,3]. Several studies have demonstrated that even some non-food by-products of artichokes, including leaves, external bracts, and stems, exhibit beneficial and therapeutic effects and are widely used as hepatoprotective [4], antioxidant [5,6], anticarcinogenic [7], hypoglycemic [8], and hypocholesterolemic [9] agents. The health-promoting properties and important nutritional values of artichokes have been extensively related to inulin, fibers, and minerals, and to the high content of some bioactive phenolic compounds, such as caffeoylquinic acid derivatives and flavonoids, showing a strong scavenging activity against reactive oxygen species (ROS) and free radicals [1,2,10].

As the artichoke is an herbaceous plant which survives in the field for several years, a large number of insects, nematodes, bacteria, fungi, and viruses can attack and invade its seeds, roots, foliage, and vascular system, causing numerous diseases [11,12]. Verticillium wilt, caused by the fungus *V. dahliae* Kleb., represents one of the greatest threats to the artichoke plantation worldwide [13–15]. This soil-borne pathogen is distributed throughout the world, and it affects over 400 plant species that belong to 14 plant families, showing a broad range of symptoms and causing significant yield losses and quality reduction for most of the host plant species [15,16]. Among these species, almond, apricot, artichoke, cabbage, cauliflower, chrysanthemum, cotton, cucurbits, olive, peach, potato, strawberry, sunflower, and tomato were defined as the most severely affected host crops [16]. *V. dahliae* causes a monocyclic disease divided into three phases: dormant, parasitic, and saprophytic [17,18]. In the presence of a host, the disease cycle begins with germination of microsclerotia that are released in the soil with the decomposition of plant materials where they can remain viable for up to 14 years. Hyphae from germinating microsclerotia can colonize and penetrate the roots of host plants, following a slow progression through the vascular (xylem) tissues, where conidia can grow and continue the colonization, leading to xylem malfunctioning and reduced movement of water and nutrients from the roots to the foliage of the infected host [17]. Over the years, several strategies have been tested to manage diseases caused by *V. dahliae*, including selection of planting site, crop rotation and manipulation of fertility and irrigation, use of healthy planting material, selection of available resistant cultivars, fungicides treatments, and soil fumigants [15,18–21].

Methyl bromide has been widely used for decades as a soil fumigant for controlling Verticillium wilt until its complete phase-out in 2005 according to the Regulation EC 2037/2000 because of its threat to the environment, being one of the major ozone depleting substances, and to humans, causing lung injury and neurological effects. Some other fumigants have been tested, such as the mixture of 1,3-dichloropropene and chloropicrin, dazomet, and metam sodium, but are not very effective [22]. Similarly, several fungicides applied as foliar sprays, soil drenches, or granular preparations have been tested, but the effectiveness was observed only with high dosages, which many a times cause phytotoxic effects [23]. Therefore, the inability of fungicide and soil fumigants treatments to control the disease successfully, the unavailability of artichoke resistant cultivars as well as the inaccessibility of *V. dahliae* during infection and its long-term persistence in the field have required alternative strategies. Furthermore, the public concern over the environment pollution, ecosystem's biodiversity, and food safety has enhanced research efforts towards eco-friendly practices for a sustainable agricultural management. To ensure that aim, beneficial microorganisms, such as AMF, could play a crucial role. AMF are symbionts, mainly belonging to the phylum Glomeromycota that form arbuscular mycorrhizal associations with the roots of over 80% of all vascular plants [24]. Numerous studies have demonstrated that AMF can improve water and mineral nutrient uptake from the soil by increasing the plant root surface area [25]. Initial stages of AMF colonization trigger an intracellular ROS burst in the host plant; however, this effect is transient and is overcome by enhanced activities of antioxidant enzymes [26]. Indeed, AMF increase the accumulation of secondary metabolites in several plants, including phenolic compounds, vitamins, and sugars [10,27], mitigate the oxidative burst of plants under abiotic stresses by increasing the activity of some antioxidants that are involved in the alleviation of oxidative damage caused by ROS [28–30], and protect host plants from pathogens, overcoming the harmful effects of abiotic and biotic stresses [31]. In particular, the efficacy of AMF in reducing the disease severity of Verticillium wilt has been demonstrated on olive [32,33], eggplant and tomato [34,35], pepper [35], oilseed rape (*Brassica napus* L. cv. Licosmos), and strawberry (*Fragaria ananassa* cv. Elsanta) [36] plants. Moreover, the presence of an autochthonous mycorrhizal consortium "Rhizolive consortium" on the early oxidative events induced in olive plants after *V. dahliae* inoculation stimulated the activity of antioxidant enzymes, reducing oxidative damage [37], and treatments with six AMF in two artichoke cultivars increased the level of total phenols and total antioxidant activity [20].

The increase in ROS is a common biochemical response to abiotic and biotic stresses in plants. Higher ROS levels in the cell could cause oxidative damage to DNA, lipids, and proteins. It is well known that the ROS level in cells is under the control of antioxidant systems, such as the ASC-GSH cycle [38], and enzymes, including SOD, catalase (CAT), and generic peroxidases (PODs), which have a pivotal role in defense mechanisms. The activity of those enzymes, with that of APX, a component of ASC-GSH cycle, is crucial for determining the steady-state level of superoxide anion and H_2O_2 in plant cells [39]. SOD dismutates superoxide anion to H_2O_2, which can be converted into oxygen and H_2O by CAT, PODs, or APX. In the ASC-GSH cycle, the APX uses two molecules of ASC to reduce H_2O_2 to water, with the concomitant generation of two molecules of monodehydroascorbate (MDHA) [40]. MDHA is a radical with a short lifetime that is rapidly reduced to ASC by MDHAR, which is a flavin enzyme that utilizes NAD(P)H as electron donors. Dehydroascorbate (DHA), the oxidized form of ASC, can be reduced back to ASC by DHA reductase (DHAR), which utilizes GSH as an electron donor, leading to the formation of glutathione disulphide (GSSG), which is in turn re-reduced to GSH by NADPH, a reaction catalyzed by the glutathione reductase (GR).

The objective of this study was to test the ability of the AMF *G. viscosum* to moderate the metabolic alterations related to oxidative stress in artichoke plants attacked by *V. dahliae*. In particular, a deeper investigation of the involvement of the defense systems in the fungi-plant interaction, by evaluating the level of the components of the ASC-GSH cycle and the antioxidant enzymes (CAT and SOD), led to a better understanding of the biochemical mechanism on the basis of this complex network among the plant, pathogen, and AMF.

2. Materials and Methods

2.1. Chemicals

All reagents used in this study were of the highest grade available, purchased from Sigma–Aldrich (Milan, Italy) and used without further purification. Ultrapure water was produced by a Milli-Q system 84 (Millipore, Bedford, MA, USA).

2.2. Plant Material and Sampling

The material analyzed was obtained from the experimental farm of "P. Martucci" of the University of Bari in Valenzano, Apulia, Italy. Plants of artichokes (*Cynara cardunculus* L. var. *scolymus* L. cv. Violetto di Provenza), obtained by micropropagation, were transplanted in pots containing a commercial peat mixture soil enriched with nutrients (organic carbon 46%, organic nitrogen 1–2%, organic matter 80%) and mixed with perlite at a 2:1 (v/v) ratio. The peat mixture was sterilized and used to fill 10 cm diameter pots. Prior to transplantation, half of the microplants were inoculated with 10 g of crude inoculum of the AMF *G. viscosum*, as described by Morone Fortunato et al. [41]. Non-mycorrhizal plants were used as controls. Acclimatization took place in greenhouse conditions at 18 ± 2 °C with mist and a relative humidity level reduced from 85–90% to 55–60% over 20 days. After 60 days, 48 non-mycorrhizal plants and 48 inoculated plants were transplanted to the open field, according to a randomized block design with treatments replicated three times. Each block consisted of eight plants, and the spacing used was 1.2 m between rows and 1 m between plants for all the treatments. A portion of the field was inoculated 15 days before the transplantation with an inert substrate enriched with mycelium, microsclerotia, and spores of *V. dahliae* isolated from naturally infected artichokes. Then, the treatments were established as follows: (i) non-mycorrhizal plants (Ctrl), (ii) non-mycorrhizal plants inoculated with *V. dahliae* (I), (iii) mycorrhizal plants (M), (iv) mycorrhizal plants inoculated with pathogen (MI).

2.3. Source of *V. dahliae* Isolates

Isolates of *V. dahliae* were recovered from symptomatic artichoke plants in a field with a known history of Verticillium wilt. Segments of 5 mm long surface-sterilized stems from infected host plants were transferred on a potato dextrose agar (PDA), (Difco, Detroit, MI, USA) supplemented with streptomycin sulphate (100 ppm) and incubated at 27 °C for 10 days in darkness. Colonies of *V. dahliae* were morphologically identified visually and microscopically, subcultured on PDA without antibiotics, and then mixed with vermiculite for the inoculation in the field. A further identity confirmation was provided by sequencing. PCR amplification and sequencing were performed using the ribosomal internal transcribed spacer region (*ITS*) as the locus, according to Inderbitzin et al. 2011 [42]. Species identification was confirmed by BLASTn against the NCBI GenBank database (http://www.ncbi.nlm.nih.gov, accessed on 14 January 2021).

2.4. Disease Assessments

A scoring metric was used to assess disease severity of the artichoke plants over time. Wilt severity was rated according to Uppal et al.'s [43] scoring system as follows: 0, no wilt symptoms; 1, inter-veinal chlorosis on the lower leaves; 2, moderate necrosis and defoliation of the lower leaves; 3, severe leaf necrosis and defoliation; and 4, severe defoliation accompanied by pronounced stunting, chlorosis, and necrosis of the remaining leaves. Furthermore, a rating scale was also established to evaluate the severity of vascular browning of artichoke stems. This scale consisted of the following grades: 0, no vascular browning; 1, trace to less than 9% of the stem cross-section showing a vascular browning; 2, 10–24% of the stem cross-section with a vascular browning; 3, 25–49% of the stem cross-section showing vascular browning; and 4, over 50% of the stem cross-section exhibiting vascular browning. In addition, the effect of mycorrhization on the growth of artichoke plants was assessed by counting the number of flower heads per plant. Data were analyzed with analysis of variance (ANOVA), and the means were compared by the Duncan test.

2.5. Determination of ASC and GSH Pool Contents

Foliar tissues (20 g) were homogenated at 4 °C with three volumes of 5% (*w/v*) metaphosphoric acid and then centrifuged for 15 min at 20,000× *g*. The resulting supernatant was used for analysis of the ASC and GSH pool content according to Zhang and Kirkham [44].

2.6. Enzyme Assays

For determination of antioxidant enzyme activities, samples were homogenized according to Mastropasqua et al. [45] with slight modifications. Briefly, twenty grams of foliar tissues were homogenized in 50 mM Tris–HCl, pH 7.8 containing 0.3 mM mannitol, 1 mM EDTA, 10 mM $MgCl_2$, 1% (*w/v*) polyvinyl-pyrrolidone (PVP), and 0.05% (*w/v*) cysteine 1%, at 4 °C. The homogenate was filtered through four layers of cheesecloth and centrifuged (20,000× *g*, 20 min, 4 °C). The supernatant was desalted by dialysis against 50 mM Tris-HCl, pH 7.8, and used for spectrophotometric analysis of the total proteins and enzymatic activities.

The total protein content of samples was measured with a Protein Assay kit from Bio-Rad (Hercules, CA, USA) with bovine serum albumin as the standard [46]. The reproducibility of the Bio-Rad kit, expressed as the coefficient of variation (%CV), is 2% approximately; the lower limit of the detection for protein molecular weight is 3000 to 5000 Daltons.

The enzymatic spectrophotometric assays for the determination of cytosolic APX (EC 1.11.1.11), DHAR (EC 1.8.5.1), CAT (EC 1.11.1.6), GR (EC 1.8.1.7), MDHAR (EC 1.6.5.4), and SOD (EC 1.15.11) were performed according to Paciolla et al. [47] and Mastropasqua et al. [48].

2.7. Electrophoretic Analyses

Native-Polyacrilamide Gel Electrophoresis (Native-PAGE) was performed on PAGE (4.3% T; 7.3% C) with a running buffer composed of 4 mM Tris-HCl pH 8.3 and 38 mM glycine. In each lane of the gel, 200 μg of total proteins were loaded. After the electrophoretic run, the gels were washed with distilled water and incubated in specific buffers for the detection of APX and CAT, as described in Paciolla et al. [49]. For the SOD, the activity on the gel was visualized by incubating it in 0.053 Tris–HCl buffer pH 8.2 containing 0.21 mM riboflavin and 0.244 mM nitro-blue tetrazolium (NBT) in the dark; after 15 min, achromatic bands on a grey background appeared, a 50% glycerol solution was used to block the reaction.

For densitometric analysis of SOD activity, the gel was acquired utilizing the Gel/ChemiDoc and Quantity One software (BioRad Laboratories Inc., Milan, Italy) to obtain information on the changes in the activity of each band due to different treatments. A relative value of 100 was assigned to the intensity of the bands of Ctrl and I samples.

2.8. Lipid Peroxidation Analysis and H_2O_2 Content

For lipid peroxidation, plant material was ground with four volumes of 0.1% (*w/v*) trichloroacetic acid (TCA). The homogenate was centrifuged at 12,000 × *g*, for 10 min, at 4 °C. One mL of the supernatant was mixed with 4 mL of 20% TCA containing 0.5% (*w/v*) thiobarbituric acid (TBA). The level of cell lipid peroxidation was evaluated in terms of malondialdehyde (MDA) content determined by the TBA reaction as described by Zhang and Kirkham [44]. Intracellular H_2O_2 concentration was evaluated according to Lee and Lee [50].

2.9. Statistical Analysis

The biochemical data presented are the means of five different experiments. Statistical analysis was done using Student's *t*-test, with level of significance for $p < 0.05$ and highly significant for $p < 0.01$; the standard deviation (SD) was calculated, and its range is shown in the figures. Data presented for disease assessments are the average of three experiments with three replicates and were analyzed with ANOVA with $p \leq 0.05$; the means were compared by the Duncan test.

3. Results

3.1. Disease Assessments

The results showed a beneficial effect of mycorrhization in containing artichoke wilt. In particular, the AMF *G. viscosum* significantly reduced the disease severity, measured by symptoms' development on leaves, on the MI treatment, while it slightly reduced the vascular browning (Table 1). In addition, the beneficial effect of mycorrhization on productivity was observed.

Table 1. Effectiveness of mycorrhizal fungus *G. viscosum* in protecting globe artichoke against Verticillium wilt. For each parameter, different letters (a,b,c) within the same column indicate that the means are significantly different at $p \leq 0.05$ according to Duncan test. The experiments were repeated three times with three replicates. Ctrl, non-mycorrhizal plants; I, non-mycorrhizal plants inoculated with *V. dahliae*; M, mycorrhizal plants; MI, mycorrhizal plants inoculated with *V. dahlia*.

Treatment	Disease Severity		Flower Heads for Plants
	Foliar Tissue	Vascular System	
Ctrl	0.0 c	0.0 b	13
I	3.9 a	3.0 a	4
M	0.0 c	0.0 b	16
MI	1.6 b	2.7 a	9

3.2. Ascorbate and Glutathione Pool Content

The mycorrhizal plants inoculated with the pathogen *V. dahliae* (MI) showed an increase ($p < 0.05$) in ASC content as compared to non-mycorrhizal inoculated plants (I), while no significant increment was observed in mycorrhizal plants (M) with respect to non-mycorrhizal plants (Ctrl) (Figure 1, Panel a). DHA did not differ significantly among the treatments (data not shown), and, hence, the ascorbate redox ratio (ASC/ASC + DHA) was higher in MI than in the other treatments (Figure 1, Panel b).

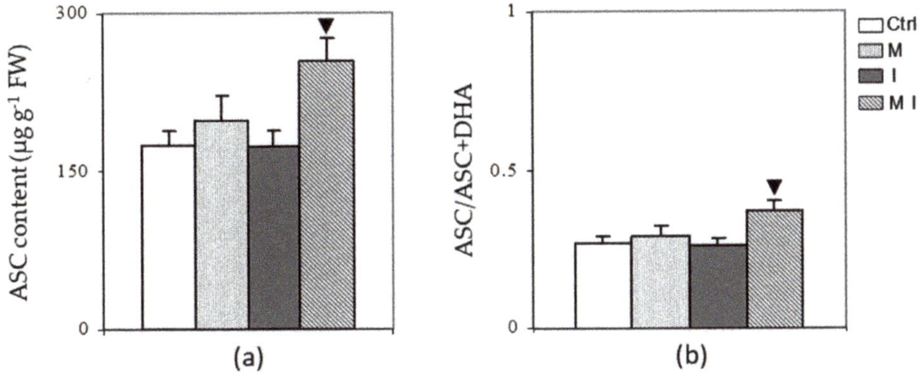

Figure 1. Ascorbate (ASC) content (**a**) and ASC redox ratio (**b**) in artichoke control plants (Ctrl), in mycorrhizal plants (M), in plants inoculated with *Verticillium dahliae* (I), and in mycorrhizal plants inoculated with *V. dahliae* (MI). The results are given as the mean values of at least five experiments ± SD; ▼ indicates values significantly different from the artichoke inoculated with *V. dahliae* (I) by the Student's *t* test with $p < 0.05$. FW, fresh weight.

In both mycorrhizal plants (M and MI), an increase ($p < 0.05$) in GSH content was observed (Figure 2, Panel a) with respect to the control and the inoculated plants (I), respectively. In addition, the GSSG content was similar in all samples (data not shown), therefore, the glutathione redox ratio (GSH/GSH + GSSG) was higher in M and MI with respect to the control and the plants inoculated with the pathogen (I), respectively (Figure 2, Panel b).

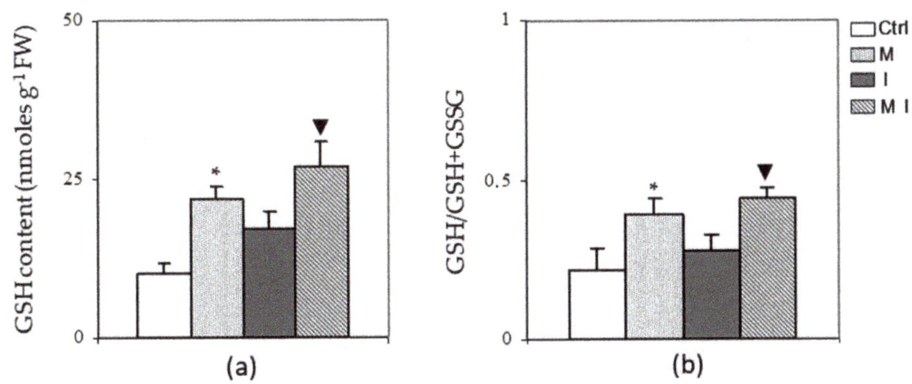

Figure 2. Glutathione (GSH) content (**a**) and glutathione redox ratio (**b**) in artichoke control plants (Ctrl), in mycorrhizal plants (M), in plants inoculated with *Verticillium dahliae* (I), and in mycorrhizal plants inoculated with *V. dahliae* (MI). The results are given as the mean values of at least five experiments ± SD; * indicates values significantly different from the control (Ctrl) by the Student's *t* test with $p < 0.05$; ▼ indicates values significantly different from the artichoke inoculated with *V. dahliae* (I) by the Student's *t* test with $p < 0.05$. FW, fresh weight.

3.3. Antioxidant Enzyme Assays

Activities of the enzymes in the ascorbate–glutathione cycle, including APX, DHAR, MDHAR, and GR, showed difference trends. The activity of APX was significantly higher in MI compared to I as shown in Figure 3 (Panel a). Similarly, M showed higher APX activity with respect to the control. These results were confirmed by Native-PAGE, which showed that enzyme activity of APX was higher in M and MI than in Ctrl and I, respectively.

Furthermore, the native electrophoretic pattern of APX showed a total number of three isoforms with same migration rate in all samples (Figure 3, Panel b).

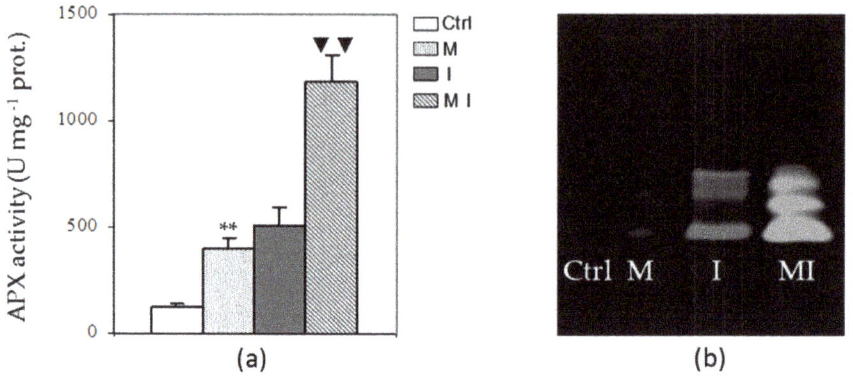

Figure 3. Spectrophotometric activity (**a**) and electrophoretic profile after Native-PAGE (**b**) of ascorbate peroxidase (APX) in the cytosolic fraction of artichoke control plants (Ctrl), mycorrhizal plants (M), plants inoculated with *Verticillium dahliae* (I), and in mycorrhizal plants inoculated with *V. dahliae* (MI); 1 U = 1 nmol of ascorbate oxidized min^{-1}; prot. = proteins. The results are given as the mean values of at least five experiments ± SD; * * and ▼▼ indicate values significantly different from the control (Ctrl) and from the artichoke inoculated with *V. dahliae*, respectively, by the Student's *t* test with $p < 0.01$.

The assay of GR and MDHAR activities revealed a trend similar to APX, showing both in M and MI a significant increment compared to Ctrl and I plants, respectively (Figure 4, Panels a and b). In contrast, the activity of DHAR remained almost unchanged among all treatments (data not shown).

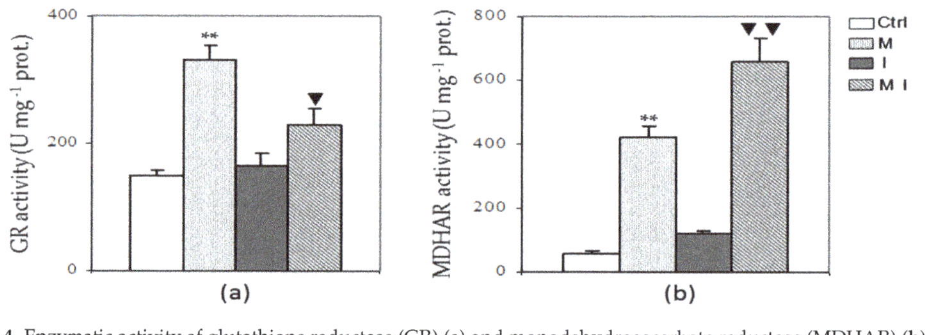

Figure 4. Enzymatic activity of glutathione reductase (GR) (**a**) and monodehydroascorbate reductase (MDHAR) (**b**) in the cytosolic fraction of artichoke control plants (Ctrl), mycorrhizal plants (M), plants inoculated with *Verticillium dahliae* (I), and in mycorrhizal plants inoculated with *V. dahliae* (MI). For GR activity, 1 U = 1 nmol of NADPH oxidized min^{-1}; for MDHAR activity, 1 U = 1 nmol of NADH oxidized min^{-1}; prot. = proteins. The results are given as the mean values of at least five experiments ± SD; * * indicates values significantly different from the control (Ctrl) by the Student's *t* test with $p < 0.01$; ▼ and ▼▼ indicate values significantly different from the artichoke inoculated with *V. dahliae* (I) by the Student's *t* test with $p < 0.05$ and 0.01, respectively.

Catalase activity was also analyzed aiming to investigate the ability of mycorrhizal plants to detoxify hydrogen peroxide. No significant changes in CAT activity were found between M and Ctrl plants, as shown in Figure 5 (Panel a). In contrast, a high significant ($p < 0.01$) decrease was observed in MI compared to I. This evidence was confirmed by native electrophoresis where I treatments showed a higher band intensity compared to the MI treatment (Figure 5, Panel b).

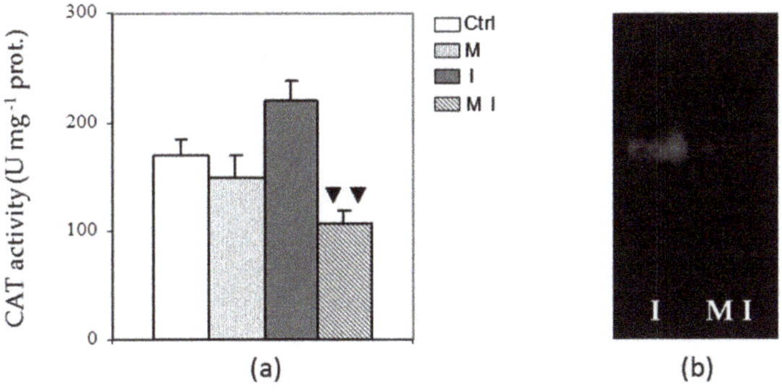

Figure 5. Spectrophotometric activity (**a**) and electrophoretic profile (**b**) in the cytosolic fraction of catalase (CAT) in artichoke control plants (Ctrl), in mycorrhizal plants (M), in plants inoculated with *Verticillium dahliae* (I), and in mycorrhizal plants inoculated with *V. dahliae* (MI); 1 U = 1 nmol H_2O_2 dismutated min^{-1}; prot. = proteins. The results are given as the mean values of at least five experiments ± SD; ▼▼ indicates values significantly different from the artichoke inoculated with *V. dahliae* (I) by the Student's *t* test with $p < 0.01$.

Regarding SOD analysis, the spectrophotometric assay (Figure 6, Panel a), the electrophoretic pattern (Figure 6, Panel b), and the densitometric analysis of the band intensity of Native-PAGE (Figure 6, Panel c), showed a significant increase in its activity in both mycorrhizal plants (M and MI), as compared to control and I plants, respectively. This increase is not due to new additional bands, as three isoforms in all samples have been observed.

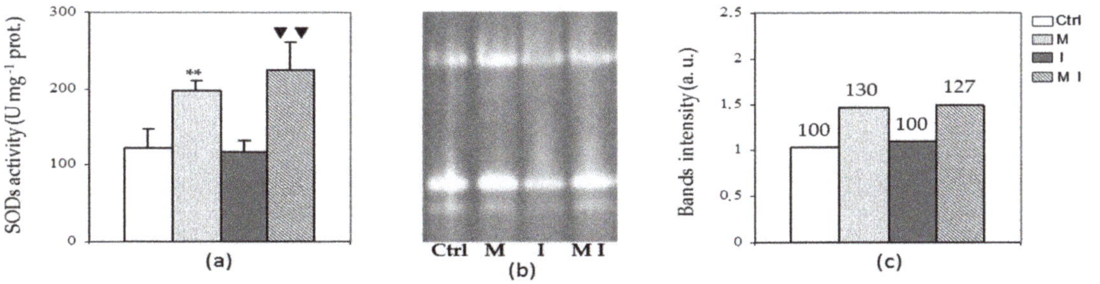

Figure 6. Spectrophotometric analysis (**a**), electrophoretic pattern of Native-PAGE (**b**), and related densitometric analysis (**c**) in the cytosolic fraction of superoxide dismutase (SOD) of artichoke control plants (Ctrl), mycorrhizal plants (M), plants inoculated with *Verticillium dahliae* (I), and in mycorrhizal plants inoculated with *V. dahliae* (MI); 1 U= the amount of enzyme required to inhibit the reduction rate of NBT by 50% at 25 °C; prot. = proteins; a.u. = arbitrary units. The results are given as the mean values of at least five experiments ± SD; * * and ▼▼ indicate values significantly different from the control (Ctrl) and from the artichoke inoculated with *V. dahliae*, respectively, by the Student's *t* test with $p < 0.01$.

3.4. H_2O_2 Content and Lipid Peroxidation Assay

The effect of mycorrhizal inoculation on ROS accumulation and preservation of membrane structure from oxidative damages after biotic interaction was evaluated by the estimation of H_2O_2 content and level of lipid peroxidation. The H_2O_2 content was significantly decreased ($p < 0.01$) in MI treatment with respect to the plants inoculated only with the pathogen (Figure 7, Panel a). Similarly, the mycorrhizal plants inoculated or not with the pathogen showed a decreased level of lipid peroxidation as compared to the inoculated and the control plants, respectively (Figure 7, Panel b).

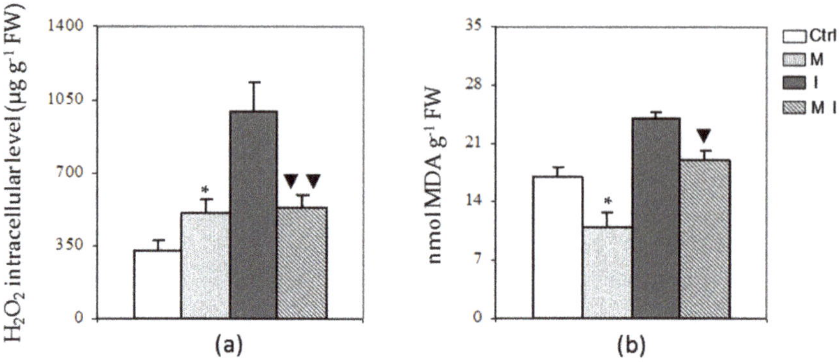

Figure 7. (**a**) Intracellular H_2O_2 content in artichoke control plants (Ctrl), in mycorrhizal plants (M), in plants inoculated with *Verticillium dahliae* (I), and in mycorrhizal plants inoculated with *V. dahliae* (MI). The results are given as the mean values of at least five experiments ± SD; * indicates values significantly different from the control (Ctrl) by the Student's *t* test with $p < 0.05$; ▼▼ indicates values significantly different from the artichoke inoculated with *V. dahliae* (I) by the Student's *t* test with $p < 0.01$; (**b**) Lipid peroxidation level in artichoke control plants (Ctrl), in mycorrhizal plants (M), in plants inoculated with *V. dahliae* (I), and in mycorrhizal plants inoculated with *V. dahliae* (MI). The results are given as the mean values of at least five experiments ± SD; * and ▼ indicate values significantly different from the control (Ctrl) and from the artichoke inoculated with *V. dahliae* (I), respectively, by the Student's *t* test with $p < 0.05$.

4. Discussion

Ensuring stable crop yields and quality while simultaneously guarding human health and the environment is a current challenge facing the farming and research communities. In recent years, the inoculation of plants with AMF has received increasing attention as an environment-friendly approach for improving plant nutrition by increasing nutrients and water availability, nutraceutical values by inducing changes in secondary metabolism, and plant tolerance to biotic and abiotic stress by selecting resistant cultivars and enhancing the activity of antioxidant enzymes [51–56]. According to this view, the present study evaluated the effectiveness of the AMF *G. viscosum* as a biocontrol agent against the soil-borne pathogen *V. dahliae* by investigating the antioxidant responses and the effects on ROS metabolism in artichokes. There is sufficient evidence in the literature to confirm that the effect of AMF varies with respect to the host plant and the fungal species [27,57–60]. Based on previous results where *G. viscosum* showed a better affinity with artichoke plantlets in terms of plant growth and physiological activities, compared with *G. intraradices* [61], we selected the former AMF in this study. Our results showed that inoculation with *G. viscosum* significantly improved productivity and ameliorated the disease severity in most of the AMF-treated plants. This is consistent with previous observations showing a beneficial effect on productivity and disease severity in cotton plants affected by Verticillium wilt inoculated with *Rhizophagus irregularis* [62], in wheat plants infected with *Fusarium pseudograminearum* and colonized by *Rhizophagus intraradices* [56], and in potato contaminated by the pathogen *Fusarium sambucinum* and inoculated with the AMF *Glomus irregular* [63]. Although *G. viscosum* had a significant effect on reducing symptoms' development on leaves, the inoculation still caused slight stem browning in most of the AMF-treated plants, confirming previous findings where several AMF treatments showed differences in efficiency towards reducing disease severity [57,64,65].

The findings of the present study highlighted that the mycorrhization process modulated the activity of the ascorbate–glutathione cycle enzymes, including APX, MDHAR, and GR, and the resulting levels of ASC and GSH. In particular, increased levels of GSH and ASC, along with enhanced activity of APX, GR, and MDHAR, were observed in mycorrhizal artichoke plants compared with non-mycorrhizal controls, while the activity of DHAR, as well as the DHA and GSSG contents, remained unchanged. The effects of AMF on the modulation of ASC-GSH cycle enzymes and antioxidative metabolism have been scarcely studied [66]. Moreover, most of the studies focused on a narrow range of fungi, such as *Trichoderma harzianum* [67], *R. intraradices* [56,68], *Glomus* spp. [69], and the AMF Rhizolive consortium [37]. Previous studies have established a correlation between the level of GSH and resistance to different biotic challenges, including plant viruses, bacteria, and fungi [66,70,71]. Furthermore, GSH represents a key metabolite in the cellular redox buffering system, protecting proteins from irreversible modifications that can be induced by oxidation, through the S-glutathionylation, a post-transcriptional protein modification, which consists in the formation of a stable mixed disulfide between GSH and a protein thiol [72,73]. In this study, the increase in GSH content, observed in both mycorrhizal treatments (M and MI), could be correlated to the higher GR enzyme activity that regenerates the GSH from GSSG, using NADPH as electron donor [74]. Due to the GSH increase, the GSH/GSSG ratio shifts toward the reduced form. Furthermore, the higher availability of NADP+ allows for accepting electrons from photosynthetic electron transport, mitigating the reduction of molecular oxygen to superoxide anion [75]. Moreover, we hypothesize that the increased levels of GSH, ASC, and GSH-dependent enzymes were related with increased mineral elements content (N, P, K, Mg, Fe, Ca, etc.), as demonstrated in previous studies where the higher activity of several antioxidant enzymes was often associated with mycorrhiza-induced increases in biomass and P or N contents [76]. On the other hand, a rise in GSH content is correlated to a higher rate of assimilation of nitrogen and sulfur [77], which are elements of the chemical structure of glutathione.

The increased content of ASC observed in MI plants compared to I, along with the significant increase in MDHAR, corroborates the key role played by MDHAR in the regeneration of ASC from MDHA for ROS scavenging. Moreover, the regeneration of ASC from its oxidative state prevents the intracellular accumulation of DHA that, at high concentrations, has been proved to be toxic for cell metabolisms [78] and to inhibit the activity of enzymes regulated by the thioredoxin–thioredoxin system [79,80]. Our results suggest that MDHAR showed a higher specific activity than DHAR, as reported in previous studies [81–83].

The remarkable increase in H_2O_2 and MDA in I plants compared with the other treatments was an indicator of oxidative stress caused by pathogen attack. Conversely, in the MI plants, we observed a strong decrease in the H_2O_2 level that could be explained by the significant increase in the APX enzyme activity of mycorrhizal plants (MI and I) compared to the other treatments. Similarly, decreased MDA concentrations in leaves of AMF, although at a lesser extent, have been observed. These findings are consistent with published results reporting the antioxidant responses in *Digitaria eriantha* plants inoculated with the AMF *R. irregularis* and subjected to drought, cold, or salinity [51] and the effect of AMF on leaf water potential, solute accumulation, and oxidative stress in soybean plants subjected to drought stress [84]. Furthermore, our findings showed a higher level of H_2O_2 in mycorrhizal plants compared to the non-mycorrhizal plants that is consistent with previous results showing that, in the early stages, the establishment of mycorrhizal symbiosis leads to an increase in ROS content followed by an enhanced defense response of the antioxidant system [26,37,62,85,86]. Accordingly, our results showed a significant stimulation of SOD and APX activity in AMF compared to non-AMF samples, indicating lower oxidative damage in the colonized plants. Conversely, the CAT showed a negative response to AMF, while its activity increased in plants inoculated with pathogen. CAT and APX activities are both involved in the scavenging of H_2O_2. Although mycorrhizal colonization has been associated with higher antioxidant enzyme activities, the response of the individual enzymes varies with respect to the host plant and the fungal species [58,60]. Moreover, previous studies showed that APX has a much higher affinity for H_2O_2 than CAT [87–90], while high concentrations of H_2O_2 induced the expression of genes involved in the synthesis of catalase gene to higher levels, and in less time, than lower H_2O_2 concentrations [91]. Furthermore, CAT, APX, and SOD are metalloenzymes depending on micronutrients' availability, so their activities may be related to the acquisition of Fe,

Cu, N, P, and Mn in the plants [58,76]. The activity of SOD, APX, and CAT was also analyzed by using electrophoretic systems. Our results showed the presence of three constitutive isoforms in SOD activity in all samples with an enhanced activity in both mycorrhizal plants, and no induction of new isoforms was detected in plants inoculated with the pathogen (I) or in the mycorrhizal plants. Similarly, three SOD isoforms were found in non-mycorrhizal plants of pepper roots affected by Verticillium wilt, although the colonization with the AMF *Glomus deserticola* induced two new isoforms with similar mobility [85].

Overall, our results demonstrated the protective role of the AMF *G. viscosum* on artichoke plants affected by Verticillium wilt through reducing disease severity and enhancing antioxidant systems and activity of investigated antioxidant enzymes. The results of this study suggest that: (1) *G. viscosum* enhances disease tolerance in artichokes; (2) the ascorbate-glutathione cycle plays a key role in maintaining redox balance and avoiding oxidative damage in contaminated artichoke plants inoculated with *G. viscosum*; (3) *G. viscosum* increases the activity of some antioxidant enzymes, such as APX and SOD, while it decreases the activity of some others (CAT), confirming that the efficiency of AMF is related to fungal and/or plant species, soil nutrient availability, and environmental factors. All those data can lend support to the applications of AM *G. viscosum* as a cost-effective and environment-friendly strategy for reducing or alleviating *V. dahliae* effects in artichoke plants.

Author Contributions: Conceptualization, C.P. and A.V.; methodology, C.P. and A.V.; data curation, C.P. and A.V.; writing—original draft preparation, C.P. and A.V.; writing—review and editing, C.P., A.V. and F.T.; supervision, C.P.; funding acquisition, C.P. All authors have read and agreed to the published version of the manuscript.

Funding: This research was funded by University of Bari Aldo Moro, grant number H96J15001610005.

Acknowledgments: The authors thank Franco Ciccarese for providing plant materials used in this study.

Conflicts of Interest: The authors declare no conflict of interest.

References

1. Lattanzio, V.; Paul, A.K.; Lansalta, V.; Cardinali, A. Globe artichoke: A functional food and source of nutraceutical ingredients. *J. Funct. Foods* **2009**, *1*, 131–144. [CrossRef]
2. Lombardo, S.; Pandino, G.; Ierna, A.; Mauromicale, G. Variation of polyphenols in a germplasm collection of globe artichoke. *Food Res. Int.* **2012**, *46*, 544–551. [CrossRef]
3. Salem, M.B.; Affes, H.; Ksouda, K.; Dhouibi, R.; Sahnoun, Z.; Hammami, S.; Zeghal, K.M. Pharmacological studies of artichoke leaf extract and their health benefits. *Plant Foods Hum. Nutr.* **2015**, *70*, 441–453. [CrossRef]
4. Fintelman, V. Therapeutic profile and mechanism of action of artichoke leaf extract: Hypolipemic, antioxidant, hepatoprotective and choleretic properties. *Phytomedica* **1996**, *201*, 50–60.
5. Liorach, R.; Espin, J.C.; Tomas-Barberan, F.A.; Ferreres, F. Artichoke byproducts as a potential source of health-promoting antioxidant phenolics. *J. Agric. Food Chem.* **2002**, *50*, 58–64.
6. Garbetta, A.; Capotorto, I.; Cardinali, A.; D'Antuono, I.; Linsalata, V.; Pizzi, F.; Minervini, F. Antioxidant activity induced by main polyphenols present in edible artichoke heads: Influence of in vitro gastro-intestinal digestion. *J. Funct. Foods* **2014**, *10*, 456–464. [CrossRef]
7. Behara, Y.B.; Pharm, B. Pharmacological studies on artichoke leaf extract –an edible herb of Mediterranean origin. *J. Pharmaceut. Biomed. Sci.* **2011**, *15*, 111–600.
8. Mohammad, A.; Sahar, N.; Effat, B. Effet of *Cynara scolymus* on fasting blood sugar of rat. *Int. J. Pharm. Ind. Res.* **2013**, *3*, 180–182.
9. Gebhardt, R. Antioxidative and protective properties of extracts from leaves of the artichoke (*Cynara scolymus* L.) against hydroperoxide-induced oxidative stress in cultured rat hepatocytes. *Toxicol. Appl. Pharmacol.* **1997**, *144*, 279–286. [CrossRef] [PubMed]
10. Rouphael, Y.; Colla, G.; Graziani, G.; Ritieni, A.; Cardarelli, M.; De Pascale, S. Phenolic composition, antioxidant activity and mineral profile in two seed-propagated artichoke cultivars as affected by microbial inoculants and planting time. *Food Chem.* **2017**, *234*, 10–19. [CrossRef]
11. Ancora, G. Globe artichoke (*Cynara scolymus* L.). In *Biotechnology in Agriculture and Forestry, 2. Crops*; Bajai, Y.P.S., Ed.; Springer: Berlin, Germany, 1986; pp. 471–484. [CrossRef]

12. Grabowska, A.; Caruso, G.; Mehrafarin, A.; Kalisz, A.; Gruszecki, R.; Kunicki, E.; Sękara, A. Application of modern agronomic and biotechnological strategies to valorise worldwide globe artichoke (*Cynara cardunculus* L.) potential-an analytical overview. *Ital. J. Agron.* **2018**, *13*, 279–289. [CrossRef]
13. Bhat, R.G.; Subbarao, K.V.; Bari, M.A. First report of *Verticillium dahliae* causing artichoke wilt in California. *Plant Dis.* **1999**, *83*, 782. [CrossRef]
14. Berbegal, M.; Ortega, A.; García-Jiménez, J.; Armengol, J. Inoculum density-disease development relationship in Verticillium wilt of artichoke caused by *Verticillium dahliae*. *Plant Dis.* **2007**, *91*, 1131–1136. [CrossRef] [PubMed]
15. Cirulli, M.; Bubici, G.; Amenduni, M.; Armengol, J.; Berbegal, M.; Jiménez-Gasco, M.D.M.; Jiménez-Díaz, R.M. Verticillium wilt: A threat to artichoke production. *Plant Dis.* **2010**, *94*, 1176–1187. [CrossRef]
16. EFSA Panel on Plant Health (PLH). Scientific opinion on the pest categorisation of *Verticillium dahliae* Kleb. *EFSA J.* **2014**, *12*, 3928.
17. Fradin, E.F.; Thomma, B.P. Physiology and molecular aspects of *Verticillium* wilt diseases caused by *V. dahliae* and *V. albo-atrum*. *Mol. Plant Pathol.* **2006**, *7*, 71–86. [CrossRef]
18. Klosterman, S.J.; Atallah, Z.K.; Vallad, G.E.; Subbarao, K.V. Diversity, pathogenicity and management of *Verticillium* species. *Annu. Rev. Phytopathol.* **2009**, *47*, 39–62. [CrossRef]
19. Cirulli, M.; Ciccarese, F.; Amenduni, M. Evaluation of Italian Clones of Artichoke for Resistance to *Verticillium dahliae*. *Plant Dis.* **1994**, *78*, 680–682. [CrossRef]
20. Pegg, G.F.; Brady, B.L. *Verticillium Wilts*; Cromwell Press: Trowbridge, UK, 2002. [CrossRef]
21. Puri, K.D.; Hu, X.; Gurung, S.; Short, D.; Sandoya Miranda, G.V.; Schild, M.; Zhang, Y.; Zhao, J.; Anchieta, A.; Klosterman, S.J.; et al. *Verticillium klebahnii* and *V. isaacii* Isolates Exhibit Host-dependent Biological Control of Verticillium Wilt Caused by *V. dahliae*. *Phytofrontiers* **2021**. [CrossRef]
22. Villarino, M.; Larena, I.; Melgarejo, P.; De Cal, A. Effect of chemical alternatives to methyl bromide on soil-borne disease incidence and fungal populations in Spanish strawberry nurseries: A long-term study. *Pest Manag. Sci.* **2021**, *77*, 766–774. [CrossRef] [PubMed]
23. Bubici, G.; Marsico, A.D.; Gaber, L.; Tsror, L. Evaluation of thiophanate-methyl in controlling Verticillium wilt of potato and artichoke. *Crop Prot.* **2019**, *119*, 1–8. [CrossRef]
24. Begum, N.; Qin, C.; Ahanger, M.A.; Raza, S.; Khan, M.I.; Ashraf, M.; Ashraf, M.; Ahmed, N.; Zhang, L. Role of arbuscular mycorrhizal fungi in plant growth regulation: Implications in abiotic stress tolerance. *Front. Plant Sci.* **2019**, *10*, 1068. [CrossRef]
25. Aggarwal, A.; Kadian, N.; Tanwar, A.; Yadav, A.; Gupta, K.K. Role of arbuscular mycorrhizal fungi (AMF) in global sustainable development. *J. Nat. Appl. Sci.* **2011**, *3*, 340–351. [CrossRef]
26. Kapoor, R.; Singh, N. Arbuscular mycorrhiza and reactive oxygen species. In *Arbuscular Mycorrhizas and Stress Tolerance of Plants*; Springer: Berlin/Heidelberg, Germany, 2017; pp. 225–243. [CrossRef]
27. Avio, L.; Maggini, R.; Ujvári, G.; Incrocci, L.; Giovannetti, M.; Turrini, A. Phenolics content and antioxidant activity in the leaves of two artichoke cultivars are differentially affected by six mycorrhizal symbionts. *Sci. Hortic.* **2020**, *264*, 109153. [CrossRef]
28. Wu, Q.S.; Zou, Y.N.; Xia, R.X. Effects of water stress and arbuscular mycorrhizal fungi on reactive oxygen metabolism and antioxidant production by citrus (*Citrus tangerine*) roots. *Eur. J. Soil Biol.* **2006**, *42*, 166–172. [CrossRef]
29. Wu, Q.S.; Zou, Y.N. Mycorrhiza has a direct effect on reactive oxygen metabolism of drought-stressed citrus. *Plant Soil Environ.* **2009**, *55*, 436–442. [CrossRef]
30. Zou, Y.N.; Wu, Q.S.; Kuča, K. Unravelling the role of arbuscular mycorrhizal fungi in mitigating the oxidative burst of plants under drought stress. *Plant Biol. J.* **2020**, *23*, 50–57. [CrossRef]
31. Boutaj, H.; Chakhchar, A.; Meddich, A.; Wahbi, S.; El Alaoui-Talibi, Z.; Douira, A.; Filali-Maltouf, A.; El Modafar, C. Bioprotection of olive tree from Verticillium wilt by autochthonous endomycorrhizal fungi. *J. Plant Dis. Prot.* **2020**, *127*, 349–357. [CrossRef]
32. Boutaj, H.; Meddich, A.; Wahbi, S.; Moukhli, A.; El Alaoui-Talibi, Z.; Douira, A.; Filali-Maltouf, A.; El Modafar, C. Effect of arbuscular mycorrhizal fungi on verticillium wilt development of olive trees caused by *Verticillium dahliae*. *Res. J. Biotechnol.* **2019**, *14*, 79–88.
33. Boutaj, H.; Meddich, A.; Chakhchar, A.; Wahbi, S.; El Alaoui-Talibi, Z.; Douira, A.; Filali-Maltouf, A.; El Modafar, C. Arbuscular mycorrhizal fungi improve mineral nutrition and tolerance of olive tree to *Verticillium* wilt. *Arch. Phytopathol. Plant Prot.* **2020**, *53*, 673–689. [CrossRef]
34. Karagiannidis, N.; Bletsos, F.; Stavropoulos, N. Effect of *Verticillium* wilt (*Verticillium dahliae* Kleb.) and mycorrhiza (*Glomus mosseae*) on root colonization, growth and nutrient uptake in tomato and eggplant seedlings. *Sci. Hortic.* **2002**, *94*, 145–156. [CrossRef]
35. Demír, S.; ¸Sensoy, S.; Ocak, E.; Tüfenkçi, ¸S.; Durak, E.D.; Erdinç, Ç.; Ünsal, H. Effects of arbuscular mycorrhizal fungus, humic acid, and whey on wilt disease caused by *Verticillium dahliae* Kleb. in three solanaceous crops. *Turk. J. Agric. For.* **2015**, *39*, 300–309. [CrossRef]
36. Berg, G.; Zachow, C.; Lottmann, J.; Go, M.; Costa, R.; Icrobiol, A.P.P.L.E.N.M. Impact of Plant Species and Site on Rhizosphere-Associated Fungi Antagonistic to *Verticillium dahliae* Kleb. *Appl. Environ. Microbiol.* **2005**, *71*, 4203–4213. [CrossRef] [PubMed]
37. Boutaj, H.; Meddich, A.; Chakhchar, A.; Wahbi, S.; Alaoui-Talibi, Z.E.; Douira, A.; Filali-Maltouf, A.; El Modafar, C. Induction of early oxidative events in mycorrhizal olive tree in response to *Verticillium* wilt. *Arch. Phytopathol. Plant Prot.* **2021**, 1–23. [CrossRef]
38. Apel, K.; Hirt, H. Reactive oxygen species: Metabolism, oxidative stress, and signal transduction. *Annu. Rev. Plant Biol.* **2004**, *55*, 373–399. [CrossRef] [PubMed]

39. De Gara, L.; de Pinto, M.C.; Tommasi, F. The antioxidant systems vis-à-vis reactive oxygen species during plant–pathogen interaction. *Plant Physiol. Biochem.* **2003**, *41*, 863–870. [CrossRef]
40. Noctor, G.; Foyer, C.H. Ascorbate and Glutathione: Keeping Active Oxygen Under Control. *Annu. Rev. Plant Physiol. Plant Mol. Biol.* **1998**, *49*, 249–279. [CrossRef]
41. Morone Fortunato, I.; Ruta, C.; Castrignanò, A.; Saccardo, F. The effect of mycorrhizal symbiosis on the development of micropropagated artichokes. *Sci. Hortic.* **2005**, *106*, 472–483. [CrossRef]
42. Inderbitzin, P.; Bostock, R.M.; Davis, R.M.; Usami, T.; Platt, H.W.; Subbarao, K.V. Phylogenetics and taxonomy of the fungal vascular wilt pathogen *Verticillium*, with the descriptions of five new species. *PLoS ONE* **2011**, *6*, e28341. [CrossRef] [PubMed]
43. Uppal, A.K.; El Hadrami, A.; Adam, L.R.; Tenuta, M.; Daayf, F. Biological control of potato *Verticillium* wilt under controlled and field conditions using selected bacterial antagonists and plant extracts. *Biol. Control* **2008**, *44*, 90–100. [CrossRef]
44. Zhang, J.; Kirkham, B. Antioxidant responses to drought in sunflower and sorghum seedlings. *New Phytol.* **1996**, *132*, 361–373. [CrossRef]
45. Mastropasqua, L.; Dipierro, N.; Paciolla, C. Effects of Darkness and Light Spectra on Nutrients and Pigments in Radish, Soybean, Mung Bean and Pumpkin Sprouts. *Antioxidants* **2020**, *9*, 558. [CrossRef]
46. Bradford, M.M. A rapid and sensitive method for the quantitation of microgram quantities of protein utilizing the principle of protein-dye binding. *Anal. Biochem.* **1976**, *72*, 248–254. [CrossRef]
47. Paciolla, C.; Ippolito, M.P.; Logrieco, A.; Dipierro, N.; Mule, G.; Dipierro, S. A different trend of antioxidant defence responses makes tomato plants less susceptible to beauvericin than to T-2 mycotoxin phytotoxicity. *Physiol. Mol. Plant Pathol.* **2008**, *72*, 3–9. [CrossRef]
48. Mastropasqua, L.; Borraccino, G.; Bianco, L.; Paciolla, C. Light qualities and dose influence ascorbate pool size in detached oat leaves. *Plant Sci.* **2012**, *183*, 57–64. [CrossRef] [PubMed]
49. Paciolla, C.; De Leonardis, S.; Zonno, M.C.; Vurro, M. Antioxidant response in *Chenopodium album* elicited by *Ascochyta caulina* mycoherbicide phytotoxins. *Phytopathol. Mediterr.* **2016**, 346–354. [CrossRef]
50. Lee, D.H.; Lee, C.B. Chilling stress-induced changes of antioxidant enzymes in the leaves of cucumber: In gel enzyme activity assays. *Plant Sci.* **2000**, *159*, 75–85. [CrossRef]
51. Pedranzani, H.; Rodríguez-Rivera, M.; Gutiérrez, M.; Porcel, R.; Hause, B.; Ruiz-Lozano, J.M. Arbuscular mycorrhizal symbiosis regulates physiology and performance of *Digitaria eriantha* plants subjected to abiotic stresses by modulating antioxidant and jasmonate levels. *Mycorrhiza* **2016**, *26*, 141–152. [CrossRef]
52. Pandino, G.; Lombardo, S.; Antonino, L.M.; Ruta, C.; Mauromicale, G. In vitro micropropagation and mycorrhizal treatment influences the polyphenols content profile of globe artichoke under field conditions. *Food Res. Int.* **2017**, *99*, 385–392. [CrossRef] [PubMed]
53. Chen, M.M.; Arato, M.M.; Borghi, L.L.; Nouri, E.; Reinhardt, D. Beneficial services of arbuscular mycorrhizal fungi—From ecology to application. *Front. Plant Sci.* **2018**, *4*, 1270. [CrossRef] [PubMed]
54. Gujre, N.; Soni, A.; Rangan, L.; Tsang, D.C.; Mitra, S. Sustainable improvement of soil health utilizing biochar and arbuscular mycorrhizal fungi: A review. *Environ. Pollut.* **2020**, 115549. [CrossRef]
55. Carillo, P.; Kyratzis, A.; Kyriacou, M.C.; Dell'Aversana, E.; Fusco, G.M.; Corrado, G.; Rouphael, Y. Biostimulatory Action of Arbuscular Mycorrhizal Fungi Enhances Productivity, Functional and Sensory Quality in 'Piennolo del Vesuvio' Cherry Tomato Landraces. *Agronomy* **2020**, *10*, 911. [CrossRef]
56. Federico, N.S.; Carmona, M.; Balestrasse, K.; Chiocchio, V.; Giacometti, R.; Lavado, R.S. The arbuscular mycorrhizal fungus *Rhizophagus intraradices* reduces the root rot caused by *Fusarium pseudograminearum* in wheat. *Rhizosphere* **2021**, 100369. [CrossRef]
57. Garmendia, I.; Goicoechea, N.; Aguirreolea, J. Effectiveness of three *Glomus* species in protecting pepper (*Capsicum annuum* L.) against verticillium wilt. *Biol. Control* **2004**, *31*, 296–305. [CrossRef]
58. Evelin, H.; Kapoor, R.; Giri, B. Arbuscular mycorrhizal fungi in alleviation of salt stress: A review. *Ann. Bot.* **2009**, *104*, 1263–1280. [CrossRef]
59. Ceccarelli, N.; Curadi, M.; Martelloni, L.; Sbrana, C.; Picciarelli, P.; Giovannetti, M. Mycorrhizal colonization impacts on phenolic content and antioxidant properties of artichoke leaves and flower heads two years after field transplant. *Plant Soil* **2010**, *335*, 311–323. [CrossRef]
60. Wang, J.; Zhai, L.; Ma, J.; Zhang, J.; Wang, G.G.; Liu, X.; Zhang, S.; Song, J.; Wu, Y. Comparative physiological mechanisms of arbuscular mycorrhizal fungi in mitigating salt-induced adverse effects on leaves and roots of *Zelkova serrata*. *Mycorrhiza* **2020**, *30*, 341–355. [CrossRef] [PubMed]
61. Campanelli, A.; Ruta, C.; Tagarelli, A.; Morone-Fortunato, I.; De Mastro, G. Effectiveness of mycorrhizal fungi on globe artichoke (*Cynara cardunculus* L. var. *scolymus*) micropropagation. *J. Plant Interact.* **2014**, *9*, 100–106. [CrossRef]
62. Zhang, Q.; Gao, X.; Ren, Y.; Ding, X.; Qiu, J.; Li, N.; Zeng, F.; Chu, Z. Improvement of *Verticillium* wilt resistance by applying arbuscular mycorrhizal fungi to a cotton variety with high symbiotic efficiency under field conditions. *Int. J. Mol. Sci.* **2018**, *19*, 241. [CrossRef]
63. Ismail, Y.; McCormick, S.; Hijri, M. The arbuscular mycorrhizal fungus, *Glomus irregulare*, controls the mycotoxin production of *Fusarium sambucinum* in the pathogenesis of potato. *FEMS Microbiol. Lett.* **2013**, *348*, 46–51. [CrossRef] [PubMed]
64. Gao, P.; Guo, Y.; Li, Y.; Duan, T. Effects of dual inoculation of AMF and rhizobium on alfalfa (*Medicago sativa*) root rot caused by *Microdochium tabacinum*. *Australas. Plant Pathol.* **2018**, *47*, 195–203. [CrossRef]

65. Campanella, V.; Miceli, C. Biological control of Fusarium wilt of Ustica landrace lentil. *Crop Prot.* **2021**, *145*, 105635. [CrossRef]
66. Hernández, J.A.; Barba-Espín, G.; Diaz-Vivancos, P. Glutathione-mediated biotic stress tolerance in plants. In *Glutathione in Plant Growth, Development, and Stress Tolerance*; Hossain, M., Mostofa, M., Diaz-Vivancos, P., Burritt, D., Fujita, M., Tran, L.S., Eds.; Springer: Cham, Switzerland, 2017; pp. 309–319. [CrossRef]
67. Bernal-Vicente, A.; Pascual, J.A.; Tittarelli, F.; Hernández, J.A.; Diaz-Vivancos, P. *Trichoderma harzianum* T-78 supplementation of compost stimulates the antioxidant defence system in melon plants. *J. Sci. Food Agric.* **2015**, *95*, 2208–2214. [CrossRef]
68. Saroy, K.; Garg, N. Relative effectiveness of arbuscular mycorrhiza and polyamines in modulating ROS generation and ascorbate-glutathione cycle in *Cajanus cajan* under nickel stress. *Environ. Sci. Pollut. Res.* **2021**, 1–18. [CrossRef]
69. Nahiyan, A.S.M.; Matsubara, Y.I. Tolerance to *Fusarium* root rot and changes in antioxidative ability in mycorrhizal asparagus plants. *HortScience* **2012**, *47*, 356–360. [CrossRef]
70. Kuzniak, E.; Skłodowska, M. Differential Implication of Glutathione, Glutathione Metabolizing Enzymes and Ascorbate in Tomato Resistance to *Pseudomonas syringae*. *J. Phytopathol.* **2004**, *152*, 529–536. [CrossRef]
71. Pieterse, C.M.; Zamioudis, C.; Berendsen, R.L.; Weller, D.M.; Van Wees, S.C.; Bakker, P.A. Induced systemic resistance by beneficial microbes. *Annu. Rev. Phytopathol.* **2014**, *52*, 347–375. [CrossRef] [PubMed]
72. Meyer, A.J.; Hell, R. Glutathione homeostasis and redox-regulation by sulfhydryl groups. *Photosynth. Res.* **2005**, *86*, 435–457. [CrossRef]
73. Noctor, G.; Mhamdi, A.; Chaouch, S.; Han, Y.; Neukermans, J.; Marquez-Garcia, B.; Queval, G.; Foyer, C.H. Glutathione in plants: An integrated overview. *Plant Cell Environ.* **2012**, *35*, 454–484. [CrossRef] [PubMed]
74. Gill, S.S.; Anjum, N.A.; Hasanuzzaman, M.; Gill, R.; Trived, D.K.; Ahmad, I.; Pereira, E.; Tuteja, N. Glutathione reductase and glutathione: A boon in disguise for plant abiotic stress defense operations. *Plant Physiol. Biochem.* **2013**, *70*, 204–212. [CrossRef] [PubMed]
75. Gamble, P.; Burke, J.J. Effect of water stress on the chloroplast antioxidant system. Alteration in glutathione reductase activity. *Plant Physiol.* **1984**, *76*, 615–621. [CrossRef] [PubMed]
76. Alguacil, M.M.; Hernández, J.A.; Caravaca, F.; Portillo, B.; Roldán, A. Antioxidant enzyme activities in shoots from three mycorrhizal shrub species afforested in a degraded semi-arid soil. *Physiol. Plant.* **2003**, *118*, 562–570. [CrossRef]
77. Kopriva, S.; Rennenberg, H. Control of sulphate assimilation and glutathione synthesis: Interaction with N and C metabolism. *J. Exp. Bot.* **2004**, *55*, 1831–1842. [CrossRef] [PubMed]
78. Paciolla, C.; De Tullio, M.C.; Chiappetta, A.; Innocenti, A.M.; Bitonti, M.B.; Liso, R.; Arrigoni, O. Short- and Long-Term Effects of Dehydroascorbate in *Lupinus albus* and *Allium cepa* Roots. *Plant Cell Physiol.* **2001**, *42*, 857–863. [CrossRef] [PubMed]
79. Morell, S.; Follmann, H.; De Tullio, M.; Häberlein, I. Dehydroascorbate and dehydroascorbate reductase are phantom indicators of oxidative stress in plants. *FEBS Lett.* **1997**, *414*, 567–570. [CrossRef]
80. Loi, M.; De Leonardis, S.; Mulè, G.; Logrieco, A.F.; Paciolla, C. A Novel and Potentially Multifaceted Dehydroascorbate Reductase Increasing the Antioxidant Systems is Induced by Beauvericin in Tomato. *Antioxidants* **2020**, *9*, 435. [CrossRef] [PubMed]
81. Mittova, V.; Tal, M.; Volokita, M.; Guy, M. Up-regulation of the leaf mitochondrial and peroxisomal antioxidative systems in response to salt-induced oxidative stress in the wild salt-tolerant tomato species *Lycopersicon pennellii*. *Plant Cell Environ.* **2003**, *26*, 845–856. [CrossRef] [PubMed]
82. Kavitha, K.; George, S.; Venkataraman, G.; Parida, A. A salt-inducible chloroplastic monodehydroascorbate reductase from halophyte *Avicennia marina* confers salt stress tolerance on transgenic plants. *Biochimie* **2010**, *92*, 1321–1329. [CrossRef] [PubMed]
83. Chen, W.; Xiangpeng, L.; Wenying, Z.; Jinggui, F. The Regulatory and Signaling Roles of Glutathione in Modulating Abiotic Stress Responses and Tolerance. In *Glutathione in Plant Growth, Development, and Stress Tolerance*; Hossain, M.A., Mostofa, M.G., Diaz-Vivancos, P., Burritt, D.J., Fujita, M., Tran, L.S.P., Eds.; Springer International Publishing: Cham, Switzerland, 2017; pp. 147–169. [CrossRef]
84. Porcel, R.; Ruiz-lozano, J.M. Arbuscular mycorrhizal influence on leaf water potential, solute accumulation, and oxidative stress in soybean plants subjected to drought stress. *J. Exp. Bot.* **2004**, *55*, 1743–1750. [CrossRef]
85. Garmendia, I.; Aguirreolea, J.; Goicoechea, N. Defence-related Enzymes in Pepper Roots During Interactions with Arbuscular Mycorrhizal Fungi and/or *Verticillium dahliae*. *Biocontrol* **2006**, *51*, 293. [CrossRef]
86. Ortega, E.P.; Noval, B.M.L.; Coca, B.M.; Noval, W.T.L.; Carmona, A.M.; Hernández, A.; León, O. Induction of defense mechanisms in mycorrhized tomato plants against the attack of *Oidiopsis taurica* (Lev.) Salm. *Cult. Trop.* **2015**, *36*, 94–102.
87. Asada, K. Ascorbate peroxidase–a hydrogen peroxide-scavenging enzyme in plants. *Physiol. Plant.* **1992**, *85*, 235–241. [CrossRef]
88. Dat, J.; Vandenabeele, S.; Vranova, E.; Van Montagu, M.; Inzé, D.; Van Breusegem, F. Dual action of the active oxygen species during plant stress responses. *Cell Mol. Life Sci.* **2000**, *57*, 779–795. [CrossRef] [PubMed]
89. Yamane, K.; Mitsuya, S.; Taniguchi, M.; Miyake, H. Transcription profiles of genes encoding catalase and ascorbate peroxidase in the rice leaf tissues under salinity. *Plant Prod. Sci.* **2010**, *13*, 164–168. [CrossRef]
90. Estrada, B.; Aroca, R.; Maathuis, F.J.M.; Barea, J.M.; Ruiz-Lozano, J.M. Arbuscular mycorrhizal fungi native from a Mediterranean saline area enhance maize tolerance to salinity through improved ion homeostasis. *Plant Cell Environ.* **2013**, *36*, 1771–1782. [CrossRef]
91. Polidoros, A.N.; Scandalios, J.G. Role of hydrogen peroxide and different classes of antioxidants in the regulation of catalase and glutathione S-transferase gene expression in maize (*Zea mays* L.). *Physiol. Plant.* **1999**, *106*, 112–120. [CrossRef]

cells

Article

Fungal Associates of Soft Scale Insects (Coccomorpha: Coccidae)

Teresa Szklarzewicz [1], Katarzyna Michalik [1], Beata Grzywacz [2], Małgorzata Kalandyk-Kołodziejczyk [3] and Anna Michalik [1,*]

[1] Department of Developmental Biology and Morphology of Invertebrates, Faculty of Biology, Institute of Zoology and Biomedical Research, Gronostajowa 9, 30-387 Kraków, Poland; teresa.szklarzewicz@uj.edu.pl (T.S.); katarzyna.michalik.bio@gmail.com (K.M.)
[2] Institute of Systematics and Evolution of Animals, Polish Academy of Sciences, Sławkowska 17, 31-016 Kraków, Poland; grzywacz@isez.pan.krakow.pl
[3] Faculty of Natural Sciences, Institute of Biology, Biotechnology and Environmental Protection, University of Silesia, Bankowa 9, 40-007 Katowice, Poland; malgorzata.kalandyk@us.edu.pl
* Correspondence: a.michalik@uj.edu.pl; Tel.: +48-12-664-5089

Abstract: *Ophiocordyceps* fungi are commonly known as virulent, specialized entomopathogens; however, recent studies indicate that fungi belonging to the Ophiocordycypitaceae family may also reside in symbiotic interaction with their host insect. In this paper, we demonstrate that *Ophiocordyceps* fungi may be obligatory symbionts of sap-sucking hemipterans. We investigated the symbiotic systems of eight Polish species of scale insects of Coccidae family: *Parthenolecanium corni*, *Parthenolecanium fletcheri*, *Parthenolecanium pomeranicum*, *Psilococcus ruber*, *Sphaerolecanium prunasti*, *Eriopeltis festucae*, *Lecanopsis formicarum* and *Eulecanium tiliae*. Our histological, ultrastructural and molecular analyses showed that all these species host fungal symbionts in the fat body cells. Analyses of ITS2 and Beta-tubulin gene sequences, as well as fluorescence in situ hybridization, confirmed that they should all be classified to the genus *Ophiocordyceps*. The essential role of the fungal symbionts observed in the biology of the soft scale insects examined was confirmed by their transovarial transmission between generations. In this paper, the consecutive stages of fungal symbiont transmission were analyzed under TEM for the first time.

Keywords: soft scale insects; *Ophiocordyceps*; symbiosis; transovarial transmission

Citation: Szklarzewicz, T.; Michalik, K.; Grzywacz, B.; Kalandyk-Kołodziejczyk, M.; Michalik, A. Fungal Associates of Soft Scale Insects (Coccomorpha: Coccidae). *Cells* **2021**, *10*, 1922. https://doi.org/10.3390/cells10081922

Academic Editor: Suleyman Allakhverdiev

Received: 28 May 2021
Accepted: 26 July 2021
Published: 29 July 2021

1. Introduction

Scale insects (coccoids) are plant sap-sucking hemipterans that are considered serious pests in agriculture, horticulture, and forestry. These insects cause direct damage to plants through sap-sucking and the injection of toxic saliva into plant tissue, which is a cause of the retardation of plant growth and recovery, and furthermore, may lead to the death of the whole or part of the plant if the infestation is severe. Scale insects are rarely known as vectors of bacterial pathogens or phytoplasmas, and only a few species are involved in virus transmission [1]. Most species of scale insects also cause indirect damage by producing a carbohydrate-rich solution, referred to as honeydew, which is a medium for the growth of saprophytic fungi known as "sooty molds", forming black superficial colonies that also reduce the host plant photosynthesis rates, further diminishing the vigor of the plant (e.g., [1–3]).

Scale insects are highly diverse in terms of the morphology of their external and internal organs, reproductive strategies and chromosome systems, as well as symbiotic systems, which makes them an interesting group of insects to study [4]. After the Diaspididae and Pseudococcidae, the Coccidae (soft scales, coccids) is the third largest family of scale insects in terms of species richness. There are 1281 described species of coccids in 176 genera. Soft scales are widely distributed in all zoogeographical regions; however, they predominantly occur in the tropics and subtropics [2,5–7]. The Coccidae, like other scale insect families, exhibit a remarkable dimorphism. The adult females are wingless and lack a well-defined head, thorax, and abdomen. The adult males are usually winged, with distinct body parts, and do not possess functional mouthparts. A large number of soft scale species are notorious plant pests that are of great economic importance to crops. Many pests of the Coccidae have been introduced into new zoogeographical regions, thus making them cosmopolitan [2].

In several insect groups, including scale insects, a mutualistic relationship with microorganisms (bacteria or fungi) evolved. The use of genomic analyses has confirmed earlier assumptions that the occurrence of symbiotic associates in the insect body is associated with the poor diet of the host-insect, e.g., plant sap-sucking hemipterans receiving amino acids, and blood-sucking insects receiving B vitamins from their mutualists [8].

As typical phloem-feeders, scale insects live in symbiotic association with microorganisms; however, in comparison with close relatives such as aphids, whiteflies, and psyllids, these insects are characterized by highly diverse symbiotic systems. Scale insects may live in mutualistic relationships with different species of bacteria or fungal symbionts. They may have only one symbiont or several species of microorganisms. Symbionts may be harbored in the fat body cells, in the midgut epithelium, in the specialized cells of mesodermal origin termed bacteriocytes, or inside the cells of other bacteria. Scale insects also developed different modes of transmission of their symbiotic associates from mother to offspring [9–12].

In contrast to other families of scale insects, symbionts of soft scale insects have not been as extensively examined through the use of modern ultrastructural and molecular methods. The results of histological studies (reviewed in [9,10,13]) have indicated that these insects are hosts to obligate fungal symbionts that may be localized freely in hemolymph or intracellularly in fat body cells, and are transovarially inherited. Studies recently conducted with the use of molecular methods and phylogenetic analyses have allowed the identification of the symbiotic associates of seven species of the Coccidae from the Mediterranean region and 28 species from southern China as the *Ophiocordyceps*-allied fungus (phylum Ascomycota) [14,15]. It is noteworthy that, for many years, fungi belonging to the genus *Ophiocordyceps*, like other members of Ophiocordycipitaceae and Cordycypitaceae, were known mainly as entomopathogenic microorganisms [16,17]. They may attack various species of insects, e.g., ants, beetles, butterflies, and hemipterans. The hyphae of these fungi penetrate the body wall and destroy their internal tissue: fat body cells, hemocytes, muscle, nerve ganglions, and the intestine. In each case, insects infected by these fungi die before beginning their reproductive phase, i.e., within 48–96 h of penetration [16]. For this reason, entomopathogenic fungi have also been tested as biological control agents for whiteflies, lepidopterans and scale insects [18,19]. The finding of a close relationship between fungal entomopathogens and symbionts has led to the hypothesis that during their co-evolution, the interaction between both of these partners shifted from parasitism to mutualism [20,21].

The aim of this study was to further explore the symbiotic systems of the Coccidae family: (1) to determine the systematic position of symbionts, (2) to verify whether symbiosis with fungi is a general rule of this family, (3) to show whether the symbiosis is the result of the single infection of the ancestor of Coccidae or multiple independent infections, (4) to describe symbiont distribution and ultrastructure as well as a mode of transmission from mother to progeny in eight species from three subfamilies of Central European origin.

2. Material and Methods

2.1. Insects

Eight species of the Coccidae family: *Parthenolecanium corni* (Bouché, 1844), *Parthenolecanium fletcheri* (Cockerell, 1893), *Parthenolecanium pomeranicum* (Kawecki, 1954), *Eriopeltis festucae* (Boyer de Fonscolombe, 1834), *Lecanopsis formicarum* Newstead, 1893, *Sphaerolecanium prunastri* (Boyer de Fonscolombe, 1834), *Eulecanium tiliae* (Linnaeus, 1758) and *Psilococcus ruber* Borchsenius, 1952 were collected in unprotected areas in Poland between the years 2017 and 2019 from their host plants. The localities, collection dates, and host plants of the investigated species have been summarized in Table 1. Species of the Coccidae family were assigned to subfamilies according to Koteja [22].

Table 1. List of investigated species with collection details and number of individuals used in each method applied.

Species	Subfamily	Place of Collection/Locality Coordinates	Date of Collection	Host Plant	No. of Individuals Examined Using Microscopic Techniques		No. of Individuals Examined Using Molecular Techniques	
					LM	TEM	PCR	FISH
Parthenolecanium corni	Coccinae	Katowice 50.245638 19.007992	V, VI 2017	*Tilia cordata*	10	2	3	1
Parthenolecanium fletcheri		Katowice, 50.245621 19.005127 Olsztyn 50.753157 19.277790	V 2018; V 2019	*Thuja* sp.	6	4	3	1
Parthenolecanium pomeranicum		Katowice 50.243978 19.001704	V, VI 2017; VI 2018	*Taxus baccata*	10	2	7	3
Eriopeltis festucae	Eriopeltinae	Olsztyn 50.749359 19.272956	VII 2017; VI 2018	*Calamagrostis epigejos*	5	2	3	1
Lecanopsis formicarum		Mikoszewo 54.341908 18.992342	VIII 2018	*Festuca ovina*	5	2	7	3
Psilococcus ruber		Mikoszewo 54.344519 18.978498	VIII 2018	*Carex* sp.	10	4	7	3
Sphaerolecanium prunastri	Filippinae	Bukowno 50.274843 19.436405	VI 2017; IV 2018	*Prunus spinosa*	15	3	7	3
Eulecanium tiliae		Olsztyn 50.749837 19.274678	V 2018	*Tilia cordata*	6	2	3	1

2.2. Light (LM) and Electron Microscopy (TEM)

Females of the investigated species, destined for detailed histological and ultrastructural analysis, were fixed in 2.5% glutaraldehyde in 0.05 M phosphate buffer (pH 7.4) and stored in a fridge for 1–4 weeks. After this time, the samples were rinsed five times in the buffer with sucrose (5.8 g/100 mL), postfixed in buffered 1% osmium tetroxide for 2 h, and then dehydrated in an ethanol series (30%, 50%, 70%, 90%, 100%) and acetone. Finally, the samples were embedded in epoxy resin Epon 812 (Serva, Heidelberg, Germany) ([23]—modified). For the histological analyses, semithin sections (1 μm thick) were stained in 1% methylene blue in 1% borax and photographed under a Nikon Eclipse 80i light microscope. For ultrastructural analyses, ultrathin sections (90 nm thick) were doubly contrasted with uranyl acetate and lead citrate and subsequently examined and photographed under a Jeol JEM 2100 at 80 kV transmission electron microscope. The number of specimens used for histological and ultrastructural analyses is summarized in Table 1.

2.3. Molecular Analyses

Specimens destined for molecular analysis were fixed in 100% ethanol. In the case of the removal of the surphase's contaminations, the specimens were placed in 5% sodium hypochlorite for 1 min and then rinsed in distilled water three times for one minute. Then the cuticle was removed, and the DNA isolated only from the fat body and internal organs. DNA extraction was performed separately from 3-7 individuals of each species (see Table 1) using the Bio-Trace DNA Purification Kit (EURx, Gdańsk, Poland) following manufacturer protocol and subsequently stored at −20 °C for further analysis.

The fungal associates of the species examined were identified and characterized based on sequences of two genes: Internal Transcribed Spacer 2 of nuclear ribosomal RNA (ITS 2) and Beta-tubulin gene using primers: ITS3/ITS4 [24] and Ophi_Btub44448F/Ophi_Btub5243R (D. Vanderpool, unpublished), respectively. The mitochondrial cytochrome c oxidase subunit I (COI) gene of soft scale insects was amplified using the primer pair PCoF1 and

HCO [25]. The conditions for all the PCR reactions were an initial denaturation step at 94 °C for 5 min, followed by 33 cycles at 94 °C for 30 s, Tm for 40 s, 70 °C for 1 min 40 s and a final extension step of 5 min at 72 °C. The PCR products were visualized on 1.5% agarose gel stained with Simply Safe (EURx, Gdańsk, Poland), purified with the Gel-Out Concentrator (A&A Biotechnology, Gdańsk, Poland) kit following manufacturer protocol, and subsequently sequenced. The sequences of the primers have been listed in Table S1. The nucleotide sequences obtained were deposited in the GenBank database under the accession numbers: ITS (MN733271-MN733277, MZ594469); COI (MN603157, MN603159-MN603160, MN603162-MN603164, MZ567176), Beta-tubulin (MN750822-MN750828, MZ576194).

2.4. Fluorescence In Situ Hybridization (FISH)

Fluorescence in situ hybridization (FISH) was conducted with the probe Hyp760 specific for the 18S rRNA gene of *Ophiocordyceps* fungi [26] (Table S1). Two individuals of each species that were preserved in 100% ethanol were rehydrated, fixed in 4% formaldehyde for two hours and dehydrated through incubation in 80%, 90% and 100% ethanol and acetone. The material was then embedded in Technovit 8100 (Kulzer, Werheim, Germany) resin and subsequently cut into sections. Hybridization was performed using a hybridization buffer containing: 1 mL 1 M Tris-HCl (pH 8.0), 9 mL 5 M NaCl, 25 µL 20% SDS, 15 mL 30% formamide and about 15 mL of distilled water. The slides were incubated in 200 µL of hybridization solution (hybridization buffer + probes) overnight at room temperature [27]. Following this, the slides were washed in PBS three times for 10 min, then dried and covered with a ProLong Gold Antifade Reagent (Life Technologies, Carlsbad, CA, USA). The hybridized slides were then examined using a confocal laser scanning microscope Zeiss Axio Observer LSM 710.

2.5. Phylogenetic and Co-Phylogenetic Analyses

The sequences were aligned in a CodonCode Aligner v.8.0 (CodonCode Corporation, www.codoncode.com, 8 March 2018). Coding regions were translated to amino acids using Mega v. X [28] in order to detect frameshift mutations and internal stop codons. The Akaike Information Criteria (AIC) in MrModeltest v. 2.2 [29] was used to estimate the best-fit substitution models. Phylogenetic trees were constructed using Bayesian inference (BI) in MrBayes v. 3.2.6 [30]. For phylogenetic trees of both scale insects and fungal symbionts, MrBayes was run for six million generations, sampling every 100 generations in order to ensure the independence of the samples. Two independent runs were performed to ensure that convergence on the same posterior distribution was reached, and if the final trees converged on the same topology. The statistical confidence in the nodes was evaluated using posterior probabilities.

Co-phylogenetic and host-switching events were tested in Jane v.4 [31] using the BI host and fungal trees as input. The analysis was performed with 100 generations, population sizes of 100 and a default event–cost scheme including "co-speciation", "duplication", "host switch", "losses", and "failure to diverge".

3. Results

3.1. Fungi Belonging to the Ophiocordycypitaceae Family Are Symbionts of the Soft Scale Insects Examined

Our histological, ultrastructural, and molecular analyses showed that all the species examined were associated with symbiotic fungi. Molecular analyses based on sequences of ITS2 and Beta-tubulin genes revealed that in all the species examined, the symbiotic fungi belonged to the Ophiocordycypitaceae family within the Ascomycota phylum (Ascomycota: Sordariomycetes: Hypocreales: Ophiocordycipitaceae).

Based on the sequences of the Beta-tubulin gene, two groups of symbiotic microorganisms may be distinguished: the first one includes symbionts of *Eulecanium tiliae*, *Parthenolecanium corni*, *Parthenolecanium pomeranicum*, and *Parthenolecanium fletcheri*, and the second one is comprised of the symbionts of remaining species. These latter sequences are almost identical (99%) and differ from the sequences in the first group by 4%. In turn, the similarity of the ITS2 sequences, which are more species-specific, ranges from 87% to 96%. Blast searches for all of the sequences obtained demonstrated the highest similarity to sequences of homologue genes of various species of *Ophiocordyceps* or its anamorphic (i.e., asexual) form—*Hirsutella*. Phylogenetic analyses also confirmed the systematic affiliation of *Hirsutella* with the genus *Ophiocordyceps*, and showed that they create a sister group, i.e., *O. cochlidiicola*, that is closely related to *H. leizhouensis*, while *O. arborescens*, *H. versilor*, and *O. xuefengensis* are closely related to *H. illustris* (Figure 1 and Figure S1). The co-phylogenetic analysis based on the ITS2 gene of fungal symbionts and the COI genes of host scale insects returned 11 potential co-speciation events, one duplication, 22 duplications with host switch, five losses, and one failure to diverge (Figure 2).

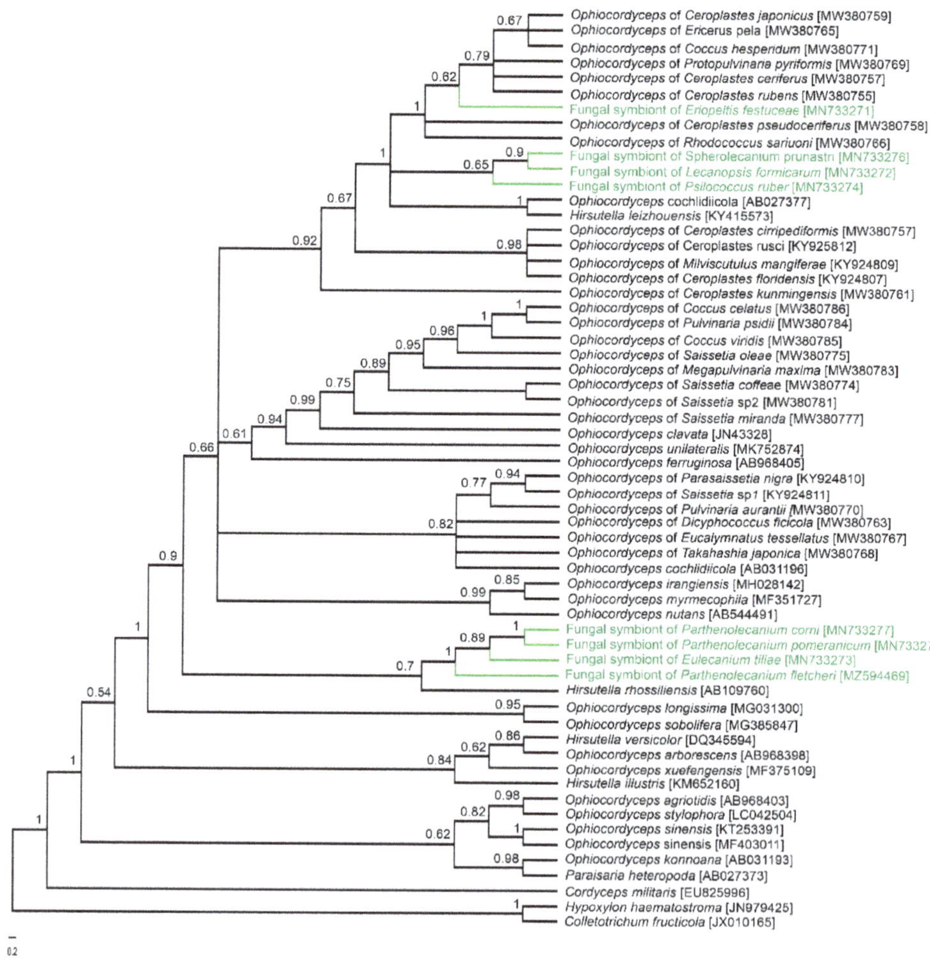

Figure 1. Phylogenetic tree showing the relationships between fungal symbionts of soft scale insect species, pathogenic, as well as free living fungi (constructed on the base of ITS2 genes). The numbers near the nodes refer to the Bayesian posterior probability.

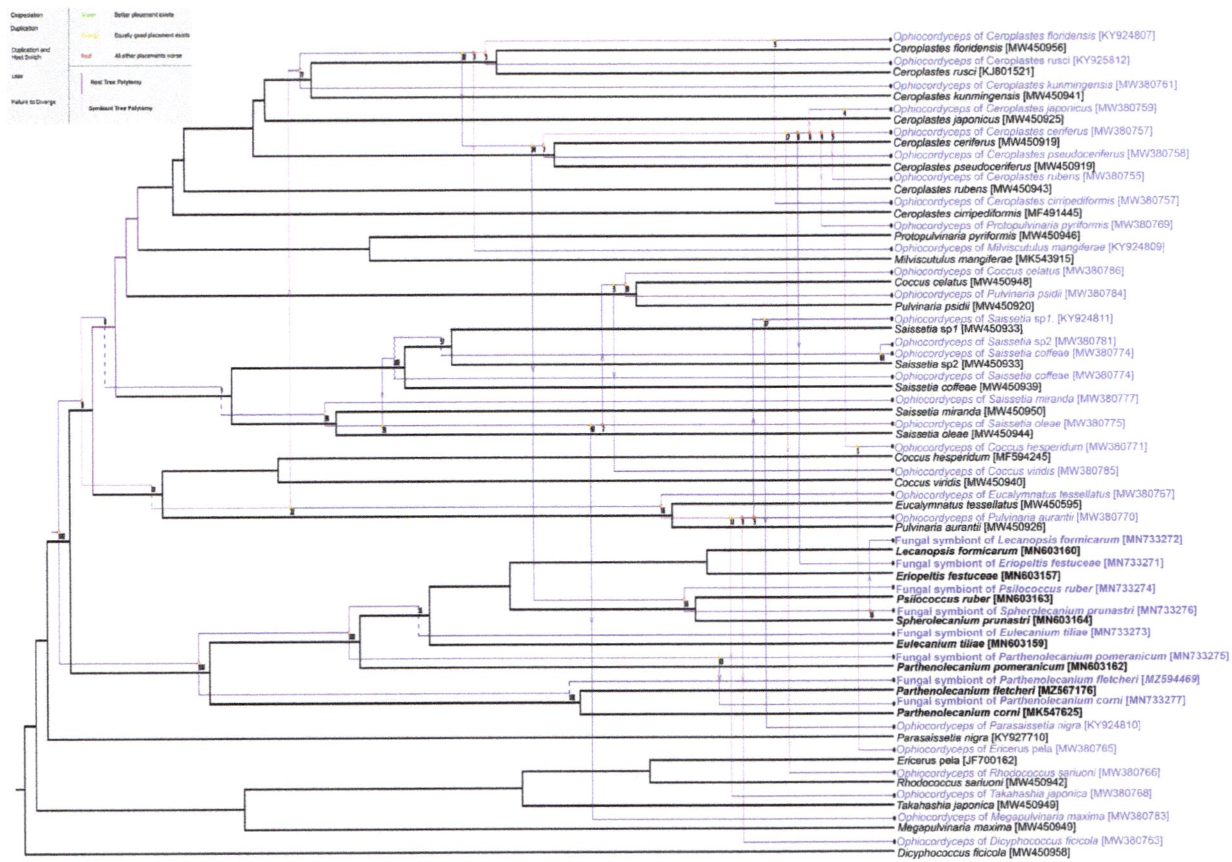

Figure 2. Co-phylogenetic analysis between the *Ophiocordyceps* symbionts' tree and their host's tree (constructed on the base of the ITS2 genes of fungal symbionts and COI genes of host scale insects). Black and blue lines indicate the phylogenies of the scale insects and *Ophiocordyceps*, respectively. Hollow red circles indicate co-speciation events, solid red and yellow circles indicate duplications, arrows indicate host events, dashed lines indicate losses, and uneven lines indicate failures to diverge.

The analysis of serial semithin sections has shown that symbiotic fungi are distributed only within the fat body cells (Figure 3). They were not observed in any other tissue except the ovaries, which is related to the transovarial transmission of these symbionts between generations (see the Results subsection, which follows). The number of symbionts and their density in the host insect body are subfamily specific and are not dependent on the stage of the insect's development (Figure 3). We observed the same amount of cells of fungi in the body cavities of larvae and mature females. The smallest number of symbionts was observed in the representatives of the Eriopeltinae subfamily (with the exception of *Psilococcus ruber*), where only single groups of fungi occur (Figure 3A–C). The highest density was observed in the members of Filippinae subfamily (Figure 3G–J). In all representatives of the Filippinae subfamily examined: *Sphaerolecanium prunasti* and *Eulecanium tiliae*, all cells of the fat body are filled with numerous symbiotic fungi (Figure 3G–J). The ultrastructural analyses showed that the cells of fungi are surrounded by a thick cell wall (Figure 3B,E,H). Large nuclei (Figure 3E,H) and vacuoles (Figure 3B) are visible in their cytoplasm.

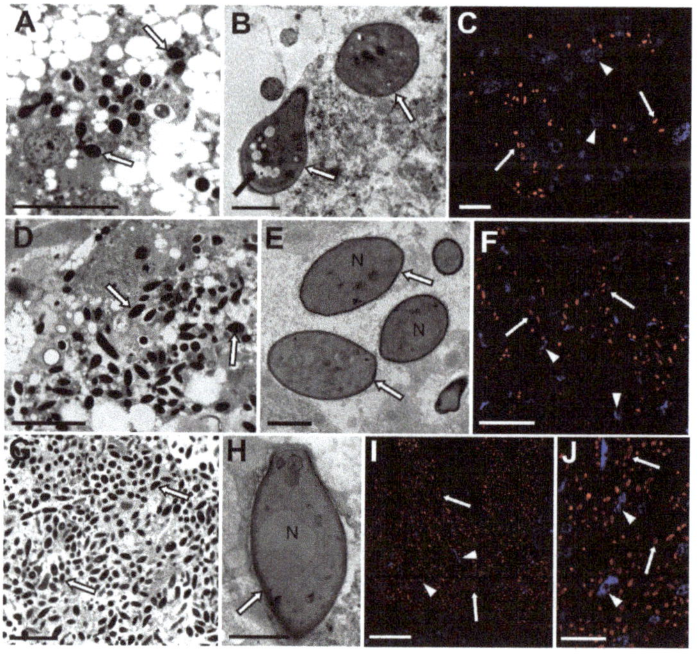

Figure 3. Distribution of symbiotic fungi in the body cavity of species examined (**A–J**). *Ophiocordyceps* fungi in the cytoplasm of the fat body cells (**A,D,G**). Light microscope, scale bar = 20 µm (**B,E,H**). TEM, scale bar = 2 µm (**C,F,I,J**). Confocal microscope, scale bar = 20 µm (**A–C**). Eriopelitnae subfamily (**A**). *Eriopeltis festucae* (**B,C**). *Lecanopsis formicarum* (**D–F**). Coccinae subfamily (**D,E**). *Parthenolecanium corni* (**F**). *Parthenolecanium pomeranicum* (**G–J**). Filippinae subfamily, *Sphaerolecanium prunastri*. N—nucleus of the fungal cell; white arrow—fungal symbiont, white arrowhead—nucleus of the fat body cell, black arrow—vacuole.

The presence of *Ophiocodyceps* fungi in the body cavity of the soft scale insects examined was also confirmed by fluorescence in situ hybridization using an *Ophiocordyceps*-specific probe (Figure 3C,F,I,J). The microscopic observations did not show any damage to the insects' tissue caused by fungi.

3.2. Fungal Associates of Soft Scale Insects Are Transovarially Transmitted between Generations

Microscopic observations revealed that fungal symbionts residing in the examined species of soft scale insects were inherited transovarially, i.e., they infect female germ cells.

The ovaries of soft scale insects are composed of numerous short telotrophic ovarioles, which are subdivided into an anterior tropharium (trophic chamber) and posterior vitellarium (Figure 4A) (for further details concerning the organization of the ovaries of scale insects, see [32]). The vitellarium houses a single oocyte, which is connected with the tropharium by means of a broad nutritive cord (Figure 4A). The oocyte is surrounded by a single-layered follicular epithelium (Figure 4A). At the time the ovarioles contain the oocytes in the stage of advanced choriogenesis (Figure 4A), the fungal symbionts begin to enter follicular cells surrounding the neck region of the ovariole (i.e., the region between the tropharium and the developing oocyte) (Figure 4B). After crossing through the cytoplasm of the follicular cells (Figure 4A,C,D), symbionts temporarily gather around the nutritive cord (Figure 4E,F). Next, symbiotic microorganisms migrate along the nutritive cord to the space between follicular epithelium and oocyte surface (Figure 4G,H). Finally, symbionts enter the oocyte cytoplasm (Figure 4I,J), where they remain until the beginning of embryonic development.

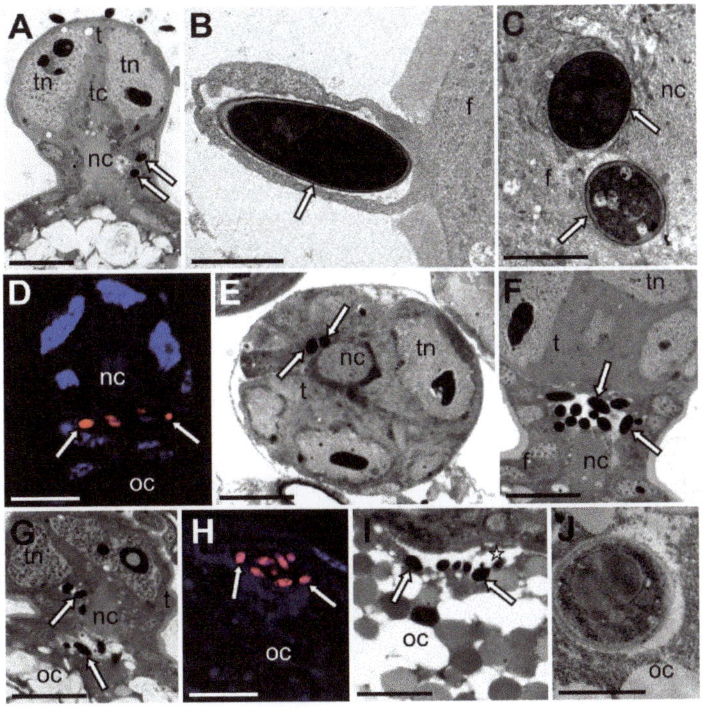

Figure 4. Consecutive stages of symbiont transmission between generations (**A**,**B**). Longitudinal section through the ovariole of *Parthenolecanium fletcheri*. Symbiotic fungi (white arrows) invade the follicular cells in the neck region of the ovariole (**A**). Light microscope, scale bar = 20 μm (**B**). TEM, scale bar = 2 μm (**C**,**D**). Symbiotic fungi in the cytoplasm of follicular cells of *Parthenolecanium pomeranicum* (**C**) and *Eulecanium tiliae* (**D**). (**C**) TEM, scale bar = 2 μm (**D**). Confocal microscope, scale bar = 20 μm (**E**,**F**). Symbiotic fungi surrounding the nutritive cord (**E**). *Parthenolecanium pomeranicum* (**F**). *Parthenolecanium fletcheri* (**E**,**F**). Light microscope, scale bar—20 μm (**G**). Symbiotic fungi move along the nutritive cord to the perivitelline space of *Parthenolecanium fletcheri*. Light microscope, scale bar = 20 μm (**H**). Symbiotic fungi gather in the invagination of the perivitelline space of *Eulecanium tiliae*. Confocal microscope, scale bar = 20 μm (**I**). Migration of symbionts from the perivitelline space to the oocyte cytoplasm of *Psilococcus ruber*. Light microscope, scale bar = 20 μm (**J**). Symbiotic microorganism in the oocyte cytoplasm of *Psilococcus ruber*. TEM, scale bar = 2 μm. f—follicular cell, nc—nutritive cord; oc—oocyte, t—trophocyte, tc—trophic core, tn—trophocyte nucleus, white arrow—fungal symbiont; asterisk—perivitelline space.

4. Discussion

The Cordycypitaceae and Ophiocordycypitaceae families of Ascomycota include several genera that are commonly occurring entomopathogens, such as *Ophiocordyceps*, *Cordyceps*, *Hirsutella*, *Lecanicilium*, and *Metarhizium* [33]. Among them, the *Ophiocordyceps* species, which usually attack ants, are best known for their ability to manipulate ant behavior. However, recent research that applies the use of molecular techniques, and concerns the interactions of insects and various microorganisms indicates that the fungi of Ascomycota phylum may also live in symbiotic associations with insects [14,26,34]. So far, fungal associates have been found and described in many representatives of Hemiptera, Coleoptera, Diptera, and Hymenoptera, as well as in some members of Isoptera, Neuroptera, and Lepidoptera; for a review, see [35]. They belong to various families of Ascomycota; however, hemipterans are usually associated with fungi from the Ophiocordycypitaceae family.

Within the Hemiptera: Auchenorrhyncha (i.e., Fulgoromorpha (planthoppers) and Cicadomorpha (leafhoppers, treehoppers, spittlebugs, and cicadas)), the presence of *Ophiocordyceps*-allied symbionts was confirmed in some leafhoppers from the Deltocephalinae subfamily [36,37] and planthoppers from the Flatidae and Delphacidae families [38–40], as well as in Japanese cicadas [26]. In Hemiptera: Sternorrhyncha, the symbiosis with fungi is not as common as in Auchenorrhyncha, and has so far only been observed in some of the Hormaphidinae aphids and members of the Coccidae, Dactylopiidae, and Kermesidae families of scale insects [14,15,34,36,41–43].

In this paper, using microscopic and molecular techniques, we investigated symbiotic systems of soft scale insects (Coccidae) belonging to three subfamilies: Coccinae, Eriopeltinae and Filippinae. Our analyses revealed that all of the species investigated were only host to fungal symbionts. Analyses of sequences of ITS2 as well as Beta-tubulin genes demonstrated that these microorganisms are representatives of the Ophiocordycypitaceae family. However, the ITS sequences, which are more species-specific than the Beta-tubulin gene, display about 87–95% similarity to each other. Fungal symbionts of members of the Coccidae family have previously been studied by Gomez-Polo and co-workers [14] and Deng and co-workers [15]. These authors investigated seven coccid species from the Ceroplastinae and Coccinae subfamilies collected in Spain, Israel, and Cyprus. Based on a high-throughput sequencing of ribosomal genes, these authors showed that the species analyzed were mainly associated with *Ophiocordyceps* fungi.

It is believed that the occurrence of fungal associates in some auchenorrhynchans and some aphids is a result of symbiont replacement [26,36,37], e.g., in deltocephalinae leafhoppers as well as in Japanese cicadas, fungi replaced the bacteria *Nasuia* and *Hodgkinia* (respectively), in Delphacidae and Flatidae planthoppers – bacteria *Sulcia* and *Vidania*, in the aphids of Hormaphidinae subfamily–bacteria *Buchnera* [26,36,37,41,44–47]. Our phylogenetic analyses showed that *Ophiocordyceps*-allied fungi in the species of soft scale insects examined form a clade (see Figure 1), which suggests that symbiosis between these insects and their microorganisms is the result of a single infection of the ancestor of extant coccids. However, the differences observed in ITS2 sequences (5–13%) may indicate their independent evolution after the initial infection.

Since symbiotic fungi have recently been observed in various hemipteran lineages, researchers continually ask themselves about the origin of these associations [14,20,34,35,40]. The literature data indicate three possible evolutionary scenarios: (1) symbiotic fungi may derive from entomopathogenic fungi; (2) they may be the descendants of non-pathogenic commensals; or (3) the ancestor of fungal symbionts may be phytopathogenic fungi [21,35,38]. Most hemipterans are plant sap-sucking insects, and due to their mode of feeding, they are also vectors of plant pathogens. Therefore, it seems probable that they may acquire fungal symbionts from the host plant. However, the results of molecular phylogenetic analyses that indicate the close relationship between fungal symbionts of hemipterans with entomopathogenic fungi favor the concept that the ancestors of the present fungal symbionts were entomopathogens that lost their virulence and shifted to a symbiotic lifestyle. The genomic analysis of the fungal symbiont of the planthopper

Nilaparvata lugens showed that it possesses a smaller genome than its free-living relatives and does not possess genes encoding enzymes responsible for penetrating the insect's cuticle, solubilizing its tissue, or genes related to sexual reproduction [40].

Ophiocordyceps fungi, in the soft scale insects examined, are localized in the cytoplasm of the fat body cells. The same localization of the fungal symbiont was observed in other Coccidae species that have previously been examined, in the scale insect *Kermes quercus* (Kermesidae) and leafhoppers *Fieberiella septentrionalis*, *Graphocraerus ventralis* and *Orientus ishidae* [9,14,34,37]. In the Deltocephalinae leafhopper *Cicadula quadrinotata*, fungal symbionts were found to be present in the cytoplasm of midgut epithelium cells, in fat body cells, and free in the hemolymph [37]. In contrast, in Japanese cicadas and planthoppers from the Flatidae and Delphacidae families, they are harbored in the cells of a specialized host's organs, termed mycetomes [26,47]. It seems probable that, similarly to the case of the bacterial associates of insects [48], the occurrence of fungal symbionts in the digestive tract represents the initial state of colonization of the insect body through microorganisms, in the hemolymph and in fat body cells represents the next (i.e., intermediate) stage, whereas their presence in the cells of mycetomes represents the most advanced condition of this association. It is worth mentioning that the occurrence in the fat body and in the specialized host's cells seems to be characteristic to fungi from the Ophiocordycypitaceae family, whereas other species of fungi found so far in insects are usually localized in a different part of the digestive tract [35].

Scale insects, like other insects living in a mutualistic relationship with microorganisms, develop stable mechanisms of transmission of these associates from one generation to the next. The results of numerous studies conducted both earlier and more recently indicate that scale insects are not only characterized by diverse species and distributions of symbionts, but also by different modes of transmission to their progeny ([9,12,49–58], this study). It should be stressed that even members of the same family may inherit symbionts differently (for further details, see [58]). It should be stressed that scale insects (i.e., all the Pseudococcidae, Eriococcidae, Coccidae, and Putoidae examined so far) are unique in that they are the only group of insects in which microorganisms invade the neck region of the ovariole ([50,55–57], this study). Until now, the course of transmission of fungal symbionts in soft scale insects had not been studied under TEM; however, Gomez-Polo and co-workers [14] reported their presence in eggs, and thus proved that these microorganisms are transported between generations transovarially. Our observations of the ovaries of the members of the Coccidae family showed that their fungal associates are transmitted between generations similarly to bacterial symbionts in Pseudococcidae, Eriococcidae, and Putoidae [50,53,55–57]. One noteworthy aspect is that, just as in the case of the infestation of ovarioles through bacterial symbionts, the time of transmission of the fungal symbionts in soft scale insects is correlated with the stage of oogenesis—the microorganisms commence the infection of the ovarioles that contain the oocytes in the stage of advanced choriogenesis. Similarly to Pseudococcidae, Eriococcidae, and Putoidae, in Coccidae, symbionts invade follicular cells surrounding the nutritive cord, because this area is the only place on the oocyte surface that is devoid of egg envelopes (see Figure 4A). After the degeneration of the nutritive cord, the symbionts may enter the oocyte cytoplasm.

Numerous molecular analyses have confirmed that bacterial symbionts co-evolve with their host insects [59–61]. The co-diversification of fungal associates and insects has previously not been tested intensively. Our co-phylogenetic analysis, through the use of Jane software, indicated that not all the species of soft scale insects examined co-evolved with their host (see Figure 2). It is a noteworthy fact that Gomez-Polo and co-workers [14] additionally revealed that co-phylogeny of Coccidae, which they tested, and their *Ophiocordyceps* symbionts were incongruent. These authors suggested that this incongruence may be a result of the independent acquisition of fungi by particular members of Coccidae. However, taking into account the fact that some of the species examined co-evolved with their symbiotic partners (this study), it may be speculated that incongruence in the co-phylogeny of some coccids and their fungal associates may result from the independent evolution of fungal symbionts or their replacement during the evolutionary history of different species.

Supplementary Materials: The following are available online at https://www.mdpi.com/article/10.3390/cells10081922/s1. Figure S1. Phylogenetic tree showing the relationships between fungal symbionts of soft scale insect species, pathogenic, as well as free-living fungi (constructed on the base of sequences of Beta-tubulin gene). The numbers near the nodes refer to the Bayesian posterior probability. Table S1. List of primers and fluorochrome-labeled probe used in this study.

Author Contributions: Conceptualization, A.M. and T.S.; methodology, A.M.; software, B.G.; investigation, A.M., K.M.; resources, M.K.-K.; data curation, A.M.; writing—original draft preparation, A.M., M.K.-K., T.S.; writing—review and editing, A.M., M.K.-K., T.S., B.G., K.M.; visualization, A.M., B.G., T.S.; supervision, A.M., T.S.; project administration, A.M.; funding acquisition, A.M. All authors have read and agreed to the published version of the manuscript.

Funding: This work was supported by the Iuventus Plus V grant IP2015050374 from the Ministry of Science and Higher Education to AM and N18/DBS/000013 to Jagiellonian University.

Institutional Review Board Statement: According to Polish Act on the Protection of Animals used for Scientific or Educational Purposes, studies on unprotected insects do not require any permission.

Informed Consent Statement: This article do not contain research involving humans.

Data Availability Statement: The data presented in this study are available on request from the corresponding author.

Acknowledgments: We are greatly indebted to Ada Jankowska for her skilled technical assistance. Ultrastructural observations were carried out using the Jeol 2100 transmission electron microscope in the Laboratory of Microscopy, Department of Cell Biology and Imaging, Institute of Zoology and Biomedical Research, Jagiellonian University. The open-access publication of this article was funded by the Priority Research Area BioSunder the program "Excellence Initiative–Research University" at the Jagiellonian University in Kraków.

Conflicts of Interest: T.S., K.M., B.G., M.K.-K. and A.M. declare no conflict of interest.

References

1. Ouvrard, D.; Kondo, T.; Gullan, P.J. Scale insects: Major pests and management. In *Encyclopedia of Pest Management*; Taylor and Francis: New York, NY, USA, 2013; pp. 1–4.
2. Hodgson, C.J. *The Scale Insect Family Coccidae: An Identification Manual to Genera*; CAB International: Wallingford, UK, 1994.
3. Camacho, E.R.; Chong, J.-H. General Biology and Current Management Approaches of Soft Scale Pests (Hemiptera: Coccidae). *J. Integr. Pest Manag.* **2015**, *6*, 17. [CrossRef] [PubMed]
4. Gullan, P.J.; Kosztarab, M. Adaptations in Scale Insects. *Annu. Rev. Entomol.* **1997**, *42*, 23–50. [CrossRef]
5. Ben-Dov, Y. *A Systematic Catalogue of the Soft Scale Insects of the World (Homoptera: Coccoidea: Coccidae)*; Sandhill Crane Press: Gainesville, FL, USA, 1993.
6. Ben-Dov, Y. The soft scale insects. Morphology, Systematics and Phylogeny. In *Soft Scale Insects: Their Biology, Natural Enemies and Control*; Elsevier: Amsterdam, The Netherlands; New York, NY, USA, 1997; Volume 7A.
7. Kozár, F.; Ben-Dov, Y. Zoogeographical considerations and status of knowledge of the family. In *Soft Scale Insects: Their Biology, Natural Enemies and Control*; Elsevier: Amsterdam, The Netherlands; New York, NY, USA, 1997; Volume 7A.
8. Douglas, A.E. The Molecular Basis of Bacterial-Insect Symbiosis. *J. Mol. Biol.* **2014**, *426*, 3830–3837. [CrossRef]
9. Buchner, P. *Endosymbiosis of Animals with Plant Microorganisms*; Interscience Publishers: New York, NY, USA, 1965.
10. Tremblay, E. Advances in Endosymbiont Studies in Coccoidea. *Va. Polytech. Inst. State Univ. Res. Div. Bull.* **1977**, *127*, 23–33.
11. Baumann, P. Biology of Bacteriocyte-Associated Endosymbionts of Plant Sap-Sucking Insects. *Annu. Rev. Microbiol.* **2005**, *59*, 155–189. [CrossRef]
12. Szklarzewicz, T.; Michalik, A.; Michalik, K. The Diversity of Symbiotic Systems in Scale Insects. In *Symbiosis: Cellular, Molecular, Medical and Evolutionary Aspects*; Kloc, M., Ed.; Springer International Publishing: Cham, Switzerland, 2020; pp. 469–495.
13. Tremblay, E. Embryonic development; oviparity and viviparity. In *Soft Scale Insects–Their Biology, Natural Enemies and Control*; Elsevier: Amsterdam, The Netherlands; New York, NY, USA, 1997; pp. 257–260.
14. Gomez-Polo, P.; Ballinger, M.J.; Lalzar, M.; Malik, A.; Ben-Dov, Y.; Mozes-Daube, N.; Perlman, S.J.; Iasur-Kruh, L.; Chiel, E. An Exceptional Family: *Ophiocordyceps*-Allied Fungus Dominates the Microbiome of Soft Scale Insects (Hemiptera: Sternorrhyncha: Coccidae). *Mol. Ecol.* **2017**, *26*, 5855–5868. [CrossRef]
15. Deng, J.; Yu, Y.; Wang, X.; Liu, Q.; Huang, X. The Ubiquity and Development-Related Abundance Dynamics of *Ophiocordyceps* Fungi in Soft Scale Insects. *Microorganisms* **2021**, *9*, 404. [CrossRef]

16. Gao, Y.; Xie, Y.P.; Xiong, Q.; Liu, W.M.; Xue, J.L. Ultrastructural Exploration on the Histopathological Change in *Phenacoccus fraxinus* Infected with *Lecanicillium lecanii*. *PLoS ONE* **2015**, *10*, e0117428. [CrossRef]
17. Sun, X.; Yan, W.; Zhang, J.; Niu, X.; Li, F.; Qin, W.; Ma, G. Frozen Section and Electron Microscopy Studies of the Infection of the Red Palm Weevil, *Rhynchophorus ferrugineus* (Coleoptera: Curculionidae) by the Entomopathogenic Fungus *Metarhizium anisopliae*. *SpringerPlus* **2016**, *5*, 1748. [CrossRef]
18. Malsam, O.; Filian, M.; Rudiger, H.; Berg, D. Fungal insecticides. In *Fungal Biotechnology*; Chapman & Hall: London, UK, 1997.
19. Assaf, L.H.; Hassan, F.R.; Ahmad, D.S. Pathogenicity Evaluation of Some Local Isolates of Entomopathogenic Fungi against the Nut Scale Insect *Eulecanium tiliae* L. *Int. J. Pure Appl. Sci. Technol.* **2013**, *19*, 37–43.
20. Suh, S.-O.; Noda, H.; Blackwell, M. Insect Symbiosis: Derivation of Yeast-like Endosymbionts within an Entomopathogenic Filamentous Lineage. *Mol. Biol. Evol.* **2001**, *18*, 995–1000. [CrossRef]
21. Drew, G.C.; Stevens, E.J.; King, K.C. Microbial Evolution and Transitions along the Parasite–Mutualist Continuum. *Nat. Rev. Microbiol.* **2021**. [CrossRef]
22. Koteja, J. Jak rozpoznawać czerwce (Homoptera, Coccinea). In *Diagnostyka Szkodników Roślin i ich Wrogów Naturalnych*; Wydawnictwo SGGW: Warszawa, Poland, 1996; Volume 2, pp. 139–231.
23. Barbosa, P.; Berry, D.L.; Kary, C.S. *Insect Histology. Practical Laboratory Techniques*; Wiley: Hoboken, NJ, USA, 2015.
24. White, T.J.; Bruns, T.; Lee, S.; Taylor, J. Amplification and direct sequencing of fungal ribosomal RNA genes for phylogenetics. In *PCR Protocols: A Guide to Methods and Applications*; Academic Press: San Diego, CA, USA, 1990; pp. 315–322.
25. Hardy, N.B.; Gullan, P.J.; Hodgson, C.J. A Subfamily-Level Classification of Mealybugs (Hemiptera: Pseudococcidae) Based on Integrated Molecular and Morphological Data. *Syst. Entomol.* **2008**, *33*, 51–71. [CrossRef]
26. Matsuura, Y.; Moriyama, M.; Łukasik, P.; Vanderpool, D.; Tanahashi, M.; Meng, X.-Y.; McCutcheon, J.P.; Fukatsu, T. Recurrent Symbiont Recruitment from Fungal Parasites in Cicadas. *Proc. Natl. Acad. Sci. USA* **2018**, *115*, E5970. [CrossRef]
27. Łukasik, P.; Newton, J.A.; Sanders, J.G.; Hu, Y.; Moreau, C.S.; Kronauer, D.J.C.; O'Donnell, S.; Koga, R.; Russell, J.A. The Structured Diversity of Specialized Gut Symbionts of the New World Army Ants. *Mol. Ecol.* **2017**, *26*, 3808–3825. [CrossRef]
28. Kumar, S.; Stecher, G.; Li, M.; Knyaz, C.; Tamura, K. MEGA X: Molecular Evolutionary Genetics Analysis across Computing Platforms. *Mol. Biol. Evol.* **2018**, *35*, 1547–1549. [CrossRef]
29. Nylander, J.A.A. *MrModeltest v2. Program Distributed by the Author*; Evolutionary Biology Centre, Uppsala University: Uppsala, Sweden, 2004.
30. Ronquist, F.; Teslenko, M.; van der Mark, P.; Ayres, D.L.; Darling, A.; Höhna, S.; Larget, B.; Liu, L.; Suchard, M.A.; Huelsenbeck, J.P. MrBayes 3.2: Efficient Bayesian Phylogenetic Inference and Model Choice across a Large Model Space. *Syst. Biol.* **2012**, *61*, 539–542. [CrossRef]
31. Conow, C.; Fielder, D.; Ovadia, Y.; Libeskind-Hadas, R. Jane: A New Tool for the Cophylogeny Reconstruction Problem. *Algorithms Mol. Biol. AMB* **2010**, *5*, 16. [CrossRef]
32. Szklarzewicz, T. The Ovaries of Scale Insects (Hemiptera, Coccinea). Morphology and Phylogenetic Conclusions. *Folia Histochem. Cytobiol.* **1998**, *36*, 157–165.
33. Araújo, J.P.M.; Hughes, D.P. Chapter One—Diversity of Entomopathogenic Fungi: Which Groups Conquered the Insect Body? In *Advances in Genetics*; Lovett, B., St. Leger, R.J., Eds.; Academic Press: Cambridge, MA, USA, 2016; Volume 94, pp. 1–39.
34. Podsiadło, E.; Michalik, K.; Michalik, A.; Szklarzewicz, T. Yeast-like Microorganisms in the Scale Insect *Kermes quercus* (Insecta, Hemiptera, Coccomorpha: Kermesidae). Newly Acquired Symbionts? *Arthropod Struct. Dev.* **2018**, *47*, 56–63. [CrossRef]
35. Vega, F.E.; Dowd, P.F. The role of yeasts as insect endosymbionts. In *Insect-Fungal Associations: Ecology and Evolution*; Oxford University Press: New York, NY, USA, 2005; pp. 211–243.
36. Nishino, T.; Tanahashi, M.; Lin, C.-P.; Koga, R.; Fukatsu, T. Fungal and Bacterial Endosymbionts of Eared Leafhoppers of the Subfamily Ledrinae (Hemiptera: Cicadellidae). *Appl. Entomol. Zool.* **2016**, *51*, 465–477. [CrossRef]
37. Kobiałka, M.; Michalik, A.; Walczak, M.; Szklarzewicz, T. Dual "Bacterial-Fungal" Symbiosis in Deltocephalinae Leafhoppers (Insecta, Hemiptera, Cicadomorpha: Cicadellidae). *Microb. Ecol.* **2018**, *75*, 771–782. [CrossRef] [PubMed]
38. Noda, H.; Nakashima, N.; Koizumi, M. Phylogenetic Position of Yeast-like Symbiotes of Rice Planthoppers Based on Partial 18S RDNA Sequences. *Insect Biochem. Mol. Biol.* **1995**, *25*, 639–646. [CrossRef]
39. Xet-Mull, A.M.; Quesada, T.; Espinoza, A.M. Phylogenetic Position of the Yeast-like Symbiotes of Tagosodes Orizicolus (Homoptera: Delphacidae) Based on 18S Ribosomal DNA Partial Sequences. *Rev. Biol. Trop.* **2004**, *52*, 777–785. [CrossRef]
40. Fan, H.-W.; Noda, H.; Xie, H.-Q.; Suetsugu, Y.; Zhu, Q.-H.; Zhang, C.-X. Genomic Analysis of an Ascomycete Fungus from the Rice Planthopper Reveals How It Adapts to an Endosymbiotic Lifestyle. *Genome Biol. Evol.* **2015**, *7*, 2623–2634. [CrossRef]
41. Fukatsu, T.; Ishikawa, H. Phylogenetic Position of Yeast-like Symbiont of *Hamiltonaphis styraci* (Homoptera, Aphididae) Based on 18S RDNA Sequence. *Insect Biochem. Mol. Biol.* **1996**, *26*, 383–388. [CrossRef]
42. Rosenblueth, M.; Sayavedra, L.; Sámano-Sánchez, H.; Roth, A.; Martínez-Romero, E. Evolutionary Relationships of Flavobacterial and Enterobacterial Endosymbionts with Their Scale Insect Hosts (Hemiptera: Coccoidea). *J. Evol. Biol.* **2012**, *25*, 2357–2368. [CrossRef]
43. Vera-Ponce de León, A.; Sanchez-Flores, A.; Rosenblueth, M.; Martínez-Romero, E. Fungal Community Associated with *Dactylopius* (Hemiptera: Coccoidea: Dactylopiidae) and Its Role in Uric Acid Metabolism. *Front. Microbiol.* **2016**, *7*, 954. [CrossRef]
44. Fukatsu, T.; Ishikawa, H. A Novel Eukaryotic Extracellular Symbiont in an Aphid, *Astegopteryx styraci* (Homoptera, Aphididae, Hormaphidinae). *J. Insect Physiol.* **1992**, *38*, 765–773. [CrossRef]

45. Fukatsu, T.; Aoki, S.; Kurosu, U.; Ishikawa, H. Phylogeny of Cerataphidini Aphids Revealed by Their Symbiotic Microorganisms and Basic Structure of Their Galls -Implications for Host-Symbiont Coevolution and Evolution of Sterile Soldier Castes. *Zoolog. Sci.* **1994**, *11*, 613–623.
46. Hongoh, Y.; Ishikawa, H. Evolutionary Studies on Uricases of Fungal Endosymbionts of Aphids and Planthoppers. *J. Mol. Evol.* **2000**, *51*, 265–277. [CrossRef] [PubMed]
47. Michalik, A.; Jankowska, W.; Szklarzewicz, T. Ultrastructure and Transovarial Transmission of Endosymbiotic Microorganisms in *Conomelus anceps* and *Metcalfa pruinosa* (Insecta, Hemiptera, Fulgoromorpha). *Folia Biol. Kraków* **2009**, *57*, 131–137. [CrossRef] [PubMed]
48. Kuechler, S.M.; Dettner, K.; Kehl, S. Characterization of an Obligate Intracellular Bacterium in the Midgut Epithelium of the Bulrush Bug *Chilacis typhae* (Heteroptera, Lygaeidae, Artheneinae). *Appl. Environ. Microbiol.* **2011**, *77*, 2869. [CrossRef] [PubMed]
49. Walczuch, A. Studien an Coccidensymbionten. *Z. für Morphol. und Ökol. der Tiere* **1932**, *25*, 623–729. [CrossRef]
50. von Dohlen, C.D.; Kohler, S.; Alsop, S.T.; McManus, W.R. Mealybug β-Proteobacterial Endosymbionts Contain γ-Proteobacterial Symbionts. *Nature* **2001**, *412*, 433–436. [CrossRef]
51. Szklarzewicz, T.; Kędra, K.; Niżnik, S. Ultrastructure and Transovarial Transmission of Endosymbiotic Microorganisms in *Palaeococcus fuscipennis* (Burmeister) (Insecta, Hemiptera, Coccinea: Monophlebidae). *Folia Biol. Kraków* **2006**, *54*, 69–74. [CrossRef]
52. Szklarzewicz, T.; Kalandyk-Kolodziejczyk, M.; Kot, M.; Michalik, A. Ovary Structure and Transovarial Transmission of Endosymbiotic Microorganisms in *Marchalina hellenica* (Insecta, Hemiptera, Coccomorpha: Marchalinidae). *Acta Zool.* **2013**, *94*, 184–192. [CrossRef]
53. Szklarzewicz, T.; Kalandyk-Kołodziejczyk, M.; Michalik, K.; Jankowska, W.; Michalik, A. Symbiotic Microorganisms in *Puto superbus* (Leonardi, 1907) (Insecta, Hemiptera, Coccomorpha: Putoidae). *Protoplasma* **2018**, *255*, 129–138. [CrossRef]
54. Niżnik, S.; Szklarzewicz, T. Structure and Development of Hermaphroditic Gonad in *Icerya purchasi* (Insecta, Hemiptera, Coccinea: Monophlebidae). *Zool. Pol.* **2007**, *52*, 71–90.
55. Michalik, K.; Szklarzewicz, T.; Kalandyk-Kołodziejczyk, M.; Jankowska, W.; Michalik, A. Bacteria Belonging to the Genus *Burkholderia* Are Obligatory Symbionts of the Eriococcids *Acanthococcus aceris* Signoret, 1875 and *Gossyparia spuria* (Modeer, 1778) (Insecta, Hemiptera, Coccoidea). *Arthropod Struct. Dev.* **2016**, *45*, 265–272. [CrossRef]
56. Michalik, A.; Schulz, F.; Michalik, K.; Wascher, F.; Horn, M.; Szklarzewicz, T. Coexistence of Novel Gammaproteobacterial and *Arsenophonus* Symbionts in the Scale Insect *Greenisca brachypodii* (Hemiptera, Coccomorpha: Eriococcidae). *Environ. Microbiol.* **2018**, *20*, 1148–1157. [CrossRef] [PubMed]
57. Michalik, A.; Michalik, K.; Grzywacz, B.; Kalandyk-Kołodziejczyk, M.; Szklarzewicz, T. Molecular Characterization, Ultrastructure, and Transovarial Transmission of *Tremblaya phenacola* in Six Mealybugs of the Phenacoccinae Subfamily (Insecta, Hemiptera, Coccomorpha). *Protoplasma* **2019**, *256*, 1597–1608. [CrossRef] [PubMed]
58. Szklarzewicz, T.; Michalik, A. Transovarial Transmission of Symbionts in Insects. In *Oocytes: Maternal Information and Functions*; Kloc, M., Ed.; Springer International Publishing: Cham, Switzerland, 2017; pp. 43–67.
59. Moran, N.A. The Coevolution of Bacterial Endosymbionts and Phloem-Feeding Insects. *Ann. Mo. Bot. Gard.* **2001**, *88*, 35–44. [CrossRef]
60. Baumann, L.; Baumann, P. Cospeciation Between the Primary Endosymbionts of Mealybugs and Their Hosts. *Curr. Microbiol.* **2005**, *50*, 84–87. [CrossRef]
61. Downie, D.A.; Gullan, P.J. Phylogenetic Congruence of Mealybugs and Their Primary Endosymbionts. *J. Evol. Biol.* **2005**, *18*, 315–324. [CrossRef] [PubMed]

cells MDPI

Article

Supra-Optimal Temperature: An Efficient Approach for Overaccumulation of Starch in the Green Alga *Parachlorella kessleri*

Vilém Zachleder [1,†], Veronika Kselíková [1,2,†], Ivan N. Ivanov [1,2], Vitali Bialevich [1], Milada Vítová [1], Shuhei Ota [3], Tsuyoshi Takeshita [4], Shigeyuki Kawano [4] and Kateřina Bišová [1,*]

1 Laboratory of Cell Cycles of Algae, Centre Algatech, Institute of Microbiology of the Czech Academy of Sciences, 37981 Třeboň, Czech Republic; zachleder@alga.cz (V.Z.); kselikova@alga.cz (V.K.); ivanov@alga.cz (I.N.I.); bialevich@alga.cz (V.B.); vitova@alga.cz (M.V.)
2 Faculty of Science, University of South Bohemia, 37005 České Budějovice, Czech Republic
3 Center for Environmental Biology and Ecosystem Studies, National Institute for Environmental Studies, Tsukuba 305 8506, Ibaraki, Japan; ota.shuhei@nies.go.jp
4 The University of Tokyo Future Center Initiative, Wakashiba 178 4 4, Kashiwa 277 0871, Chiba, Japan; takeshita@algalbio.co.jp (T.T.); kawano@edu.k.u-tokyo.ac.jp (S.K.)
* Correspondence: bisova@alga.cz; Tel.: +420-384-340-480
† These authors contributed equally to the work.

Abstract: Green algae are fast-growing microorganisms that are considered promising for the production of starch and neutral lipids, and the chlorococcal green alga *Parachlorella kessleri* is a favorable model, as it can produce both starch and neutral lipids. *P. kessleri* commonly divides into more than two daughter cells by a specific mechanism—multiple fission. Here, we used synchronized cultures of the alga to study the effects of supra-optimal temperature. Synchronized cultures were grown at optimal (30 °C) and supra-optimal (40 °C) temperatures and incident light intensities of 110 and 500 µmol photons m^{-2} s^{-1}. The time course of cell reproduction (DNA replication, cellular division), growth (total RNA, protein, cell dry matter, cell size), and synthesis of energy reserves (net starch, neutral lipid) was studied. At 40 °C, cell reproduction was arrested, but growth and accumulation of energy reserves continued; this led to the production of giant cells enriched in protein, starch, and neutral lipids. Furthermore, we examined whether the increased temperature could alleviate the effects of deuterated water on *Parachlorella kessleri* growth and division; results show that supra-optimal temperature can be used in algal biotechnology for the production of protein, (deuterated) starch, and neutral lipids.

Keywords: microalgae; *Parachlorella kessleri*; starch; supra-optimal temperature; cell cycle; energy reserves; growth processes; reproduction events; deuterium; deuterated starch; deuterated lipid

Citation: Zachleder, V.; Kselíková, V.; Ivanov, I.N.; Bialevich, V.; Vítová, M.; Ota, S.; Takeshita, T.; Kawano, S.; Bišová, K. Supra-Optimal Temperature: An Efficient Approach for Overaccumulation of Starch in the Green Alga *Parachlorella kessleri*. *Cells* **2021**, *10*, 1806. https://doi.org/10.3390/cells10071806

Academic Editor: Suleyman Allakhverdiev

Received: 14 June 2021
Accepted: 12 July 2021
Published: 16 July 2021

1. Introduction

Light and temperature are two crucial factors affecting algal growth and division, both in natural habitats and in biotechnological applications. For autotrophically grown organisms, there is a physiological range of light intensities and temperatures that support growth and division, and light intensities and temperatures below the threshold are not sufficient. On the other hand, light intensities and temperatures above this range are increasingly stressful, and will inhibit cell division and/or growth. The duration of light, along with its intensity and spectral composition will, in autotrophically grown algae, affect growth rates [1–11], but will not affect light-independent processes such as cell division, which uses internal energy stores for its support [1,5–9,11]. In contrast, temperature is much less specific, as it will affect all metabolic processes, including "dark" ones such as cell division; thus, its effects reach beyond setting growth rates. In principle, within the physiological range, a 10 °C increase in temperature will increase metabolic rate twofold [12]. Increasing temperature will thus speed up growth, as well as individual processes leading to cell division, and consequently shorten the duration of the cell cycle, as shown in different algae such as *Chlorella ellipsoidea* [11], *Chlamydomonas reinhardtii* [13], *Chlamydomonas eugametos* [14], and *Desmodesmus quadricauda* [15]. However, individual metabolic processes are differentially sensitive to various stresses [16], including temperature [17], which significantly affects how cells react to changes in temperature. In particular, cell division seems to be more sensitive to temperature increases than growth does [17,18]; this has peculiar consequences; there are threshold temperatures that will only slightly or not at all affect cell growth, but will completely block cell division. The effects of such supra-optimal temperatures were described in the green algae *Chlamydomonas reinhardtii* [17] and *Chlorella vulgaris* [18].

The chlorococcal alga *Parachlorella kessleri* (formerly *Chlorella kessleri* [19]) is a biotechnologically promising algal species [20–29] that divides by multiple fission. Its cell cycle consists of multiple rounds of DNA replication that, after completion of the last one, are followed by successive nuclear divisions. Each of the nuclear divisions is immediately followed by cell division. Morphologically, the cells become sequentially polyploid but not polynuclear [30]. Within a single cell cycle, the mother cell can give rise to 2, 4, 8, or 16 cells; this is consistent with the established scheme of multiple fission, where a single mother cell can divide into 2^n daughter cells—where n is the number of rounds of started reproductive sequences (DNA replication, nuclear division, cell division) [31]. Increasing light intensity will increase both growth rate and the number of reproductive sequences started, and thus, the number of daughter cells being formed from a single cell [30].

The primary energy and carbon store of *P. kessleri* is starch, which is produced under optimal growth conditions in complete nutrient medium and at physiological values of light intensity and temperature [23,32]. Starch is exclusively located in the chloroplasts in the form of starch grains of different numbers and sizes. In the autotrophically grown algal cell, starch reserves serve mostly as a buffer to supply a stable carbon and energy source. This is indispensable, as in nature the energy supply from photosynthesis varies with sunlight intensity during the day, and is absent during nights. The buffering role of starch is crucial for cell reproduction. The processes of DNA replication, along with nuclear and cellular divisions, are of importance in the life of a cell and so they depend on a constant and reliable supply of energy and carbon. Indeed, the majority of starch produced by cells is degraded exclusively for cell reproduction [33], and this is true even for cells growing in continuous light [34]. This was evidenced in synchronized cultures of different algal species, where net starch increased to some maximal value during the cell cycle, and subsequently was nearly completely degraded during nuclear and cellular division at the end of the cell cycle [33,35,36]. The cellular starch content at any given time is the net result of starch synthesis and utilization; thus, decreasing starch content can be caused by lower starch synthesis, by increased consumption, or by a combination of both factors. It seems that a decrease in starch content at the time of cell division is caused by a combination of increased starch spending for cell reproduction and a minimum photosynthetic rate specific for this period [35]. Furthermore, a blockage of cell division processes as the primary starch consumers will lead to starch (over)accumulation. This has been established for different stress conditions, such as the application of inhibitors [36], nutrient starvation [24,25,32,36], high light intensity [32], the presence of a high concentration of CO_2 [37], or supra-optimal temperatures [18], and it should be explored biotechnologically [36,38]. If stress conditions are prolonged, starch reserves will gradually start to be replaced by neutral lipids—the secondary energy store of *P. kessleri* [23,24,32,39]. Although neutral lipids under optimal growth conditions are maintained at a relatively low level, at 1–10% of dry matter (DM) [23,39,40], they can also (over)accumulate under stress conditions similar to those inducing starch accumulation, such as nitrogen, sulfur, or phosphorus depletion [24,26,32,39,41], dilution of all the nutrients in the medium [25,39] or salt stress [42]. The ability to (over)produce both starch and neutral lipids makes *P. kessleri* a useful model for applications in biotechnology. A peculiar source of stress for *P. kessleri* is cultivation in deuterium, which leads predominantly to the accumulation of starch [30], but also that of neutral lipids [43]. Deuterium—the stable isotope of hydrogen—is known to have the highest kinetic effect among the stable isotopes of biogenic elements [43–45]. Deuterium in deuterated water is known to increase cell stress via several mechanisms [43,44], including greater bond dissociation and activation energies [46]. Its presence affects the entire metabolism, and its specific effects range from disrupting signaling [47] and energy production in mitochondria and chloroplasts [48,49] to disrupting cell division [50,51]. Given the very low concentrations of deuterium (or deuterated water) in nature, its application in high concentrations is artificial, and is also quite expensive. However, deuterated molecules are used extensively as analytical standards and for metabolic labeling [43,52,53], and deuterated compounds are exploited in pharmacology [44,54–57]. Therefore, production of algal-derived deuterated compounds is justified for the production of fine (bio)chemicals with very high added value. Indeed, algal-derived deuterated biomolecules such as sugars, proteins, carotenoids, lipids, and starch can be used commercially [43,52,53,58–60].

Here, we supplement the extensive knowledge that has accumulated on starch and neutral lipid production under different conditions in *P. kessleri* by providing baseline information on the effects of supra-optimal temperature on cell reproduction (DNA replication, cell division) and growth (RNA, protein, dry matter, cell volume). Particular focus has been placed on the production of starch and lipids as the primary and secondary energy stores. Furthermore, we analyze the effects of a high concentration of deuterium on cell growth and reproduction, as well as the production of deuterated starch and lipids. The results of this study can be exploited as a starting point for further optimization in larger scale biotechnological production, but will also widen the *P. kessleri* portfolio as a model for basic research.

2. Materials and Methods

2.1. Organism and Culture

The green unicellular microalga *Parachlorella kessleri* (Trebouxiophyceae, Chlorophyta) (strain CCALA 255) was obtained from the Culture Collection of Autotrophic Organisms at the Institute of Botany, Czech Academy of Sciences in Třeboň, Czech Republic (CCALA; https://ccala.butbn.cas.cz/, accessed on 7 June 2021).

For routine sub-culturing, the cultures were streaked every three weeks onto nutrient medium (see below) solidified by agar (1.5%), and grown on a light shelf at an incident light intensity of 100 μmol photons m $^{-2}$ s $^{-1}$ of photosynthetically active radiation.

2.2. Mineral Nutrient Medium

The mineral medium was based on the mean content of P, N, K, Mg, and S in algal biomass [61], and had the following initial composition (in mg/L): 1100 KNO_3, 237 KH_2PO_4, 204 $MgSO_4.7H_2O$, 40 $C_{10}H_{12}O_8N_2NaFe$, 88 $CaCl_2$, 0.83 H_3BO_3, 0.95 $CuSO_4.5H_2O$, 3.3 $MnCl_2.4H_2O$, 0.17 $(NH_4)_6Mo_7O_{24}.4H_2O$, 2.7 $ZnSO_4.7H_2O$, 0.6 $CoSO_4.7H_2O$, and 0.014 NH_4VO_3 in distilled water [36]. For the preparation of the medium, 100x concentrated stock solutions of macroelements and microelements were used. All components were diluted in distilled water and autoclaved for 30 min at 121 °C. The pH was adjusted to 7 with 1 M NaOH.

2.3. Culture Conditions

Two types of cultivation units were used for experiments: (1) glass cylinders (inner diameter 36 mm, height 500 mm, volume of suspension 300 mL), and (2) flat and rectangular glass vessels (inner dimensions 400 × 300 × 20 mm, volume of suspension 2500 mL). The first type of the units was used for routine growth, synchronization, and small-scale experiments with deuterated water; the second type of the units was used for large-scale experiments. Culture units were placed in a thermostatic water bath, where the temperature was set and automatically maintained at the values stated in the Results section for each experiment. The vessels were illuminated from one side with a panel of dimmable fluorescent lamps (DULUX L55W/950 Daylight, OSRAM, Munich, Germany).

For the experiments, the incident light intensity was set either to 110 or 500 µmol photons m $^{-2}$ s $^{-1}$ of photosynthetically active radiation. The cultures were vigorously mixed with air bubbles containing 2% carbon dioxide (*v*/*v*) dispersed through a 200-µL micropipette tip fixed in a glass tube placed at the bottom of the cylinder. The aeration in the flat vessels was performed via bubble stream from a perforated stainless steel tube located at the bottom of the vessel. The flow rate of the aeration mix was 60 L/h.

2.4. Synchronization of Cultures

Cultures were synchronized under optimal conditions: incident light intensity 500 µmol photons m $^{-2}$ s $^{-1}$, temperature 30 °C, 2% CO_2 (*v*/*v*) in aeration mixture. The cultures were initiated by inoculating algal cells directly from the culture plates into the liquid medium. Such cultures were synchronized by alternation of 18 h light and 7 h of dark, as established earlier [30]. During the synchronization procedure, the cell density was kept below 1×10^6 cells mL^{-1} by dilution at the end of the dark period, in order to prevent cell shading. This regime was maintained for several days (about 3 cycles) until the required culture synchrony was reached. Once the cultures were synchronized, they were grown for approximately one more cycle at the same cell density until the volume of the culture required for the start of the experiment was reached. The synchronized daughter cells were again diluted to the initial cell density of approximately 1×10^6 cells mL^{-1}, and used as inocula for experimental cultures.

2.5. Management of Deuterated Cultures

The synchronized inocula for the deuterated water experiments were centrifuged at 3000× *g* for 5 min and resuspended in a medium containing the desired concentration of deuterated water. The cultures were cultivated in a semi-batch mode, controlled by the optical density at 750 nm (OD_{750}). Upon reaching an OD_{750} of 0.4, the cultures were diluted with fresh medium with corresponding D_2O content to the initial value of OD_{750}. This way, growth unlimited by nutrients and light was achieved.

2.6. Measurement of Light Intensity

The dimmable fluorescent tubes were used for the adjustment of the incident irradiance. The light intensity was measured using a quantum/radiometer/photometer (LI-COR, Inc., Lincoln, NE, USA). Incident light intensity (I_i) was measured at the surface of the culture vessel, while the transmitted light intensity was measured at the rear side of the culture vessel (I_t). The different values of incident (I_i) and transmitted (I_t) light intensities at different optical densities given by cell size and concentrations were used to calculate the mean light intensity (I_m)—i.e., light energy absorbed by a layer of cell suspension—according to the Lambert–Beer formula: $I_m = (I_i - I_t)/\ln(I_i/I_t)$.

2.7. Assessment of Cell Division Curves

The proportion of mother cells and daughter cells was determined by light microscopy in cells fixed in Lugol's solution (1 g I, 5 g KI, 100 mL H_2O) at a final concentration of 10 µL of Lugol's solution per 1 mL of cell suspension.

2.8. Dry Matter Determination

Biomass was separated from the medium by centrifugation of 4 mL of the cell suspension in pre-weighed microtubes at 3000× *g* for 5 min; the sediment was dried at 105 °C for 12 h and weighed on an analytical balance (TE214S-0CE, Sartorius, Goettingen, Germany) [36].

2.9. Cell Volume and Number

Cell volume and number were measured using a Beckman Coulter Multisizer 4 (Beckman Coulter Life Sciences, Indianapolis, IN, USA) by diluting 50 µL of fixed (0.2% glutaraldehyde) cell suspension into 10 mL of 0.9% NaCl (*w*/*v*) electrolyte solution.

2.10. Quantum Yield Measurement

Aliquots of 2 mL were withdrawn from the culture and placed into 10 × 10-mm plastic cuvettes for 30 min in the dark. Quantum yield was measured using an AquaPen-C 100 (Photon Systems Instruments, Drasov, Czech Republic).

2.11. Neutral Lipids and Starch Staining

For neutral lipids and starch staining, 1 mL of culture was sampled and stored for up to several hours at 4 °C until analysis. Staining was conducted on microscopy slides prior to observation. For starch staining, 20 μL of culture was combined with 0.5 μL of Lugol's solution (1 g I, 5 g KI, 100 mL H_2O) and observed immediately via light microscope (model BX51, Olympus, Tokyo, Japan). Neutral lipids were stained with the addition of 0.5 μL of freshly prepared Nile red dye (0.5 mg/mL in DMSO, catalog no. 72485, Sigma-Aldrich, Prague, Czech Republic) to 20 μL of culture, directly on the microscopic slide. Such slides were incubated at room temperature for 20 min in the dark. Stained cells were observed using a microscope (model BX51, Olympus, Tokyo, Japan) equipped with a mercury lamp and filter combination U-MNU2 (360–370 nm excitation and >515 nm barrier filter). Microphotographs were taken with a DP72 camera (Olympus, Tokyo, Japan).

2.12. Neutral Lipids Quantification

Neutral lipids content was measured spectrophotometrically in a microplate format following the modified procedure of Takeshita, et al. [62]. Aliquots (100 μL) of the cultures were combined with 5 μL of freshly prepared Nile red dye (0.5 mg/mL in DMSO, catalog no. 72485, Sigma-Aldrich, Prague, Czech Republic) in a 96-well plate. The same amount of Nile red dye was added to a sample blank consisting of 100 μL of H_2O. The plate was incubated at room temperature in the dark for 15 min. Fluorescence was measured using an Infinite 200 PRO microplate reader (Tecan, Männedorf, Switzerland) equipped with a 485-nm excitation filter and a 595-nm emission filter. The fluorescence intensity of the samples was normalized using the fluorescence intensity of unstained samples and a blank. Standard curves produced from a commercial lipid standard—triolein (catalog no. Y0001113, Sigma-Aldrich, St. Luis, MO, USA)—were used to quantify neutral lipids.

2.13. Estimation of Bulk RNA, DNA, and Proteins

2.13.1. Total Nucleic Acids Extraction

The procedure of Wanka [63], as modified by Lukavský, et al. [64], was used for the acidic extraction of total nucleic acids. After the removal of small-molecular-weight contaminants, the total nucleic acids were hydrolyzed with 0.5 M perchloric acid, and the absorbance of total nucleic acids in the supernatant was read at 260 nm (A_{260}). Total nucleic acid concentration was calculated based on calibration with a DNA standard of known concentration treated using the same procedure, and normalized to the number of cells in the sample.

2.13.2. DNA and RNA Determination

The light-activated reaction of diphenylamine with hydrolyzed DNA, as described by Decallonne and Weyns [65], was used with the modifications of Zachleder [66]. The concentrations of DNA within the samples were set by comparison to the A_{600} and A_{700} nm values of the sample with known DNA concentration treated using the same procedure. The values were normalized to the number of cells extracted; the RNA content was calculated as the difference between the total nucleic acid and DNA contents.

2.13.3. Protein Determination

The sediment remaining after nucleic acid extraction was used for protein determination; it was hydrolyzed with 1M NaOH for 1 h at 70 °C. The protein concentration in the supernatant after centrifugation of the hydrolysate (15 min, 5300× *g*, room temperature) was estimated by BCA assay (cat. no. 23225, Thermo Fisher Scientific, Waltham, MA, USA) according to the manufacturer's specifications. The same procedure was carried out with a calibration curve set by different concentrations of bovine serum albumin.

2.14. Starch Analyses

Net starch content was determined as described by Zachleder, et al. [30]. In short, the cells were disintegrated by vortexing with zirconium beads, and the pigments were removed via the addition of 80% (v/v) ethanol to the pellet, followed by incubation for 15 min at 68 °C. The procedure was repeated until the pellets were completely discolored. After that, 1 mL of α-amylase from porcine pancreas (Sigma-Aldrich, St. Luis, MO, USA), (0.5 mgl^{-1} (w/v) in 0.1 M sodium phosphate buffer (pH 6.9)) was added to the samples, and they were incubated for 1 h at 37 °C (FTC 90i, VELP Scientifica, Usmate Velate MB, Italy). Starch content was determined in the supernatant using DNSA color reaction for the quantification of reducing sugars [67]. The concentration of starch was estimated through a calibration curve of potato starch (Lach-Ner, Neratovice, Czech Republic) digested with α-amylase.

2.15. Statistical Analysis

Experiments were performed in at least three biological replicates, and the mean values were used to construct the graphs in SigmaPlot version 11 (Jandel Scientific Software, Erkrath, Germany). Regression curves of various orders were applied to fit the data. If not stated otherwise, all results are presented as means and standard deviation (n = 3). MS Excel 2016 was used to compute one- and two-way ANOVA. A p-value < 0.05 was considered to be significant.

3. Results

3.1. Growth in Normal Water

Synchronized cultures of *Parachlorella kessleri* were grown at optimal (30 °C) and supra-optimal (40 °C) temperatures, at low (110 μmol photons m^{-2} s^{-1}) and high (500 μmol photons m^{-2} s^{-1}) incident light intensities. The reproductive events (DNA replication, cellular division) (Figure 1), growth processes (RNA and protein synthesis, dry matter, and cell volume) (Figure 2), and accumulation of energy reserves (starch and lipids) (Figure 3) were followed within the time interval corresponding to two consequent cell cycles (48 h) in the control culture.

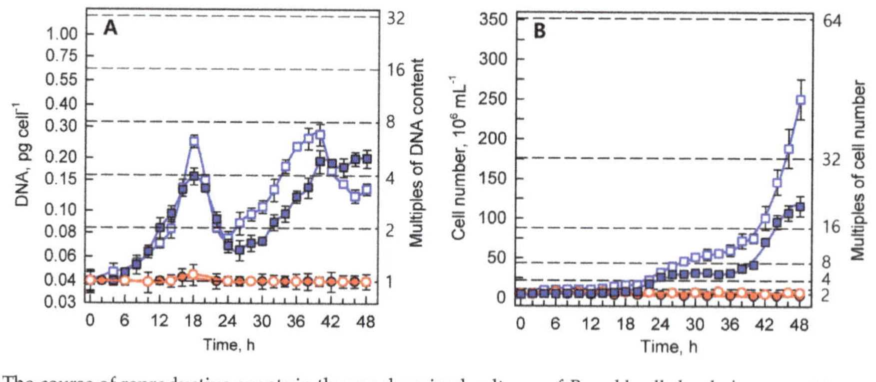

Figure 1. The course of reproductive events in the synchronized cultures of *Parachlorella kessleri* grown at temperatures of 30 °C and 40 °C (blue and red symbols and lines, respectively), and at incident light intensities of 500 and 110 μmol photons m^{-2} s^{-1} (empty and full symbols, respectively). (**A**) DNA (pg cell^{-1}), and (**B**) cell number (10^6 mL^{-1}). Horizontal dashed lines indicate the number of doublings of the initial values at the beginning of the cell cycle (0 h). Multiples of these values are given on the right ordinate. Data are presented as means ± SE. The differences between 30 °C and 40 °C are statistically significant at p < 0.001 (two-way ANOVA) in both A and B.

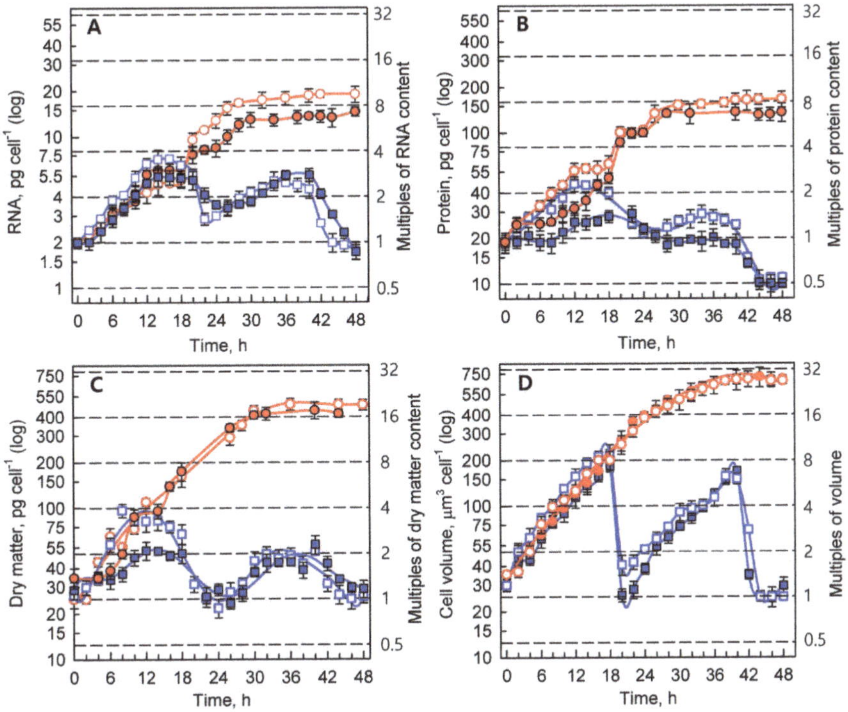

Figure 2. Time course of growth events in synchronized cultures of *Parachlorella kessleri* grown at 30 °C and 40 °C (blue and red symbols and lines, respectively), and at incident light intensities of 500 and 110 µmol photons m^{-2} s^{-1} (empty and full symbols, respectively). (**A**) RNA (pg $cell^{-1}$), (**B**) protein (pg $cell^{-1}$), (**C**) dry matter (pg $cell^{-1}$), and (**D**) cell volume (μm^3 $cell^{-1}$). Horizontal dashed lines indicate the number of doublings of the initial values at the beginning of the cell cycle (0 h). Multiples of these values are given at right ordinate. Data are presented as means ± SE. The differences between 30 °C and 40 °C are statistically significant at $p < 0.001$ (two-way ANOVA) in both A and B.

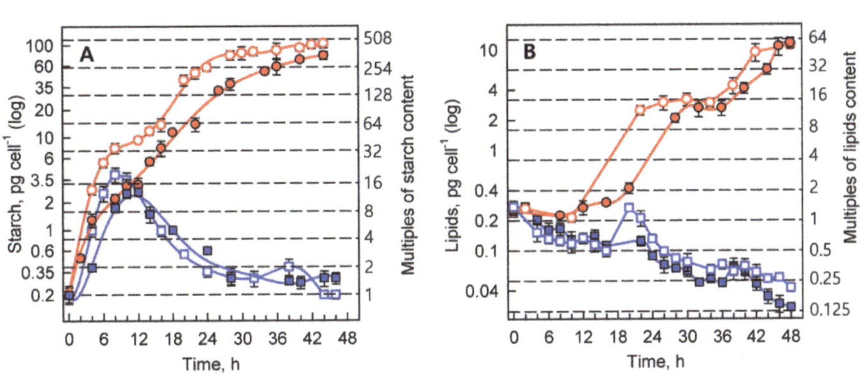

Figure 3. The course of changes in energy reserves in synchronized cultures of *Parachlorella kessleri* at temperatures of 30 °C and 40 °C (blue and red symbols and lines, respectively), and at incident light intensities of 500 and 110 µmol photons m^{-2} s^{-1}(empty and full symbols, respectively). (**A**) Starch (pg $cell^{-1}$), and (**B**) lipids (pg $cell^{-1}$). Horizontal dashed lines indicate the number of doublings of the initial values at the beginning of the cell cycle (0 h). Multiples of these values are given on the right ordinate. Data are presented as means ± SE. The differences between 30 °C and 40 °C are statistically significant at $p < 0.001$ (two-way ANOVA) in both A and B. The differences between different light intensities in A are statistically significant at $p < 0.05$ (two-way ANOVA).

3.1.1. Reproductive Events

Daughter cells that were released during the dark period were uninucleate, containing about 0.04 pg of DNA per nucleus. At the optimal temperature, the first replication round started after about 8 h and was completed at the 12th h of growth in light at both light intensities. The second replication round was terminated by the 18th h, when cell division started. The extent of multiplication differed between the two light intensities, as the DNA content increased about fourfold at low light intensity, and almost eightfold at high light intensity. The content of DNA per cell became correspondingly reduced during cell division between the 20th and 24th hours, when the cell numbers increased about eightfold and fourfold at high and low light intensities, respectively (Figure 1A). In the second cell cycle, the completion of DNA replication again shortly preceded the division of cells, attaining a maximum DNA content per cell just before the release of the daughter cells (Figure 1A,B). At high light intensity, DNA replication occurred earlier and to a greater extent than at the low light intensity. Similarly to the first cell cycle, nearly three replication rounds of DNA occurred at high light intensity, followed by division into mostly eight daughters, while at lower light intensity only two replication rounds and division into four daughter cells occurred during the cell cycles (compare Figure 1A,B).

The reproductive events were completely inhibited in the cultures grown at the supra-optimal temperature (40 °C), which was demonstrated by constant values of DNA content (Figure 1A), and number of cells (Figure 1B) corresponding to the initial values at the beginning of the first cell cycle.

3.1.2. Growth Processes

Growth was characterized as changes in the cellular content (pg cell^{-1}) of total RNA (Figure 2A), proteins (Figure 2B), dry matter (Figure 2C), and cell volume (μm^3 cell^{-1}) (Figure 2D). Total RNA content (Figure 2A) in the first cell cycle increased approximately threefold at the low light intensity and almost fourfold at high light intensity within 18 h of growth in cells grown at 30 °C, reaching a maximum just before cellular division, and then decreasing with time as the cells divided to a value corresponding to the RNA content of the daughter cells (Figure 2A). A similar behavior was also noted in the second cell cycle, with the maximum being reached after 36 h, i.e., 18 h after cell division started. However, the rate of RNA content increase was slower in the second cell cycle than in the first one, probably due to a decrease in the mean light intensity (Table 1) due to an increase in cell concentration by the end of the first cell cycle (Figure 2A).

Table 1. Changes in mean light intensities experienced by the cultures during the experiment at different incident light intensities and temperatures.

		Mean Light Intensity (μmol photons m^{-2} s^{-1})			
Incident Light Intensity		**110 μmol photons m^{-2} s^{-1}**		**500 μmol photons m^{-2} s^{-1}**	
Temperature		**30 °C**	**40 °C**	**30 °C**	**40 °C**
Time (h)	0	70 ± 5	75 ± 5	332 ± 15	332 ± 12
	24	56 ± 5	54 ± 7	274 ± 10	261 ± 10
	42	50 ± 5	50 ± 5	186 ± 10	220 ± 10
	48	50 ± 5	50 ± 5	177 ± 10	220 ± 10

Similar kinetics were also observed in the time courses of protein content (Figure 2B), dry matter (Figure 2C), and cell volume (Figure 2D) at 30 °C. The growth characteristics (RNA, protein, and dry matter) were slowed, and attained a lower maximum content at the lower light intensity of 110 μmol photons m^{-2} s^{-1} than at the higher intensity of 500 μmol photons m^{-2} s^{-1}. Within the second cell cycle, the differences between these variants were less apparent, and growth rate was suppressed by an increase in cell concentration, and a consequent decrease in mean light intensity (Table 1). Cell volume multiplied about eightfold (Figure 2D) in proportion to the number of daughter cells liberated at the end of both cell cycles at both light intensities (Figure 1B).

Completely different kinetics of growth processes occurred in cultures grown at 40 °C. As described above, the reproductive processes (events) at this temperature were inhibited (Figure 1), but the growth processes continued undisturbed for the duration of two cell cycles of the control cultures grown at 30 °C (Figure 2). The cells reached a giant size corresponding to about five doublings (32-fold increase) in cell volume within 36 h (Figure 2D). Within the same time period, protein content attained three doublings (8-fold increase; Figure 2B), dry matter attained four doublings (16-fold increase, Figure 2C), and RNA content increased 8-fold (Figure 2A). The proportions of different components of the cells differed compared to control cells.

3.1.3. Energy Reserves

The biochemical analyses of starch content solely detect net starch, i.e., the total amount of starch produced in light minus starch consumed in light. Cells of the control culture grown at high light intensity increased their starch content to about 16 times the daughter cells' content within 12 h of growth in the first cell cycle (Figure 3A). At the end of the cell cycle, the starch was extensively degraded, until it reached the values present in the daughter cells. At lower light intensity (110 µmol photons m^{-2} s^{-1}), accumulation of starch to its maximal value during the growth phase was slower, and it attained a lower maximum than in the culture grown at high light intensity (500 µmol photons m^{-2} s^{-1}) (Figure 3A). In the second cell cycle, the synthesis of starch decreased due to an increase in cell concentration caused by mother cell divisions (Figure 1B). This slowed the accumulation of starch so that it only covered the requirements of metabolism, but did not accumulate (Figure 3A).

Due to inhibited reproductive processes at 40 °C, the net starch content did not decrease at the time of cell division in controls. Consequently, starch overaccumulated to between 508- and 254-fold compared to the initial content in the daughter cells, and to about 32-fold the maximum values found in the control cells (Figure 3A) The rate of accumulation at the lower light intensity was slightly slower, and the final content of overaccumulated starch was lower (77 vs. 104 pg $cell^{-1}$) (Figure 3A).

The secondary energy reserve, neutral lipids, did not accumulate at the optimal temperature of 30 °C. The lipid content remained fairly constant, and decreased during the second cell cycle (Figure 3B). Contrary to growth at 30 °C, lipid reserves at 40 °C increased from the middle of the first cell cycle until the end of the second cell cycle (Figure 3B). Lipid accumulation was delayed at low light intensity, but the final values —about a 64-fold increase compared to the initial values—were comparable between both light intensities (Figure 3B).

3.2. Growth in Deuterated Water

Stress caused by growth at the supra-optimal temperature affected the cell composition, and significantly increased the content of both starch and lipids. Since increased temperature increases kinetic energy—leading to a seeming decrease in both dissociation and activation energies, which are both affected by the presence of deuterium—we tested whether the combination of increased temperature and the presence of deuterium would affect cell growth and accumulation of energy reserves. Synchronized cultures of *P. kessleri* were grown at the optimal temperature of 30 °C and two higher temperatures: 35 °C, and 39 °C. In contrast to the experiments described above, for the experiments in deuterated water, the temperature of 39 °C was used as the supra-optimal one; this was due to the fact that combination of the higher temperature (40 °C) and deuterated water strongly affected cell vitality and viability, but the effect of 39 °C on deuterated cultures (and control cultures) was similar to the effect of 40 °C. As high light intensity, in combination with the presence of deuterium, increases cellular stress [30], only a low light intensity of 110 µmol photons m^{-2} s^{-1} was used for these experiments. Due to the inhibitively high price of deuterated water, only cellular division was followed, and growth was assessed by changes in optical density at 750 nm (OD_{750}) and cell volumes (Figure 4). The stress caused to the photosynthetic apparatus was estimated as quantum yield.

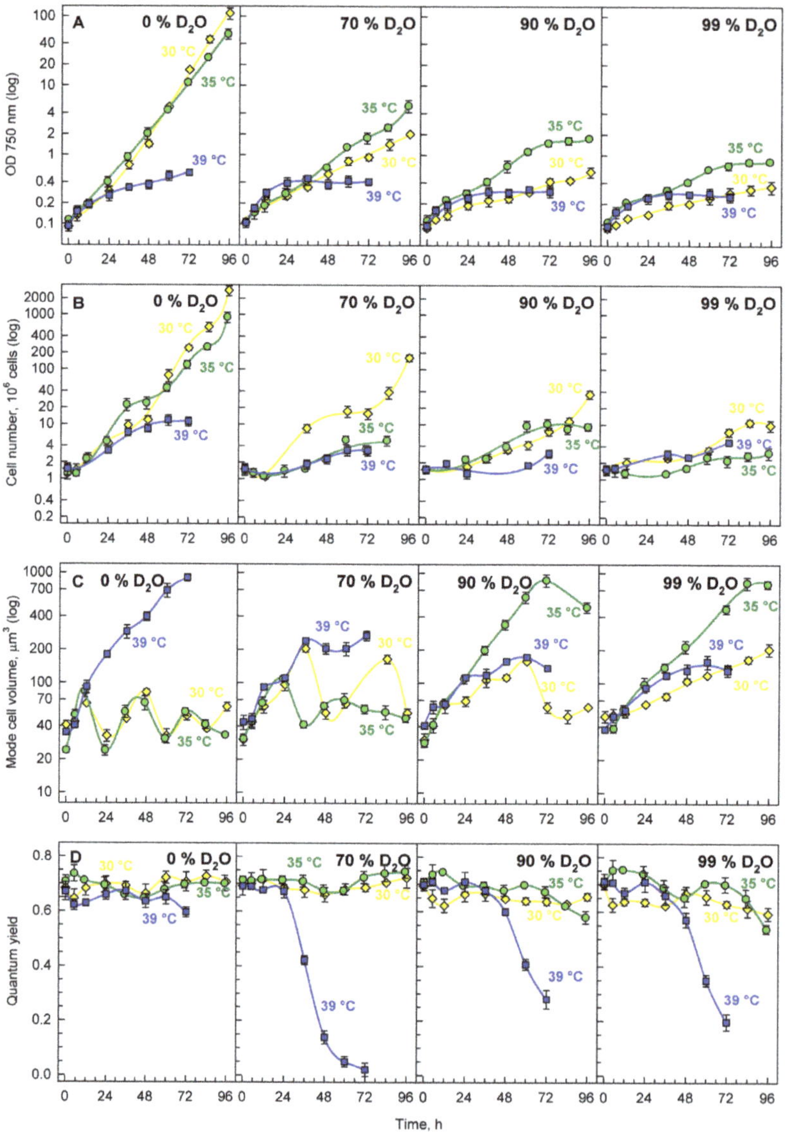

Figure 4. Time course of cell division and growth in synchronized cultures of *Parachlorella kessleri* grown at an incident light intensity of 110 µmol photons m^{-2} s^{-1} and temperatures of 30 °C (yellow diamonds), 35 °C (green circles), and 39 °C (blue squares), with different concentrations of deuterated water in the medium (0, 70, 90, and 99%). (**A**) Optical density at 750 nm, (**B**) cell number (10^6 mL^{-1}), (**C**) modal cell volume, and (**D**) quantum yield. Note the logarithmic scale on the Y-axis for (**A**–**C**). Note that the cultures were maintained in semi-batch mode. However, both panels (**A**) and (**B**) show recalculated values multiplied by individual dilution factors. Thus, they represent a projection of the growth in semi-batch cultivation mode to the continuous growth. This way, the plots remain easily readable.

3.2.1. Growth and Division

Synchronous cultures of *P. kessleri* grown in deuterated water showed a concentration-dependent decrease in their growth rate (Figure 4A). Apart from decreased growth rates, changes in cell numbers (Figure 4B) indicated decreased division rates in deuterated cultures. This was further supported by observed changes in mode cell volume (Figure 4C), which reflected cell cycle progression in synchronous cultures. At 30 °C, the culture in 0% D_2O (Figure 4C, first panel, yellow line) completed three full cell cycles, as illustrated by three sequences of cell volume increase and decrease, while there were only two cell cycles completed in 70% D_2O (Figure 4C, second panel, yellow line), one full cell cycle in 90% D_2O (Figure 4C, third panel, yellow line), and only growth—not followed by cell division—which was characteristic for the majority of cells in 99% D_2O (Figure 4C, fourth panel, yellow line).

Interestingly, the effects of temperature varied between different deuterated cultures. In the control culture, temperatures of 30 and 35 °C showed similar performance, both in terms of growth (Figure 4A, first panel) and cell cycle progression (Figure 4B,C, first panels), while at a temperature of 39 °C cell division was clearly inhibited, as documented by the occurrence of abnormally enlarged cells at 39 °C (Figure 4C, first panel, blue line), accompanied by significantly lower cell numbers (Figure 4B, first panel, blue line). This phenotype recapitulates what is described in more detail above. In deuterated cultures, the best performance, in terms of growth measured as OD_{750}, was obtained at 35 °C (Figure 4A, green lines). However, such a performance was mainly caused by an increase in cell volume under such conditions—especially in 90% and 99% D_2O (Figure 4C, green lines). At the same time, division was more adversely affected by cultivation at 35 °C (Figure 4B, green lines), as fewer cells were produced during cultivation. The adverse effect of cultivation at 39 °C was especially pronounced in quantum yields (Figure 4D, blue lines), which points to severe impairment of photosynthetic efficiency in such cultures.

In order to quantify and compare the effects of different deuterium concentrations (and their combination with tested temperatures), mass doubling time and cell number doubling time were calculated (Table 2). Mass doubling time increased with increasing D_2O content in the growth medium, thus proving the adverse effect of D_2O on growth. At the same time, cell number doubling time also increased with increasing D_2O content in the medium, thus proving the adverse effect of deuterium on division. This trend applied to all temperatures tested; however, at 39 °C, it was the weakest, as such a temperature already poses severe stress—even for the control, undeuterated culture. Moreover, the effect of increasing D_2O concentration on growth seemed to be more pronounced than the effect on cell division at 30 °C, but not at 35 and 39 °C (Table 2).

3.2.2. Energy Reserves

The combined effects of different deuterium concentrations with optimal and supra-optimal temperatures affected both starch and neutral lipid accumulation. After 24 h of cultivation, the starch content was increased in all variants except for the control culture grown at 30 °C (Figure 5, compare A with B–D and I–L). The degree of starch accumulation at 30 °C generally increased with increasing D_2O concentration—except for 99% D_2O, which contained the lowest amount of starch among cultures grown at 30 °C (Figure 5, compare panels A–D); at 39 °C, the net starch content was comparable in all variants (Figure 5I–L). The neutral lipids content was negligible in the control cultures grown in 0% D_2O (Figure 5E), but it started to accumulate in some of the cells grown at 39 °C (Figure 5M), thus repeating the behavior described above (Figure 3). The deuterium-treated cultures had a higher content of neutral lipids, and their content increased with increases in both temperature and deuterium concentration from 70% D_2O to 90% D_2O (Figure 5). The cells grown at the highest deuterium concentration (99% D_2O) contained slightly less neutral lipids than those grown in 90% D_2O, but both of the cultures contained neutral lipids in almost all cells at 39 °C, which was in contrast to 70% D_2O, where only about half of the cells contained detectable neutral lipids (Figure 5, compare O and P with N).

Table 2. Comparison of mass and cell number doubling times for cultures grown in media containing different concentrations of deuterated water at different temperatures. The average of at least three experiments is shown.

Concentration of Deuterated Water (%)	Temperature (°C)	Mass Doubling Time (h)	% of Control	Cell Number Doubling Time (h)	% of Control
0	30	9.58	100.00	8.85	100.00
	35	10.66	100.00	9.99	100.00
	39	28.18	100.00	24.84	100.00
70	30	22.97	239.67	14.18	160.25
	35	17.01	159.54	28.90	289.35
	39	36.48	129.46	63.35	255.00
90	30	32.23	336.30	21.00	237.37
	35	24.01	225.27	36.35	363.91
	39	43.92	155.86	73.63	296.38
99	30	42.34	441.87	35.35	399.54
	35	35.32	331.37	92.33	924.32
	39	50.61	179.58	63.51	255.66

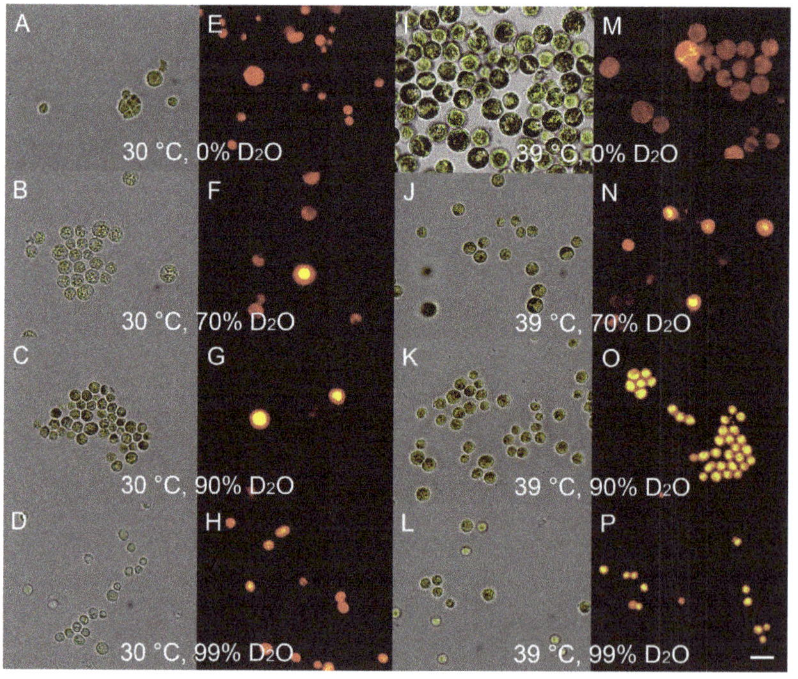

Figure 5. Light and fluorescence photomicrographs of synchronized cultures of *Parachlorella kessleri* grown for 24 h at an incident light intensity of 110 µmol photons m $^{-2}$ s $^{-1}$, temperatures of 30 °C and 39 °C, and different concentrations of deuterated water in the media (0, 70, 90, and 99%). (**A,E,I,M**) control cultures grown in 0% D_2O, (**B,F,J,N**) cultures grown in 70% D_2O, (**C,G,K,O**) cultures grown in 90% D_2O, and (**D,H,L,P**) cultures grown in 99% D_2O. In (**A–D**) and (**I–L**) the cells were stained with Lugol's solution to visualize starch; for (**E–H**) and (**M–P**) the cells were stained with Nile red to visualize neutral lipids. Scale is 10 µm.

4. Discussion

Parachlorella kessleri is an alga with emerging biotechnological potential, particularly because it can produce both starch as a primary energy store and neutral lipids as a secondary one [23,32]. Here, we used synchronized cultures of *P. kessleri* to analyze the effects of supra-optimal temperature on the production of starch and neutral lipids, both in the presence of normal (hydrogenated) water and of deuterated (heavy) water. We chose to use synchronized cultures where all of the cells within the cultures were in the same phase of the cell cycle. This allowed us to identify any subtle differences in the behavior of the supra-optimal-temperature-treated cultures and, thus, better characterize the effects.

The application of different light intensities to the control culture at 30 °C led to different degrees of multiplication of DNA, and to the production of different numbers of daughter cells: four daughter cells at low light intensity, and eight daughter cells at high light intensity (Figure 1). Interestingly, the number of daughter cells produced was not significantly affected by the decreasing light intensities in the second cycle. At the low light intensity, about 70 µmol photons m^{-2} s^{-1} at the beginning of the first and 56 µmol photons m^{-2} s^{-1} at the beginning of the second cell cycle were both sufficient for division into four daughter cells (Table 1). Similarly, for division into eight cells, a sufficient mean light intensity was 332 µmol photons m^{-2} s^{-1} at the beginning of the first cell cycle and 274 µmol photons m^{-2} s^{-1} at the beginning of the second. This ability to divide into multiple daughter cells, even at low light intensities, allows the alga to grow and divide at high cell densities when the majority of incoming light is shaded [68,69], making it a highly efficient species for algal biotechnology [70,71]. Although the number of daughter cells produced was not affected, the different incident light intensities affected the timing of the reproduction processes, so that DNA replication and cell divisions in the second cell cycle were delayed by 4 and 2 h, respectively (Figure 1). Moreover, the daughter cells produced at the end of the second cell cycle were smaller than those at the end of the first cycle, and contained fewer proteins than the cells at the beginning of the experiment (Figure 2). As the RNA and starch contents per cell stayed approximately the same at 30 °C (Figures 2 and 3), the cell composition was clearly altered by light intensity. Furthermore, cells at high light intensity were, in both cell cycles, richer in protein compared to the low-light-grown cells (Figure 2B). This phenomenon might be exploited when a protein-rich algal biomass is required [68,72]. Altogether, this documents the plasticity of the multiple fission lifestyle, which allows the cells to slightly prolong the growth phase in order to be able to reach a critical cell size for starting another reproductive sequence, and to maintain the reproduction potential at the expense of producing slightly smaller daughter cells with altered cell compositions. Such growth and metabolic plasticity of algae allows them to thrive both at low light intensities [68,69], and in the harsh conditions of a desert, with very high light intensity and extreme fluctuating temperatures [73–76].

Growth at the supra-optimal temperature led to a blockage of DNA replication and cell division (Figure 1), an approximately 30-fold increase in cell size (Figure 2D), and overaccumulation of both starch and lipids (Figure 3). The cell cycle arrest at supra-optimal temperatures has been established both in the related alga *Chlorella vulgaris* [18] and in the distantly related *Chlamydomonas reinhardtii* [17,77]. However, the nature of the blockage seems to be different in *P. kessleri*, since no DNA replication was detected (Figure 1), which is in contrast to both *Chlorella vulgaris* and *Chlamydomonas reinhardtii*, where (at least some) DNA replication occurred and the cell cycle blockage took place at the level of nuclear (and cellular) division(s) [17,18,77]. Indeed, it has been established that individual cell reproduction processes are sensitive to stress in an order opposite to their occurrence in the cell cycle, with cell division being the most sensitive process [16]. Thus, the observed behavior may also be caused by the supra-optimal temperature used, and if a lower temperature is used, DNA replication might not be affected. In the experiments presented, both supra-optimal temperatures of 39 °C and 40 °C caused the same blockage of cell division (Figures 1 and 4), suggesting that the temperature of 40 °C is well above the threshold required to cause the cell cycle blockage. However, the effects of temperature are very precise and subtle, so a 1 °C variation in temperature can have the quite distinct effect of only a partial blockage [17,77]. Of the three effects of supra-optimal temperature treatment, the most striking was the approximately 32-fold overaccumulation of starch compared to the highest content reached in the control cells, and the approximately 508-fold increase compared to the initial cell content. The accumulation was very fast, as close to maximal values were reached within about 24 h at the high light intensity, and within about 36 h at the low light intensity (Figure 3A). At the optimal temperature, the net accumulation of starch was slower, and reached lower levels at the low light intensity; this phenomenon was even more pronounced at the supra-optimal temperature (Figure 3). This stresses the importance of light intensity for starch (over)production, and is consistent with the established knowledge in the field [38]. The starch overaccumulation at supra-optimal temperatures was consistent with the previous experiments on the related *Chlorella* sp. K [78], *Chlorella vulgaris* [18] and distantly related *Chlamydomonas reinhardtii* [17,77]. The overaccumulation of neutral lipids was delayed by about 16 hours compared with that of starch (Figure 3). This fits with the two-step production of starch, followed by neutral lipids, as stress conditions are prolonged [23,24,32,39]. Lipid production also proved to be light-dependent, as it was about six hours delayed at the low light intensity (Figure 3B). Although the lipids overaccumulated to about 64 times the initial values, their final cell content was much lower than that of starch (Figure 3), making starch the predominant component of the cell. Indeed, cell composition was strikingly affected by the supra-optimal temperatures. At the high light intensity, the cell volume (Figure 2D) increased about 30-fold, dry matter (Figure 2C) increased about 19-fold, RNA content more than 8-fold (Figure 2A), and protein content by about 32-fold (Figure 2B). Together with a massive 508-fold increase in starch, this made the biomass significantly enriched in starch and proteins. In general, starch (over)production can be attained either through increasing starch production by improving growth conditions (high light, high CO_2 concentration) [36,37] or by decreasing starch consumption—mainly by blocking cell reproduction as the biggest starch consumer under nutrient-limiting conditions [24–26,32,39,41]. Indeed, a blockage of cell division has emerged as an effective tool to (over)produce starch [79]. Improved starch production by optimizing growth conditions, combined with decreasing its consumption due to inhibition of the major consumers, should lead to superior starch (over)accumulation. Rapid and extensive starch (over)production by supra-optimal temperature treatment was shown in the green alga *Chlamydomonas reinhardtii* in both laboratory-scale [17] and pilot-scale photobioreactors [38]. Compared to *Chlamydomonas reinhardtii*, *P. kessleri* grows at lower light intensities, and is able to reach very high cell densities [70,71]. Together with the ability to produce significant amounts of neutral lipids via a simple alteration of conditions, this makes it an excellent organism for algal biotechnology. The experiments presented here provide a means for the fast and reliable overproduction of starch by increasing the temperature to a supra-optimal one. Such a treatment is very simple, but a rapid and controllable increase in temperature, and maintenance of that temperature, might be complicated at a very large scale. Alternatively, the treatment can be used in closed pilot-scale reactors [38], where temperature control is more feasible. Such production would only be justifiable for high-added-value compounds, such as stable isotopically labeled ones.

The experiments on the combined effects of deuterium and temperature had a double rationale: Firstly, from a basic research point of view, it is intriguing to analyze the combination of deuterium, which increases bond dissociation and activation energies [46], with increased temperature, which is known to increase kinetic energy and, thus, speed up general metabolism. Secondly, (over)production of deuterated compounds such as starch, proteins, and/or neutral lipids is biotechnologically relevant, as these are very high-added-value products that can be exploited both as biochemical standards and in stable isotope-labeling experiments [43,52,53]. When starting the experiments, it was not clear which of the two phenomena would prevail. We assumed that the temperature increase might decrease the effect of the presence of deuterium, and at the same time that the deuterium might prevent the blockage of cell division observed in hydrogenated water at supra-optimal temperatures. As it turned out, some of our assumptions were confirmed, while others were not. Cultivation in increasing concentrations of deuterated water showed a concentration-dependent decrease in growth rates and cell division (Figure 4A,B). Interestingly, although the difference in growth rates between 30 °C and 35 °C in the control was negligible (Figure 4A, first panel), the cultivation at 35 °C promoted growth in all deuterated cultures (Figure 4A, second to fourth panels, green circles). This partially confirmed the prediction that increased temperature might alleviate the effects of deuterium on the cells. However, the effect was specific only to growth, and was not reflected in cell division (Figure 4B, green circles), as the cells remained large (Figure 4C, green circles). This implies that although the effects of deuterium on metabolism in general can be partially compensated for by temperature, the deuterium-induced problems with cell division are either more sensitive or more specific. A further increase in temperature to 39 °C inhibited both cell growth and division (Figure 4A,B, blue squares). The effect of supra-optimal temperatures was more detrimental to the deuterated cultures than to the controls, suggesting that the alleviating effect of 35 °C was replaced with a strictly stress effect. Due to the combination of the two stresses, the same temperature considered to be supra-optimal in control cultures might be above this threshold for highly deuterated cultures. Thus, temperature in deuterated cultures showed aspects of hormesis—a phenomenon where low concentrations of a compound are beneficial but higher levels cause stress [80,81]. The significant stress caused to the deuterated cells by the supra-optimal temperatures was further supported by a rapid decrease in the Fv:Fm ratio in all deuterated cultures after about 36 h of the experiment (Figure 4D, second to fourth panels, blue squares). This was in contrast to the control cultures at the same temperatures (Figure 4D, blue squares), which were not significantly affected. The cultures of *P. kessleri* showed striking resilience to the presence of deuterium, as they were not significantly stressed by growth in very high deuterated water concentrations for more than three days (Figure 4D, yellow and green lines). Only on the fourth day of cultivation did the cultures at the two highest deuterated water concentrations (90% and 99%) start to be stressed (Figure 4D, third and fourth panels). However, the absence of cell division at the higher deuterated water concentrations allowed (over)accumulation of starch, although the temperature did not seem to have a promoting or inhibiting effect in such cultures (Figure 5, compare B–D with J–L). The combination of supra-optimal temperatures and growth in deuterated water promoted the accumulation of neutral lipids (Figure 5, panels N–P). These experiments confirmed that the effect of supra-optimal temperatures is a general phenomenon not only in different organisms [17,18,77], but also in different growth conditions. If desired, such treatment can be used to (over)produce deuterated starch and, interestingly, also deuterated neutral lipids. Due to limitations in maintaining supra-optimal temperatures and the inhibitively high prices of deuterated water, such treatment is conceivable only at small scales for the production of stable isotopically labeled compounds [43,52,53]. This is consistent with the proposed algal production of deuterated sugars and amino acids [58], proteins [59], chlorophylls [52,60], and carotenoids [52]. Supra-optimal temperature seems to be a promising tool for a rapid increase in starch (and neutral lipid) content, compared to the slower effect of nutrient starvation [43]. Furthermore, it offers recovery of other deuterated cell components.

5. Conclusions

We describe the effects of supra-optimal temperatures on synchronized cultures of the green alga *P. kessleri*. Supra-optimal temperatures block DNA replication and cell division, but do not affect cell growth. Moreover, due to the absence of cell division, starch and neutral lipids overaccumulate to levels 508-fold and 64-fold those of the initial values in the daughter cells. This leads to the production of large cells with altered cell compositions, consisting mostly of proteins and starch. We tested whether increased cultivation temperatures could alleviate the effects of deuterium in heavy water. Temperature showed a hormetic effect, as a temperature increase over optimum improved cell growth, but the supra-optimal temperatures caused extensive stress. The cultures were strikingly resilient to the stress caused by heavy water, and were able to grow and divide even at 90% D_2O, although they were only able to grow but not divide at 99% D_2O. The supra-optimal temperature treatment, when applied to cultures grown in heavy water, had similar effects as in normal water, and led to the production of cells with increased starch and neutral lipid contents. Our results establish new treatments for the overaccumulation of starch (and proteins) in *P. kessleri*, and they also prove that such treatment can be used to produce deuterated starch, proteins, and neutral lipids for biotechnological purposes.

Author Contributions: Conceptualization, V.Z., and K.B.; methodology, I.N.I., V.K., V.B., S.O. and T.T.; validation, K.B., M.V., S.K. and V.Z.; formal analysis, I.N.I., M.V., V.K., V.B., S.O. and T.T.; investigation, V.Z., I.N.I., M.V., V.K., V.B., S.O., T.T., S.K. and K.B.; resources, K.B. and S.K.; data curation, V.Z., I.N.I., V.K., V.B., S.O., T.T. and K.B.; writing—original draft preparation, V.Z. and K.B.; writing—review and editing, V.Z., I.N.I., M.V., V.K., V.B., S.O., T.T., S.K. and K.B.; visualization, V.Z., I.N.I., V.K., M.V., S.O., T.T. and K.B.; supervision, V.Z., S.K. and K.B.; project administration, V.Z. and K.B.; funding acquisition, K.B. All authors have read and agreed to the published version of the manuscript.

Funding: This work was supported by Grantová Agentura České Republiky, grant no. 17-06264S, and by Institutional Research Concept no. AVOZ61388971.

Institutional Review Board Statement: Not applicable.

Informed Consent Statement: Not applicable.

Data Availability Statement: All data presented in this study are available within this article. There are no special databases associated with this manuscript.

Acknowledgments: We are obliged to the technical staff of the Laboratory of Cell Cycles of Algae for their excellent technical support. We also thank J. D. Brooker for critical reading and language editing of the text.

Conflicts of Interest: The authors declare no conflict of interest.

References

1. Zachleder, V.; Bišová, K.; Vítová, M.; Štěpán, K.; Hendrychová, J. Variety of cell cycle patterns in the alga *Scenedesmus quadricauda* (*Chlorophyta*) as revealed by application of illumination regimes and inhibitors. *Eur. J. Phycol.* **2002**, *37*, 361–371. [CrossRef]
2. Oldenhof, H.; Zachleder, V.; van den Ende, H. Blue- and red-light regulation of the cell cycle in *Chlamydomonas reinhardtii* (Chlorophyta). *Eur. J. Phycol.* **2006**, *41*, 313–320. [CrossRef]
3. Oldenhof, H.; Zachleder, V.; van den Ende, H. Blue light delays commitment to cell division in *Chlamydomonas reinhardtii*. *Plant Biol.* **2004**, *6*, 689–695. [CrossRef]
4. Munzner, P.; Voigt, J. Blue light regulation of cell division in *Chlamydomonas reinhardtii*. *Plant Physiol.* **1992**, *99*, 1370–1375. [CrossRef] [PubMed]
5. Voigt, J.; Munzner, P. The *Chlamydomonas* cell-cycle is regulated by a light dark- responsive cell-cycle switch. *Planta* **1987**, *172*, 463–472. [CrossRef]
6. McAteer, M.; Donnan, L.; John, P.C.L. The timing of division in *Chlamydomonas*. *New Phytol.* **1985**, *99*, 41–56. [CrossRef]
7. Donnan, L.; Carvill, E.P.; Gilliland, T.J.; John, P.C.L. The cell-cycles of *Chlamydomonas* and *Chlorella*. *New Phytol.* **1985**, *99*, 1–40. [CrossRef]
8. Donnan, L.; John, P.C.L. Cell cycle control by timer and sizer in *Chlamydomonas*. *Nature* **1983**, *304*, 630–633. [CrossRef]
9. Zachleder, V.; Šetlík, I. Timing of events in overlapping cell reproductive sequences and their mutual interactions in the alga *Scenedesmus quadricauda*. *J. Cell Sci.* **1990**, *97*, 631–638. [CrossRef]
10. Los, D.A.; Zachleder, V.; Kuptsova, E.S.; Ksenofontov, A.L.; Markelova, A.G.; Shapiguzov, Y.M.; Semenenko, V.E. Effect of the spectral composition of light on replication of chloroplast DNA and division of chloroplast nucleoids in the green-alga *Dunaliella salina*. *Russ. Plant Physiol.* **1990**, *37*, 791–798.
11. Morimura, Y. Synchronous culture of *Chlorella*. I. Kinetic analysis of the life cycle of *Chlorella ellipsoidea* as affected by changes of temperature and light intensity. *Plant Cell Physiol.* **1959**, *1*, 49–62.
12. Bisova, K.; Zachleder, V. Cell-cycle regulation in green algae dividing by multiple fission. *J. Exp. Bot.* **2014**, *65*, 2585–2602. [CrossRef]
13. Vítová, M.; Bišová, K.; Hlavová, M.; Kawano, S.; Zachleder, V.; Čížková, M. *Chlamydomonas reinhardtii*: Duration of its cell cycle and phases at growth rates affected by temperature. *Planta* **2011**, *234*, 599–608. [CrossRef]

14. Zachleder, V.; van den Ende, H. Cell cycle events in the green alga *Chlamydomonas eugametos* and their control by environmental factors. *J. Cell Sci.* **1992**, *102*, 469–474. [CrossRef]
15. Zachleder, V.; Ivanov, I.; Vítová, M.; Bišová, K. Effects of cyclin-dependent kinase activity on the coordination of growth and the cell cycle in green algae at different temperatures. *J. Exp. Bot.* **2019**, *70*, 845–858. [CrossRef]
16. Bišová, K.; Hendrychová, J.; Cepák, V.; Zachleder, V. Cell growth and division processes are differentially sensitive to cadmium in *Scenedesmus quadricauda*. *Folia Microbiol.* **2003**, *48*, 805–816. [CrossRef]
17. Zachleder, V.; Ivanov, I.; Vítová, M.; Bišová, K. Cell cycle arrest by supraoptimal temperature in the alga *Chlamydomonas reinhardtii*. *Cells* **2019**, *8*, 1237–1257. [CrossRef] [PubMed]
18. Šetlík, I.; Zachleder, V.; Doucha, J.; Berková, E.; Bartoš, J. The nature of temperature block in the sequence of reproductive processes in *Chlorella vulgaris* BEIJERINCK. *Arch. Hydrobiol. Suppl. 49 Algol. Stud.* **1975**, *14*, 70–104.
19. Krienitz, L.; Hegewald, E.H.; Hepperle, D.; Huss, V.A.R.; Rohr, T.; Wolf, M. Phylogenetic relationship of *Chlorella* and *Parachlorella* gen nov (Chlorophyta, Trebouxiophyceae). *Phycologia* **2004**, *43*, 529–542. [CrossRef]
20. Řezanka, T.; Podojil, M. The very long-chain fatty-acids of the green-alga *Chlorella kessleri*. *Lipids* **1984**, *19*, 472–473. [CrossRef]
21. Ota, S.; Matsuda, T.; Takeshita, T.; Yamazaki, T.; Kazama, Y.; Abe, T.; Kawano, S. Phenotypic spectrum of *Parachlorella kessleri* (Chlorophyta) mutants produced by heavy-ion irradiation. *Bioresour. Technol.* **2013**, *149*, 432–438. [CrossRef]
22. Ota, S.; Yoshihara, M.; Yamazaki, T.; Takeshita, T.; Hirata, A.; Konomi, M.; Oshima, K.; Hattori, M.; Bišová, K.; Zachleder, V.; et al. Deciphering the relationship among phosphate dynamics, electron-dense body and lipid accumulation in the green alga *Parachlorella kessleri*. *Sci. Rep.* **2016**, *6*, 25731. [CrossRef] [PubMed]
23. Fernandes, B.; Teixeira, J.; Dragone, G.; Vicente, A.A.; Kawano, S.; Bišová, K.; Přibyl, P.; Zachleder, V.; Vítová, M. Relationship between starch and lipid accumulation induced by nutrient depletion and replenishment in the microalga *Parachlorella kessleri*. *Bioresour. Technol.* **2013**, *144*, 268–274. [CrossRef] [PubMed]
24. Mizuno, Y.; Sato, A.; Watanabe, K.; Hirata, A.; Takeshita, T.; Ota, S.; Sato, N.; Zachleder, V.; Tsuzuki, M.; Kawano, S. Sequential accumulation of starch and lipid induced by sulfur deficiency in *Chlorella* and *Parachlorella* species. *Bioresour. Technol.* **2013**, *129*, 150–155. [CrossRef]
25. Takeshita, T.; Ivanov, I.N.; Oshima, K.; Ishii, K.; Kawamoto, H.; Ota, S.; Yamazaki, T.; Hirata, A.; Kazama, Y.; Abe, T.; et al. Comparison of lipid productivity of *Parachlorella kessleri* heavy-ion beam irradiation mutant PK4 in laboratory and 150-L mass bioreactor, identification and characterization of its genetic variation. *Algal Res.* **2018**, *35*, 416–426. [CrossRef]
26. Taleb, A.; Legrand, J.; Takache, H.; Taha, S.; Pruvost, J. Investigation of lipid production by nitrogen-starved *Parachlorella kessleri* under continuous illumination and day/night cycles for biodiesel application. *J. Appl. Phycol.* **2017**. [CrossRef]
27. Rathod, J.P.; Prakash, G.; Pandit, R.; Lali, A.M. Agrobacterium-mediated transformation of promising oil-bearing marine algae *Parachlorella kessleri*. *Photosynth. Res.* **2013**, *118*, 141–146. [CrossRef] [PubMed]
28. Sato, N.; Tsuzuki, M.; Kawaguchi, A. Glycerolipid synthesis in *Chlorella kessleri* 11h - I. Existence of a eukaryotic pathway. *Biochim. Biophys. Acta* **2003**, *1633*, 27–34. [CrossRef]
29. Saleh, M.M.; Matorin, D.N.; Zayadan, B.K.; Todorenko, D.A.; Lukashov, E.P.; Gaballah, M.M. Differentiation between two strains of microalga *Parachlorella kessleri* using modern spectroscopic method. *Bot. Stud.* **2014**, *55*, 53. [CrossRef]
30. Zachleder, V.; Ivanov, I.N.; Kselíková, V.; Bialevich, V.; Vítová, M.; Ota, S.; Takeshita, T.; Kawano, S.; Bišová, K. Characterization of growth and cell cycle events as affected by light intensity in the green alga *Parachlorella kessleri*, as a new model for cell cycle research. *Biomolecules* **2021**, *11*, 891. [CrossRef] [PubMed]
31. Lien, T.; Knutsen, G. Synchronous growth of *Chlamydomonas reinhardtii* (Chlorophyceae): A review of optimal conditions. *J. Phycol.* **1979**, *15*, 191–200. [CrossRef]
32. Takeshita, T.; Ota, S.; Yamazaki, T.; Hirata, A.; Zachleder, V.; Kawano, S. Starch and lipid accumulation in eight strains of six *Chlorella* species under comparatively high light intensity and aeration culture conditions. *Bioresour. Technol.* **2014**, *158*, 127–134. [CrossRef] [PubMed]
33. Torres-Romero, I.; Kong, F.; Legeret, B.; Beisson, F.; Peltier, G.; Li-Beisson, Y. *Chlamydomonas* cell cycle mutant crcdc5 over-accumulates starch and oil. *Biochimie* **2019**. [CrossRef]
34. Hirokawa, T.; Hata, M.; Takeda, H. Correlation between the starch level and the rate of starch synthesis during the development cycle of *Chlorella ellipsoidea*. *Plant Cell Physiol.* **1982**, *23*, 813–820.
35. Sorokin, C. Changes in photosynthetic activity in the course of cell development in *Chlorella*. *Physiol. Plant.* **1957**, *10*, 659–666. [CrossRef]
36. Brányiková, I.; Maršálková, B.; Doucha, J.; Brányik, T.; Bišová, K.; Zachleder, V.; Vítová, M. Microalgae-novel highly efficient starch producers. *Biotechnol. Bioeng.* **2011**, *108*, 766–776. [CrossRef] [PubMed]
37. Izumo, A.; Fujiwara, S.; Oyama, Y.; Satoh, A.; Fujita, N.; Nakamura, Y.; Tsuzuki, M. Physicochemical properties of starch in *Chlorella* change depending on the CO_2 concentration during growth: Comparison of structure and properties of pyrenoid and stroma starch. *Plant Sci.* **2007**, *172*, 1138–1147. [CrossRef]
38. Ivanov, I.N.; Zachleder, V.; Vítová, M.; Barbosa, M.J.; Bišová, K. Starch production in *Chlamydomonas reinhardtii* through supraoptimal temperature in a pilot-scale photobioreactor. *Cells* **2021**, *10*. [CrossRef]
39. Li, X.; Přibyl, P.; Bišová, K.; Kawano, S.; Cepák, V.; Zachleder, V.; Čížková, M.; Brányiková, I.; Vítová, M. The microalga *Parachlorella kessleri*—a novel highly efficient lipid producer. *Biotechnol. Bioeng.* **2013**, *110*, 97–107. [CrossRef]

40. Přibyl, P.; Cepák, V.; Zachleder, V. Production of lipids in 10 strains of *Chlorella* and *Parachlorella*, and enhanced lipid productivity in *Chlorella vulgaris*. *Appl. Microbiol. Biotechnol.* **2012**, *94*, 549–561. [CrossRef] [PubMed]
41. Gao, Y.; Feng, J.; Lv, J.; Liu, Q.; Nan, F.; Liu, X.; Xie, S. Physiological changes of *Parachlorella kessleri* ty02 in lipid accumulation under nitrogen stress. *Int. J. Environ. Res. Public Health* **2019**, *16*. [CrossRef]
42. You, Z.; Zhang, Q.; Peng, Z.; Miao, X. Lipid droplets mediate salt stress tolerance in *Parachlorella kessleri*. *Plant Physiol.* **2019**, *181*, 510–526. [CrossRef] [PubMed]
43. Zachleder, V.; Vítová, M.; Hlavová, M.; Moudříková, Š.; Mojzeš, P.; Heumann, H.; Becher, J.R.; Bišová, K. Stable isotope compounds - production, detection, and application. *Biotechnol. Adv.* **2018**, *36*, 784–797. [CrossRef] [PubMed]
44. Yang, J. *Deuterium: Discovery and applications in organic chemistry*; Elsevier: Amsterdam, The Netherlands, 2016.
45. Lehmann, W.D. A timeline of stable isotopes and mass spectrometry in the life sciences. *Mass Spectrom. Rev.* **2017**, *36*, 58–85. [CrossRef]
46. Kinetic isotope effects. Available online: https://chem.libretexts.org/@go/page/1685 (accessed on 10 June 2021).
47. Hirakura, Y.; Sugiyama, T.; Takeda, M.; Ikeda, M.; Yoshioka, T. Deuteration as a tool in investigating the role of protons in cell signaling. *Biochim. Biophys. Acta* **2011**, *1810*, 218–225. [CrossRef]
48. Salomonsson, L.; Branden, G.; Brzezinski, P. Deuterium isotope effect of proton pumping in cytochrome c oxidase. *Biochim. Biophys. Acta* **2008**, *1777*, 343–350. [CrossRef]
49. De Kouchkovsky, Y.; Haraux, F.; Sigalat, C. Effect of hydrogen-deuterium exchange on energy-coupled processes in thylakoids. *FEBS Lett.* **1982**, *139*, 245–249. [CrossRef]
50. Evans, B.R.; Bali, G.; Reeves, D.T.; O'Neill, H.M.; Sun, Q.; Shah, R.; Ragauskas, A.J. Effect of D_2O on growth properties and chemical structure of annual ryegrass (*Lolium multiflorum*). *J. Agr. Food Chem.* **2014**, *62*, 2595–2604. [CrossRef]
51. Sacchi, G.A.; Cocucci, M. Effects of deuterium oxide on growth, proton extrusion, potassium influx, and in vitro plasma membrane activities in maize root segments. *Plant Physiol.* **1992**, *100*, 1962–1967. [CrossRef]
52. Saha, S.K.; Hayes, J.; Moane, S.; Murray, P. Tagging of biomolecules with deuterated water (D_2O) in commercially important microalgae. *Biotechnol. Lett.* **2013**, *35*, 1067–1072. [CrossRef] [PubMed]
53. Gireesh, T.; Jayadeep, A.; Rajasekharan, K.N.; Menon, V.P.; Vairamany, M.; Tang, G.; Nair, P.P.; Sudhakaran, P.R. Production of deuterated b-carotene by metabolic labelling of *Spirulina platensis*. *Biotechnol. Lett.* **2001**, *23*, 447–449. [CrossRef]
54. Cargnin, S.; Serafini, M.; Pirali, T. A primer of deuterium in drug design. *Future Med. Chem.* **2019**, *11*, 2039–2042. [CrossRef]
55. DeWitt, S.H.; Maryanoff, B.E. Deuterated drug nolecules: Focus on FDA-approved deutetrabenazine. *Biochemistry* **2018**, *57*, 472–473. [CrossRef]
56. Raffa, R.B.; Pergolizzi, J.V.; Taylor, R. The First Approved "Deuterated" Drug: A Short Review of the Concept. *Pharmacol. Pharm.* **2018**, *09*, 440–446. [CrossRef]
57. Schmidt, C. First deuterated drug approved. *Nat. Biotechnol.* **2017**, *35*, 493–494. [CrossRef] [PubMed]
58. Blake, M.I.; Crespi, H.L.; Mohan, V.; Katz, J.J. Isolation of fully deuterated metabolites from *Scenedesmus obliquus* grown in deuterium oxide. *J. Pharm. Sci.* **1961**, *50*, 425–429. [CrossRef]
59. Hattori, A.; Crespi, H.L.; Katz, J.J. Effect of side-chain deuteration on protein stability. *Biochemistry* **1965**, *4*, 1213–1225. [CrossRef]
60. Closs, D.L.; Katz, J.J.; Pennington, M.R.; Thomas, H.R.; Strain, J. Hydrogen exchange at methine and C-10 positions in chlorophyll. *Amer. Chem. Soc.* **1963**, *85*, 3809. [CrossRef]
61. Doucha, J.; Lívanský, K. Productivity, CO_2/O_2 exchange and hydraulics in outdoor open high density microalgal (*Chlorella* sp.) photobioreactors operated in a Middle and Southern European climate. *J. Appl. Phycol.* **2006**, *18*, 811–826. [CrossRef]
62. Takeshita, T.; Takeda, K.; Ota, S.; Yamazaki, T.; Kawano, S. A simple method for measuring the starch and lipid contents in the cell of microalgae. *Cytologia* **2015**, *80*, 475–481. [CrossRef]
63. Wanka, F. Die bestimmung der nucleinsäuren in *Chlorella pyrenoidosa*. *Planta* **1962**, *58*, 594–619. [CrossRef]
64. Lukavský, J.; Tetík, K.; Vendlová, J. Extraction of nucleic acid from the alga *Scenedesmus quadricauda*. *Arch. Hydrobiol. Suppl. 41 Algol. Stud.* **1973**, *9*, 416–426.
65. Decallonne, J.R.; Weyns, C.J. A shortened procedure of the diphenylamine reaction for measurement of deoxyribonucleic acid by using light activation. *Anal. Biochem.* **1976**, *74*, 448–456. [CrossRef]
66. Zachleder, V. Optimization of nucleic acids assay in green and blue-green algae: Extraction procedures and the light-activated reaction for DNA. *Arch. Hydrobiol. Suppl. 67 Algol. Stud.* **1984**, *36*, 313–328. [CrossRef]
67. Miller, G.L. Use of dinitrosalicylic acid reagent for determination of reducing sugar. *Anal. Chem.* **1959**, *31*, 426–428. [CrossRef]
68. Metsoviti, M.N.; Papapolymerou, G.; Karapanagiotidis, I.T.; Katsoulas, N. Effect of light intensity and quality on growth rate and composition of *Chlorella vulgaris*. *Plants* **2019**, *9*. [CrossRef]
69. Barnett, J.Z.; Foy, J.; Malone, R.; Rusch, K.A.; Gutierrez-Wing, M.T. Impact of light quality on a native Louisiana *Chlorella vulgaris/Leptolyngbya sp.* co-culture. *Eng. Life. Sci.* **2017**, *17*, 678–685. [CrossRef]
70. Borowitzka, M.A. Energy from microalgae: A short history. In *Algae for biofuels and energy*; Borowitzka, M.A., Moheimani, N.R., Eds.; Springer: Dordrecht, The Netherlands; New York, NY, USA; London, UK, 2013; Volume Developments in applied phycology 5, pp. 1–16.
71. Borowitzka, M.A.; Raven, J.A.; Beardall, J. *The physiology of microalgae*, 6 ed.; Borowitzka, M.A., Raven, J.A., Beardall, J., Eds.; Springer International Publishing Switzerland: Cham, Germany; Heidelberg, Germany; New York, NY, USA; Dordrecht, The Netherlands; London, UK, 2016; p. 681. [CrossRef]

72. Schuler, L.; Greque de Morais, E.; Trovao, M.; Machado, A.; Carvalho, B.; Carneiro, M.; Maia, I.; Soares, M.; Duarte, P.; Barros, A.; et al. Isolation and characterization of novel *Chlorella vulgaris* mutants with low chlorophyll and improved protein contents for food applications. *Front. Bioeng. Biotechnol.* **2020**, *8*, 469. [CrossRef] [PubMed]
73. Al Jabri, H.; Taleb, A.; Touchard, R.; Saadaoui, I.; Goetz, V.; Pruvost, J. Cultivating microalgae in desert conditions: Evaluation of the effect of light-temperature summer conditions on the growth and metabolism of *Nannochloropsis* QU130. *Appl. Sci.* **2021**, *11*. [CrossRef]
74. Levin, G.; Kulikovsky, S.; Liveanu, V.; Eichenbaum, B.; Meir, A.; Isaacson, T.; Tadmor, Y.; Adir, N.; Schuster, G. The desert green algae *Chlorella ohadii* thrives at excessively high light intensities by exceptionally enhancing the mechanisms that protect photosynthesis from photoinhibition. *Plant J.* **2021**. [CrossRef] [PubMed]
75. Treves, H.; Raanan, H.; Kedem, I.; Murik, O.; Keren, N.; Zer, H.; Berkowicz, S.M.; Giordano, M.; Norici, A.; Shotland, Y.; et al. The mechanisms whereby the green alga *Chlorella ohadii*, isolated from desert soil crust, exhibits unparalleled photodamage resistance. *New Phytol.* **2016**. [CrossRef] [PubMed]
76. Treves, H.; Raanan, H.; Finkel, O.M.; Berkowicz, S.M.; Keren, N.; Shotland, Y.; Kaplan, A. A newly isolated *Chlorella sp.* from desert sand crusts exhibits a unique resistance to excess light intensity. *FEMS Microbiol. Ecol.* **2013**, *86*, 373–380. [CrossRef] [PubMed]
77. Hemme, D.; Veyel, D.; Muhlhaus, T.; Sommer, F.; Juppner, J.; Unger, A.K.; Sandmann, M.; Fehrle, I.; Schonfelder, S.; Steup, M.; et al. Systems-wide analysis of acclimation responses to long-term heat stress and recovery in the photosynthetic model organism *Chlamydomonas reinhardtii*. *Plant Cell* **2014**, *26*, 4270–4297. [CrossRef] [PubMed]
78. Semenenko, V.E.; Vladimirova, M.G.; Orleanskaya, O.B. Physiological characteristics of *Chlorella* sp. K under conditions of high extremal temperatures I. Uncoupling effect of extreme temperatures on the cellular functions of *Chlorella*. *Russian J. Plant Physiol.* **1967**, *14*, 612–625.
79. Zachleder, V.; Brányiková, I. Starch overproduction by means of algae. In *Algal biorefineries*; Bajpai, R.K., Prokop, A., Zappi, M., Eds.; Springer: Dordrecht, The Netherlands; Heidelberg, Germany; London, UK; New York, NY, USA, 2014; Volume 1: Cultivation of cells and products, pp. 217–240.
80. Cedergreen, N.; Streibig, J.C.; Kudsk, P.; Mathiassen, S.K.; Duke, S.O. The occurrence of hormesis in plants and algae. *Dose Response* **2006**, *5*, 150–162. [CrossRef]
81. Calabrese, E.J.; Mattson, M.P. How does hormesis impact biology, toxicology, and medicine? *NPJ Aging Mech. Dis.* **2017**, *3*, 13. [CrossRef]

cells

MDPI

Article

Cytokinin Modulates Cellular Trafficking and the Cytoskeleton, Enhancing Defense Responses

Lorena Pizarro [1,*], Daniela Munoz [1], Iftah Marash [2,3], Rupali Gupta [2], Gautam Anand [2], Meirav Leibman-Markus [2] and Maya Bar [2,*]

1 Institute of Agri-Food, Animal and Environmental Sciences, Universidad de O'Higgins, Rancagua 2820000, Chile; daniela.munoz@uoh.cl
2 Department of Plant Pathology and Weed Research, Institute of Plant Protection, ARO, Volcani Institute, Rishon LeZion 7505101, Israel; iftah.marash@mail.huji.ac.il (I.M.); rupaligupta862@gmail.com (R.G.); gautaming@gmail.com (G.A.); meiravleibman@gmail.com (M.L.-M.)
3 School of Plant Sciences and Food Security, Faculty of Life Sciences, Tel Aviv University, Tel Aviv 6997801, Israel
* Correspondence: lorena.pizarro@uoh.cl (L.P.); mayabar@volcani.agri.gov.il (M.B.)

Abstract: The plant hormone cytokinin (CK) plays central roles in plant development and throughout plant life. The perception of CKs initiating their signaling cascade is mediated by histidine kinase receptors (AHKs). Traditionally thought to be perceived mostly at the endoplasmic reticulum (ER) due to receptor localization, CK was recently reported to be perceived at the plasma membrane (PM), with CK and its AHK receptors being trafficked between the PM and the ER. Some of the downstream mechanisms CK employs to regulate developmental processes are unknown. A seminal report in this field demonstrated that CK regulates auxin-mediated lateral root organogenesis by regulating the endocytic recycling of the auxin carrier PIN1, but since then, few works have addressed this issue. Modulation of the cellular cytoskeleton and trafficking could potentially be a mechanism executing responses downstream of CK signaling. We recently reported that CK affects the trafficking of the pattern recognition receptor LeEIX2, influencing the resultant defense output. We have also recently found that CK affects cellular trafficking and the actin cytoskeleton in fungi. In this work, we take an in-depth look at the effects of CK on cellular trafficking and on the actin cytoskeleton in plant cells. We find that CK influences the actin cytoskeleton and endomembrane compartments, both in the context of defense signaling—where CK acts to amplify the signal—as well as in steady state. We show that CK affects the distribution of FLS2, increasing its presence in the plasma membrane. Furthermore, CK enhances the cellular response to flg22, and flg22 sensing activates the CK response. Our results are in agreement with what we previously reported for fungi, suggesting a fundamental role for CK in regulating cellular integrity and trafficking as a mechanism for controlling and executing CK-mediated processes.

Keywords: cytokinin; endocytosis; cytoskeleton; actin; plant immunity; induced resistance

Citation: Pizarro, L.; Munoz, D.; Marash, I.; Gupta, R.; Anand, G.; Leibman-Markus, M.; Bar, M. Cytokinin Modulates Cellular Trafficking and the Cytoskeleton, Enhancing Defense Responses. *Cells* **2021**, *10*, 1634. https://doi.org/10.3390/cells10071634

Academic Editor: Suleyman Allakhverdiev

Received: 31 May 2021
Accepted: 23 June 2021
Published: 29 June 2021

1. Introduction

The plant hormone cytokinin (CK) is central to plant life, regulating many processes including embryogenesis, cell division and differentiation, stem cell maintenance, growth and branching of roots and shoots, leaf senescence, nutrient balance, and stress tolerance. CHASE domain-containing AHK (Arabidopsis histidine kinase) receptors, which are located primarily in the endoplasmic reticulum (ER), perceive CK and commence the resultant signaling cascades, which mediate CK function [1–3]. Based on the subcellular localization of AHKs, the endoplasmic reticulum (ER) has been considered to be the primary CK perception site [4,5]. Upon perception through AHKs, AHPs (Arabidopsis histidine phosphotransferases) transfer a phosphate from AHKs to downstream type B ARRs (Arabidopsis response regulators), initiating transcriptional reprogramming as a result of CK perception [6]. CK signaling can also start at the plasma membrane (PM) [7]. Extracellular CKs were shown to bind to cell-surface receptors and elicit signaling [8]. ER-localized CK receptors were also demonstrated to be transported to the PM [9]. In that work, AHK4 was shown to reside in the PM and in vesicles, in addition to the ER. Moreover, CK was shown to accumulate in BFA compartments, attesting to AHKs being trafficked between the ER and PM on endocytic vesicles. Thus, initiation sites of cytokinin perception occur at both the PM and the ER, and CK receptors undergo endosomal trafficking between these locales. It has been suggested that the dual CK response locales may contribute to the flexibility of the CK response [8].

CK regulates some developmental processes through crosstalk with the auxin pathway. CK was previously shown to regulate the transcription of auxin pathway genes [10–12]. Importantly, CK was also demonstrated to modulate auxin activity by regulating the endocytic trafficking of the auxin carrier PIN1 [13]. CK regulates the recycling of the auxin efflux carrier PIN1 to the plasma membrane, by directing it to the vacuole for degradation, in a rapid regulatory pathway that does not require transcriptional reprogramming [13]. Thus, through the endocytic redirection of PIN1, CK controls auxin fluxes that are required for lateral root organogenesis, and it was postulated that this endocytic regulation may also be involved in other CK-regulated developmental processes. However, CK was found not to influence the stability of two additional membranal auxin carrier proteins, prompting the authors to suggest that CK affects trafficking in a protein-specific manner [13].

CKs were shown to mediate disease resistance through induction of host immunity. Transgenic plants with high CK levels had increased resistance to *Pseudomonas syringae* [14], while transgenes with low levels of CK or Arabidopsis histidine kinase (AHK) receptor mutants displayed enhanced pathogen susceptibility [14,15]. CKs were also found to mediate enhanced resistance in tobacco [16]. In Arabidopsis, it was suggested that CK-mediated resistance depends on salicylic acid (SA) [14], and that CK signaling enhances SA-mediated immunity [17]. In tobacco, an SA-independent, phytoalexin-dependent mechanism was reported [16]. We recently reported that CK induces systemic immunity in tomato, promoting resistance to fungal and bacterial pathogens in an SA- and ethylene (ET)-dependent mechanism [18,19]. Until recently, other than the evidence that SA is required for CK-induced immunity [14,15,17], and that phytoalexins are involved [16], no additional cellular mechanisms related to CK-mediated immunity had been reported.

Pattern recognition receptors (PRRs) are the first line of defense and immune activation in plant cells. Since CK was shown to modulate endocytic trafficking of PIN1 in its regulation of auxin [13], in our recent work, we examined whether CK might also affect the trafficking of immune receptors as a possible mechanism for promoting immune responses and disease resistance. We were able to show that CK modulates the cellular trafficking of the PRR LeEIX2, which mediates immune responses to Xyn11 family xylanases [18]. We found that CK enhances both the endosomal presence and the vesicular size of LeEIX2 endosomes, without affecting the total cellular content of the proteins [18]. Furthermore, asking whether this CK-mediated enhancement of PRR endosomal presence acts as a mechanism for increased disease resistance, we used a SlPRA1A-overexpressing line, which has a decreased presence of receptor-like protein (RLP)-type PRRs in the cell plasma membrane, due to receptor degradation [20]. Indeed, we found that in plants overexpressing SlPRA1A, which have decreased levels of RLP-type PRRs, CK-mediated disease resistance to *Botrytis cinerea* is compromised [18].

Recently, we have found that CK directly affects the growth, development, and virulence of fungal plant pathogens [21]. *Botrytis cinerea* (*Bc*) growth, sporulation, and spore germination were all inhibited by CK, with some fungal developmental processes being inhibited by plant physiological CK concentrations [21]. We found similar effects in a variety of plant pathogenic fungi. We also found that CK affects both budding and fission yeast in a similar manner. Transcriptome profiling revealed that cell cycle and DNA replication, cytoskeleton integrity, and endocytosis are all inhibited in *Bc* by CK [21]. We directly confirmed these results, demonstrating that CK affects the cell cycle and DNA replication, cytoskeleton distribution, and cellular trafficking in fungi and yeasts [21]. CK interfered with actin localization in growing hyphae and lowered endocytic rates and endomembrane compartment sizes, likely underlying the reduced growth and development we observed [21].

In this work, we take an in-depth look at the effects of CK on cellular trafficking and on the cytoskeleton in plant cells, both in the context of plant immunity and in general. We have found that in addition to the xylanase receptor-like protein (RLP) LeEIX2, CK also affects the distribution of the flagellin receptor-like kinase (RLK) flagellin-sensing 2 (FLS2), increasing its presence in the plasma membrane. FLS2, first characterized in Arabidopsis, acts as the PRR for the bacterial PAMP (pathogen-associated molecular pattern) flagellin [22] in several plant species [23–25]. Furthermore, CK enhances the cellular response to the 22-amino-acid, flagellin-derived peptide flg22, and flg22 sensing activates the CK response. Examining cellular trafficking compartments and the cytoskeleton in general, we also show that CK affects endosome distribution and increases the amount of endomembrane compartments. CK also caused disorganization and reduction in actin filaments, but not in tubulin. Our results are in agreement with what we previously reported for fungi, suggesting a fundamental role for CK in regulating cellular integrity and trafficking as a mechanism for controlling and executing CK-mediated processes.

2. Results

2.1. CK Affects PRR Trafficking

We previously reported that CK modulates the trafficking of the PRR LeEIX2 and enhances its defense signaling, and that PRR-RLPs are required to achieve CK-mediated disease resistance [18]. To examine whether this is a general effect on PRRs, we tested whether CK influences the signaling of the LRR-receptor-like kinase (RLK)-type PRR FLS2. FLS2 is very different from LeEIX2 in its structure, specificity, and regulation [20,22,23]. Figure 1 demonstrates that CK affects cellular distribution of FLS2, increasing its presence at the PM.

2.2. CK Response Is Activated during Immunity Elicitation

We and others have previously demonstrated that pathogenesis processes can activate the CK pathway [17–19], raising the possibility that CK pathway activation might be required for pathogen resistance. Interestingly, we previously found that *B. cinerea* infection activates the synthetic CK response promoter TCSv2 [18]. To test whether defense elicitation activates the CK pathway, we examined whether TCSv2 could respond to flg22 elicitation. As seen in Figure 2, flg22 treatment in Arabidopsis roots significantly enhanced TCSv2-driven Venus expression, indicating that flg22 activates the CK response.

Our previous work has shown that in addition to affecting LeEIX2 trafficking, CK enhances defense signals elicited by its ligand, EIX [18]. Since we observed that CK also affected FLS2 cellular distribution, increasing its presence at the PM, we tested whether CK also enhanced defense signaling elicited by flg22. As seen in Figure 3, CK strongly enhances flg22 elicited reactive oxygen species (ROS) production. CK also has a small but significant inductive effect on ROS production on its own (Figure 3B). The CK trans-zeatin was previously reported to induce ROS production in Arabidopsis guard cells [26].

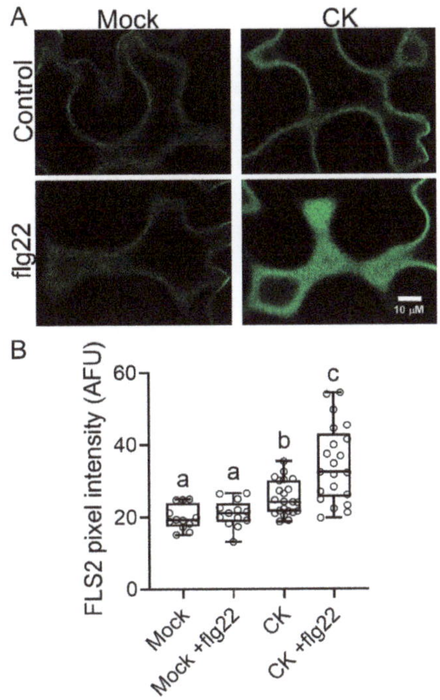

Figure 1. Cytokinin increases the membranal presence of the PRR FLS2. *N. benthamiana* epidermal cells transiently expressing FLS2-GFP were mock treated, or treated with cytokinin (CK, 100 μM 6-benzylaminopurine), for 4 h, and subsequently treated with flg22 (30 min). Membranal plane images (1 μm) were captured in 3 experiments, and images were analyzed using Fiji-ImageJ. (**A**) Representative confocal microscopy images. Contrast was uniformly adjusted; scale bar = 10 μm. (**B**) Twelve images were analyzed. Data are presented as boxplots with inner quartile ranges (boxes), outer quartile ranges (whiskers), and medians (lines in boxes), all points displayed. Letters represent statistical significance in one-way ANOVA with Dunnett's post-hoc test ($p < 0.035$).

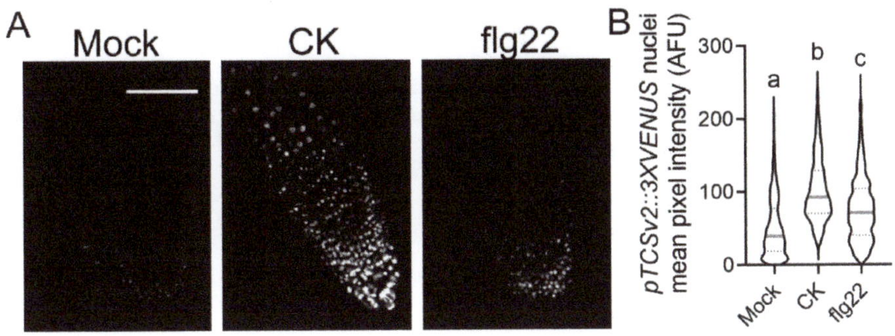

Figure 2. flg22 activates cytokinin response. Transgenic *A. thaliana* roots stably expressing the cytokinin (CK) response marker *pTCSv2::3XVENUS* were treated with mock (1 mM NaOH), CK (100 μM 6-benzylaminopurine), or flg22 (1 μM), for 12 h. Images were captured in 3 experiments using at least 15 plants per treatment, and analyzed using Fiji-ImageJ; scale bar = 100 μm. (**A**) Representative confocal microscopy images. Contrast was uniformly adjusted; scale bar = 100 μm. (**B**) Images from at least 15 plants were analyzed. Mean fluorescence intensity for each nucleus in each image was quantified, N > 770 nuclei. Data are presented as violin plots—solid gray lines indicate medians, and dotted lines indicate quartile ranges. Letters represent statistical significance in a Kruskal–Wallis test with Dunn's multiple comparisons post-hoc test ($p < 0.0001$).

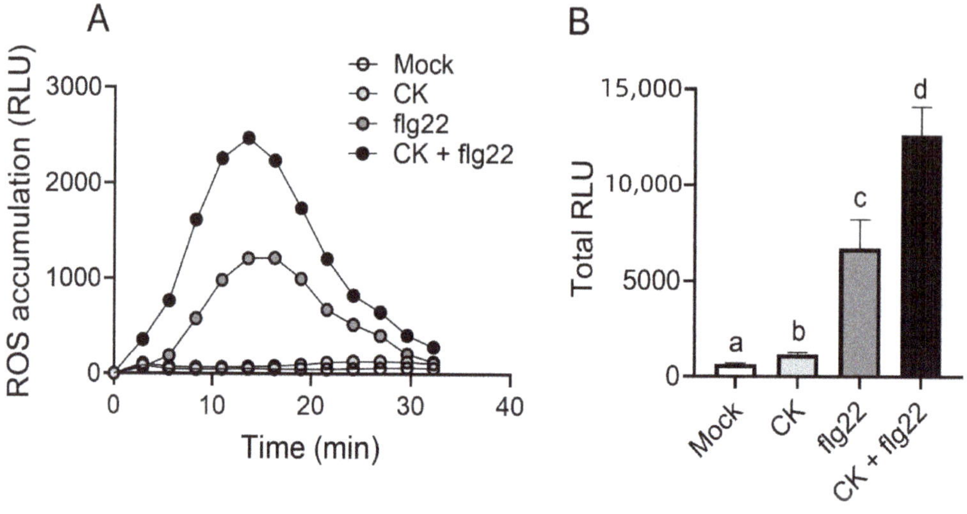

Figure 3. Cytokinin enhances flg22-mediated defense. *S. lycopersicum* cv. M82 leaves were treated for 12 h with mock or with cytokinin (CK, 100 μM 6-benzylaminopurine). After 12 h, leaf discs were harvested and treated with flg22 (1 μM). Reactive oxygen species (ROS) production was measured every 2.5 min for 35 min, using the HRP-luminol method. (**A**) Kinetics of ROS accumulation are plotted; RLU = relative light units; N = 24. (**B**) Total ROS production, area under the graph in A. Average ± SEM is shown, N = 24. Different letters represent statistically significant differences in one-way ANOVA, $p < 0.0001$, with a Dunnett's post-hoc test, $p < 0.044$.

2.3. CK Treatment Modulates Cellular Trafficking

As a developmental hormone, CK regulates various processes that require growth and rapid membrane modeling. These processes could be controlled through cellular trafficking. We have recently found that in fungi, CK attenuates cellular trafficking and causes mislocalization in the actin cytoskeleton [21]. Interestingly, investigation of the effect of CK on the cellular trafficking of the auxin carrier PIN1 showed that CK does not affect the cellular localization of two additional auxin carriers [13].

To test whether the effect of CK on cellular trafficking is general or specific to defense receptors, we examined the distribution of ARA6 and FM-4-64 endosomes following CK treatment. As seen in Figure 4, CK treatment significantly increases the density and size of both FM4-64 compartments (Figure 4A–C) and ARA6 endosomes (Figure 4A,D,E). ARA6 is considered to be a late endosome marker [27], while FM4-64 is present in all endomembrane compartments at the timepoint tested [28].

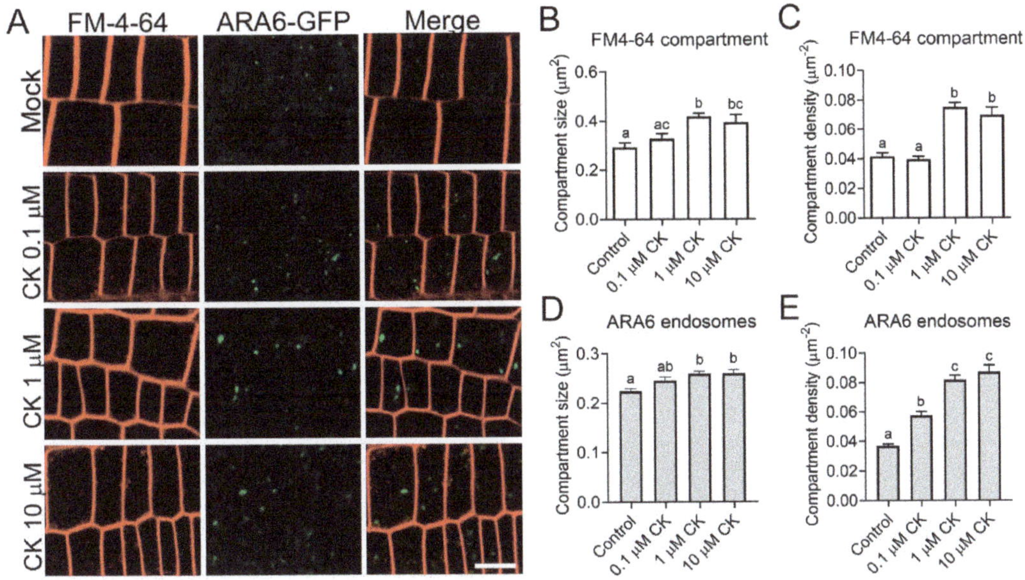

Figure 4. Cytokinin increases endosomal size and density. Transgenic *A. thaliana* root roots stably expressing the endosomal marker ARA6-GFP were grown vertically on 0.5X MS-agar plates, mock or supplemented with cytokinin (CK, 6-benzylaminopurine) at indicated concentrations, and stained with 5 µM FM-4-64 for 5 min on ice. Images were captured in 4 experiments using at least 10 plants per treatment, and analyzed using Fiji-ImageJ. (**A**) Representative confocal microscopy images. Contrast was uniformly adjusted; scale bar = 10 µm. (**B–D**) A total of at least 40 cells from at least 10 images per treatment were captured in 4 separate experiments and analyzed. Compartment size and density for FM4-64 (**B**,**C**) and ARA6-GFP (**D**,**E**) are presented as mean ± SEM. Letters represent statistically significant differences in a one-way ANOVA with Tukey's post-hoc test. (**B**) N > 425, $p < 0.0076$. (**C**) N > 40, $p < 0.0001$. (**D**) N > 500, $p < 0.0037$. (**E**) N > 60, $p < 0.0001$.

2.4. CK Treatment Affects the Cellular Cytoskeleton

Growth requires changes to both the composition and the orientation of the cytoskeleton [29]. Furthermore, observed effects on vesicular trafficking can relate to the cytoskeleton, since trafficking relies on an intact cytoskeleton [30]. We previously demonstrated that CK can affect F-actin distribution in fungi [21]. We therefore examined whether CK treatment could affect the plant cytoskeleton. Figure 5 shows that CK affects actin distribution, reducing the quantity and organization of actin in the cell (Figure 5A–C). CK did not affect tubulin (Figure 5D–F).

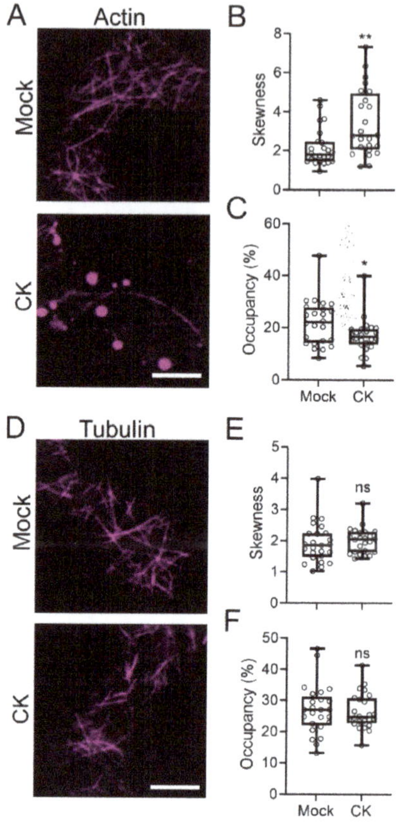

Figure 5. Cytokinin decreases actin organization and density. *Nicotiana benthamiana* epidermal cells transiently expressing Actin-mCherry (**A**–**C**) or mCherry-Tubulin (**D**–**F**) were mock treated, or treated with cytokinin (CK, 100 µM 6-benzylaminopurine) for 4 h. Membranal plane images (1 µm) were captured in 3 experiments, and images were analyzed using Fiji-ImageJ. (**A**,**D**) Representative confocal microscopy images. Contrast was uniformly adjusted; scale bar = 10 µm. (**B**,**C**,**E**,**F**) 25 images per treatment were analyzed. Box plots represent inner quartile ranges (boxes), outer quartile ranges (whiskers), and medians (lines in boxes), all points shown. Asterisks represent statistically significant differences in an unpaired two-tailed *t*-test with Welch's correction, N = 25. (**B**) ** p = 0.0031. (**C**) * $p < 0.028$. (**E**,**F**) ns = not significant.

3. Discussion

CK mediates plant immunity and disease resistance [14–19]. Traditionally a plant growth hormone with various developmental roles, CK supports juvenility and delays senescence [31,32], indicating that perhaps commonalities beyond the execution of the cell death program may exist between age-related cell death and pathogen-related cell death, both of which CK can delay or prevent.

In addition to its role in controlling auxin distribution by regulating its trafficking [13], we recently reported that CK can influence cellular trafficking in plants—where it enhances defense signaling of EIX through modulation of the trafficking of its PRR LeEIX2 [18]—and in fungi, where CK treatment lowers growth and development by attenuating trafficking [21]. Furthermore, we observed that CK-mediated disease resistance was lost in a transgenic line with reduced PRR expression [18]. These reports suggest that control of cellular trafficking could be a general mechanism by which CK exerts its varied effects on development and defense.

In this work, we affirmed the role we previously reported for CK in controlling receptor trafficking, demonstrating that CK enhances flagellin signaling (Figure 2), possibly by increasing the presence of its receptor—FLS2—in the membrane (Figure 1), from which it may transmit part of its defense signals [33]. Interestingly, though ligand-activated FLS2 is endocytosed after about 80–90 min [34], ROS responses occur within 5 min (Figure 3), suggesting that an increase in FLS2 presence at the membrane could underlie the increase in ROS observed upon CK treatment. Furthermore, we observed that flagellin activates CK response (Figure 2). Flagellin was previously reported to increase expression of the CK receptor AHK3 and the CK response regulator ARR2 in Arabidopsis guard cells [26]. We also previously observed a similar mode of action—activation of the CK pathway—in the pathogenesis of *B. cinerea* [18] and bacterial pathogens [19].

The fact that CK enhances signals emanating from both the fungal response RLP LeEIX2 [18] and the bacterial response RLK FLS2 (Figures 1 and 2), and augments defense elicited by both of their respective ligands, EIX [18]—which is derived from *Trichoderma*, which mostly signals through the jasmonic acid pathway [35], and flg22 (Figure 2), which mostly signals through the salicylic acid pathway [36]—indicates that the modulation of the cellular distribution of PRRs is likely a general mechanism by which CK affects cellular defense.

Our recent report [21] observed attenuation of cellular trafficking and disorganization of the actin cytoskeleton in response to CK treatment in *B. cinerea*, suggesting that perhaps control of cellular trafficking and cytoskeletal integrity could be a general mechanism employed to execute CK-mediated responses in plants as well. We examined endocytosis and actin distribution in response to CK treatment in Arabidopsis and *N. benthamiana*. We found that CK significantly increased the size and density of endomembrane compartments, as observed with both the general FM-4-64 dye and the late-endosome-specific ARA6 (Figure 4). Interestingly, the effects of CK on endomembrane compartments did not appear to be dose-dependent in most examined cases, with a cutoff of 1 μM observed, though this effect could be specific to 6-BAP in Arabidopsis, and testing of additional CK concentrations is required in order to definitively make this assertion.

Taken together, our results indicate that one of the early mechanisms for the activation of CK-mediated plant immunity is likely through cellular trafficking. CK may enhance defense signals, both by specifically increasing PRR occupancy in the cellular compartments required for signaling, and by increasing cellular trafficking compartments in general. flg22 is known to induce the entry of FLS2 to endosomes. flg22-induced FLS2-GFP endosomes were shown to co-localize to ARA6 compartments, though there was no difference in the amount of ARA6 endosomes at time points between 30 min and 105 min after flg22 application [37]. This could support the notion that the increased immunity observed upon flg22 and CK co-treatment (Figure 3) stems both from the increased presence of FLS2 in signaling compartments due to flg22 activation, and from a general increase in the available amount of endomembrane compartments, due to CK-mediated activity.

We found that CK inhibits the actin cytoskeleton, though not the tubulin cytoskeleton, in plant cells (Figure 5). We observed similar inhibition of the actin cytoskeleton in *B. cinerea*, underlying the mechanism for growth and development inhibition in fungi [21]. Actin inhibition likely also underlies CK-mediated growth inhibition observed in plants, as exemplified in the inhibition of root growth [38], and also perhaps in processes that maintain plant juvenility, e.g., inhibition of changes to cellular trafficking or degradation of the cytoskeleton required for senescence and cell death. The cytoskeleton is known to be an important component of cellular trafficking. The effect of CK on the actin cytoskeleton could result in altered trafficking rates and paths for endomembrane compartments—which could, in turn, affect the distribution of defense-related cargo and result in altered defense signaling.

Interestingly, our findings on the effects of CK on cellular trafficking differ somewhat between plants (Figure 4) and fungi [21]. While in both cases we observed a decrease in both the content and organization of actin filaments, in fungi—in contrast to plants—we observed a decrease in endomembrane compartments upon CK treatment. These differences could relate to differences in both the assay methodologies and the cell biology attributes of different species. Notably, the endocytosis pathway was significantly downregulated upon CK treatment in *B. cinerea* [21].

In summary, we found that CK regulates actin distribution, endocytosis, and PRR trafficking. A model summarizing this work is provided in Figure 6. This is an interesting starting point for future work, which will elucidate how the dual role of CK in growth and defense is regulated, and which additional proteins and feedback circuits are involved, perhaps also elucidating how some of these functionalities evolved in both plants and fungi.

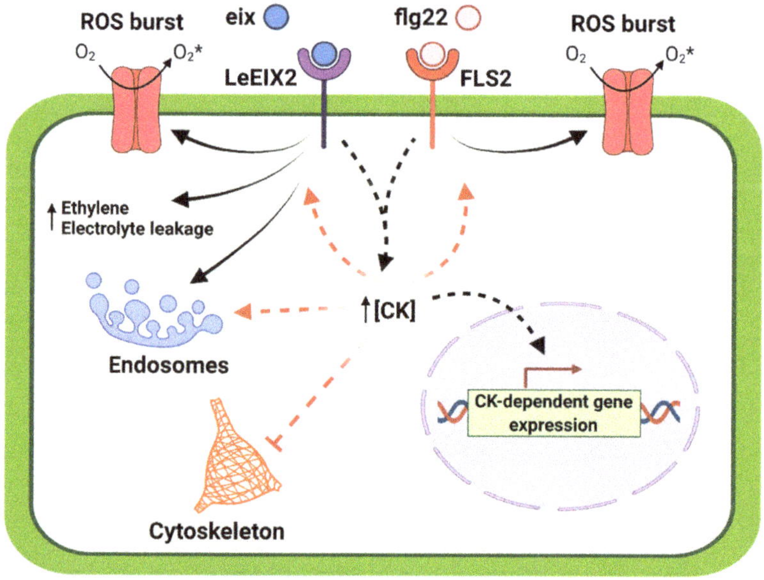

Figure 6. Model for the effects of cytokinin on plant cell compartments and defense signaling. Cytokinin (CK) increases endosomal compartment size and density, and decreases actin content and organization. CK promotes defense signaling mediated by both FLS2 and LeEIX2, resulting in increases in reactive oxygen species (ROS) in response to flg22 and EIX, as well as increases in ethylene and electrolyte leakage in response to EIX. CK also primes ROS on its own, and induces gene expression. LeEIX2 and FLS2 elicitor activation induces CK signaling pathways, increasing CK levels and/or activating CK-dependent gene expression. Model created with BioRender.com, accessed on 31 May 2021.

4. Materials and Methods

4.1. Plant Growth Conditions

Nicotiana benthamiana and *Solanum lycopersicum* cv. M82 were grown from seeds in soil (Green Mix; Even Ari, Ashdod, Israel) in a growth chamber at 24 °C, under long-day conditions (16 h:8 h, light:dark).

A. thaliana cv. Columbia transgenic plants overexpressing *pTCS::3XVENUS* [39] or *pARA6::ARA6-Venus* [27] were germinated on 1/2 MS (Duchefa M0225) agar plates containing 50 ug/mL kanamycin, and grown upright in a growth chamber at 22 °C under short-day conditions (8 h:16 h, light:dark).

4.2. Transient Expression

Four-week-old *N. benthamiana* plant leaves were abaxially infiltrated with a needleless syringe, with an *Agrobacterium tumefaciens* (strain GV3101) harboring the following previously described constructs: *pAtFLS2::AtFLS2-3xmyc-GFP* [40] (Addgene plasmid#86157; http://n2t.net/addgene:86157, accessed on 31 May 2021; RRID:Addgene_86157), *pCMU-ACTLr* (AtUBQ10::F-actin LifeAct-mCherry; Addgene plasmid #61193; http://n2t.net/addgene:61193, accessed on 31 May 2021; RRID:Addgene_61193) [41], and *pCMU-MTUBr* (AtUBQ10::mCherry-MAP4-MBD; Addgene plasmid #61196; http://n2t.net/addgene, accessed on 31 May 2021; RRID:Addgene_61196) [41]. Treatments were applied 40 h after infiltration. Transgenic lines and constructs used in this study are detailed in Table 1.

Table 1. Transgenic lines and constructs used in this study.

Line/Construct	Source	Organism/Use	Fig#
pAtFLS2::AtFLS2−3xmyc-GFP [40]	Addgene plasmid#86157; http://n2t.net/addgene:86157, accessed on 31 May 2021; RRID:Addgene_86157	Transient expression in *N. benthamiana*	1
pTCS::3XVENUS [39]	David Weiss, Faculty of Agriculture, HUJI.	Stable transgenic *A. thaliana*, col. background	2
pARA6:ARA6-Venus [27]	Takashi Ueda.	Stable transgenic *A. thaliana*, col. background	4
pCMU-ACTLr [41]	Addgene plasmid #61193; http://n2t.net/addgene:61193, accessed on 31 May 2021;	Transient expression in *N. benthamiana*	5
pCMU-MTUBr [41]	Addgene plasmid #61196; http://n2t.net/addgene:, accessed on 31 May 2021; RRID:Addgene_61196	Transient expression in *N. benthamiana*	5

4.3. Chemical Treatments

The CK 6-benzylaminopurine (6-BAP, Benzyladenine, sigma) was dissolved in 10 mM NaOH. *pARA6:ARA6-Venus* Arabidopsis plants were grown in 0.5X MS-agar media and transferred 3 days after germination to 0.5X MS-agar media supplemented with 6-BAP at indicated concentrations. *pTCS::3XVENUS* Arabidopsis plants, 5 days after germination, were incubated for 16 h in 0.5X MS media supplemented with 6-BAP at indicated concentrations (Figures 2 and 4). In the transient expression assays, 6-BAP water solutions were infiltrated into *N. benthamiana* leaves, and images were captured 6 h after infiltration (Figures 1 and 5). flg22 (1 μM, PhytoTech Labs #P6622) was dissolved in DMSO and applied as described for CK.

FM-4-64 (Invitrogen # T13320) was dissolved in water and applied to Arabidopsis roots in solution at 5 μM. Arabidopsis roots were incubated with this solution for 5 min on ice prior to live-cell imaging.

4.4. Confocal Microscopy

Confocal microscope images were acquired using a Zeiss LSM780 confocal microscope with Objective C-Apochromat 40 ×/1.2 W Corr M27 (Figure 1, Figure 2, and Figure 5) or Objective C-Apochromat 63×/1.2 W Corr (Figure 4). GFP was excited using a laser of 488 nm (5% power), and emission was collected in the range of 493–535 nm. mCherry fluorescence was excited using a laser of 561 nm (3% power), and emission was collected in the range of 588–641 nm. FM4-64 was excited using a laser of 514 nm (3% power), and emission was collected in the range of 650–750 nm. Images of 8 bits were acquired using a pixel dwell time of 1.27 μs, pixel averaging of 4, and pinhole of 1 airy unit. Image analysis was performed using Fiji-ImageJ with the raw images [42]. Endosome density and size measurements were conducted with the 3D Object counter tool, pixel intensity was measured using the measurement analysis tool, cytoskeleton organization was measured using the skew tool, and occupancy was assessed using the 3D Object tool [42].

4.5. Reactive Oxygen Species Burst Measurement

ROS measurement was performed as previously described [43]. Leaf disks of 0.5 cm in diameter were harvested from leaves 4 to 6 of 5–6-week-old M82 plants, treated for 12 h with mock (1 mM NaOH) or CK (100 µM 6-benzylaminopurine). Disks were floated in a white 96-well plate (SPL Life Sciences, Korea) containing 250 µL distilled water for 4–6 h at room temperature. After incubation, water was removed, and an ROS measurement reaction containing 1 µM flg22 was added. Light emission was immediately measured using a luminometer (Spark, Teacan, SL).

4.6. Data Analysis

Differences between two groups were analyzed for statistical significance using an unpaired two-tailed *t*-test, with Welch's correction where samples were found to have unequal variances, or with Holm–Sidak correction where multiple *t*-tests were applied. Differences among three groups or more were analyzed for statistical significance with a one-way ANOVA, or with a Kruskal–Wallis test. For ANOVA analyses, regular ANOVA was used for groups with equal variances, and Welch's ANOVA for groups with unequal variances. When a significant result for a group was returned, differences between the means of different samples in the group were assessed using a post-hoc test. Tukey's test was employed for ANOVA conducted on samples with equal variances, Dunnett's test was employed for ANOVA conducted on samples with unequal variances, and Dunn's multiple comparisons post-hoc test was employed for Kruskal–Wallis tests. All statistical analyses were conducted using Prism9.

Author Contributions: Conceptualization, L.P., M.L.-M., and M.B.; methodology, L.P. and M.B.; validation, L.P., D.M., I.M., R.G., G.A., and M.L.-M.; formal analysis, L.P., D.M., I.M., M.L.-M., and M.B.; writing—original draft preparation, M.B.; writing—review and editing, L.P. and M.B. All authors have read and agreed to the published version of the manuscript.

Funding: This research was partially supported by grant PAI 77190027 to LP. The APC was funded by M.B.

Institutional Review Board Statement: Not applicable.

Informed Consent Statement: Not applicable.

Data Availability Statement: The authors declare that the data supporting the findings of this study are available within the paper. Raw data are available from the corresponding author upon reasonable request.

Acknowledgments: We thank Takashi Ueda for providing seeds of *A. thaliana* col. *pARA6:ARA6-Venus*, Silke Robatzek for *pAtFLS2:FLS2-3xmyc-GFP*, Maria Harrison for *pCMU-ACTLr* and *pCMU-MTUBr*, and Adi Avni for assistance with microscopy. We thank all members of the Pizarro and Bar groups for helpful discussions and support.

Conflicts of Interest: The authors declare no conflict of interest.

References

1. Kieber, J.J.; Schaller, G.E. Cytokinins. *Arab. Book* **2014**, *12*, e0168. [CrossRef]
2. D'Agostino, I.B.; Deruère, J.; Kieber, J.J. Characterization of the Response of the Arabidopsis Response Regulator Gene Family to Cytokinin. *Plant Physiol.* **2000**, *124*, 1706–1717. [CrossRef]
3. Hutchison, C.E.; Li, J.; Argueso, C.; Gonzalez, M.; Lee, E.; Lewis, M.W.; Maxwell, B.B.; Perdue, T.D.; Schaller, G.E.; Alonso, J.; et al. The Arabidopsis Histidine Phosphotransfer Proteins Are Redundant Positive Regulators of Cytokinin Signaling. *Plant Cell* **2006**, *18*, 3073–3087. [CrossRef]
4. Wulfetange, K.; Lomin, S.N.; Romanov, G.A.; Stolz, A.; Heyl, A.; Schmülling, T. The Cytokinin Receptors of Arabidopsis Are Located Mainly to the Endoplasmic Reticulum. *Plant Physiol.* **2011**, *156*, 1808–1818. [CrossRef]

5. Romanov, G.A.; Schmülling, T. Opening Doors for Cytokinin Trafficking at the ER Membrane. *Trends Plant Sci.* **2021**, *26*, 305–308. [CrossRef] [PubMed]
6. Müller, B.; Sheen, J. Arabidopsis Cytokinin Signaling Pathway. *Sci. STKE* **2007**, *2007*, cm5. [CrossRef] [PubMed]
7. Zürcher, E.; Müller, B. Cytokinin Synthesis, Signaling, and Function—Advances and New Insights. *Int. Rev. Cell Mol. Biol.* **2016**, *324*, 1–38. [CrossRef] [PubMed]
8. Antoniadi, I.; Novák, O.; Gelová, Z.; Johnson, A.; Plíhal, O.; Simerský, R.; Mik, V.; Vain, T.; Mateo-Bonmatí, E.; Karady, M.; et al. Cell-surface receptors enable perception of extracellular cytokinins. *Nat. Commun.* **2020**, *11*, 4284. [CrossRef] [PubMed]
9. Kubiasová, K.; Montesinos, J.C.; Šamajová, O.; Nisler, J.; Mik, V.; Semerádová, H.; Plíhalová, L.; Novák, O.; Marhavý, P.; Cavallari, N.; et al. Cytokinin fluoroprobe reveals multiple sites of cytokinin perception at plasma membrane and endoplasmic reticulum. *Nat. Commun.* **2020**, *11*, 4285. [CrossRef]
10. Müller, B.; Sheen, J. Cytokinin and auxin interaction in root stem-cell specification during early embryogenesis. *Nature* **2008**, *453*, 1094–1097. [CrossRef]
11. Ioio, R.D.; Nakamura, K.; Moubayidin, L.; Perilli, S.; Taniguchi, M.; Morita, M.T.; Aoyama, T.; Costantino, P.; Sabatini, S. A genetic framework for the control of cell division and differentiation in the root meristem. *Science* **2008**, *322*, 1380–1384. [CrossRef]
12. Zhao, Z.; Andersen, S.U.; Ljung, K.; Dolezal, K.; Miotk, A.; Schultheiss, S.J.; Lohmann, J.U. Hormonal control of the shoot stem-cell niche. *Nature* **2010**, *465*, 1089–1092. [CrossRef]
13. Marhavy, P.; Bielach, A.; Abas, L.; Abuzeineh, A.; Duclercq, J.; Tanaka, H.; Pařezová, M.; Petrášek, J.; Friml, J.; Kleine-Vehn, J.; et al. Cytokinin Modulates Endocytic Trafficking of PIN1 Auxin Efflux Carrier to Control Plant Organogenesis. *Dev. Cell* **2011**, *21*, 796–804. [CrossRef] [PubMed]
14. Choi, J.; Huh, S.U.; Kojima, M.; Sakakibara, H.; Paek, K.-H.H.; Hwang, I. The cytokinin-activated transcription factor ARR2 promotes plant immunity via TGA3/NPR1-dependent salicylic acid signaling in arabidopsis. *Dev. Cell* **2010**, *19*, 284–295. [CrossRef]
15. Argueso, C.T.; Ferreira, F.J.; Epple, P.; To, J.P.C.; Hutchison, C.E.; Schaller, G.E.; Dangl, J.L.; Kieber, J.J. Two-component elements mediate interactions between cytokinin and salicylic acid in plant immunity. *PLoS Genet.* **2012**, *8*, e1002448. [CrossRef]
16. Grosskinsky, D.K.; Naseem, M.; Abdelmohsen, U.R.; Plickert, N.; Engelke, T.; Griebel, T.; Zeier, J.; Novák, O.; Strnad, M.; Pfeifhofer, H.; et al. Cytokinins Mediate Resistance against Pseudomonas syringae in Tobacco through Increased Antimicrobial Phytoalexin Synthesis Independent of Salicylic Acid Signaling. *Plant Physiol.* **2011**, *157*, 815–830. [CrossRef]
17. Naseem, M.; Philippi, N.; Hussain, A.; Wangorsch, G.; Ahmed, N.; Dandekar, T. Integrated Systems View on Networking by Hormones in Arabidopsis Immunity Reveals Multiple Crosstalk for Cytokinin. *Plant Cell* **2012**, *24*, 1793–1814. [CrossRef] [PubMed]
18. Gupta, R.; Pizarro, L.; Leibman-Markus, M.; Marash, I.; Bar, M. Cytokinin response induces immunity and fungal pathogen resistance, and modulates trafficking of the PRR LeEIX2 in tomato. *Mol. Plant Pathol.* **2020**, *21*, 1287–1306. [CrossRef]
19. Gupta, R.; Leibman-Markus, M.; Pizarro, L.; Bar, M. Cytokinin induces bacterial pathogen resistance in tomato. *Plant Pathol.* **2021**, *70*, 318–325. [CrossRef]
20. Pizarro, L.; Leibman-Markus, M.; Schuster, S.; Bar, M.; Avni, A. SlPRA1A/RAB attenuate EIX immune responses via degradation of LeEIX2 pattern recognition receptor. *Plant Signal. Behav.* **2018**, *13*, e1467689. [CrossRef]
21. Gupta, R.; Anand, G.; Pizarro, L.; Laor, D.; Kovetz, N.; Sela, N.; Yehuda, T.; Gazit, E.; Bar, M. Cytokinin inhibits fungal development and virulence by targeting the cytoskeleton and cellular trafficking. *bioRxiv* **2020**. [CrossRef]
22. Chinchilla, D.; Zipfel, C.; Robatzek, S.; Kemmerling, B.; Nürnberger, T.; Jones, J.D.G.; Felix, G.; Boller, T. A flagellin-induced complex of the receptor FLS2 and BAK1 initiates plant defence. *Nature* **2007**, *448*, 497–500. [CrossRef]
23. Roberts, R.; Liu, A.E.; Wan, L.; Geiger, A.M.; Hind, S.R.; Rosli, H.G.; Martin, G.B. Molecular characterization of differences between the tomato immune receptors flagellin sensing 3 and flagellin sensing 2. *Plant Physiol.* **2020**, *183*, 1825–1837. [CrossRef]
24. Wahlig, T.A.; Bixler, B.J.; Valdés-López, O.; Mysore, K.S.; Wen, J.; Ané, J.M.; Kaspar, C.W. Salmonella enterica serovar Typhimurium ATCC 14028S is tolerant to plant defenses triggered by the flagellin receptor FLS2. *FEMS Microbiol. Lett.* **2019**, *366*, fny296. [CrossRef]
25. Shi, Q.; Febres, V.J.; Jones, J.B.; Moore, G.A. A survey of FLS2 genes from multiple citrus species identifies candidates for enhancing disease resistance to *Xanthomonas citri* ssp. *citri*. *Hortic. Res.* **2016**, *3*, 16022. [CrossRef]
26. Arnaud, D.; Lee, S.; Takebayashi, Y.; Choi, D.; Choi, J.; Sakakibara, H.; Hwang, I. Cytokinin-mediated regulation of reactive oxygen species homeostasis modulates stomatal immunity in *Arabidopsis*. *Plant Cell* **2017**, *29*, 543–559. [CrossRef]
27. Ebine, K.; Fujimoto, M.; Okatani, Y.; Nishiyama, T.; Goh, T.; Ito, E.; Dainobu, T.; Nishitani, A.; Uemura, T.; Sato, M.H.; et al. A membrane trafficking pathway regulated by the plant-specific RAB GTPase ARA6. *Nat. Cell Biol.* **2011**, *13*, 853–859. [CrossRef]
28. Rigal, A.; Doyle, S.M.; Robert, S. Live Cell Imaging of FM4-64, a Tool for Tracing the Endocytic Pathways in Arabidopsis Root Cells. In *Plant Cell Expansion*; Estevez, J.M., Ed.; Methods in Molecular Biology; Springer Nature: Basingstoke, UK, 2015; Volume 1242, pp. 93–103.
29. Ketelaar, T.; Emons, A.M.C. The cytoskeleton in plant cell growth: Lessons from root hairs. *New Phytol.* **2001**, *152*, 409–418. [CrossRef] [PubMed]
30. Šamaj, J.; Baluška, F.; Voigt, B.; Schlicht, M.; Volkmann, D.; Menzel, D. Endocytosis, Actin Cytoskeleton, and Signaling. *Plant Physiol.* **2004**, *135*, 1150–1161. [CrossRef] [PubMed]

31. Kurakawa, T.; Ueda, N.; Maekawa, M.; Kobayashi, K.; Kojima, M.; Nagato, Y.; Sakakibara, H.; Kyozuka, J. Direct control of shoot meristem activity by a cytokinin-activating enzyme. *Nature* **2007**, *445*, 652–655. [CrossRef] [PubMed]
32. Gordon, S.P.; Chickarmane, V.S.; Ohno, C.; Meyerowitz, E.M. Multiple feedback loops through cytokinin signaling control stem cell number within the Arabidopsis shoot meristem. *Proc. Natl. Acad. Sci. USA* **2009**, *106*, 16529–16534. [CrossRef]
33. Smith, J.M.; Salamango, D.J.; Leslie, M.E.; Collins, C.A.; Heese, A. Sensitivity to Flg22 Is Modulated by Ligand-Induced Degradation and de Novo Synthesis of the Endogenous Flagellin-Receptor FLAGELLIN-SENSING2. *Plant Physiol.* **2014**, *164*, 440–454. [CrossRef] [PubMed]
34. Loiseau, J.; Robatzek, S. Detection and Analyses of Endocytosis of Plant Receptor Kinases. In *Plant Receptor Kinases*; Methods in Molecular Biology; Humana Press: New York, NY, USA, 2017; Volume 1621, pp. 177–189.
35. Shoresh, M.; Yedidia, I.; Chet, I. Involvement of Jasmonic Acid/Ethylene Signaling Pathway in the Systemic Resistance Induced in Cucumber by Trichoderma asperellum T203. *Phytopathology* **2005**, *95*, 76–84. [CrossRef]
36. Tateda, C.; Zhang, Z.; Shrestha, J.; Jelenska, J.; Chinchilla, D.; Greenberg, J.T. Salicylic Acid Regulates Arabidopsis Microbial Pattern Receptor Kinase Levels and Signaling. *Plant Cell* **2014**, *26*, 4171–4187. [CrossRef]
37. Beck, M.; Zhou, J.; Faulkner, C.; MacLean, D.; Robatzek, S. Spatio-Temporal Cellular Dynamics of the Arabidopsis Flagellin Receptor Reveal Activation Status-Dependent Endosomal Sorting. *Plant Cell* **2012**, *24*, 4205–4219. [CrossRef] [PubMed]
38. Werner, T.; Motyka, V.; Strnad, M.; Schmülling, T. Regulation of plant growth by cytokinin. *Proc. Natl. Acad. Sci. USA* **2001**, *98*, 10487–10492. [CrossRef] [PubMed]
39. Steiner, E.; Israeli, A.; Gupta, R.; Shwartz, I.; Nir, I.; Leibman-Markus, M.; Tal, L.; Farber, M.; Amsalem, Z.; Ori, N.; et al. Characterization of the cytokinin sensor *TCSv2* in *arabidopsis* and tomato. *Plant Methods* **2020**, *16*, 152. [CrossRef]
40. Robatzek, S.; Chinchilla, D.; Boller, T. Ligand-induced endocytosis of the pattern recognition receptor FLS2 in Arabidopsis. *Genes Dev.* **2006**, *20*, 537–542. [CrossRef]
41. Ivanov, S.; Harrison, M.J. A set of fluorescent protein-based markers expressed from constitutive and arbuscular mycorrhiza-inducible promoters to label organelles, membranes and cytoskeletal elements in *Medicago truncatula*. *Plant J.* **2014**, *80*, 1151–1163. [CrossRef]
42. Schindelin, J.; Arganda-Carreras, I.; Frise, E.; Kaynig, V.; Longair, M.; Pietzsch, T.; Preibisch, S.; Rueden, C.; Saalfeld, S.; Schmid, B.; et al. Fiji: An open-source platform for biological-image analysis. *Nat. Methods* **2012**, *9*, 676–682. [CrossRef]
43. Leibman-Markus, M.; Schuster, S.; Avni, A. LeEIX2 Interactors' Analysis and EIX-Mediated Responses Measurement. In *Plant Pattern Recognition Receptors*; Shan, L., He, P., Eds.; Methods in Molecular Biology; Humana Press: New York, NY, USA, 2017; Volume 1578, pp. 167–172.

cells

MDPI

Article

Identification of Putative Virulence Genes by DNA Methylation Studies in the Cereal Pathogen *Fusarium graminearum*

Francesco Tini [1], Giovanni Beccari [1], Gianpiero Marconi [1,*], Andrea Porceddu [2], Micheal Sulyok [3], Donald M. Gardiner [4], Emidio Albertini [1] and Lorenzo Covarelli [1,†]

1 Department of Agricultural, Food and Environmental Sciences, University of Perugia, Borgo XX Giugno, 74, 06121 Perugia, Italy; francensco.tini@collaboratori.unipg.it (F.T.); giovanni.beccari@unipg.it (G.B.); emidio.albertini@unipg.it (E.A.); lorenzo.covarelli@unipg.it (L.C.)
2 Department of Agriculture, University of Sassari, Viale Italia, 39a, 07100 Sassari, Italy; aporceddu@uniss.it
3 Department of Agrobiotechnology (IFA-Tulln), University of Natural Resources and Applied Life Sciences, Vienna (BOKU), Konrad Lorenz Strasse, 20, A-3430 Tulln, Austria; michael.sulyok@boku.ac.at
4 Commonwealth Scientific and Industrial Research Organisation, Agriculture and Food, 306 Carmody Road, St Lucia, QLD 4067, Australia; donald.gardiner@csiro.au
* Correspondence: gianpiero.marconi@unipg.it
† Current co-address: Centre for Crop and Disease Management, School of Molecular and Life Sciences, Curtin University, Bentley, Perth, WA 6102, Australia.

Abstract: DNA methylation mediates organisms' adaptations to environmental changes in a wide range of species. We investigated if a such a strategy is also adopted by *Fusarium graminearum* in regulating virulence toward its natural hosts. A virulent strain of this fungus was consecutively sub-cultured for 50 times (once a week) on potato dextrose agar. To assess the effect of subculturing on virulence, wheat seedlings and heads (cv. A416) were inoculated with subcultures (SC) 1, 23, and 50. SC50 was also used to re-infect (three times) wheat heads (SC50×3) to restore virulence. *In vitro* conidia production, colonies growth and secondary metabolites production were also determined for SC1, SC23, SC50, and SC50×3. Seedling stem base and head assays revealed a virulence decline of all subcultures, whereas virulence was restored in SC50×3. The same trend was observed in conidia production. The DNA isolated from SC50 and SC50×3 was subject to a methylation content-sensitive enzyme and double-digest, restriction-site-associated DNA technique (ddRAD-MCSeEd). DNA methylation analysis indicated 1024 genes, whose methylation levels changed in response to the inoculation on a healthy host after subculturing. Several of these genes are already known to be involved in virulence by functional analysis. These results demonstrate that the physiological shifts following sub-culturing have an impact on genomic DNA methylation levels and suggest that the ddRAD-MCSeEd approach can be an important tool for detecting genes potentially related to fungal virulence.

Keywords: DNA methylation; *Fusarium graminearum; in vitro* subcultures; virulence reduction; ddRAD-MCSeEd; virulence genes

Citation: Tini, F.; Beccari, G.; Marconi, G.; Porceddu, A.; Sulyok, M.; Gardiner, D.M.; Albertini, E.; Covarelli, L. Identification of Putative Virulence Genes by DNA Methylation Studies in the Cereal Pathogen *Fusarium graminearum*. *Cells* **2021**, *10*, 1192. https://doi.org/10.3390/cells10051192

Academic Editor: N. Louise Glass

Received: 19 March 2021
Accepted: 10 May 2021
Published: 13 May 2021

1. Introduction

Fusarium Head Blight (FHB) is one of the most widespread and damaging diseases of cereal crops, such as bread and durum wheat and barley, capable of strongly impairing not only yield but also quality, by contaminating grains with mycotoxins. The disease is caused by several members of the *Fusarium* species complex [1]. *F. graminearum* is globally considered the most dangerous FHB pathogen due to its aggressiveness and diffusion worldwide [2–4]. The pathogen is able to biosynthesize mycotoxins belonging to the type B trichothecenes, such as deoxynivalenol (DON) and nivalenol (NIV) [5] as well as the DON acetylated forms: 3-acetyl-deoxynivalenol (3-ADON) and 15-acetyl-deoxynivalenol (15-ADON) [4,6].

All living organisms can adopt strategies to enable rapid adaptation to new environmental conditions, without changing the DNA sequence [7,8]. For example, to respond to the environmental changes or biotic and abiotic stresses, some organisms adjust physiological and development machinery by gene expression regulation. DNA methylation and demethylation of cytosine play key roles in such a strategy [9–12]. Cytosine methylation is conventionally classified in CG, CHG, and CHH sequence contexts, where H is adenine, cytosine, or thymine. DNA methylation involves the addition of a methyl group to cytosine to produce 5-methylcytosine (5mC). Furthermore, the addition of the same group to adenine has been recently explored (N6-methyladenine, 6 mA), [13,14]. Methylation changes on cytosine residues are important for transposon silencing, epigenetic regulation, and genome expression [15–17]. Generally, methylation is related to the silencing of genes and transposable elements, whereas demethylation is correlated to active transcription [18], even if the reverse has also been described [15]. DNA methylation is catalysed by a conserved set of proteins called DNA methyltransferases (MTases) [19]. DNA MTases in eukaryotes belong to five different groups based on their structure and functions [20–22]. DNA MTase homologs have been identified in many fungal pathogens, including *F. graminearum* [23]. In pioneering studies, the histone proteins DIM-2, DIM-5, and HP1 were defined to be essential for DNA methylation in *Neurospora* [24–26]. Homologues for these three proteins are present in all sequenced *Fusarium* species [27], demonstrating that the DNA methylation machinery is present in this genus [28].

Some pathogenic fungi may partially lose or attenuate their virulence in response to environmental changes [29] as well as in response to other external factors, such as prolonged subculturing on artificial rich media [30]. Virulence may be restored if the attenuated strains are re-inoculated onto healthy host tissues. In the present work, we analysed whether virulence changes due to subculturing in different media/hosts was associated with DNA methylation changes and if these changes affected genes known to be involved in virulence regulation toward different hosts.

The loss of aggressiveness, following subculturing in artificial rich media, was evaluated by (i) execution of *in planta* virulence assays, (ii) characterization of secondary metabolites biosynthesis, and (iii) determination of *in vitro* fungal development and conidiation. Furthermore, the colony that had undergone consecutive transfers for one year on an artificial, rich, nutrient medium was used to infect healthy bread wheat heads. The DNA extracted from the last *in vitro* subculture was compared to the DNA isolated from mycelia sampled on infected wheat heads. Several genes affected by methylation level changes have already been demonstrated to be involved in virulence toward the host.

2. Materials and Methods

2.1. Fungal Strain and Subculturing

F. graminearum strain FG8 (15-ADON producer) from the fungal collection of the Department of Agricultural, Food, and Environmental Sciences (University of Perugia, Perugia, Italy) was used for all experiments. FG8 was isolated from durum wheat grain, molecularly identified and characterized for the *in vitro* mycotoxigenic profile [31]. The experimental design is shown in Figure 1.

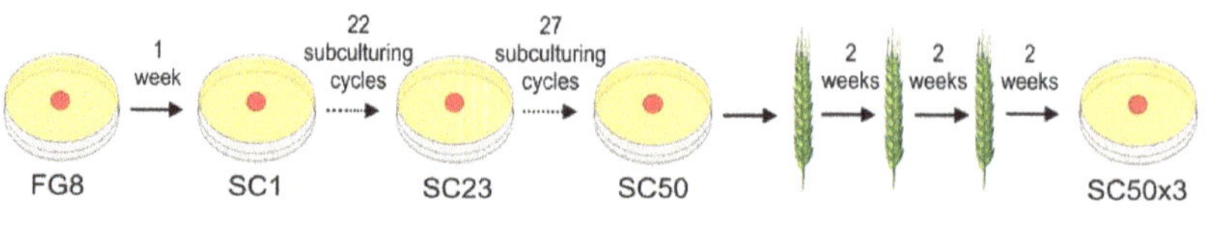

Figure 1. Experimental design followed throughout the experiment.

To prepare the subcultures' inoculum, FG8 was cultured on potato dextrose agar (PDA, Biolife Italiana, Milan, Italy) for 50 weeks. Briefly, a piece of fungal mycelium was cultured on PDA in a 9-cm Petri dish at 22 °C. After one week, one mycelium plug (0.5 cm of diameter) was used to inoculate a sterile PDA plate that was incubated for another week at 22 °C, whereas the rest of the mycelium was cut in small pieces and stored into 2-mL plastic tubes (Eppendorf, Hamburg, Germany) at −80 °C and represented the SC1 inoculum. The same actions were repeated every week for 50 weeks, obtaining a total of 50 subcultures stored at −80 °C (from SC1 to SC50). Sterile PDA plates were inoculated with mycelium plugs deriving from SC1, SC23, and SC50 samples for further analyses.

SC50×3 subcultures were obtained from sterile PDA plates inoculated with mycelium derived from three head-to-head passages, as described in Section 2.2.2.

2.2. Virulence Assays

2.2.1. Crown Rot Assay

The virulence assay on the stem base of bread wheat was carried out following the method previously described [32–34]. The mycelium of SC1, SC23, and SC50 was cut in small squares and homogenised with 12 mL of sterile water with a Mixer Mill MM400 (Retsch, Haan, Germany) to obtain a gel for pipetting. Bread wheat seeds (cv. A416, an Italian cultivar with well-known susceptibility to FHB) were previously surface sterilized with a solution composed of 7% sodium hypochlorite (8% *v/v*), 98% ethanol (10% *v/v*), and sterile, deionised water (82% *v/v*) for 5 min and rinsed three times with sterile, deionised water. Surface-sterilized seeds were sown in 6 × 8 × 8 cm pots (10 seeds per pot), and filled with a sterile soil mix (50% sand and 50% peat). Pots were incubated at 22 °C with a 15/9 h day/night light cycle. A 3-cm-long PVC collar (3-mm internal diameter) was placed around the emerging coleoptiles. When the second leaf was fully expanded, plants were inoculated by injecting 700 µL of inoculum into the space between seedling and the PVC collar. PDA macerated with sterile water was used as a control treatment. Three replicates (corresponding to three different pots, 10 plants per pot/replicate) for each FG8 subculture and for the control were realized for a total of 12 pots (120 plants). The inoculated seedlings were covered by plastic bags for 3 days to keep the moisture high. Seedlings were maintained for 25 days at 22 °C with a 15/9 h day/night light cycle.

Stem-base infections were evaluated by measuring the length (cm) of the necrotic area on the first leaf and the presence/severity of necrosis across leaf sheaths with a 0–17 arbitrary scale (clean = 0; coleoptile = 1–2; 1st leaf = 3–4–5; 2nd leaf = 6–7–8; 3rd leaf = 9–10–11; 4th = 12–13–14; 5th leaf = 15–16–17).

The fungal subcultures virulence toward the bread wheat stem base was evaluated using crown rot disease index (DI). DI was calculated as the product between the average length of the necrotic area on the first leaf (cm) and the average value (0–17) of necrosis across leaf sheaths of 10 plants for 3 replicates.

2.2.2. Fusarium Head Blight Assay

Flasks containing 300 mL of mung bean broth were inoculated with a SC50 mycelium plug. Mung bean broth was prepared by boiling 1 L of sterile water and adding 40 g of mung beans for 10 min. Subsequently, beans were removed from the broth by filtering with cheesecloth and the broth was autoclaved. The inoculated flasks were shaken on an orbital shaker at 150 rpm for 10 days at room temperature and 12/12 h light/dark. The fungal broth was filtered through Miracloth (Millipore Corporation Billerica, MA, USA) and the conidia suspension concentration was adjusted to 1×10^6 conidia mL^{-1} using a haemocytometer to count the cells. Sterilised seeds were incubated in Petri dishes for one day in the dark at 4 °C on water-soaked filter paper and three days in the dark at room temperature for germination. Germinated seeds were transplanted into 9 × 9 × 13 cm pots (one seed per pot) filled with peat and placed in a growth chamber at 23 °C with a photoperiod of 16 h. At mid-anthesis, wheat heads were point inoculated with macroconidia of SC50 by pipetting 10 µL of conidial suspension, containing approximately 10^4 conidia. The inoculum was injected between the glumes of a central spikelet. Heads were covered in plastic bags for 7 days to increase moisture content. Two weeks after inoculation, a little piece of mycelium was scraped from inoculated heads and used for inoculating other healthy wheat heads. This was repeated for three consecutive direct head-to-head transfers to obtain a sample named SC50×3. A portion of scraped mycelium of SC50×3 was cultured on PDA and stored at −80 °C for DNA extraction or used to inoculate mung bean flasks to obtain SC50×3 conidia inoculum, as described for the FHB assay. At mid-anthesis, heads were point inoculated as mentioned above, with 10 µL of conidial suspension containing approximately 10^4 conidia. A total of 15 heads per subculture were inoculated (5 heads per replicate) for a total of 75 heads, including control (sterile-water inoculation). After inoculation, heads were covered for 72 h in plastic bags to maintain a high humidity level and promote the infection. Inoculated plants were placed into a growth chamber with a photoperiod of 16 h at 23 °C. Symptoms caused by the different subcultures were assessed at 14 days post-inoculation (dpi), determining the proportion of spikelets of each head that displays browning symptoms.

2.3. *In Vitro Growth Rate Assay*

One mycelium plug of 0.5 cm diameter of SC1, SC23, and SC50 was taken from the edge of a 4-day-old colony and placed in the middle of the plates containing PDA. Six replicates per subculture were realized for a total of 18 plates. The growth rate was evaluated measuring the mycelial diameters in the two perpendicular directions, as previously described [35]. The diameters of the colonies were measured alongside the two axes. The growth value was calculated as the average of the measures taken from the two axes for three replicates per subculture and expressed in centimeters.

2.4. *In Vitro Conidial Production*

Three PDA Petri dishes per sample (three replicates) were inoculated with one mycelium plug of each subculture (diameter of 0.5 cm) and incubated for 4 weeks at room temperature, under near-UV light for 12 h per day. At the end of the incubation period, 15 mL of sterile water were added with a pipette to each incubated plate and the mycelium was scraped and mixed to the added water with a sterile spatula. The conidia were separated from the mycelium by Miracloth filtration. Conidia concentration was estimated with a haemocytometer and conidia production was calculated as the average of three replicates per sample.

2.5. *Determination of Secondary Metabolites Biosynthesized In Vitro by F. graminearum Subcultures*

2.5.1. *F. graminearum* Subcultures Preparation

Ten milliliters of deionized, sterile water were added to 20 g of rice kernels and placed into 100-mL glass flasks, autoclaved three times on alternate days, and inoculated with one mycelium plug per sample. Flasks were incubated for 4 weeks at 22 °C in the dark and the developed cultures were milled with mortar and pestle and stored at −80 °C. Three replicates per sample were realized. Three non-inoculated flasks with rice kernels were used as controls.

2.5.2. Extraction and Analysis of Secondary Metabolites

Five grams of each ground sample were extracted using 20 mL of extraction solvent (acetonitrile-water-acetic acid, 79:20:1, *v/v/v*) followed by a 1 + 1 dilution using acetonitrile-water-acetic acid (20:79:1, *v/v/v*) and direct injection of 5 µL of diluted extract. Concentrations exceeding the linear range of the detector were quantified by reanalysis of the extracts after further dilution steps (1:50 and 1:1000, respectively). LC-MS/MS screening of target fungal metabolites was performed with a QTrap 5500 LC-MS/MS System (Applied Biosystems, Foster City, CA, USA) equipped with a Turbo Ion Spray electrospray ionization (ESI) source and a 1290 Series HPLC System (Agilent, Waldbronn, Germany). Chromatographic separation was performed at 25 °C on a Gemini® C18-column, 150 × 4.6 mm i.d., 5-µm particle size, equipped with a C18 4 × 3 mm i.d. security guard cartridge (all from Phenomenex, Torrance, CA, USA). The chromatographic method as well as chromatographic and mass spectrometric parameters are described in Reference [36]. Confirmation of positive analyte identification was obtained by the acquisition of two MRMs per analyte (with the exception of MON and three nitro propionic acids that exhibit only one fragment ion), which yielded 4.0 identification points according to the commission decision (Commission Decision, 2002). In addition, the liquid chromatography retention time and the intensity ratio of the two MRM transitions agreed with the related values of an authentic standard within 0.03 min and 30% rel., respectively. Quantification was performed via external calibration using serial dilutions of a multi-analyte stock solution. Results were corrected for apparent recoveries obtained for wheat [36]. The accuracy of the method is verified on a continuous basis by regular participation in proficiency testing schemes.

2.6. DNA Extraction

To proceed with methylation analysis, DNA from SC50 and SC50×3 was extracted. In detail, SC50 was grown in Petri dishes containing PDA. Fungal mycelium was scraped using a spatula, and placed into 2-mL sterile plastic tubes with a steel bead at −80 °C. The SC50×3 mycelium developed on host tissues after three head-to-head consecutive transfers (Section 2.2.2) was collected with tweezers and stored in 2-mL plastic tubes with one steel bead at −80 °C. Mycelium samples (SC50 and SC50×3) were freeze-dried (Heto Power Dry LL3000) and reduced to a fine powder whit using a Mixer Mill MM400 (Retsch). Genomic DNA of the four samples was obtained using a PureLink™ Plant Total DNA Purification Kit (Thermo Fisher Scientific, Walthman, MA, USA) according to the manufacturer's instruction. Extracted DNA concentration was quantified with a Qubit® 3.0 Fluorometer (Thermo Fisher Scientific), using a dsDNA High Sensitivity (HS) Assay (Thermo Fisher Scientific) kit, following the manufacturer's protocol.

2.7. DNA Methylation Analysis

The library set-up protocol was performed according to Reference [37]. Three specific enzyme combinations were chosen to infer the CG (*Aci*I/MseI), CHG (*Sex*AI/*Mse*I), and CHH (*Eco*T22I/*Mse*I), methylation contexts, respectively. Briefly, for each library, 150 ng DNA were double-digested with one of these enzyme combinations following the protocol previously described [37,38]. The libraries were then pooled, purified using magnetic beads (Agencourt AMPure XP, Beckman Coulter, MA, USA), size selected by gel electrophoresis, and purified using QIAquick Gel Extraction kits (Qiagen, Hilden, Germany) for fragments ranging from 250 bp to 600 bp. Size-selected libraries were quantified using a Qubit® 3.0 Fluorometer (Thermo Fisher Scientific), and a normalized DNA amount (15 ng) was amplified with a primer that introduced an Illumina index (at the Y common adapter site) for demultiplexing. Following PCR with uniquely indexed primers, multiple samples were pooled and subjected to PCR-enrichment, as previously described [37]. The grouped libraries were pooled in an equimolar fashion, and the final library was Illumina-sequenced using 150-bp single-end chemistry. Raw reads from the Illumina sequencing of the CG, CHG, and CHH libraries were analyzed following the protocol and the pipeline previously described [37].

The relative methylation levels at each site were calculated following a described procedure [37] and the DMPs (Differentially Methylated Positions) were called following the methyl kit's manual best practices [39]. The mapping of the DMPs in the same scaffold and as closer than a given threshold provided their clustering together to identify the DMRs (Differentially Methylated Regions), as previously reported [37].

Synteny Block and Statistical Analysis

Synteny block analysis was performed with MCSCANX with default settings. *F. graminearum* proteins were used as a query against a database of *F. verticillioides* proteins for BLASTP homology searches. The BLASTP results were exported in a tabular format (m 8). The criteria for synteny block analysis were: match score 50, match size >5, gap_penalty of −1, and max gaps of 25. The chromosomes of *F. graminearum* were partitioned in an adjacent window of 20 kb and, for each of these regions, the proportion of mapping genes collinear to *F. verticillioides* was calculated. Chromosomal windows with a portion of collinear genes below 50% were considered to be non-conserved. The relative abundance of DMPs and DMR mapping in conserved and not conserved (NC) regions were compared with a permutation test.

The effect of subculturing on pathogen virulence, mycelium growth, conidia production, and metabolite biosynthesis were tested by one-way ANOVA and Duncan's multiple comparison tests, as implemented in the program DSAASTAT [40].

3. Results

3.1. Subculturing Reduces Fungal Virulence, but Passaging Can Rescue These Defects

To assess the effect of continuous subculturing on virulence, three subcultures of the FG8 strain (SC1, SC23, and SC50) were selected for virulence assays toward bread wheat with inoculations performed on stem bases.

A different aggressiveness between SC1, SC23, and SC50 was observed (Figure 2A). All three subcultures showed the ability to cause the typical necrotic lesions on the first leaves and across leaf sheaths of soft wheat plants. In detail, crown rot disease index (DI), calculated as the product between the average length of the necrotic area on the first leaf (cm) and the average value (0–17), was 53.8, 24.5, and 17.4 in plants inoculated with SC1, SC23, and SC50, respectively. The DI decrease was significant ($p < 0.05$) and followed the gradient SC1 > SC23 > SC50.

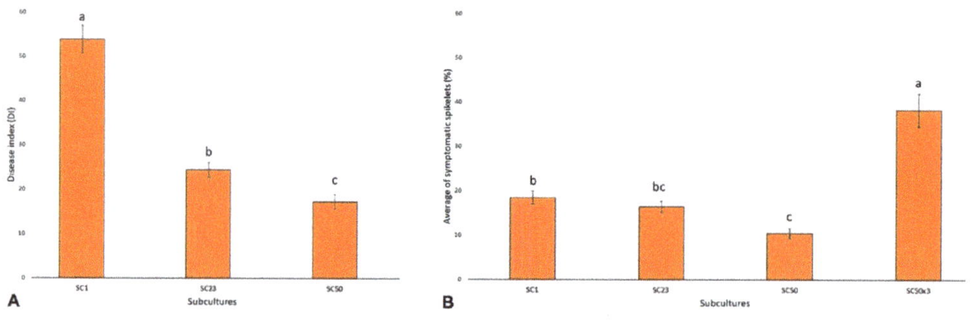

Figure 2. (**A**) Stem base Disease Index (DI) of soft wheat inoculated with subcultures SC1, SC23, and SC50 of *F. graminearum* strain FG8. Columns represent the average of three replicates (± SE) with each composed of 10 plants. Values with different letters are significantly different based on Duncan's multiple comparison tests ($p < 0.05$). (**B**) Average percentage of symptomatic spikelets (%) of heads point-inoculated with the FG8 different subcultures. Columns represent the average of three replicates (±SE), in which each is composed of five heads. Values with different letters are significantly different based on Duncan's multiple comparison tests ($p < 0.05$).

The aggressiveness of the same subcultures was also evaluated toward bread wheat heads. All three subcultures were able to induce the typical FHB bleached spikelets with aggressiveness decreasing by the subculturing time (Figure 2B). These subcultures were compared with SC50×3, which was obtained from mycelia derived from three head-to-head passages, as described in MM. The aggressive average levels described a trend with SC50×3 > SC1 ≥ SC23 ≥ SC50 (Figure 2B). In detail, the initial subculture (SC1) caused 18.5% of symptomatic spikelets whereas the virulence of SC50 was significantly ($p < 0.05$) reduced when compared to the first one, with only 10% of spikelets showing symptoms. SC23, with an average of 16% infected spikelets, showed an intermediate aggressiveness in comparison to SC1 and SC50. These results showed that continuous subculturing on PDA caused a pathogen virulence decrease as a consequence of its adaptation to a nutrient-rich medium while three transfers of SC50 on wheat heads fully restored virulence (38% of bleached spikelets) to a degree even higher than that observed for SC1 ($p < 0.05$).

3.2. Subculturing Affects Conidial Production but Not an In Vitro Growth Rate

Mycelium growth rates from subcultures SC1, SC23, and SC50 were measured on PDA plates. Despite the three subcultures having different adaptation times on PDA (1, 23, or 50 weeks), no significant differences in growth rate were observed ($p = 0.42$) after 5 days (Figure 3A). In detail, SC1 showed an average growth of 5.1 cm along the axes, whereas SC23 and SC50 showed an average growth of 5.4 cm.

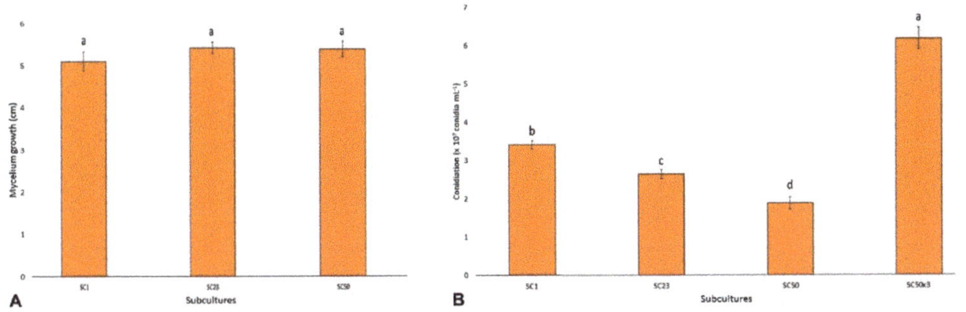

Figure 3. (**A**) Subcultures development (cm) along the diameters of Petri dishes containing PDA. Columns represent the average of three replicates (±SE). Values with the same letters are not significantly different ($p > 0.05$) based on Duncan's multiple comparison tests. (**B**) Average conidia production by SC1, SC23, SC50, and SC50×3 developed on PDA under near-UV light for 28 days. Columns represent the average of three replicates (±SE). Values with different letters are significantly different based on Duncan's multiple comparison tests ($p < 0.05$).

Conversely, significantly different conidia productions by the varying subcultures were detected (Figure 3B). In detail, conidiation followed the significant ($p < 0.05$) gradient: SC50×3 > SC1 > SC23 > SC50. In addition, after *in vivo* passages of the last subculture (SC50) for three times onto wheat head tissues, this strain produced a high number of conidia, which resulted in an even higher number ($p < 0.05$) than that observed in the first subculture (SC1).

3.3. Subculturing Does Not Affect In Vitro Secondary Metabolite Production

The accumulation of the main secondary metabolites detected in SC1, SC23, SC50, and SC50×3 subcultures are reported in Table S1.

In general, no significant differences in secondary metabolite biosynthesis between the four subcultures were detected. In detail, SC1 produced 19,900 μg kg^{-1} of 15-ADON in addition to 36,600 μg kg^{-1} of DON. A very low production of 3-ADON (700 μg kg^{-1}), NIV (114 μg kg^{-1}), and sambucinol (412 μg kg^{-1}) was also detected. In addition, SC1 biosynthesised 56,000 μg kg^{-1} of zearalenone (ZEN) and some ZEN derivatives, such as 1700 μg kg^{-1} of alpha-zearalenol and 12,000 μg kg^{-1} of beta-zearalenol. Other metabolites such as culmorin (3400 μg kg^{-1}), 15-hydroxyculmorin (4200 μg kg^{-1}), and fusarin C (54,000 μg kg^{-1}) completed the SC1 *in vitro* mycotoxigenic profile. All the reported values are the average of three replicates. As mentioned before, the subculturing process performed on PDA as well as the following passages on the host head tissues did not significantly alter the ability of the fungus to produce secondary metabolites. For example, the production of DON showed an increase (even if without significant differences, $p > 0.05$) going from SC1 to SC23 (60,000 μg kg^{-1}) and to SC50 (74,000 μg kg^{-1}). A partial, non-significant decrease was detected for SC50×3 (53,000 μg kg^{-1}).

The 15-ADON biosynthesis levels did not show significant changes between subcultures ($p = 0.91$). A very similar amount was detected in all subcultures (SC1 = 19,900 μg kg^{-1}, SC23 = 16,600 μg kg^{-1}, SC50 = 18,100 μg kg^{-1}, SC50×3 = 19,400 μg kg^{-1}).

Similarly, 3-ADON biosynthesis was nearly the same in all subcultures ($p = 0.37$). In detail, 1100 μg kg^{-1} of 3-ADON were produced by SC23 and 1400 μg kg^{-1} by SC50 and

SC50×3. NIV and sambucinol were detected with similar levels for SC1, SC23, SC50, and SC50×3 ($p = 0.81$ and $p = 0.70$, respectively). Culmorin levels showed a decreasing trend that was followed by a restoration after host re-infection, whereas 15-hydroxiculmorin increased from SC1 to SC50 and then decreased again on SC50×3, but both without significant differences ($p = 0.96$ for culmorin and $p = 0.57$ for 15-hydroxiculmorin).

Subculturing induced a very small increase of ZEN production while plant re-infection brought it back to the SC1 levels, even if no significant differences were detected ($p = 0.32$). In addition, alpha-zearelenol and beta-zearalenol did not show any significant differences ($p = 0.87$ and $p = 0.57$). Among other *Fusarium* metabolites, fusarin C showed a decrease in the total amount produced following the three passages but, again, without significant differences ($p = 0.34$).

Finally, no significant variations were detected for butenolid ($p = 0.59$), gibepyron D ($p = 0.22$), aurofusarin ($p = 0.60$), and rubrofusarin ($p = 0.24$) biosynthesis among the different subcultures.

3.4. DNA Methylation Analysis

3.4.1. Identification of Differentially Methylated Positions and Differentially Methylated Regions

MCSeEd (Methylation Context Sensitive Enzyme ddRAD) [37,38] was used to investigate DNA methylation changes induced by subculturing. To this end, next-generation sequencing (NGS) libraries from genomic DNA purified from PDA plates (SC50) and wheat heads (SC50×3) were constructed. Therefore, a total of 12 libraries were produced by double restriction ligations with each using *Mse*I in combination with one of the three methylation-sensitive enzymes *Aci*I, *Sex*AI, and *Eco*T22I, for the CG, CHG, and CHH contexts, respectively (Supplementary Table S2).

After quality control, a mean of 822,679 thousand 150-bp-long reads from each library were obtained and aligned to the *F. graminearum* PH-1 reference genome [41]. Only reads mapped at unique genomic positions were retained. Thus, considering the three different contexts, a total of 2,899,673, 4,755,848, and 1,516,029 reads were mapped uniquely on the reference genome (92.6% of the total reads, with a minimum of 86.8% for *Aci*I, and a maximum of 96.6% for *Sex*AI) and were classified as MCSeEd loci (Supplementary Table S3).

Therefore, a total of 138,119 loci containing cytosines (120,439 in symmetric, and 17,680 in asymmetric contexts) (Supplementary Table S4) were identified.

The mapping location of each MCSeEd locus was investigated to determine whether it fell within a gene window that included the region within 0.5 kb upstream of the transcription start site (TSS), the transcribed region (i.e., the gene body), and the region within 0.5 kb downstream of the transcription termination site (TTS). Furthermore, 92% (*Aci*I), 91% (*Sex*AI), and 87% (*Eco*T22I) of the identified MCSeEd loci were included within these gene windows (Supplementary Figure S1).

After normalization of the MCSeEd loci, the sites covered by a total number of reads <4 or showing excessive read-count variation among the replicates (standard deviation of 5 for CG and 10 for GHG and CHH) were discarded. The remaining sites were used to estimate a total of 13,899 DMPs, out of the 138,119 MCSeEd loci, with significantly altered methylation levels between the SC50×3 and SC50 samples (false discovery rate, $\leq$0.05). Of these, 12,326 belonged to symmetric contexts, and 1573 belonged to asymmetric contexts (Supplementary Table S4).

Principal component analysis was used to graphically portray the samples based on the DMPs' methylation levels (Supplementary Figure S2).

The first latent component (PC1) accounted for 71.6%, 78.4%, and 71.8% of the total variance for the CG, CHG, and CHH, contexts, respectively, and clearly discriminated between SC50×3 and SC50, indicating that the head-to-head transfer of mycelium induced genome-wide methylation changes. Accordingly, complete linkage clustering of the methylation levels at DMPs clearly separated the SC50×3 and SC50 (Supplementary Figure S2).

Considering all of the methylation changes being induced by host colonization in the replicates, we observed 1.4-fold (CG) to 1.15-fold (CHH) more methylation decreases than increases in response to healthy heads' infection whereas, for CHG, the proportion of methylation changes was 1.12 in the other direction.

Genomic regions with co-regulated methylation changes upon subculturing, known as DMRs, were identified. In total, 932 DMRs were scored for CG (874), CHG (19), and CHH (39) contexts (Supplementary Table S5).

The estimated relative methylation level of the DMPs belonging to each DMR were hierarchically clustered and, as expected, clustered according to the treatment, as SC50×3 or SC50 (Figure 4 and Supplementary Figure S3). In particular, for all three contexts, the number of DMRs with higher methylation levels in the SC50×3 samples (relative to the SC50 samples) was lower than the number of DMRs that showed a lower level in the SC50×3 samples (Figure 4 and Supplementary Figure S3).

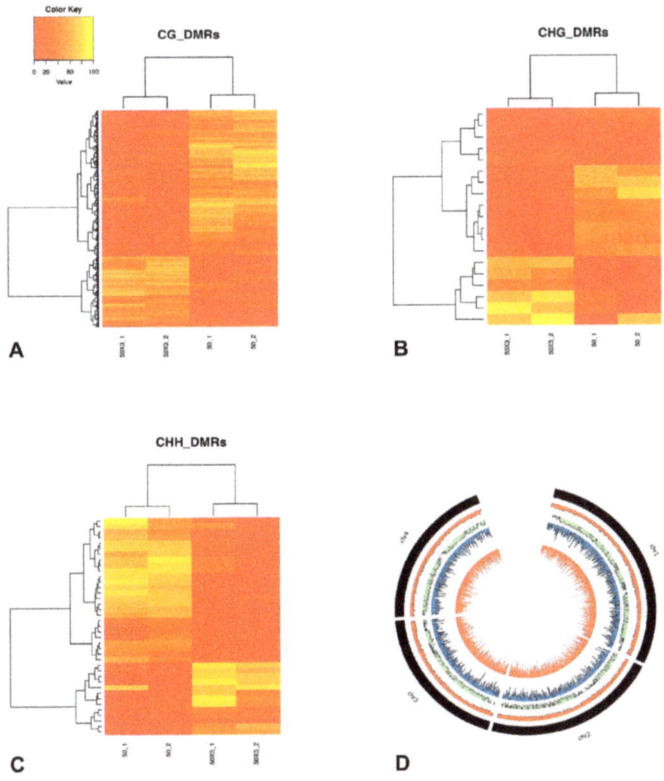

Figure 4. Relative methylation frequencies of differentially methylated regions as identified from the comparison between SC50 and SC50×3 subcultures. Relative methylation frequencies of the differentially methylated positions contained in each differentially methylated region (**A–C**) for CG, CHG, and CHH, respectively) were averaged and used in complete linkage clustering analysis of samples derived from SC50 and SC50×3 based on differentially methylated regions. (**D**) Circos plot. From outer inward: number of genes in adjacent genomic chromosome regions of 20 kb. Ratio of the number of *F. graminearum-F. verticillioides* collinear genes and total number of *F. graminearum* genes in each region. DMPs (CG context) and DMRs (CG context).

Next, we analysed the distribution of DMPs and DMRs along *F. graminearum* chromosomes. Several studies proposed that *F. graminearum* genome can be partitioned in a core portion enriched for housekeeping genes and a dispensable portion with a high frequency of pathogenesis and virulence-related genes [42]. The dispensable genomic portions can be identified as regions of low gene collinearity (hereafter, referred to as not conserved (NC) regions) between *F. graminearum* and *F. verticillioides* or *F. oxysporum*. The distributions of both DMPs and DMRs along chromosomes were visualized as the total number of these features for each adjacent genomic window of 20 kb. The black histograms of Figure 3 indicate the location of NC regions in four *F. graminearum* chromosomes. For all the analysed contexts, no preferential accumulation of either DMPs or DMRs between NC regions and other chromosome regions were observed ($p > 0.05$ of a non-parametric permutation test).

3.4.2. Differentially Methylated Genes

DMP and DMR distributions were analysed in relation to the coding and regulatory genomic sequences. In particular, we compared the distribution of DMPs and DMRs in transcribed genic regions extended by 0.5 kb at both ends (extended gene bodies, EGBs) (Supplementary Figure S1) and found that DMRs mapped preferentially to EGBs.

In addition, we plotted the distribution of significant DMPs along the EGBs for CG (Figure 5), CHG, and CHH (Supplementary Figure S4) contexts. The main differences in CG relative methylation levels between SC50 and SC50×3 were observed in the regulative regions and in proximity of TSS and TTS sites. In particular, a decreased, relative, methylation level of the samples grown on plants as compared to artificial substrate was found. Conversely, along the gene body, no changes were highlighted. In the other two contexts, the low number of significant DMPs (Supplementary Table S4) were not able to properly reveal differences along EGB's genomic region (Supplementary Figure S4).

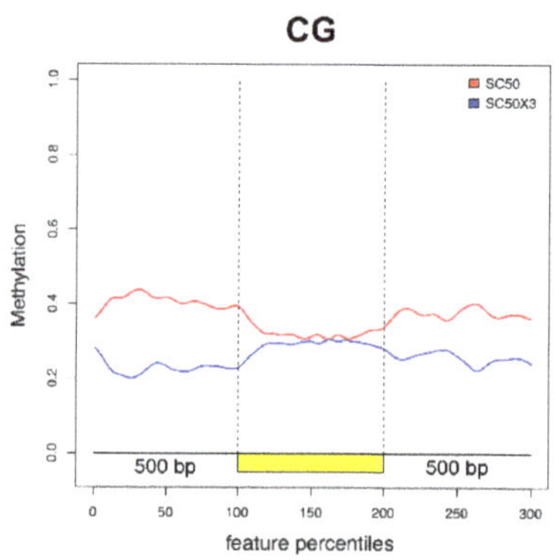

Figure 5. Plotted DMPs along the EGBs (coding region, in yellow, with the regions 500 bp upstream and downstream) for the CG context.

In particular 930, 13, and 40 EGBs were overlapped at least once by 932 DMRs in the 0.5-kb windows upstream of TSS, within the gene body, or in the 0.5-kb windows downstream of TTS, respectively. The genes belonging to these EGBs were defined as differentially methylated genes (DMGs, Supplementary Table S6).

Analysis of Gene Ontologies' enrichment demonstrated that DMGs are enriched for GO terms in relation to transcriptional regulation (GO_0006355) and chitin metabolism (GO:0006032). More DMGs than expected by chance were involved in zinc ion (GO:0008270) and DNA (0003677) binding. Next, we assigned DMGs to gene families based on the presence of PFAM domains within the encoded proteins. The PFAM domains are significantly more abundant than expected by chance in the DMGs' dataset are reported in Table S7 (Fisher exact test $p < 0.05$). Several of these domains have been identified in genes known to be linked directly or indirectly to virulence.

The top five PFAM domains enriched in the DMGs were associated with isoprenes and carbohydrate metabolism (PF08544.13, PF00180.20), endonuclease and exonuclease activity (PF003372), vacuolar 14 fab1-binding (PF12755.7), and ureo-hydrolase (PF00491.21).

4. Discussion

Fungal pathogens are the predominant causal agents of plant diseases, causing yield and quality losses [43–45]. To successfully infect plants, fungal pathogens use different strategies to exert their virulence during the infection process, such as when adjusting the activity of various molecules, which may be effectors or extracellular factors [46,47], by regulating their transcription levels [20]. Some virulence factors are upregulated to facilitate host colonization and infection, whereas others are downregulated to mitigate host responses [48,49]. Furthermore, the genome plasticity of fungal plant pathogens allows the adaptation of the metabolism and of the reproductive strategies to variable environmental conditions, such as light cycle, temperature, substrate type, and the presence/absence of hosts [50,51]. DNA methylation is a basic modification of genomic DNA in eukaryotes with significant effects on gene expression, genomic imprinting, and transposon silencing involving gene promoter regions, transposable elements, repeat sequences, and transcribed regions of genes [52–56]. In filamentous fungi, transcriptome and methylome studies have shown that DNA methylation is linked to gene expression and to the silencing of transposable elements [52,57]. For these reasons, the present study was based on the DNA methylation approach to reveal hypothetical genes involved in the virulence of *F. graminearum*, which is the most important FHB pathogen of wheat worldwide. In fact, DNA methylation studies can be considered useful tools to identify novel genes associated with the aggressiveness of fungal pathogens subject to environmental modifications [58] and/or stresses, such as adapting to an artificial substrate or to host tissues. Changes of DNA methylation patterns in response to environmental stresses were observed in plants, during their growth and development [53,59,60].

In the present work, a high virulent strain of *F. graminearum* (FG8) was preliminarily stressed when performing subculturing (50 times for 50 weeks) on an artificial substrate (PDA) to verify if the adaptation to a nutrient-rich medium induced an attenuation of virulence. To assess the effect of the *in vitro* subculturing process on fungal virulence, the aggressiveness of three selected subcultures (SC1–SC23–SC50) was examined. Stem base crown rot virulence assays showed a progressive, aggressive decline related to subculturing time toward this bread wheat tissue. A similar result was also observed on bread wheat heads. Furthermore, the mycelium from the subculture developed for 50 weeks on PDA (SC50) was used to inoculate for three consecutive head-to-head passages (SC50×3) bread wheat heads with the objective of restoring the native virulence of the FG8 strain. In fact, SC50×3 exhibited a strong aggressiveness on heads, that is even higher than SC1, likely due to the repeated inoculation of healthy heads in a short time window (6 weeks).

These results are in line with previous studies that showed a virulence decline of different entomopathogenic fungal species or isolates (e.g., *Metarhizium anisopliae, Beauveria bassiana, Beauveria densa, Nomuraea rileyi, Paecilomyces farinosus, Verticillium lecanii*), caused by artificial *in vitro* subculturing on nutrient-reach media and long-term routine maintenance [61–64], even if this decline is not always reported, such as in some strains of *Paecilomyces fumosoroseus, P. farinosus*, and *B. bassiana* [65–67]. As observed in this study, other researchers reported that virulence can be restored when a pathogen passes from an artificial media to a suitable host [68–71]. In addition to the subculturing process, the simple *in vitro* growth on different artificial media can influence conidial germination, growth, and virulence of fungal pathogens [72–74]. The *in vitro* growth on artificial media could also induce genetic modifications, as already demonstrated for a *F. verticillioides* strain subject to a subculturing process, in which the accumulation of about 14 genetic variants was observed [75]. The present work also shows that consecutive *in vitro* subcultures of *F. graminearum* caused a considerable decline of conidiation passing from SC1 to SC50, confirming previous studies on *B. bassiana, M. anisopliae*, and *Metarhizium brunneum* strains [76,77]. With the passage on a healthy host, a significant increase of conidia production from SC50 to SC50×3 was observed. Even if some phenotypic changes, such as the reduction in the growth rate, may be typically associated with *in vitro* degeneration [69]. No differences were observed comparing the measures of the diameters of the subcultures, as also found in *B. bassiana* [76].

Finally, a very large spectrum of *F. graminearum* secondary metabolites biosynthesis following prolonged subculturing was investigated. Even if the lower biosynthesis of some secondary metabolites is known to be linked to subculturing processes in different fungal genera, such as *Periconia* sp., *Fusarium* spp., *Galactomyces* sp., and *Phomopsis* sp. [78–80], the exact mechanisms that cause this attenuation are still unclear. They may be attributed to the absence of a host stimulus or to gene silencing occurring in axenic cultures [81,82]. However, this phenomenon was not observed in this study.

DNA methylation and transposon activity have already been investigated to be at the base of virulence loss and conidiation ability as a consequence of serial *in vitro* subcultures as well as at the origin of virulence restoration following host re-inoculation [69]. In the present study, we detected a lower level of relative methylation of SC50×3 when compared to SC50 in all contexts, suggesting that this genomic strategy was employed by *F. graminearum* in order to restore an efficient virulence. No significant differences in accumulating methylation changes were observed between genomic compartments. This finding suggests that, to achieve the phenotypic changes described in this study, both genomic regions hosting genes involved in basal metabolism and those regulating virulence are equally important. At a lower scale, the regulatory regions of genes were mainly affected by the previously mentioned methylation changes, especially for the CG context. Moreover, the different number of significant DMPs (Supplementary Table S4) revealed that CG was the methylation context mainly affected by the subculturing process. Differentially methylated region distributions in relation to coding and regulatory genomic sequences identified a total of 1024 genes, which are putatively regulated by DNA methylation.

Some of these genes have already been investigated for their crucial role in *F. graminearum* aggressiveness. Hereafter, some examples of genes that showed different methylation levels between SC50 and SC50×3 with our analyses and that have been previously described in other studies for their role in *F. graminearum* virulence are discussed. The genes FGSG_06675 (FgLeu2A) and FGSG_10671 (FgLeu2B) are known to be involved in the leucine metabolic pathway [83] of *F. graminearum*, and their importance in pathogenicity (in particular of FGSG_06675) and DON production (both FGSG_06675 and FGSG_10671) is well-known [84]. In the present experiment, other genes with different methylation levels during wheat infection are involved in the DON biosynthetic pathway: FGSG_05912 (mevalonate kinase) and FGSG_09764 (5′-phosphomevalonate kinase). These enzymes are responsible for the transformation of mevalonate in 5′-phosphomevalonate and, subsequently, in 5′-pyrphosphomevlonate during the chemical conversion of the acetyl-CoA in farnesyl pyrophosphate (FPP), which is the main substrate for DON biosynthesis [85]. Considering that DON is a well-known virulence factor of *F. graminearum* [86,87], likely FGSG_05912 and FGSG_09764 could be indirectly involved in fungal virulence due to their crucial role in DON production. Another interesting gene differentially methylated between SC50 and 50SCx3 is FGSG_07896. Even if not directly implicated in fungal virulence, this gene encodes for a trichothecene 3-O-acetyltransferase (TRI101) involved in the DON self-protection mechanism of *F. graminearum* [88,89]. In addition, the target of rapamycin (TOR) kinase gene (FGSG_08133) showed different methylation between the two samples. The protein FgTOR encoded by this gene is a key component of the TOR complex. The TOR signaling pathway of *F. graminearum* plays critical roles in regulating vegetative differentiation and virulence [90]. Other differentially methylated genes are FGSG_07067 and FGSG_06944, which encode for two transcription factors. *F. graminearum* mutants with the deletion of these genes showed a significant loss of virulence toward wheat heads [91]. Varied methylation levels of FGSG_07593, encoding a glycoside hydrolase, were observed between SC50×3 and SC50. This gene is usually upregulated at the beginning of the host colonization process [92]. Another gene differentially methylated comparing the two samples is FGSG_11955. The gene was previously identified like the velvet gene (FgVe1 or FgVeA) of *F. graminearum*, a domain conserved in various genera of filamentous fungi. The velvet gene regulates the trichothecene biosynthesis and pathogenicity against wheat heads and affects fungal development [93,94]. In addition to this gene, FGSG_01362 and FGSG_06774 belonged to the velvet gene domain, and, in this study, have shown different methylation levels. FGSG_01973, FGSG_09917, and FGSG_06175 encode for phospholipid hydrolases (phospholipase D, PLD) of *F. graminearum* (FgPLD1, FgPLD2, and FgPLD3). FgPLD1 is involved in the virulence toward flowering wheat heads and the mutant lacking this gene also showed reduced DON production, whereas FgPLD2 and FgPLD3 are not primarily involved in plant infection [95]. The differentially methylated FGSG_05902 gene between SC50×3 and SC50, which is almost identical with FGL15 cloned in previous research, encodes for a lipase known to be an important virulence factor [96]. Again, gene FGSG_04580, encoding a pleiotropic drug resistance class ABC transporter, is known to play a role in *F. graminearum* virulence toward different wheat tissues [97]. Furthermore, the ATP-binding cassette transporter Abc1, encoded by this gene, may be involved in DON release [98]. Several other ABC-G family transporters are highly expressed during host infection, such as FGSG_08309, which showed a different methylation between the two subcultures investigated [97]. The gene FGSG_09329, encoding an ABC-2 family transporter protein, showed a high expression during barley heads and wheat coleoptile colonization [97]. Recently, another ATP-binding cassette transporter (FgArb1) encoded by FGSG_04181, differentially methylated between SC50 and SC50×3, proved to have a function in pathogenesis and DON production [99]. In general, the ABC transporters family has a crucial role in *F. graminearum* pathogenicity [97,100]. Comparing SC50×3 to SC50, FGSG_01964 (GzCHS5), which encodes for a chitin synthase, is indispensable for perithecia formation and pathogenicity as well as for normal sept formation and hyphal growth [101]. Another chitin synthase gene, which is known to be involved both in DON synthesis and pathogenicity [102], is FGSG_06550 that showed different methylation levels between the two explored subcultures, such as FGSG_03538 (transcription factor Tri10) that is essential for DON production, and regulates the expression of the entire Tri-cluster [103,104]. FGSG_00352, differentially methylated between SC50 and SC50×3, is the orthologous protein of Hap2, which is one of the three subunits composing the heme activator protein (HAP), also known as a nuclear factor Y (NF-Y) or CCAAT-binding factor (CBF). *F. graminearum* has eight different genes encoding for CCAAT-binding factors. The deletion of FGSG_00352 did not significantly affect fungal mycelium growth, sexual development, mycotoxin production, and virulence [91], but other CCAAT-binding factors (FGSG_01182 and FGSG_05304) are involved in trichothecene production and virulence [105]. Thus, further studies on all the genes of the CCAAT-binding complex are necessary to reveal the relationship among them during host colonization. Another aggressiveness-associated gene (FGSG_08010), that is, a regulatory virulence [106], and is usually reported to be up-regulated during infection [107], showed a different level of methylation in the present work. A different methylation between the two analyzed subcultures was observed in FGSG_00332, encoding for a beta transducing-like (WD-40 repeat) protein, that has been demonstrated to be essential for pathogenicity in wheat [108], and in FGSG_01665 (FSR1) that regulates *F. graminearum* virulence by acting as a scaffold for a signal transduction pathway [109]. Comparing SC50 to SC50×3, a different methylation level was observed in FGSG_06798. Recently, this gene has been identified to encode for an acetyltransferase (FgHAT2) involved in regulating vegetative growth, conidiation, DNA damage repair, DON production, and virulence in the pathogen [110]. Another highlighted gene previously proven to be involved in pathogenicity was FGSG_00416, belonging to a major facilitator superfamily (MFS) [111]. Furthermore, the deletion of FGSG_03716 (Famfs1), which belongs to the MFS gene family, affected fungal development and virulence [112]. FGSG_03541 (Tri12), with different methylations between the two subcultures, is required for DON production and fungal virulence [113].

In addition, it has been demonstrated that FGSG_03168 has 90% similarity to FST1 of *F. verticillioides* (putative hexose transporter gene), which is functional in pathogenesis during the colonization of living maize kernels [114].

In the present study, we demonstrated that *F. graminearum* exhibits reduced virulence on bread wheat stem bases and heads after a prolonged subculturing process. However, the virulence on head tissue of bread wheat can be restored with the in planta transfer. Additionally, an innovative approach, based on the relative methylation level analysis, was used to explore novel putative virulence genes, comparing the pathogen after three generations of mycelium growth on bread wheat heads to the same fungus after approximately one-year of an *in vitro* subculturing process. Some of the genes that showed different methylation levels have been previously studied and were revealed to be related to *Fusarium* aggressiveness toward hosts. This suggests that the approach of the present study could be a promising tool in the study of *F. graminearum* genes associated with virulence on bread wheat tissues. In the future, it will be interesting to verify the function and possible involvement in virulence of all the other genes that have shown different methylation levels in this study (listed in Supplementary Table S7) but for which evidence about their implication in *F. graminearum* aggressiveness are not currently available. Finally, the present approach may be an important tool to use for other fungal pathogens to explore the pool of genes that could be involved in their virulence toward host species.

Supplementary Materials: The following materials are available online at https://www.mdpi.com/article/10.3390/cells10051192/s1. Table S1: *In vitro* biosynthesis of secondary metabolites (μg kg^{-1}) produced by subcultures of *F. graminearum* strain FG8 (SC1–SC23–SC50–SC50×3) as detected by LC-MS/MS analysis. Per each mycotoxin, the standard error (SE) is reported. Statistical differences were detected ($p > 0.05$) by one-way ANOVA. Table S2: Experimental design. Table S3: Sequencing data summary. Table S4: Methylation sequencing statistics. Uniquely mapped loci (MCSeEd loci) were normalized, filtered, and then analysed with the methyl kit to infer the number of differentially methylated positions (DMPs). Figure S1: Abundance of MCSeEd loci, DMPs, and DMRs in genomic regions (CDS, Introns, Regulatory Intergenic). Figure S2: Principal component analysis and clustering for the relative methylation levels at the differentially methylated positions obtained across all of the replicates in each sequence context: CG, CHG, and CHH. Number in brackets indicate the fraction of overall variance explained by the respective component (Dim1, Dim2). Table S5: List of significant DMRs identified in each context. Figure S3: Principal component analysis and boxplot for the relative methylation levels at the differentially methylated positions obtained across all of the replicates in each sequence context: CG, CHG, and CHH. Numbers in brackets indicate the fraction of overall variance explained by the respective component (Dim1, Dim2). Figure S4: Plotted DMPs along the EGBs (coding region, in yellow, with the regions 500 bp upstream and downstream) for CHG and CHH context. Table S6: Differentially methylated genes (DMGs) identified by intersecting DMRs. Table S7: Enrichment tests of PFAM domains within DMGs and *F. graminearum* loci. First column reports the PFAM domain, the second column reports the odds ratio, and the third column shows the *p*-value calculated for the Fisher exact test for an odd ratio equal to 1 (alternative hypothesis odds ratio >1).

Author Contributions: Conceived and designed the experiments, L.C., F.T., E.A., and G.M.; performed the experiments, F.T., G.B., G.M., and M.S.; statistical analysis, F.T.; data analysis, F.T., G.M., and A.P.; resources, L.C. and E.A.; wrote the original draft of the paper, F.T. and G.M.; critical revision of the manuscript, L.C., G.B., G.M., E.A., and D.M.G.; supervised the experiments, L.C. and G.M. All authors have read and agreed to the published version of the manuscript.

Funding: This research was supported by the "University of Perugia fund for basic research 2014", project title "DNA methylation analysis on the adaptation of fungi (yeasts and phytopathogenic fungi) grown outside their preferred habitat".

Institutional Review Board Statement: Not applicable.

Informed Consent Statement: Not applicable.

Acknowledgments: The authors wish to thank M.V. Consalvi, L. Ceccarelli, L. Bonciarelli, and M. Orfei for technical assistance.

Conflicts of Interest: The authors declare no conflict of interest.

References

1. O'Donnell, K.; Rooney, A.P.; Proctor, R.H.; Brown, D.W.; McComick, S.P.; Ward, T.J.; Frandsen, R.J.N.; Lysoe, E.; Rehner, S.A.; Aoki, T.; et al. Phylogenetic analyses of RPB1 and RPB2 support a middle Cretaceous origin for a clade comprising all agriculturally and medically important fusaria. *Fungal Genet. Biol.* **2013**, *52*, 20–31. [CrossRef]
2. Goswami, R.S.; Kistler, H.C. Heading for disaster: *Fusarium graminearum* on cereal crops. *Mol. Plant Pathol.* **2004**, *5*, 515–525. [CrossRef]
3. Kazan, K.; Gardiner, D.M.; Manners, J.M. On the trace of a cereal killer: Recent advances in *Fusarium graminearum* pathogenomics and host resistance. *Mol. Plant Pathol.* **2012**, *13*, 399–413. [CrossRef] [PubMed]
4. Mielniczuk, E.; Skwarylo-Bednarz, B. Fusarium head blight, mycotoxins and strategies for their reduction. *Agronomy* **2020**, *10*, 509. [CrossRef]
5. Geraldo, M.R.F.; Tessmann, D.J.; Kemmelmeier, C. Production of mycotoxins by *Fusarium graminearum* isolated from small cereal (wheat, triticale and barley) affected with scab disease on Southern Brazil. Braz. *J. Microbiol.* **2006**, *37*, 58–63. [CrossRef]
6. Ward, T.J.; Bielawski, J.P.; Kistler, H.C.; Sullivan, E.; O'Donnel, K. Ancestral polymorphism and adaptive evolution in the trichothecene mycotoxin gene cluster of phytopathogenic Fusarium. *Proc. Natl. Acad. Sci. USA* **2002**, *99*, 9278–9283. [CrossRef] [PubMed]
7. Feil, R.; Fraga, M.F. Epigenetics and the environment: Emerging patterns and implications. *Nat. Rev. Genet.* **2012**, *13*, 97–109. [CrossRef] [PubMed]
8. Causevic, A.; Delaunay, A.; Ounnar, S.; Righezza, M.; Delmotte, F.; Brignolas, F.; Hagege, D.; Maury, S. DNA methylating and demethylating treatments modify phenotype and cell wall differentiation state in sugar beet cell lines. *Plant Physiol. Biochem.* **2005**, *43*, 681–691. [CrossRef]
9. Cao, D.; Gao, X.; Liu, J.; Wang, X.; Geng, S.; Yang, C.; Liu, B.; Shi, D. Root-specific DNA methylation in *Chloris virgata*, a natural alkaline-resistant halophyte, in response to salt and alkaline stresses. *Plant Mol. Biol. Rep.* **2012**, *30*, 1102–1109. [CrossRef]
10. Angers, B.; Castonguay, E.; Massicotte, R. Environmentally induced phenotypes and DNA methylation: How to deal with unpredictable conditions until the next generation and after. *Mol. Ecol.* **2010**, *19*, 1283–1295. [CrossRef]
11. Lu, Y.; Rong, T.; Cao, M. Analysis of DNA methylation in different maize tissues. *J. Genet. Genom.* **2008**, *35*, 41–48. [CrossRef]
12. Lizal, P.; Relichova, J. The effect of day length, vernalization and DNA demethylation on the flowering time in *Arabidopsis thaliana*. *Physiol. Plant.* **2001**, *113*, 121–127. [CrossRef]
13. Seidl, M.F. Adenine N6-methylation in diverse fungi. *Nat. Genet.* **2017**, *49*, 823–824. [CrossRef] [PubMed]
14. Mondo, S.J.; Dannebaum, R.O.; Kuo, R.C.; Louie, K.B.; Bewick, A.J.; Labutti, K.; Haridas, S.; Keu, A.; Salamov, A.; Ahrendt, S.R.; et al. Widespread adenine N6- methylation of active genes in fungi. *Nat. Genet.* **2017**, *49*, 964–968. [CrossRef] [PubMed]
15. Lang, Z.; Wang, Y.; Tang, K.; Tang, D.; Datsenka, T.; Cheng, J.; Zhang, J.; Handa, A.K.; Zhu, J.K. Critical roles of DNA demethylation in the activation of ripening-induced genes and inhibition of ripening-repressed genes in tomato fruit. *Proc. Natl. Acad. Sci. USA* **2017**, *114*, E4511–E4519. [CrossRef]
16. Zhang, X.; Yazaki, J.; Sundaresan, A.; Cokus, S.; Chan, S.W.L.; Chen, H.; Henderson, I.R.; Shinn, P.; Pellegrini, M.; Jacobsen, S.E.; et al. Genome-wide high-resolution mapping and functional analysis of DNA methylation in Arabidopsis. *Cells* **2006**, *126*, 1189–1201. [CrossRef]
17. Goll, M.G.; Bestor, T.H. Eukaryotic cytosine methyltransferases. *Annu. Rev. Biochem.* **2005**, *74*, 481–514. [CrossRef]
18. Paszkowski, J.; Whitham, S.A. Gene silencing and DNA methylation processes. *Curr. Opin. Plant Biol.* **2001**, *4*, 123–129. [CrossRef]
19. Suzuki, M.M.; Bird, A. DNA methylation landscapes: Provocative insights from epigenomics. *Nat. Rev. Genet.* **2008**, *9*, 465–476. [CrossRef]
20. He, C.; Zhang, Z.; Li, B.; Tian, S. The pattern and function of DNA methylation in fungal plant pathogens. *Microorganisms* **2020**, *8*, 227. [CrossRef] [PubMed]
21. Freitag, M.; Williams, R.L.; Kothe, G.O.; Selker, E.U. A cytosine methyltransferase homologue is essential for repeat-induced point mutation in *Neurospora crassa*. *Proc. Natl. Acad. Sci. USA* **2002**, *99*, 8802–8807. [CrossRef]
22. Malagnac, F.; Wendel, B.; Goyon, C.; Faugeron, G.; Zickler, D.; Rossignol, J.L.; Noyer-Weidner, M.; Vollmayr, P.; Trautner, T.A.; Walter, J. A gene essential for de novo methylation and development in *Acobolus* reveals a novel type of eukaryotic DNA methyltransferase structure. *Cell* **1997**, *91*, 281–290. [CrossRef]
23. Bewick, A.J.; Hofmeister, B.T.; Powers, R.A.; Mondo, S.J.; Grigoriev, I.V.; James, T.Y.; Stajich, J.E.; Schmitz, R.J. Diversity of cytosine methylation across the fungal tree life. *Nat. Ecol. Evol.* **2019**, *3*, 479–490. [CrossRef]
24. Freitag, M.; Hickey, P.C.; Khlafallah, T.K.; Read, N.D.; Selker, E.U. HP1 is essential for DNA methylation in *Neurospora*. *Mol. Cell* **2004**, *13*, 427–434. [CrossRef]
25. Tamaru, H.; Selker, E.U. Synthesis of signals for de novo DNA methylation I *Neurospora crassa*. *Mol. Cell Biol.* **2003**, *23*, 2379–2394. [CrossRef] [PubMed]
26. Kouzminova, E.; Selker, E.U. dim-2 encode a DNA methyltransferase responsible for all known cytosine methylation in *Neurospora*. *EMBO J.* **2001**, *20*, 4309–4323. [CrossRef]

27. Ma, L.J.; van der Does, H.C.; Borkovich, K.A.; Coleman, J.J.; Daboussi, M.J.; Di Pietro, A.; Dufresne, M.; Freitag, M.; Grabherr, M.; Henrissat, B.; et al. Comparative genomics reveals mobile pathogenicity chromosomes in *Fusarium*. *Nature* **2010**, *464*, 367–373. [CrossRef] [PubMed]
28. Pomraning, K.R.; Connolly, L.R.; Whalen, J.P.; Smith, K.M.; Freitag, M. Repeat-induced point mutation, DNA methylation and heterochromatin in *Gibberella zeae* (Anamorph: *Fusarium graminearum*. In *Fusarium Genomics and Molecular Cellular Biology*; Brown, D., Proctor, R.H., Eds.; Horizon Scientific Press: Norwich, UK, 2013; p. 93.
29. Gijezen, M.; Ishmael, C.; Shrestha, D. Epigenetic control of effectors in plant pathogens. *Front. Plant Sci.* **2014**, *5*, 638.
30. Kim, D.H. Induced change in DNA methylation of *Fusarium oxysporum* f. sp. niveum due to successive transfer. *J. Biochem. Mol. Biol.* **1997**, *30*, 216–221.
31. Covarelli, L.; Beccari, G.; Prodi, A.; Generotti, S.; Etruschi, F.; Juan, C.; Ferrer, E.; Mañes, J. *Fusarium* species, chemotype characterization and trichothecene contamination of durum and soft wheat in an area of central Italy. *J. Sci. Food Agric.* **2015**, *95*, 540–551. [CrossRef]
32. Simpson, D.R.; Rezanoor, H.N.; Parry, D.W.; Nicholson, P. Evidence for differential host preference in *Microdochium nivale* var. *majus*, *M. nivale* var. *nivale*. *Plant Pathol.* **2000**, *49*, 261–268.
33. Beccari, G.; Covarelli, L.; Nicholson, P. Infection processes and soft wheat response to root rot and crown rot caused by *Fusarium culmorum*. *Plant Pathol.* **2011**, *60*, 671–684. [CrossRef]
34. Covarelli, L.; Beccari, G.; Steed, A.; Nicholson, P. Colonization of soft wheat following infection of the stem base by *Fusarium culmorum* and translocation of deoxynivalenol to the head. *Plant Pathol.* **2012**, *61*, 1121–1129. [CrossRef]
35. Brunner, K.; Lichtenauer, A.M.; Kratochwill, K.; Delic, M.; Mach, R.L. Xyr1 regulates xylanase but not cellulase formation in the head blight fungus *Fusarium graminearum*. *Curr. Genet.* **2007**, *52*, 213–220. [CrossRef] [PubMed]
36. Sulyok, M.; Stadler, D.; Steiner, D.; Krska, R. Validation of an LC-MS/MS-based diluted-and-shoot approach for the quantification of >500 mycotoxins and other secondary metabolites in food crops: Challenges and solutions. *Anal. Bioanal. Chem.* **2020**, *412*, 2607–2620. [CrossRef] [PubMed]
37. Marconi, G.; Capomaccio, S.; Comino, C.; Acquadro, A.; Portis, E.; Porceddu, A.; Albertini, E. Methylation content sensitive enzyme ddRAD (MCSeEd): A reference-free, whole genome profiling system to address cytosine/adenine methylation changes. *Sci. Rep.* **2019**, *9*, 14864. [CrossRef]
38. Di Marsico, M.; Cerruti, E.; Comino, C.; Porceddu, A.; Acquadro, A.; Capomaccio, S.; Marconi, G.; Albertini, E. MCSeEd (Methylation Context Sensitive Enzyme ddRAD): A new method to analyze DNA methylation. In *Plant Epigenetics and Epigenomics. Methods in Molecular Biology*; Spillane, C., McKeown, P., Eds.; Humana: New York, NY, USA, 2020; Volume 2093, pp. 47–64.
39. Akalin, A.; Kormaksson, M.; Li, S.; Garrett-Bakelman, F.E.; Melnick, A.; Mason, C.E. MethylKit: A comprehensive R package for the analysis of genome-wide DNA methylation profiles. *Genome Biol.* **2012**, *13*, R87. [CrossRef]
40. Onofri, A.; Pannacci, E. Spreadsheet tools for biometry classes in crop science programmes. *CBCS* **2014**, *9*, 43–53.
41. Cuomo, C.A.; Guldener, U.; Xu, J.R.; Trail, F.; Turgeon, B.G.; Di Pietro, A.; Walton, J.D.; Baker, S.E.; Rep, M.; Adam, G.; et al. The *Fusarium graminearum* genome reveals a link between localized polymorphism and pathogen specialization. *Science* **2007**, *317*, 1400–1402. [CrossRef] [PubMed]
42. Zhao, C.; Waalwijk, C.; de Wit, P.J.G.M.; Tang, D.; van der Lee, T. RNA-Seq analysis reveals new gene models and alternative splicing in the fungal pathogen *Fusarium graminearum*. *BMC Genom.* **2014**, *14*, 21. [CrossRef]
43. Li, B.Q.; Zong, Y.Y.; Du, Z.L.; Shang, Y.J.; Chen, Y.; Zhang, Z.Q.; Qin, G.Z.; Zhao, W.M.; Tian, S.P. Genomic characterization reveals insights into patulin biosynthesis and pathogenicity in *Penicillium* species. *Mol. Plant Microbe Interact.* **2015**, *28*, 635–647. [CrossRef] [PubMed]
44. Malcom, G.M.; Kuldau, G.A.; Gugino, B.K.; Jimenez-Gasco, M.D.M. Hidden host plant associations of soilborne fungal pathogens: An ecological perspective. *Phytopathology* **2013**, *103*, 538–544. [CrossRef] [PubMed]
45. Dean, R.; Kan, J.A.V.; Pretorius, Z.A.; Hammond-Kosack, K.E.; Pietro, A.D.; Spanu, P.D.; Rudd, J.J.; Dickman, M.; Kahmann, R.; Ellis, J.; et al. The top 10 fungal pathogens in molecular plant pathology. *Mol. Plant Pathol.* **2012**, *13*, 414–430. [CrossRef]
46. Presti, L.L.; Lanver, D.; Schweizer, G.; Tanaka, S.; Liang, L.; Tollot, M.; Zuccaro, A.; Reissmann, S.; Kahmann, R. Fungal effectors and plant susceptibility. *Annu. Rev. Plant Biol.* **2015**, *66*, 513–545. [CrossRef]
47. Presti, L.L.; Kahmann, R. How filamentous plant pathogen effectors are translocated to host cells. *Curr. Opin. Plant Biol.* **2017**, *38*, 19–24. [CrossRef]
48. Oliveira-Garcia, E.; Valent, B. How eukaryotic filamentous pathogens evade plant recognition. *Curr. Opin. Microbiol.* **2015**, *26*, 92–101. [CrossRef]
49. Shalaby, S.; Horwitz, B.A. Plant phenolic compounds and oxidative stress: Integrated signals in fungal-plant interactions. *Curr. Genet.* **2015**, *61*, 347–357. [CrossRef] [PubMed]
50. Reverberi, M.; Punelli, M.; Scala, V.; Scarpari, M.; Uva, P.; Mentzen, W.I.; Dolezal, A.L.; Woloshuk, C.; Pinzari, F.; Fabbri, A.A.; et al. Genotypic and phenotypic versatility of *Aspergillus flavus* during maize exploitation. *PLoS ONE* **2013**, *8*, e68735. [CrossRef]
51. Slepecky, R.A.; Starmer, W.T. Phenotypic plasticity in fungi: A review with observations on *Aureobasidium pullulans*. *Mycologia* **2009**, *101*, 823–832. [CrossRef] [PubMed]
52. Bartels, A.; Han, Q.; Nair, P.; Stacey, L.; Gaynier, H.; Mosley, M.; Huang, Q.Q.; Pearson, J.K.; Hsieh, T.F.; An, Y.C.; et al. Dynamic DNA methylation in plant growth and development. *Int. J. Mol. Sci.* **2018**, *19*, 2144. [CrossRef] [PubMed]

53. Zhong, S.; Fei, Z.; Chen, Y.R.; Zheng, Y.; Huang, M.; Vrebalov, J.; McQuinn, R.; Gapper, N.; Liu, B.; Xiang, J.; et al. Single-base resolution methylomes of tomato fruit development reveal epigenome modifications associated with ripening. *Nat. Biotechnol.* **2013**, *31*, 154–159. [CrossRef] [PubMed]
54. Moore, L.D.; Le, T.; Fan, G.P. DNA methylation and its basic function. *Neuropsychopharmacology* **2013**, *38*, 23–38. [CrossRef]
55. Su, Z.X.; Han, L.; Zhao, Z.M. Conservation and divergence of DNA methylation in eukaryotes: New insights from single base resolution DNA methylomes. *Epigenetics* **2011**, *6*, 134–140. [CrossRef]
56. Bird, A. DNA methylation patterns and epigenetic memory. *Genes Dev.* **2002**, *16*, 6–21. [CrossRef] [PubMed]
57. Howlett, B.J.; Lowe, R.G.T.; Marcroft, S.J.; van de Wouw, A.P. Evolution of virulence in fungal plant pathogens: Exploiting fungal genomics to control plant disease. *Mycologia* **2015**, *107*, 441–451. [CrossRef]
58. Jeon, J.; Choi, J.; Lee, G.W.; Park, S.Y.; Huh, A.; Dean, R.; Lee, Y.H. Genome-wide profiling of DNA methylation provides insights into epigenetic regulation of fungal development in a plant pathogenic fungus, *Magnaporthe oryzae*. *Sci. Rep.* **2015**, *5*, 8567. [CrossRef]
59. An, Y.C.; Goettel, W.; Han, Q.; Bartels, A.; Liu, Z.; Xiao, W. Dynamic changes of genome-wide DNA methylation during soybean seed development. *Sci. Rep.* **2017**, *7*, 12263. [CrossRef] [PubMed]
60. Cokus, S.J.; Feng, S.H.; Zhang, X.Y.; Chen, Z.G.; Merriman, B.; Haudenschild, C.D.; Pradhan, S.; Nelson, S.F.; Pellegrini, M.; Jacobsen, S.E. Shotgun bisulphite sequencing of the *Arabidopsis* genome reveals DNA methylation patterning. *Nature* **2008**, *452*, 215–219. [CrossRef]
61. Nahar, P.B.; Kulkarani, S.A.; Kulye, M.S.; Chavan, S.B.; Kulkarani, G.; Rajendran, A.; Yadav, P.D.; Shouche, Y.; Deshpande, M.V. Effect of repeated *in vitro* subculturing on the virulence of *Metarhizium anisopliae* against *Helicoverpa armigera* (Lepidoptera: Noctuidae). *Biocontrol Sci. Technol.* **2008**, *18*, 337–355. [CrossRef]
62. Shah, F.A.; Allen, N.; Wright, C.J.; Butt, T.M. Repeated *in vitro* subculturing alters spore surface properties and virulence of *Metarhizium anisopliae*. *FEMS Microbiol. Lett.* **2007**, *276*, 60–66. [CrossRef]
63. Morrow, B.J.; Boucias, D.G.; Heath, M.A. Loss of virulence in an isolate of an entomopathogenic fungus, *Nomuraea rileyi*, after serial *in vitro* passages. *J. Econ. Entomol.* **1989**, *82*, 404–407. [CrossRef]
64. Ignoffo, C.M.; McIntosh, A.H.; Garcia, C.; Kroha, M.; Johnson, J.M. Effects of successive *in vitro* and *in vivo* passages on the virulence of the entomopathogenic fungus, *Nomuraea rileyi*. *Entomophaga* **1982**, *27*, 371–378. [CrossRef]
65. Vandenberg, J.D.; Cantone, F.A. Effect of serial transfer of three strains of *Paecilomyces fumosoroseus* on growth *in vitro*, virulence and host specificity. *J. Invertebr. Pathol.* **2004**, *85*, 40–45. [CrossRef] [PubMed]
66. Brownbridge, M.; Costa, S.; Jaronski, S.T. Effects of *in vitro* passage of *Beauveria bassiana* on virulence to *Bemisia argentifolii*. *J. Invertebr. Pathol.* **2001**, *77*, 280–283. [CrossRef] [PubMed]
67. Hayden, T.P.; Bidochka, M.J.; Khachatourians, G.C. Entomopathogenicity of several fungi toward the English grain aphid (Homoptera: Aphididae) and enhancement of virulence with host passage of *Paecilomyces farinosus*. *J. Econ. Entomol.* **1992**, *85*, 58–64. [CrossRef]
68. Hussain, A.; Tian, M.Y.; He, Y.R.; Lei, Y.Y. Differential fluctuation in virulence and VOC profiles among different cultures of entomopathogenic fungi. *J. Invertebr. Pathol.* **2010**, *104*, 166–171. [CrossRef]
69. Butt, T.M.; Wang, C.S.; Shah, F.A.; Hall, R. Degeneration of entomogenous fungi. In *An Ecological and Societal Approach to Biological Control*; Eilenberg, J., Hokkanen, H.M.T., Eds.; Springer: Cham, The Netherlands, 2006; pp. 213–226.
70. Shah, F.A.; Wang, C.S.; Butt, T.M. Nutrition influences growth and virulence of the insect-pathogenic fungus *Metarhizium anisopliae*. *FEMS Microbiol. Lett.* **2005**, *251*, 259–266. [CrossRef]
71. Butt, T.M.; Goettel, M. Bioassays of entomopathogenic fungi. In *Bioassays of Entomopathogenic Microbes and Nematodes*; Navon, A., Ascher, K.R.S., Eds.; CAB International: Wallingford, UK, 2000; pp. 141–195.
72. Hussain, A.; Tian, M.Y.; He, Y.R.; Ruan, L.; Ahmed, S. *In vitro* and in vivo culturing impacts on the virulence of characteristics of serially passed entomopathogenic fungi. *J. Food Agric. Environ.* **2010**, *8*, 481–487.
73. Hutwimmer, S.; Wagner, S.; Affenzeller, M.; Burgstaller, W.; Strasser, H. Algorithmbased design of synthetic growth media stimulating virulence properties of *Metarhizium anisopliae* conidia. *J. Appl. Microbiol.* **2008**, *105*, 2026–2034. [CrossRef]
74. Rangel, D.; Braga, G.; Flint, S.D. Variations in UV-B tolerance and germination speed of *Metarhizium anisopliae* conidia produced on insects and artificial substrates. *J. Invertebr. Pathol.* **2004**, *87*, 77–83. [CrossRef]
75. Scala, V.; Grottoli, A.; Cigliano, R.A.; Anzar, I.; Beccacioli, M.; Fanelli, C.; Dall'Asta, C.; Battilani, P.; Reverberi, M.; Sanseverino, W. Carefully with that axe, gene, genome perturbation after a PEG-mediated protoplast transformation in *Fusarium verticillioides*. *Toxins* **2017**, *9*, 183. [CrossRef] [PubMed]
76. Safavi, S.A. Attenuation of the entomopathogenic fungus *Beauveria bassiana* following serial *in vitro* transfers. *Biologia* **2012**, *67*, 1062–1068. [CrossRef]
77. Ansari, M.A.; Butt, T.M. Effects of successive subculturing on stability, virulence, conidial yield, germination and shelf-life of entomopathogenic fungi. *J. Appl. Microbiol.* **2011**, *110*, 1460–1469. [CrossRef]
78. Kusari, S.; Zuelke, S.; Spiteller, M. Effect of artificial reconstitution of the interaction between the plant *Camptotheca acuminata* and the fungal endophyte *Fusarium solani* on camptothecin biosynthesis. *J. Nat. Prod.* **2011**, *74*, 764–775. [CrossRef] [PubMed]
79. Gurudatt, P.S.; Priti, V.; Sweta, S.; Ramesha, B.T.; Ravikanth, G.; Vasudeva, R.; Amna, T.; Deepika, S.; Ganeshaiah, K.N.; Shaanker, U.R.; et al. Attenuation of camptothecin production and negative relation between hyphal biomass and camptothecin content in endophytic fungal strains isolated from *Nothapodytes nimmoniana* Grahm (Icacinaceae). *Curr. Sci.* **2010**, *98*, 1006–1009.

80. Li, J.; Sidhu, R.S.; Ford, E.J.; Long, D.M.; Hess, W.M.; Strobel, G.A. The induction of taxol production in the endophytic fungus-*Periconia* sp. from *Torreya grandifolia*. *J. Ind. Microbiol. Biotechnol.* **1998**, *20*, 259–264. [CrossRef]
81. Priti, V.; Ramesha, B.T.; Shweta, S.; Ravikanth, G.; Ganeshaiah, K.N.; Suryanarayanan, T.S.; Shaanker, U.R. How promising are endophytic fungi as alternative sources of plant secondary metabolites. *Curr. Sci.* **2009**, *97*, 4–6.
82. Sachin, N.; Manjunatha, B.L.; Kumara, M.P.; Ravikanth, G.; Shweta, S.; Suryanarayanan, T.S.; Ganeshaiah, K.N.; Shaanker, U.R. Do endophytic fungi possess pathway genes for plant secondary metabolites? *Curr. Sci.* **2013**, *104*, 178–182.
83. Subramaniam, R.; Narayanan, S.; Walkowiak, S.; Wang, L.; Joshi, M.; Rocheleau, H.; Ouellet, T.; Harris, L.J. Leucine metabolism regulates TRI6 expression and affects deoxynivalenol production and virulence in *Fusarium graminearum*. *Mol. Microbiol.* **2015**, *98*, 760–769. [CrossRef] [PubMed]
84. Liu, X.; Han, Q.; Wang, J.; Wang, X.; Xu, J.; Shi, J. Two FgLEU2 genes with different roles in leucine biosynthesis and infection-related morphogenesis in *Fusarium graminearum*. *PLoS ONE* **2016**, *11*, e0165927. [CrossRef]
85. Zheng, X.; Zhang, X.; Zhao, L.; Apaliya, M.T.; Yang, Q.; Sun, W.; Zhang, X.; Zhang, H. Screening of deoxynivalenol producing strains and elucidation of possible toxigenic molecular mechanism. *Toxins* **2017**, *9*, 184. [CrossRef] [PubMed]
86. Bai, G.H.; Desjardins, A.E.; Plattner, R.D. Deoxynivalenol-nonproducing *Fusarium graminearum* causes initial infection, but does not cause disease spread in wheat spikes. *Mycopathologia* **2002**, *153*, 97–98. [CrossRef]
87. Eudes, F.; Comeau, A.; Rioux, S.; Collin, J. Impact of trichothecenes on Fusarium head blight (*Fusarium graminearum*) development in jspring wheat (*Triticum aestivum*). *Can. J. Plant Pathol.* **2001**, *23*, 318–322. [CrossRef]
88. Kimura, M.; Kaneko, I.; Komiyama, M.; Takatsuki, A.; Koshino, H.; Yoneyama, K.; Yamaguchi, I. Trichothecene 3-O-acetyltransferase protects both the producing organism and transformed yeast from related mycotoxins. Cloning and characterization of Tri101. *J. Biol. Chem.* **1998**, *273*, 1654–1661. [CrossRef]
89. McCormick, S.P.; Alexander, N.J.; Trapp, S.E.; Hohn, T.M. Disruption of TRI101, the gene encoding trichothecene 3-O-acetyltransferase, from *Fusarium sporotrichioides*. *Appl. Environ. Microbiol.* **1999**, *65*, 5252–5256. [CrossRef] [PubMed]
90. Yu, F.; Gu, Q.; Yun, Y.; Yin, Y.; Xu, J.R.; Shim, W.B.; Ma, Z. The TOR signaling pathway regulates vegetative development and virulence in *Fusarium graminearum*. *New Phytol.* **2014**, *203*, 219–232. [CrossRef] [PubMed]
91. Son, H.; Seo, Y.S.; Min, K.; Park, A.R.; Lee, J.; Jin, J.M.; Lin, Y.; Cao, P.; Hong, S.Y.; Kim, E.K.; et al. A phenome-based functional analysis of transcription factors in the cereal head blight fungus, *Fusarium graminearum*. *PLoS Pathog.* **2011**, *7*, e1002310. [CrossRef] [PubMed]
92. Puri, K.D.; Yan, C.; Leng, Y.; Zhong, S. RNA-Seq revealed differences in transcriptomes between 3ADON and 15ADON populations of *Fusarium graminearum in vitro* and *in planta*. *PLoS ONE* **2016**, *11*, e0163803.
93. Merhej, J.; Urban, M.; Dufresne, M.; Hammond-Kosack, K.E.; Richard-Forget, F.; Barreau, C. The velvet gene, FgVe1, affects fungal development and positively regulates trichothecene biosynthesis and pathogenicity in *Fusarium graminearum*. *Mol. Plant Pathol.* **2011**, *13*, 363–374. [CrossRef]
94. Jiang, J.; Liu, X.; Yin, Y.; Ma, Z. Involvement of a velvet protein FgVeA in the regulation of asexual development, lipid and secondary metabolisms and virulence in *Fusarium graminearum*. *PLoS ONE* **2011**, *6*, e28291. [CrossRef] [PubMed]
95. Ding, M.; Zhu, Q.; Liang, Y.; Li, J.; Fan, X.; Yu, X.; He, F.; Xu, H.; Ling, Y.; Yu, J. Differential roles of three FgPLD genes in regulating development and pathogenicity in *Fusarium graminearum*. *Fungal Genet. Biol.* **2017**, *109*, 46–52. [CrossRef]
96. Niu, X.W.; Zheng, Z.Y.; Feng, Y.G.; Guo, W.Z.; Wang, X.Y. The *Fusarium graminearum* virulence factor FGL targets an FKBP12 immunophilin of wheat. *Gene* **2013**, *525*, 77–83. [CrossRef]
97. Gardiner, D.M.; Stephens, A.E.; Munn, A.L.; Manners, J.M. An ABC pleiotropic drug resistance transporter of *Fusarium graminearum* with a role in crown and root diseases of wheat. *FEMS Microbiol. Lett.* **2013**, *348*, 36–45. [CrossRef] [PubMed]
98. O'Mara, S.P.; Broz, K.; Boenisch, M.; Zhong, Z.; Dong, Y.; Kistler, H.C. The *Fusarium graminearum* t-SNARE Sso2 is involved in growth, defense, and DON accumulation and virulence. *Mol. Plant-Microbe Int.* **2020**, *33*, 888–901. [CrossRef] [PubMed]
99. Yin, Y.; Wang, Z.; Cheng, D.; Chen, X.; Chen, Y.; Ma, Z. The ATP-binding protein FgArb1 is essential for penetration, infectious and normal growth of *Fusarium graminearum*. *New Phytol.* **2018**, *219*, 1447–1466. [CrossRef] [PubMed]
100. Ammar, G.A.; Tryono, R.; Doll, K.; Karlovsky, P.; Deising, H.B.; Wirsel, S.G.R. Identification of ABC transporter genes of *Fusarium graminearum* with roles in azole tolerance and/or virulence. *PLoS ONE* **2013**, *8*, e79042. [CrossRef]
101. Kim, J.E.; Lee, H.J.; Lee, J.; Kim, K.W.; Yun, S.H.; Shim, W.B.; Lee, Y.W. *Gibberella zeae* chitin synthase genes, GzCHS5 and GzCHS7, are required for hyphal growth, perithecia formation, and pathogenicity. *Curr. Genet.* **2009**, *55*, 449–459. [CrossRef]
102. Zhang, Y.Z.; Chen, Q.; Liu, C.H.; Liu, Y.B.; Yi, P.; Niu, K.X.; Wang, Y.Q.; Wang, A.Q.; Yu, H.Y.; Pu, Z.E.; et al. Chitin synthase gene FgCHS8 affects virulence and fungal cell wall sensitivity to environmental stress in *Fusarium graminearum*. *Fungal Biol.* **2016**, *120*, 764–774. [CrossRef]
103. Seong, K.Y.; Pasquali, M.; Zhou, X.; Song, J.; Hilburn, K.; McCormick, S.; Dong, Y.; Xu, J.R.; Kistler, H.C. Global gene regulation by *Fusarium* transcription factors Tri6 and Tri10 reveals adaptations for toxin biosynthesis. *Mol. Microbiol.* **2009**, *72*, 354–367. [CrossRef]
104. Guo, L.; Ji, M.; Ye, K. Dynamic network inference and association computation discover gene modules regulating virulence, mycotoxin and sexual reproduction in *Fusarium graminearum*. *BMC Genom.* **2020**, *21*, 176. [CrossRef]
105. Kim, J.E.; Nam, H.; Park, J.; Choi, G.J.; Lee, Y.W.; Son, H. Characterization of the CCAAT-binding transcription factor complex in the plant pathogenic fungus *Fusarium graminearum*. *Sci. Rep.* **2020**, *10*, 4898. [CrossRef]

106. Talas, F.; Kalih, R.; Miedaner, T.; McDonald, B.A. Genome-wide association study identifies novel candidate genes for aggressiveness, deoxynivalenol production, and azole sensitivity in natural field populations of *Fusarium graminearum*. *Mol. Plant-Microbe Int.* **2016**, *29*, 417–430. [CrossRef]
107. Lawler, K.; Hammond-Kosack, K.; Brazma, A.; Coulson, R.M.R. Genomic clustering and co-regulation of transcriptional networks in the pathogenic fungus *Fusarium graminearum*. *BMC Syst. Biol.* **2013**, *7*, 52. [CrossRef]
108. Ding, S.; Mehrabi, R.; Koten, C.; Kang, Z.; Wei, Y.; Seong, K.; Kistler, H.C.; Xu, J.R. Transduction beta-like gene FTL1 is essential for pathogenesis in *Fusarium graminearum*. *Eukaryot. Cell* **2009**, *8*, 867–876. [CrossRef]
109. Shin, W.B.; Sagaram, U.S.; Choi, Y.E.; So, J.; Wilkinson, H.H.; Lee, Y.W. FSR1 is essential for fungal virulence and female fertility in *Fusarium verticillioides* and in *Fusarium graminearum*. *Mol. Plant Microbe Int.* **2006**, *19*, 725–733. [CrossRef]
110. Lu, W.Y.; Yang, N.; Xu, Z.; Dai, H.; Tang, S.; Wang, Z.Y. FgHAT2 is involved in regulating vegetative growth, conidiation, DNA damage repair, DON production and virulence in *Fusarium graminearum*. *J. Integr. Agric.* **2020**, *19*, 1813–1824. [CrossRef]
111. Dufresne, M.; van der Lee, T.; M'Barek, S.B.; Xu, X.; Zhang, X.; Liu, T.; Waalwijk, C.; Zhang, W.; Kema, G.H.J.; Daboussi, M.J. Transposon-tagging identifies novel pathogenicity genes in *Fusarium graminearum*. *Fungal Genet. Biol.* **2008**, *45*, 1552–1561. [CrossRef]
112. Ren, W.; Zhao, H.; Shao, W.; Ma, W.; Wang, J.; Zhou, M.; Chen, C. Identification of a novel phenamacril-resistance-related gene by the cDNA-RAPD method in *Fusarium asiaticum*. *Pest Manag. Sci.* **2016**, *72*, 1558–1565. [CrossRef]
113. Menke, J.; Dong, Y.; Kistler, H.C. *Fusarium graminearum* Tri12p influences virulence to wheat and trichothecene accumulation. *Mol. Plant Microbe Int.* **2012**, *25*, 1408–1418. [CrossRef]
114. Kim, H.; Woloshuk, C.P. Functional characterization of fst1 in *Fusarium verticillioides* during colonization of maize kernels. *Mol. Plant Microbe Int.* **2010**, *24*, 18–24. [CrossRef]

cells

MDPI

Article

Starch Production in *Chlamydomonas reinhardtii* through Supraoptimal Temperature in a Pilot-Scale Photobioreactor

Ivan N. Ivanov [1,2], Vilém Zachleder [1], Milada Vítová [1], Maria J. Barbosa [3] and Kateřina Bišová [1,*]

1 Laboratory of Cell Cycles of Algae, Centre Algatech, Institute of Microbiology of the Czech Academy of Sciences, 379 81 Třeboň, Czech Republic; ivanov@alga.cz (I.N.I.); zachleder@alga.cz (V.Z.); vitova@alga.cz (M.V.)
2 Faculty of Science, University of South Bohemia, Branišovská 1760, 370 05 České Budějovice, Czech Republic
3 Bioprocess Engineering Group, AlgaePARC, Wageningen University & Research, P.O. Box 16, 6700 AA Wageningen, The Netherlands; maria.barbosa@wur.nl
* Correspondence: bisova@alga.cz; Tel.: +420-384-340-485

Abstract: An increase in temperature can have a profound effect on the cell cycle and cell division in green algae, whereas growth and the synthesis of energy storage compounds are less influenced. In *Chlamydomonas reinhardtii*, laboratory experiments have shown that exposure to a supraoptimal temperature (39 °C) causes a complete block of nuclear and cellular division accompanied by an increased accumulation of starch. In this work we explore the potential of supraoptimal temperature as a method to promote starch production in *C. reinhardtii* in a pilot-scale photobioreactor. The method was successfully applied and resulted in an almost 3-fold increase in the starch content of *C. reinhardtii* dry matter. Moreover, a maximum starch content at the supraoptimal temperature was reached within 1–2 days, compared with 5 days for the control culture at the optimal temperature (30 °C). Therefore, supraoptimal temperature treatment promotes rapid starch accumulation and suggests a viable alternative to other starch-inducing methods, such as nutrient depletion. Nevertheless, technical challenges, such as bioreactor design and light availability within the culture, still need to be dealt with.

Keywords: microalgae; *Chlamydomonas reinhardtii*; starch; supraoptimal temperature; cell cycle; pilot-scale production

Citation: Ivanov, I.N.; Zachleder, V.; Vítová, M.; Barbosa, M.J.; Bišová, K. Starch Production in *Chlamydomonas reinhardtii* through Supraoptimal Temperature in a Pilot-Scale Photobioreactor. *Cells* **2021**, *10*, 1084. https://doi.org/10.3390/cells10051084

Academic Editor: Suleyman Allakhverdiev

Received: 6 April 2021
Accepted: 29 April 2021
Published: 1 May 2021

1. Introduction

Together with light and nutrient availability, temperature is one of the major abiotic factors affecting growth of microalgae [1–4]. Temperature has been found to affect individual metabolic processes in various ways. Cell division and the duration of the cell cycle are particularly susceptible to changes in temperature while other parts of cellular metabolism, such as growth and other related synthetic processes, appear to be less influenced by such changes [5,6].

In green algae dividing by multiple fission, a gradual increase in temperature results in a physiological response in which cells at first increase their growth rate and shorten their cell cycle. Upon a further increase in temperature, an optimal point is reached at which growth rate is at its maximum and cell cycle duration is at its minimum. However, after exceeding this optimal temperature, the duration of the cell cycle is gradually prolonged while growth rates remain unaffected [1]. Eventually, after reaching a certain temperature (hereafter referred to as supraoptimal), the cell cycle is completely blocked while growth and metabolism remain seemingly further unaffected [1,6]. This effect of supraoptimal temperature was first observed in the 1960s and 1970s during small-scale laboratory experiments conducted with *Chlorella* sp. (Chlorophyta) [6–9]. It was determined that the specific supraoptimal temperature that causes cell cycle arrest varies between species of microalgae and must be controlled within a very narrow range. Otherwise, the cells will not achieve cell cycle arrest (at a temperature lower than supraoptimal) or will have their metabolism strongly affected which might lead to cell death (at a temperature higher than supraoptimal) [6].

An inherent property of cell division is that it is an energy-demanding process, consuming the majority of the cell's energy reserves [10]. A simple block of cell division leads to accumulation of starch and/or lipids in microalgal cultures grown in nitrogen (and other nutrient) starvation or limiting conditions [11–15]. A combination of cell cycle arrest and unaltered growth metabolism, as is the case of supraoptimal temperature treatment, leads to the build-up of surplus energy reserves [6,9]. For starch producing green algae, the accumulation of starch under supraoptimal temperature can be extensive and it can reach levels considerably higher than in cells cultivated at the optimal growth temperature and hence, it can be utilized as an approach to increase starch productivity.

Chlamydomonas reinhardtii has served as a well-established model for a number of years [13,16]. This green alga benefits from a wide array of readily available molecular tools for genetic engineering and strain optimization [16–18]. However, in spite of these benefits, the adoption of *C. reinhardtii* as a biotechnology platform has been limited. Only recently, attempts were made to increase the starch content of *C. reinhardtii* by utilizing techniques such as nutrient deprivation and temperature stress [6,10,13,19–22]. Although nutrient deprivation is an effective technique that can increase the starch content of *C. reinhardtii* to almost 49% (*w*/*w*), the build-up is rather slow and it can take weeks until the maximum concentration is reached, which reduces overall productivity [22]. In contrast, temperature stress can provide a rapid method for starch accumulation within short periods of time. In laboratory-scale experiments with synchronized cultures of *C. reinhardtii*, the cells at 30 °C initially accumulated starch as they grew in size, but this starch was spent for cell division. The cells at 39 °C grew in size similar to those at 30 °C, but they did not divide. Instead, they continued to increase their cell size and after 24 h, their total starch content was more than two-fold higher than the maximum at 30 °C [6]. Although promising, these results were obtained only under controlled laboratory conditions that utilized synchronized cultures with relatively low biomass densities and were exposed to abundant light intensities. Hence, the applicability of the supraoptimal temperature method for industrial production of starch is still largely unknown.

In the present study, we examine the potential for pilot-scale starch production in *C. reinhardtii* by supraoptimal temperature, a method that has already been proven to cause a rapid 2-fold increase in starch yields under laboratory conditions [6]. In doing so we investigate whether and how biomass density affects starch productivity, the possibility of culture recovery and reuse after the supraoptimal temperature treatment, as well as potential practical challenges and limitations of the method. To our knowledge, the experiments described here are the first attempt to employ supraoptimal temperature in the production of starch in microalgae at pilot-scale.

2. Materials and Methods

2.1. Microorganism and Culturing Conditions

The algal strain used in these experiments was the unicellular alga *Chlamydomonas reinhardtii* wild type 21gr (CC-1690), obtained from the Chlamydomonas Resource Center at the University of Minnesota (St. Paul, MN, USA). For routine subculturing, the strains were streaked onto culture plates containing standard high salt (HS) medium [23] solidified with agar every three weeks.

For the purpose of the experiments, a starting culture was cultivated in a bench-top flat-panel airlift photobioreactor (Algaemist, Technical Development Studio, Wageningen University, Wageningen, The Netherlands) in the following manner: 400 mL of liquid HS medium was inoculated directly from the culture plates and was cultivated at 30 °C and under constant incident light intensity of 500 µmol photons m^{-2} s^{-1} of photosynthetically active radiation (PAR) provided by light-emitting diode lamps (BXRA W1200, Bridgelux, Fremont, CA, USA). The cultures were aerated with a mixture of air and CO_2 (2%, *v*/*v*) at a flow rate of 0.63 VVM in order to provide a carbon source and mixing of the cell suspension.

2.2. Culture Medium for the Pilot-Scale Cultivation

All experiments were performed under photoautotrophic conditions. The culture medium used for the pilot-scale experiments was based on a HS medium but was modified in order to facilitate high biomass yields with the NH_4Cl concentration being increased 5-fold. This resulted in a growth medium with the following final composition: 250 g L^{-1} NH_4Cl, 2 $CaCl_2{\cdot}2H_2O$, 20 g L^{-1} $MgSO_4{\cdot}7H_2O$, 1.84 g L^{-1} $C_{10}H_{12}FeN_2NaO_8$, 0.05 g L^{-1} $Na_2MoO_4{\cdot}2H_2O$, 3.09 g L^{-1} H_3BO_3, 1.18 g L^{-1} $MnSO_4{\cdot}7H_2O$, 1.40 g L^{-1} $CoSO_4{\cdot}7H_2O$, 1.24 g L^{-1} $CuSO_4{\cdot}5H_2O$, 1.43 g L^{-1} $ZnSO_4{\cdot}7H_2O$, 72 g L^{-1} KH_2PO_4, 134 g L^{-1} K_2HPO_4. For the preparation of the medium, 100× concentrated stock solutions of macro elements and microelements were used. All components, excluding phosphates, were diluted in distilled H_2O and autoclaved for 20 min at 121 °C. After cooling, the sterile autoclaved phosphates were added. The medium used for cultivation in the pilot-scale flat-panel photobioreactor was not sterilized and tap water was used to dilute the stock solutions. In the course of the experiments, pH was monitored daily and was maintained at 7.0 ± 0.1 by the addition of 2 M NaOH. Foam formation in the reactor vessels was controlled with the help of 10× diluted antifoam silicone Snapsil RE 20 containing 30% active compound (Product code: 84538.290, VWR International, LLC, Radnor, PA, USA).

2.3. Pilot-Scale Flat-Panel Photobioreactor

A flat-panel Algae-Germ photobioreactor with two cultivation vessels, each of a total volume of 25 L (20 L of culture volume) (Figure 1), was used in all experiments described here. The photobioreactor was situated at 51°59′45.6″ N, 5°39′25.7″ E in Wageningen, Netherlands and was placed within a greenhouse with panels facing 240° SW. Each of the cultivation vessels had the following dimensions: length: 70 cm, height: 72 cm, width (optical path: 5 cm). Cooling and heating of the microalgal suspension culture was provided by two refrigerating/heating circulators (Julabo GmbH, Seelbach, Germany) that circulated water through temperature control coils, which were submerged in the culture suspension. A simple aeration system provided a constant flow of a mixture of air and CO_2 (2%, v/v) and ensured mixing of the suspension culture. Both cultivation vessels were constantly illuminated by a panel of luminescent lamps (Master TL-D 58W/840, Philips, Amsterdam, the Netherlands) delivering 50 µmol photons m^{-2} s^{-1} of PAR (measured at the vessel surface). However, the majority of PAR delivered to the cultures was through exposure to natural sunlight.

Figure 1. Changes of optical densities of *C. reinhardtii* cultures starting at the same cell density (day 6, left photobioreactor) and grown at the same light intensity for 5 days at temperatures of 30 °C and 39 °C (day 11, right photobioreactor). For description of the photobioreactor see Materials and Methods in Section 2.3.

2.4. Experimental Approach

All the experiments were performed in the time-span between July 17th and September 17th, 2018 (Table 1). Each experiment consisted of two phases. During the biomass accumulation phase, one cultivation vessel was filled with 20 L of HS medium and was inoculated with 0.8 L of starting inoculum of cell concentration approximately 3.5×10^7 cells mL^{-1}. The resulting culture, with an initial cell concentration of approximately 1.4×10^6 cells mL^{-1}, was then incubated at 30 °C for 4 to 6 days. The supraoptimal temperature phase started upon reaching a biomass concentration exceeding 1.0 g L^{-1}. At this point the culture was diluted with HS medium and was separated into two cultivation vessels, which were then transferred to 30 °C or 39 °C.

Table 1. Overview of the experiments performed including date and duration. Temperature and initial biomass concentration after dilution of the culture from the biomass accumulation phase are shown. All experiments were performed in the summer of 2018.

Experiment Date	Total Duration (days)	Temperature Treatment (°C)	DM at the Beginning of Experiment (g L^{-1})
17.07–31.07	14	39 39	0.2 0.8
13.08–23.08	10	39 39	0.1 0.2
23.08–03.09	11	39 39	0.1 0.8
03.09–17.09	12	30 39	0.1 0.1

2.5. Light Measurements

The photon flux density (Iph, μmol photons m^{-2} s^{-1}) was measured with a LI-COR 190-SA 2π PAR (400–700 nm) quantum sensor (LiCor, Lincoln, NE, USA). Continuous light data logging was made with a sensor from the same model and manufacturer, mounted outdoors and facing the sky, parallel to the ground.

To obtain a measure of light energy absorbed by the cell suspension grown at different concentrations of cells, the mean light intensity (I) was calculated according to the Lambert-Beer law:

$$I = (I_i - I_t)/\ln(I_i/I_t),$$

where I_i is the incident light intensity at the surface of the culture vessel and I_t is the transmitted light intensity measured at the rear side of the culture vessel. The mean light (μmol $cell^{-1}$) was calculated by dividing the mean light intensity during 24 h (obtained by continuous light data logging) by the number of viable cells for that period.

2.6. Cell Size and Cell Number Measurements

One milliliter aliquots of culture suspension were taken, fixed with 10 μL of iodine solution (1 g I, 5 g KI, 100 mL H_2O), and stored at 4 °C. Cell diameter was measured on microphotographs taken with an Olympus Camedia C-5050 Zoom digital camera. The microphotographs were then analyzed using ImageJ image processing and analysis software (U. S. National Institute of Health, Rockville Pike, MD, USA). The cell diameter was recalculated to volume by a basic formula assuming a spherical cell shape. Cell number was determined by means of a Bürker counting chamber (Meopta, Prerov, Czech Republic).

2.7. Dry Matter Measurements

Aliquots of culture suspension (50 mL) were taken and centrifuged (1580R, Labogene ApS, Lillerød, Denmark) for 10 min at 4000 rpm. The supernatant was then removed, and the remaining pellet was transferred to a 2 mL preweighed test tube and dried for 24 h at 105 °C. After cooling for 2 h, the test tube with the pellet was weighed on an analytical balance (CP224S-OCE, Sartorius AG, Göttingen, Germany) and the weight of the pellet was determined by subtracting the weight of the empty test tube.

2.8. Starch Analysis

A starch-specific enzymatic method was used to accurately estimate the biomass starch content. Aliquots of culture suspension (10 mL) were harvested and centrifuged (1580R, Labogene ApS, Lillerød, Denmark) for 10 min at 4000 rpm. The supernatant was then discarded, and the resulting pellets were stored at −20 °C. After thawing, the cells in the pellets were disintegrated by adding 300 μL of glass beads (0.7 mm in diameter) and vortexing vigorously (MS3, IKA-Werke GmbH & Co. KG, Staufen, Germany) for 15 min. Depigmentation of the samples was carried out by adding 1 mL of 80% (*v*/*v*) ethanol to the pellet and incubating in a water bath for 15 min at 68 °C after which the samples were centrifuged (1580R, Labogene ApS, Lillerød, Denmark) for 2 min at 14,000 rpm and the supernatant was removed. The depigmentation procedure was repeated 3 to 4 times (or until the pellet was completely discolored). After that, 1 mL of α-amylase (porcine pancreas, Sigma–Aldrich, St. Louis, MO, USA) solution (0.5 $g \cdot L^{-1}$ *w*/*v* in 0.1 M sodium phosphate buffer (pH 6.9)) was added to the samples and incubated for 1 h at 37 °C. The samples were centrifuged (1580R, Labogene ApS, Lillerød, Denmark) for 2 min at 14,000 rpm, after which the supernatant was used for the quantification of reducing sugars through the dinitrosalicylic acid (DNSA) color reaction, as described by Miller [24]. Briefly, 500 μL of supernatant were mixed with 500 μL DNSA solution (1% (*w*/*v*) DNSA, 30% (*w*/*v*) potassium sodium tartrate tetrahydrate, 20% (*v*/*v*) 2 M sodium hydroxide) and incubated for 5 min at 105 °C on a heat block. Following a cooling period of 10 min at room temperature the mixture was diluted five-fold with distilled water, after which the absorbance of the samples was measured at 570 nm. The concentration of starch was estimated through a calibration curve of potato starch (Lach-Ner s.r.o., Neratovice, Czech Republic) digested with α-amylase.

2.9. Microscopic Observations and Starch Staining

Microscopic observations during the course of the experiments were carried out daily on a Leica Laborlux S microscope. Staining of starch was with the same iodine solution that was used for fixing cell counting samples in a 1:10 volume ratio of staining solution to sample.

3. Results

Starch accumulation in *C. reinhardtii* can be induced by nutrient depletion [12,13,22]. In order to exclude such effect, a biomass accumulation phase in fully supplemented medium at the optimal temperature was included prior to the supraoptimal temperature phase. The purpose of the biomass accumulation phase was to demonstrate that the cultures were not limited by nutrients and to estimate the typical starch content in *C. reinhardtii* cultures during exponential growth under the optimal growth temperature. Please, refer to Table 1 for an overview of the time span of the experiments as well as combinations of biomass densities and temperature treatments applied.

3.1. The Effects of Supraoptimal Temperature

To assess the effects of supraoptimal temperature on cell growth and division, as well as starch accumulation, the culture behavior at two parallel treatments of 30 °C (control) and 39 °C was compared. At first a *C. reinhardtii* culture was cultivated at 30 °C for 6 days. After reaching a biomass concentration of 1.0 $g\ L^{-1}$, the culture was diluted with HS medium to 0.1 $g\ L^{-1}$ and split into two cultures which were then cultivated at 30 °C and 39 °C, respectively (Figure 1).

The control culture cultivated at 30 °C had a similar pattern of biomass accumulation to that before dilution and returned to the pre-dilution biomass concentration within 6 days (reaching a maximum of nearly 1 $g\ L^{-1}$) (Figure 2A). In contrast, biomass accumulation of the culture at 39 °C stopped after 2 days, reaching a maximum of only 0.4 g L^{-1}. The biomass starch content in terms of percentage of starch within the dry matter (DM) was much greater in the culture cultivated at 39 °C, reaching 18% of DM as opposed to 8% of DM at 30 °C (Figure 2C). Moreover, the volumetric starch concentration in mg mL^{-1} was faster in the culture cultivated at 39 °C, reaching a maximum of 0.07 g L^{-1} in only 2 days as opposed to 5 days and 0.06 g L^{-1} in the culture cultivated at 30 °C (Figure 2D).

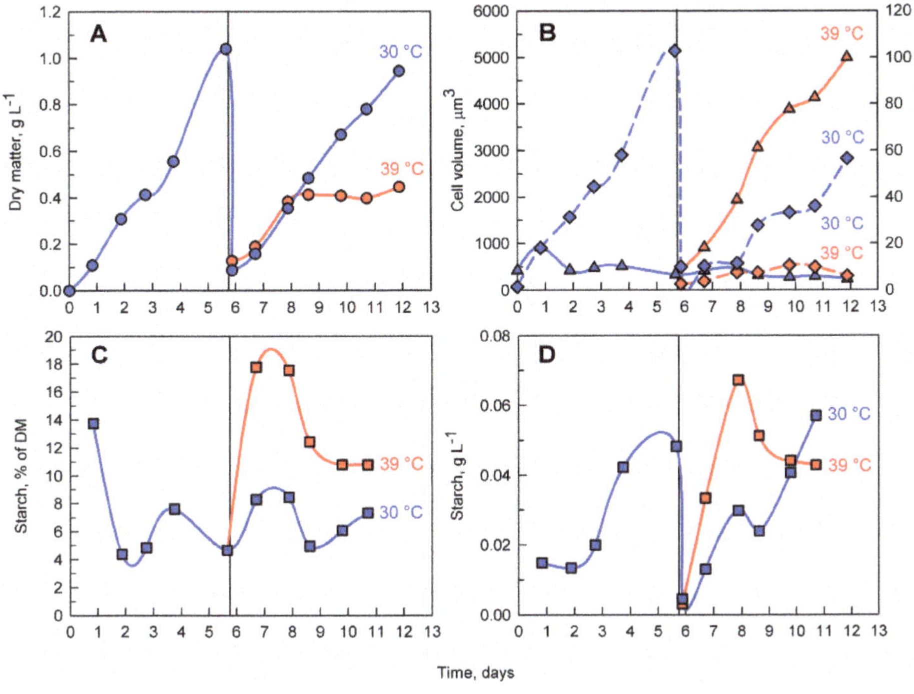

Figure 2. Effect of supraoptimal temperature on dry matter accumulation (**A**), average cell volume (solid line, triangles) and cell number (dashed line, diamonds) (**B**), biomass starch content (**C**) and volumetric starch concentration in the culture (**D**). The vertical line on day 6 represents the shift from biomass accumulation phase to supraoptimal temperature phase. Blue lines represent cultivation at 30 °C while red lines represent cultivation at 39 °C. During the biomass accumulation phase a single *C. reinhardtii* culture was cultivated at 30 °C. After 6 days the culture was diluted to a biomass concentration of 0.1 g L^{-1} and divided into two separated cultures. One of the cultures was cultivated at 39 °C (supraoptimal temperature treatment) while the other one was cultivated at 30 °C (control). Panel B represents the median volume of cells and an estimation of the mean cell number within a defined volume of culture suspension.

Microscopic observations of the culture cultivated at 30 °C did not reveal any change in the pattern of cell division with cells having a median cell volume of 422 μm^3 and cell division occurring, as expected, during the course of the experiment with mother cells dividing predominantly into eight daughter cells (Figures 2B and 3). In contrast, the cells transferred to 39 °C largely stopped dividing and the few dividing cells formed mostly two or four daughter cells. The inhibition of cell division was also reflected in the median cell volume, which increased eight-fold at 39 °C reaching 3479 μm^3.

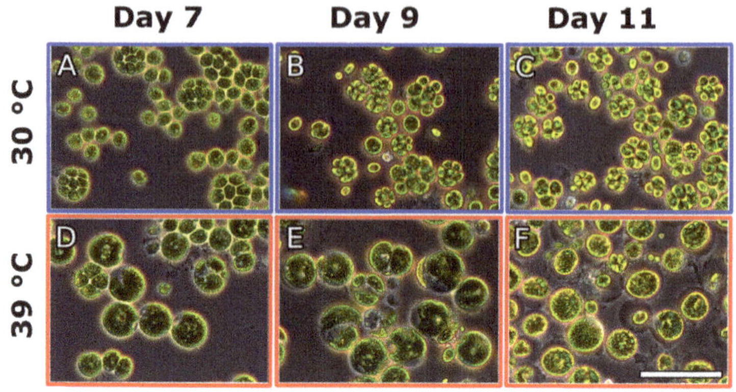

Figure 3. Effect of supraoptimal temperature on the cell division pattern and cell size in *C. reinhardtii*. The algal culture was cultivated for 6 days at 30 °C after which it was split, transferred to 30 °C (**A–C**) and 39 °C (**D–F**) and monitored for another 5 days. Day 7 corresponds to the first day after the split. Size of bar: 50 μm.

3.2. The Effects of Biomass Density

Light availability within the culture suspension itself is a function of biomass density and incident light intensity. To study the effect of light availability on starch accumulation at a supraoptimal temperature, two parallel cultures of different biomass concentrations and identical incident light intensity were compared. To do so, a *C. reinhardtii* culture was cultivated at 30 °C for 6 days. After reaching a biomass concentration of 1.5 g L^{-1} the culture was diluted with a HS medium to avoid nutrient limitation and split into two cultures, with an initial biomass concentration of 0.2 g L^{-1} (less dense culture) and 0.8 g L^{-1} (dense culture); these were then cultivated at 39 °C for 8 more days (Figure 4A). During the first three days after dilution and transfer to 39 °C, both cultures increased in biomass and reached a maximum of 0.6 g L^{-1} and 1.1 g L^{-1}, respectively. After this, the biomass concentration in the less dense culture remained constant while the biomass in the dense culture started to decline rapidly. The difference in biomass concentrations was also reflected in the mean light availability in the cultures, with the cells in the less dense culture being exposed to notably more light than the ones in the dense culture (Figure 4B). The increase in biomass relative starch content was remarkably rapid and was much more pronounced in the less dense culture, reaching a maximum of 13.2 % of DM within the first day of the transfer to 39 °C (Figure 4C). This rapid 3-fold increase from the initial culture represents a striking contrast to the starch values within the dense culture, which did not increase when compared to the pre-treatment phase. When comparing the different biomass concentrations in the cultures, the volumetric starch concentration (g L^{-1}) in the less dense culture was about 20% higher than in the dense culture. However, the maximum volumetric starch concentration reached at 39 °C was similar (in the case of the less dense culture) or lower (in the case of the dense culture) than that reached during the biomass accumulation phase at 30 °C (Figure 4D). Similar experiments with similar results were also performed at two lower initial cell densities in 39 °C, 0.1 g L^{-1} and 0.2 g L^{-1} (Figure S1).

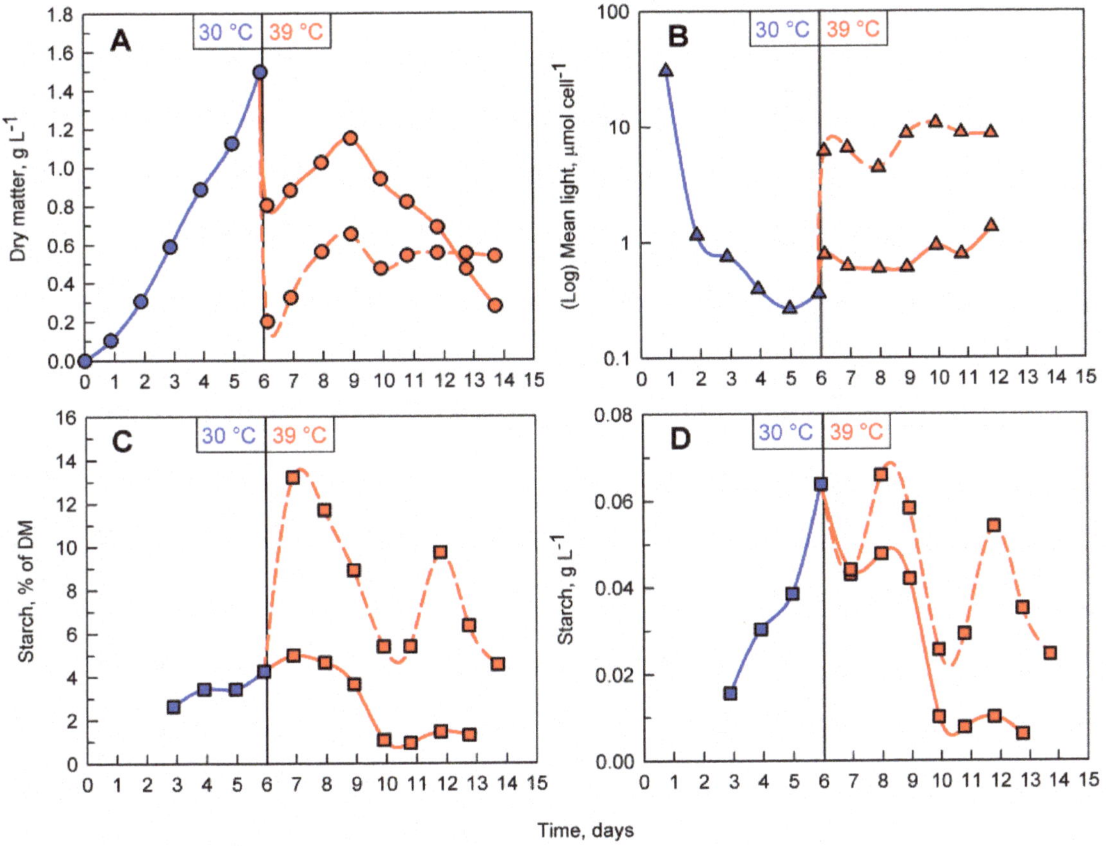

Figure 4. Effect of different biomass concentrations at the supraoptimal temperature on the time course of dry matter accumulation (**A**), mean light availability (**B**), biomass starch content (**C**), and volumetric starch concentration (**D**) in the culture. The vertical line at day 6 represents the shift between biomass accumulation and supraoptimal temperature phases. Blue lines and markers indicate cultivation at 30 °C and red lines and markers indicate cultivation at 39 °C. During the biomass accumulation phase a single *C. reinhardtii* culture was cultivated at 30 °C. After 6 days, the culture was split into two and diluted to 0.2 g L^{-1} (dashed red line) and 0.8 g L^{-1} (solid red line) then transferred to 39 °C.

3.3. Transfer Back to Optimal Temperature

As previously demonstrated, the supraoptimal temperature treatment allows rapid accumulation of starch, during which time, the maximum biomass starch content is reached rapidly within 1 to 2 days. However, due to the nature of the temperature block on cell division, a biomass accumulation phase is required before the treatment can be applied. Thus, a possible starch production process on an industrial scale will involve a repeated-batch culture which is treated with consecutive alterations of temperature between 30 and 39 °C. To investigate whether the cells are viable, and their cell cycle block can recover after the supraoptimal temperature phase, the *C. reinhardtii* culture was cultivated at 30 °C for 4 days. After reaching a biomass concentration of 1.2 g L^{-1}, the culture was diluted with a HS medium and split into two cultures with initial biomass concentrations of 0.1 g L^{-1} and 0.8 g L^{-1}, respectively (Figure 5A). Immediately after dilution, both cultures were transferred to 39 °C for a period of three days. During the transfer to supraoptimal temperature, biomass accumulation in both cultures decreased with the decrease being much more pronounced in the 0.8 g L^{-1} culture. When the culture was moved back to

30 °C, biomass accumulation resumed in the 0.1 g L^{-1} culture while in the 0.8 g L^{-1} culture, biomass concentrations declined gradually. The difference in initial biomass concentration after dilution was also reflected in the mean light availability in the cultures, with the cells in the less dense culture being exposed to notably more light than the ones in the dense culture (Figure 5B). Starch synthesis was much more pronounced in the 0.1 g L^{-1} culture with biomass starch content increasing 4-fold and reaching nearly 20% of DM within the first day of transfer to supraoptimal temperature (compared to only 8% of DM in the 0.8 g L^{-1} culture) (Figure 5C). Although the biomass starch content in terms of percent of DM was higher in the 0.1 g L^{-1} culture, the volumetric starch concentration (g L^{-1}) in both cultures was nearly identical due to the difference in biomass concentrations within the two cultures (Figure 5D).

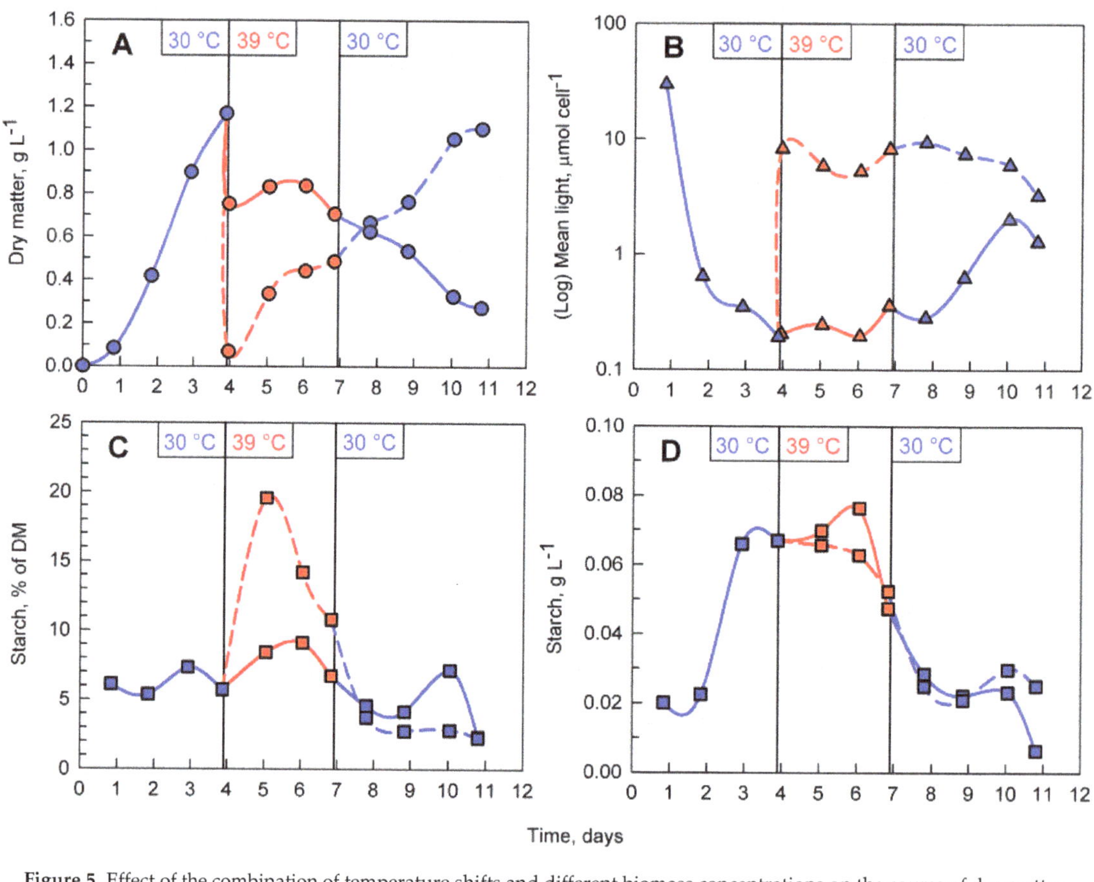

Figure 5. Effect of the combination of temperature shifts and different biomass concentrations on the course of dry matter accumulation (**A**), mean light availability (**B**), biomass starch content (**C**), volumetric starch concentration in the culture (**D**). A single *C. reinhardtii* culture was cultivated at 30 °C. After 4 days the culture was split into two, transferred to 39 °C and diluted to 0.1 mg mL^{-1} (dashed red lines) and 0.8 mg mL^{-1} (solid red lines), respectively. At day 7, the 0.1 mg mL^{-1} culture (dashed blue lines) and the 0.8 mg mL^{-1} culture (solid blue line) were transferred to 30 °C. Vertical lines at days 4 and 7 indicate those shifts in temperature (from 30 °C to 39 °C on day 4 and from 39 °C to 30 °C on day 7).

Microscopic observations and analysis of the cell cycle confirmed that *C. reinhardtii* cells were dividing during the biomass accumulation phase at 30 °C (Figure 6A) and transfer to supraoptimal temperature resulted in a block of cell division (Figure 6B,C,E,F). The cells in the 0.1 g L^{-1} culture were larger and rounder as opposed to cells in the 0.8 g L^{-1}

culture, which were smaller (Figure 6B,E). Staining with iodine revealed the presence of large amounts of starch granules in the chloroplasts of the cells in both cultures one day after the transfer (Figure 6C,F). Upon transfer back to 30 °C cell cycle progression was restored in the 0.1 g L^{-1} culture within one day (Figure 6D) as opposed to the cells in the 0.8 g L^{-1} which did not recover (Figure 6G).

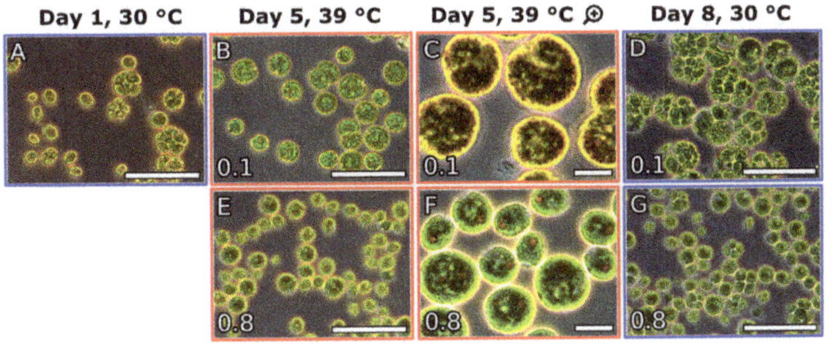

Figure 6. Microscopic observations. (**A**): Cells during day one of the biomass accumulation phase at 30 °C. (**B**,**E**): Cells one day after transfer to 39 °C and dilution to 0.1 g L^{-1} and 0.8 g L^{-1}, respectively. (**C**,**F**): magnification and iodine staining of starch granules in the cells from panels (**B**,**E**), respectively. (**D**,**G**): cells one day after transfer back to 30 °C of the cultures originally diluted to 0.1 g L^{-1} and 0.8 g L^{-1}. Size of bar on panels (**A**,**B**,**D**,**E**,**G**): 50 µm, panels (**C**,**F**): 10 µm.

4. Discussion

4.1. Effects of Supraoptimal Temperature on Starch Accumulation in C. reinhardtii at Pilot-Scale

The transfer of *C. reinhardtii* cultures from optimal to supraoptimal temperature proved to have a pronounced effect on both cell cycle progression and the accumulation of energy reserves in the form of starch. Upon 39 °C treatment, cell division was inhibited, mean cell size was increased, and the biomass starch content was enhanced more than 2-fold (Figures 2C, 4C, and 5C, Figure S1C and Table 2). These observations are in agreement with results from supraoptimal temperature experiments conducted with synchronized cultures of *C. reinhardtii* in laboratory-scale experiments [6] and resemble the effect of supraoptimal temperature on *Chlorella* sp. [7,8]. The increase in starch as an energy storage molecule during a period of inhibited cell division supports the inverse relationship between chemical energy storage and energy expenditure for the normal operation of the cell cycle under optimal growth conditions. Increased starch accumulation in green algae is often linked to a block in the cell cycle. It was observed not only as an effect of supraoptimal temperature treatment but also as a response to nutrient deprivation [13,20,22,25,26] and for cell cycle gene mutants [10]. The starch (over)accumulation is genetically linked to mutations in *phosphoglucomutase* 1, an enzyme involved in both glycolysis and starch biosynthesis [19]. Furthermore, *C. reinhardtii* mutants in DYRK kinase were shown to hyper-accumulate both starch and oil [21].

Table 2. Comparison of the effect of supraoptimal temperature in combination with various initial biomass concentrations on the maximum biomass starch content and the time when the maximum volumetric starch concentration was attained.

Initial DM (g L^{-1})	Temperature (°C)	Max. Biomass Starch Content (% of DM)	Max. Volumetric Starch Concentration (g L^{-1})	Time Required to Achieve Max. Volumetric Starch Concentration (Days)
0.1	39	21 *	0.067 *	1.3 *
0.2	39	14 **	0.069 **	2 **
0.8	39	7 **	0.060 **	2 **
0.1	30	8	0.057	5

* Shown value is an average of three cultivations; ** Shown value is an average of two cultivations.

Starch synthesis in response to nitrogen depletion in laboratory-scale experiments with *C. reinhardtii* cultures has been reported to increase starch content to up to 70 μg 10^{-6} cells (i.e., 70 pg $cell^{-1}$). However, this type of treatment required 7 days of nitrogen starvation until the maximum starch content was reached [26]. Similarly, starch levels induced by sulfur depletion reached up to 49% of DM after 20 days. However, during the upscaling of the process, the maximum starch content which was reached was nearly two-fold lower at around 25% of DM [22]. Thus, the results produced under stable laboratory conditions are often difficult to directly extrapolate to an industrial scale. This is also well documented on the comparison of starch production in supraoptimal temperature in laboratory conditions [6] and the results described here, i.e., 80 pg $cell^{-1}$ versus 16 pg $cell^{-1}$. Notwithstanding, the limitation of pilot-scale cultivation, the starch accumulation in experiments presented here was significantly faster (only 1 to 2 days) compared to the nutrient limitation conditions [22,26]. Such a significant decrease in cultivation time might prove of notable importance for the economic viability of industrial-scale production of microalgal based starch.

4.2. The Importance of Light Availability

A modern biotechnological process requires high productivity and cost efficiency [27]. This can only be achieved through rapid accumulation and high volumetric concentrations of the desired product. In order for microalgal starch production to be economically viable, starch yields per volume of culture must be high. This is only possible through increased starch content per cell at high biomass concentrations. Yet, increase in biomass density led to a decrease in mean light availability within the culture (Figures 4B, 5B and Figure S1B). This is likely due to light scattering and self-shading of the microalgal cells which have been found to cause sharp reductions in light availability with depth [28–30]. This limited light availability within the culture led to a reduction in the starch content and ultimately limited the effectiveness of the supraoptimal temperature treatment (Table 2). Similarly, recent findings in *Nannochloropsis* sp. showed a strong link between photosynthetic efficiency and the accumulation rate of lipids which are the primary energy storage compound in this microalga [31]. As a result, although the supraoptimal treatment led to a more rapid accumulation of starch, the maximum volumetric starch concentration achieved during cultivation at 39 °C did not show any notable increase over the maximum volumetric starch concentrations reached at 30 °C.

Furthermore, in the cultivation system used in this study, the combination of high cell density (leading to low light availability) and supraoptimal temperature seemed to have a negative effect on *C. reinhardtii* cultures after a certain period of time. In both cultivations at a starting biomass density of 0.8 mg mL^{-1}, biomass decline was observed after 3 days of exposure to supraoptimal temperature (Figures 4A and 5A), and the cultures did not recover when transferred to the optimal growth temperature (Figure 5A). In addition, light penetration into the high biomass cultures was further hindered by excessive biofilm formation (Figure S2).

When synthesis of a substance such as starch is linked to the photosynthetic capacity of the cell, and consequently to light availability, production in large-scale requires efficient utilization of light. In this regard, an effective reactor design that ensures proper light distribution within a high biomass culture is essential [28,29,32]. This can be achieved by decreasing the optical path that light has to travel within the culture and carefully controlling the biomass concentration at the time when supraoptimal temperature is applied.

4.3. Perspectives

The main advantage of the experiments presented here, compared to the other treatments, is the speed of starch accumulation (Table 2). In the context of large-scale cultivation of microalgae, short turnaround times are important not only because they allow for higher productivity but also because they reduce the risk of biological contamination by a competitive microalgal species or fortuitous grazers. This makes the supraoptimal temperature method a viable option for increasing starch productivity in microalgae.

Based on the results described here, a tentative industrial-scale starch production strategy based on the use of supraoptimal temperature can employ both batch and repeated batch modes of operation. This is made possible by the fact that the culture can be recovered and reused as an inoculum after the temperature treatment (Figures 5A and 6). The production flow can consist of alternating biomass accumulation (6–7 days) and temperature treatment phases (1–2 days). Moreover, large and heavy cells that are filled with starch should enable better downstream processing of the biomass. However, the economic viability of the process depends greatly on improving bioreactor design and reducing the associated energy and labor costs.

5. Conclusions

The method of supraoptimal temperature treatment was successfully applied in pilot-scale and resulted in a considerable, nearly 3-fold, enhancement of starch content in *C. reinhardtii* at low biomass densities. Moreover, starch synthesis was faster, with the maximum being reached within only 1–2 days, compared to five days at the optimal temperature. Thus, the supraoptimal temperature treatment provides a viable alternative to other starch stimulating methods, such as nutrient depletion. This is especially true when times required for starch accumulation are taken into account. However, technical challenges, such as bioreactor design and improved light availability per cell, still need to be dealt with.

Supplementary Materials: The following are available online at https://www.mdpi.com/article/10.3390/cells10051084/s1, Figure S1: Effect of the combination of supraoptimal temperature and different biomass concentrations (A) on % of starch in the microalgal biomass (B), mean light availability (C) and starch concentration in the culture (D). Figure S2: Cultivation vessels where *C. reinhardtii* cultures treated at a supraoptimal temperature and with different starting biomass densities have been cultured.

Author Contributions: Conceptualization, K.B., M.J.B., V.Z. and M.V.; methodology, I.N.I.; validation, K.B., M.J.B., V.Z. and M.V.; formal analysis, I.N.I. and K.B.; investigation, I.N.I., M.V., V.Z., M.J.B. and K.B.; resources, V.Z., M.J.B. and K.B.; data curation, I.N.I., M.V., V.Z., M.J.B. and K.B.; writing—original draft preparation, I.N.I., V.Z. and K.B.; writing—review and editing, I.N.I., M.V., V.Z., M.J.B. and K.B.; visualization, I.N.I., M.V., V.Z. and K.B.; supervision, M.J.B. and K.B.; project administration, M.J.B. and K.B.; funding acquisition, M.J.B., K.B. and I.N.I. All authors have read and agreed to the published version of the manuscript.

Funding: This work was supported by the Czech Science Foundation (grant no. 15-09231S) and the Ministry of Education, Youth and Sports of the Czech Republic (project no. EF16_027/0007990).

Institutional Review Board Statement: Not applicable.

Informed Consent Statement: Not applicable.

Data Availability Statement: All data presented in this study are available within this article or Supplementary Materials. There are no special databases associated with this manuscript.

Acknowledgments: We would like to express our sincere gratitude to Rick Wieggers at AlgaeParc™ for his indispensable help in the organization of the experiments and the operation of the microalgae cultivation equipment. We also thank J. D. Brooker for critical reading and language editing of the text.

Conflicts of Interest: The authors declare no conflict of interest.

References

1. Vítová, M.; Bišová, K.; Hlavová, M.; Kawano, S.; Zachleder, V.; Čížková, M. *Chlamydomonas reinhardtii*: Duration of its cell cycle and phases at growth rates affected by temperature. *Planta* **2011**, *234*, 599–608. [CrossRef]
2. Vítová, M.; Bišová, K.; Umysová, D.; Hlavová, M.; Kawano, S.; Zachleder, V.; Čížková, M. *Chlamydomonas reinhardtii*: Duration of its cell cycle and phases at growth rates affected by light intensity. *Planta* **2011**, *233*, 75–86. [CrossRef]

3. Ras, M.; Steyer, J.-P.; Bernard, O. Temperature effect on microalgae: A crucial factor for outdoor production. *Rev. Environ. Sci. Biotechnol.* **2013**, *12*, 153–164. [CrossRef]
4. Singh, S.P.; Singh, P. Effect of temperature and light on the growth of algae species: A review. *Renew. Sustain. Energy Rev.* **2015**, *50*, 431–444. [CrossRef]
5. Zachleder, V.; Ivanov, I.; Vítová, M.; Bišová, K. Effects of cyclin-dependent kinase activity on the coordination of growth and the cell cycle in green algae at different temperatures. *J. Exp. Bot.* **2019**, *70*, 845–858. [CrossRef]
6. Zachleder, V.; Ivanov, I.; Vítová, M.; Bišová, K. Cell cycle arrest by supraoptimal temperature in the alga *Chlamydomonas reinhardtii*. *Cells* **2019**, *8*, 1237. [CrossRef] [PubMed]
7. Semenenko, V.E.; Vladimirova, M.G.; Opleanskaja, O.B.; Raikov, N.I.; Kovanova, E.S. Physiological characteristics of *Chlorella* sp. K under conditions of high extremal temperatures II. Changes in biosynthesis, ultrastructure and activity of photosynthetic apparatus of *Chlorella* at uncoupling cellular functions at extreme temperature. *Russian J. Plant Physiol.* **1969**, *16*, 210–220.
8. Semenenko, V.E.; Vladimirova, M.G.; Orleanskaya, O.B. Physiological characteristics of *Chlorella* sp. K under conditions of high extremal temperatures I. Uncoupling effect of extreme temperatures on the cellular functions of *Chlorella*. *Russian J. Plant Physiol.* **1967**, *14*, 612–625.
9. Šetlík, I.; Zachleder, V.; Doucha, J.; Berková, E.; Bartoš, J. The nature of temperature block in the sequence of reproductive processes in *Chlorella vulgaris* BEIJERINCK. *Arch. Hydrobiol. Algol. Stud.* **1975**, *14*, 70–104.
10. Torres-Romero, I.; Kong, F.; Legeret, B.; Beisson, F.; Peltier, G.; Li-Beisson, Y. *Chlamydomonas* cell cycle mutant crcdc5 over-accumulates starch and oil. *Biochimie* **2019**. [CrossRef]
11. Lacour, T.; Sciandra, A.; Talec, A.; Mayzaud, P.; Bernard, O. Neutral lipid and carbohydrate productivities as a response to nitrogen status in *Isochrysis* sp. (T-iso; Haptophyceae): Starvation vs. limitation. *J. Phycol.* **2012**, *48*, 647–656. [CrossRef]
12. Ball, S.G.; Dirick, L.; Decq, A.; Martiat, J.C.; Matagne, R.F. Physiology of starch storage in the monocellular alga *Chlamydomonas reinhardtii*. *Science* **1990**, *66*, 1–9. [CrossRef]
13. Burlacot, A.; Peltier, G.; Li-Beisson, Y. Subcellular energetics and carbon storage in *Chlamydomonas*. *Cells* **2019**, *8*, 1154. [CrossRef] [PubMed]
14. Chen, H.; Hu, J.; Qiao, Y.; Chen, W.; Rong, J.; Zhang, Y.; He, C.; Wang, Q. Ca^{2+}-regulated cyclic electron flow supplies ATP for nitrogen starvation-induced lipid biosynthesis in green alga. *Sci. Rep.* **2015**, *5*, 15117. [CrossRef]
15. Peltier, G.; Schmidt, G.W. Chlororespiration: An adaptation to nitrogen deficiency in *Chlamydomonas reinhardtii*. *Proc. Natl. Acad. Sci. USA* **1991**, *88*, 4791–4795. [CrossRef]
16. Salome, P.A.; Merchant, S.S. A series of fortunate events: Introducing *Chlamydomonas* as a reference organism. *Plant Cell* **2019**. [CrossRef] [PubMed]
17. Sasso, S.; Stibor, H.; Mittag, M.; Grossman, A.R. From molecular manipulation of domesticated *Chlamydomonas reinhardtii* to survival in nature. *eLife* **2018**, *7*, e39223. [CrossRef] [PubMed]
18. Zhang, M.-P.; Wang, M.; Wang, C. Nuclear transformation of *Chlamydomonas reinhardtii*: A review. *Biochimie* **2021**, *181*, 1–11. [CrossRef] [PubMed]
19. Koo, K.M.; Jung, S.; Lee, B.S.; Kim, J.B.; Jo, Y.D.; Choi, H.I.; Kang, S.Y.; Chung, G.H.; Jeong, W.J.; Ahn, J.W. The mechanism of starch over-accumulation in *Chlamydomonas reinhardtii* high-starch mutants identified by comparative transcriptome analysis. *Front. Microbiol.* **2017**, *8*, 858. [CrossRef] [PubMed]
20. Gifuni, I.; Olivieri, G.; Pollio, A.; Franco, T.T.; Marzocchella, A. Autotrophic starch production by *Chlamydomonas* species. *J. Appl. Phycol.* **2016**, *29*, 105–114. [CrossRef]
21. Schulz-Raffelt, M.; Chochois, V.; Auroy, P.; Cuiné, S.; Billon, E.; Dauvillée, D.; Li-Beisson, Y.; Peltier, G. Hyper-accumulation of starch and oil in a *Chlamydomonas* mutant affected in a plant-specific DYRK kinase. *Biotechnol. Biofuels* **2016**, *9*, 55. [CrossRef]
22. Mathiot, C.; Ponge, P.; Gallard, B.; Sassi, J.F.; Delrue, F.; Le Moigne, N. Microalgae starch-based bioplastics: Screening of ten strains and plasticization of unfractionated microalgae by extrusion. *Carbohydr. Polym.* **2019**, *208*, 142–151. [CrossRef]
23. Sueoka, N. Mitotic replication of deoxyribonucleic acid in *Chlamydomonas reinhardtii*. *Proc. Natl. Acad. Sci. USA* **1960**, *46*, 83–91. [CrossRef]
24. Miller, G.L. Use of dinitrosalicylic acid reagent for determination of reducing sugar. *Anal. Chem.* **1959**, *31*, 426–428. [CrossRef]
25. Fernandes, B.; Teixeira, J.; Dragone, G.; Vicente, A.A.; Kawano, S.; Bišová, K.; Přibyl, P.; Zachleder, V.; Vítová, M. Relationship between starch and lipid accumulation induced by nutrient depletion and replenishment in the microalga *Parachlorella kessleri*. *Bioresour. Technol.* **2013**, *144*, 268–274. [CrossRef]
26. Siaut, M.; Cuine, S.; Cagnon, C.; Fessler, B.; Nguyen, M.; Carrier, P.; Beyly, A.; Beisson, F.; Triantaphylides, C.; Li-Beisson, Y.; et al. Oil accumulation in the model green alga *Chlamydomonas reinhardtii*: Characterization, variability between common laboratory strains and relationship with starch reserves. *BMC Biotechnol.* **2011**, *11*, 7. [CrossRef] [PubMed]
27. Nitsos, C.; Filali, R.; Taidi, B.; Lemaire, J. Current and novel approaches to downstream processing of microalgae: A review. *Biotechnol. Adv.* **2020**, *45*, 107650. [CrossRef] [PubMed]
28. Gifuni, I.; Pollio, A.; Marzocchella, A.; Olivieri, G. New ultra-flat photobioreactor for intensive microalgal production: The effect of light irradiance. *Algal Res.* **2018**, *34*, 134–142. [CrossRef]
29. Carvalho, A.P.; Silva, S.O.; Baptista, J.M.; Malcata, F.X. Light requirements in microalgal photobioreactors: An overview of biophotonic aspects. *Appl. Microbiol. Biotechnol.* **2011**, *89*, 1275–1288. [CrossRef] [PubMed]

30. Murphy, T.E.; Berberoğlu, H. Effect of algae pigmentation on photobioreactor productivity and scale-up: A light transfer perspective. *J. Quant. Spectrosc. Radiat. Transf.* **2011**, *112*, 2826–2834. [CrossRef]
31. Li, T.; Wang, W.; Yuan, C.; Zhang, Y.; Xu, J.; Zheng, H.; Xiang, W.; Li, A. Linking lipid accumulation and photosynthetic efficiency in *Nannochloropsis* sp. under nutrient limitation and replenishment. *J. Appl. Phycol.* **2020**, *32*, 1619–1630. [CrossRef]
32. Huang, Q.; Jiang, F.; Wang, L.; Yang, C. Design of photobioreactors for mass cultivation of photosynthetic organisms. *Engineering* **2017**, *3*, 318–329. [CrossRef]

cells

MDPI

Article

Overexpression of an *Agave* Phospho*enol*pyruvate Carboxylase Improves Plant Growth and Stress Tolerance

Degao Liu [1,2,†], Rongbin Hu [1,†], Jin Zhang [1,2], Hao-Bo Guo [3], Hua Cheng [1], Linling Li [1], Anne M. Borland [1,4], Hong Qin [3], Jin-Gui Chen [1,2], Wellington Muchero [1,2], Gerald A. Tuskan [1,2] and Xiaohan Yang [1,2,*]

[1] Biosciences Division, Oak Ridge National Laboratory, Oak Ridge, TN 37831, USA; liudegao909@gmail.com (D.L.); rhu1@utk.edu (R.H.); zhang007jin@163.com (J.Z.); chenghua1437@126.com (H.C.); lilinling1437@126.com (L.L.); anne.borland@newcastle.ac.uk (A.M.B.); chenj@ornl.gov (J.-G.C.); mucherow@ornl.gov (W.M.); tuskanga@ornl.gov (G.A.T.)
[2] The Center for Bioenergy Innovation (CBI), Oak Ridge National Laboratory, Oak Ridge, TN 37831, USA
[3] Department of Computer Science and Engineering, SimCenter, University of Tennessee Chattanooga, Chattanooga, TN 37403, USA; guohaobo@gmail.com (H.-B.G.); hong-qin@utc.edu (H.Q.)
[4] School of Natural and Environmental Science, Newcastle University, Newcastle upon Tyne NE1 7RU, UK
* Correspondence: yangx@ornl.gov; Tel.: +1-865-241-6895; Fax: +1-865-576-9939
† These authors contribute equally to this work.

Abstract: It has been challenging to simultaneously improve photosynthesis and stress tolerance in plants. Crassulacean acid metabolism (CAM) is a CO_2-concentrating mechanism that facilitates plant adaptation to water-limited environments. We hypothesized that the ectopic expression of a CAM-specific phospho*enol*pyruvate carboxylase (PEPC), an enzyme that catalyzes primary CO_2 fixation in CAM plants, would enhance both photosynthesis and abiotic stress tolerance. To test this hypothesis, we engineered a CAM-specific *PEPC* gene (named *AaPEPC1*) from *Agave americana* into tobacco. In comparison with wild-type and empty vector controls, transgenic tobacco plants constitutively expressing *AaPEPC1* showed a higher photosynthetic rate and biomass production under normal conditions, along with significant carbon metabolism changes in malate accumulation, the carbon isotope ratio $\delta^{13}C$, and the expression of multiple orthologs of CAM-related genes. Furthermore, *AaPEPC1* overexpression enhanced proline biosynthesis, and improved salt and drought tolerance in the transgenic plants. Under salt and drought stress conditions, the dry weight of transgenic tobacco plants overexpressing *AaPEPC1* was increased by up to 81.8% and 37.2%, respectively, in comparison with wild-type plants. Our findings open a new door to the simultaneous improvement of photosynthesis and stress tolerance in plants.

Keywords: *Agave americana*; crassulacean acid metabolism; genetic engineering; *Nicotiana sylvestris*; phospho*enol*pyruvate carboxylase; photosynthesis; drought tolerance; salt tolerance

Citation: Liu, D.; Hu, R.; Zhang, J.; Guo, H.-B.; Cheng, H.; Li, L.; Borland, A.M.; Qin, H.; Chen, J.-G.; Muchero, W.; et al. Overexpression of an *Agave* Phospho*enol*pyruvate Carboxylase Improves Plant Growth and Stress Tolerance. *Cells* **2021**, *10*, 582. https://doi.org/10.3390/cells10030582

Academic Editor: Suleyman Allakhverdiev

Received: 2 February 2021
Accepted: 18 February 2021
Published: 6 March 2021

1. Introduction

Human population increases, global climate changes and natural resources reductions could seriously impact food and energy security in the future [1–8]. The drylands, with precipitation amounts that are inadequate for most present-day food and bioenergy crops, covers approximately 40% of the world's land [9,10]. Moreover, around 20% of irrigated areas in the world are under salt stress which has become a big constraint limiting agricultural production [11–13]. In addition, the phytotoxic effects of nanoparticles could impact plant growth [14,15]. To address these challenges, tremendous efforts have been put into improving photosynthesis, drought tolerance and salt tolerance in crop plants through breeding and genetic engineering over the past 50 years, though there has been limited success in simultaneously enhancing both biomass production and stress tolerance [16]. One common problem shared by previous breeding and genetic engineering approaches is that they have focused on drought- or salt-responsive genes in C_3 or C_4 species, with narrow genetic diversity for the tolerance of abiotic stresses.

A potential solution to this problem can be obtained from the desert plants that employ crassulacean acid metabolism (CAM) for photosynthesis (e.g., *Agave americana*), as these plants maximize water-use efficiency (WUE) by shifting all or part of their net atmospheric carbon dioxide (CO_2) uptake to the nighttime, when the evapotranspiration rates are generally lower than that in the daytime [1,2,8,9,17]. The nocturnal CO_2 taken up in CAM plants is intracellularly fixed, usually leading to the generation of C_4 organic acid (e.g., malate) as storage intermediates, which accumulate in the vacuole during the dark period before being gradually catabolized during the day [2,9]. High rates of malate decarboxylation behind closed stomata during the day can concentrate CO_2 from 2- to 60-fold around ribulose-1–5-bisphosphate carboxylase/oxygenase (Rubisco) [9,18]. This effectively creates a 'CO_2 pump' that can strongly favor the carboxylase activity of Rubisco and potentially suppress photorespiration, a process that can reduce photosynthesis by up to 40% in C_3 photosynthesis plants [1,2,9,19–21]. In addition, as an obligate CAM species, *Agave americana* is highly tolerant to salinity with an electrical conductivity (EC) level of 9.4 dS m^{-1} [22]. Additionally, *A. americana* has been reported to achieve high biomass productivity and drought tolerance when grown as a crop on arid and semi-arid lands [23].

CAM photosynthesis can be divided into nocturnal and day-time reactions [24–27]: At night atmospheric CO_2 is taken up through open stomata and primary fixation of CO_2 (as HCO_3^-) by phospho*enol*pyruvate carboxylase (PEPC) produces oxaloacetate (OAA) that is subsequently converted to malate by malate dehydrogenase (MDH). Malate is transported by tonoplast-localized aluminum-activated malate transporter (ALMT) into the vacuole, where it is stored as malic acid throughout the dark period. During the light period, Rubisco fixes the CO_2 released from NAD(P)-malic enzyme (ME)- or PEP carboxykinase (PEPCK)-mediated malate decarboxylation when stomata are closed [9,19]. It has been shown that PEPC is a highly abundant enzyme in leaves of *A. americana* and its transcript level and protein abundance are correlated with CAM [28]. In this study, we identified a CAM-isoform *PEPC* through genome-wide analysis of the *PEPC* gene family in *A. americana*, as well as protein structure modeling analysis and molecular dynamics (MD) simulation studies. We then determined the impacts of the overexpression of the *Agave PEPC* on the photosynthetic carbon metabolism, and abiotic stress tolerance in the model C_3 plant tobacco.

2. Materials and Methods

2.1. Genome-Wide Analysis of the PEPC Gene Family

Arabidopsis thaliana PEPC sequences (AT1G68750.1, AT1G53310.1, AT3G14940.1, and AT2G42600.2) were retrieved and used as queries in BLAST searches against *Agave americana* transcriptomics data [28] to identify potential *PEPC* genes. An expectation (E) value of $<1 \times 10^{-10}$ was used to obtain the homologous protein sequences of the predicted *PEPC* family members. The expression data of the *Agave* PEPCs were also obtained from the *A. americana* transcriptomics data [28]. For gene expression patterns of *Agave PEPC*, the log10 transformed FPKM values and z-score normalized relative expression were used for heatmap analysis.

2.2. Phylogenetic Analysis

Multiple alignment of PEPC proteins was performed using the MAFFT online service [29]. The maximum likelihood (ML) phylogenetic tree was constructed using W-IQ-TREE [30]. The sequences used for the analysis were: foxtail millet (AY491400), foxtail millet C_4 (AF495586), maize C_3 (X61489), maize C_4 (X15642), maize root (AB012228), rice C_3 (Os08g0366000, Os09g0315700, and Os01g0758300), rice root (Os02g0244700), sorghum C_4 (X63756), sugarcane C_3 (M86661), sugarcane C_4 (AY135709), and wheat (AJ007705) [31]. The bootstrap values were calculated as percentages for 1000 replications.

2.3. Structural Modeling and Molecular Dynamics Simulation

The *A. americana* PEPC model was constructed using the I-TASSER v. 5.1 protein structure modeling toolkit [32]. Structure averaging from multiple MD simulations or a single long time-scale MD simulation could effectively refine the predicted structures [33]. Here, starting with the best I-TASSER model, the rotamer states of Asn, Gln, and His residues as well as protonation states of titratable residues were validated by MolProbity [34]. The HBUILD module in CHARMM [35] was employed to add missing hydrogen atoms to the model. A water box with the size of 118 × 106 × 102 Å3 (at least 15 Å to the edge of the protein) was used, and sodium and chloride ions were added to neutralize the net charge. The final model contained 120,210 atoms with 34,893 water molecules. NAMD [36] was used to perform the MD simulations with the CHARMM protein force field [37] and TIP3P water model [38]. Using the SHAKE algorithm to fix all bond lengths involving hydrogen atoms, a time step of 2-fs was applied to all MD simulations. A 50k-step energy minimization was conducted, followed by a "natural" heating to 300 K with the rate of 0.001 K/step for 300 k steps. An NPT ensemble maintained by Langevin piston controls was used in the MD simulations with a system pressure of 1 atm and a temperature of 300 K. A cutoff switching between 9 and 11 Å was applied for non-bonded interactions. The particle mesh Ewald (PME) summation was applied for long-range electrostatic interactions with a grid spacing of 1.35 Å. MD simulation was performed for 200 ns, and analysis was carried out on the final 50 ns of the MD trajectory. The PDBsum online tool [39] was used to plot the cartoon topology of the protein structure.

2.4. Plasmid Construction

A 2943-bp DNA fragment containing the coding sequence of *AaPEPC1* (Aam080248) [28] fused to two FLAG epitope tags [40] was chemically synthesized by Integrated DNA Technology (Coralville, IA, USA) and used to produce a chimeric gene construct, p35S:*FLAG-AaPEPC1*/pNOS: *nptII*. The vector contains the CaMV35S promoter which drives *FLAG-AaPEPC1*, and the nopaline synthase (NOS) promoter which drives the *nptII* gene for kanamycin resistance as a selection marker. The vector was delivered into the GV3101 *Agrobacterium tumefaciens* strain via the freeze-thaw method [41] for plant transformation.

2.5. Plant Transformation

A tobacco (*Nicotiana sylvestris*) cultivar (USNGC TW136, PI555569) was used for genetic transformation. The generation and culturing of transgenic plants were carried out as previously described [42]. The transgenic lines were based on single copy lines with a segregation ratio of approximately 3:1 (kanamycin resistance vs. sensitivity) in the T1 generation, and homozygous lines were presumed if there was no segregation in the T2 and T3 generations (n > 100).

2.6. Measurement of Photosynthesis

The photosynthetic rates of leaves of transgenic and wild-type (WT) plants that were well-watered and grown in pots for 6 weeks was measured as previously described [11]. Mature leaves from 3 individual replicate plants were used for gas exchange analysis by an LI-COR Portable Photosynthesis System (LI-COR Inc., Lincoln, NE, USA). The relative chlorophyll content (SPAD value in fresh leaves) was analyzed using an SPAD-502 Chlorophyll Meter (Minolta, Japan) [11].

2.7. Analysis of Malate, Glucose and Proline Content

Mature leaves were sampled and frozen in liquid nitrogen at the indicated times and stored at −80 °C until use. The leaf samples were ground to a fine powder and assayed for malate and glucose content using the standard enzyme-linked spectrophotometric methods according to the manufacturer's instructions (Sigma-Aldrich, St. Louis, MO, USA), respectively.

The proline content was measured as described by He, et al. [43].

2.8. Carbon Isotope Ratio Analysis

Plants were well watered throughout the growing period. Mature leaves were harvested from 6-week-old plants and dried for 1 week at 50 °C. Finely ground dry powder was placed in capsules and then analyzed at the University of California Davis Stable Isotope Facility (http://stableisotopefacility.ucdavis.edu, accessed on 25 February).

2.9. Salt and Drought Stress Treatment

For salt tolerance analysis, the transgenic plants and controls were watered with 200 mM NaCl solution every other day for 4 weeks as previously described [11]. For drought tolerance analysis, the transgenic and WT plants were exposed to progressive drought stress by withholding water until a nearly lethal effect of dehydration was observed on the WT. A recovery study was carried out for plants under drought stress by re-irrigating with water [44]. After salt or drought treatment, all plants were dried for 48 h in an oven at 80 °C and weighed [11]. All treatments were performed in triplicate.

2.10. Expression Levels Analysis of the Related Genes

The expression levels of related genes in the transgenic and WT plants were analyzed by qRT-PCR as previously described [11]. Total leaf RNA was extracted from transgenic and WT plants using the Spectrum Plant Total RNA Kit (Sigma-Aldrich, St. Louis, MO, USA). RNA samples were reverse-transcribed using a High-Capacity cDNA Reverse Transcription Kit (Applied Biosystems, Foster City, CA, USA). The cDNA solution was used as a template for qPCR amplification using SYBR Green Master Mix (Applied Biosystems, Foster City, CA, USA) with the specific primers designed for each gene (Supplementary Table S1). The tobacco *actin* gene was used as an internal control (Supplementary Table S1). The quantification of the gene expression level was performed with comparative CT method [45].

2.11. Statistical Analysis

The data presented as the mean ± SD were analyzed by a one-way ANOVA analysis with post-hoc Tukey honestly significant difference (HSD) test. p-values of <0.05 or <0.01 were considered to be statistically significant.

3. Results

3.1. CAM-Specific PEPC in Agave americana

Through a tBLASTn search against *A. americana* transcriptomics data [28], using the PEPC protein sequences of *Arabidopsis thaliana* as queries, we identified a total of 21 transcripts encoding PEPC in *A. americana*. Several types of PEPC are present in plants, including plant-type PEPCs (PTPCs) and one bacterial-type PEPC (BTPC) [46,47]. The plant-type PEPCs studied so far have been classified into four groups: C_3, C_4, and CAM-types from photosynthetic tissues and root-type from non-photosynthetic tissue [31]. In order to gain insight into the evolutionary relationships among PTPCs, we constructed a phylogenetic tree using the 21 predicted *Agave PEPC* transcripts and the C_3-, C_4-, CAM- and root-type PEPCs from rice (*Oryza sativa*), maize (*Zea mays*), wheat (*Triticum aestivum*), sugarcane (*Saccharum* spp.), sorghum (*Sorghum bicolor*), foxtail millet (*Setaria italica*) and orchid (*Phalaenopsis equestris*) [24,31]. A phylogenetic analysis of the plant-type PEPCs indicated that Aam080248 [28] (named AaPEPC1) belongs to CAM-type PEPC (Figure 1a). Our results also showed the CAM- and C_4-type PEPCs belong to different clades, suggesting that these two PEPC types evolved independently.

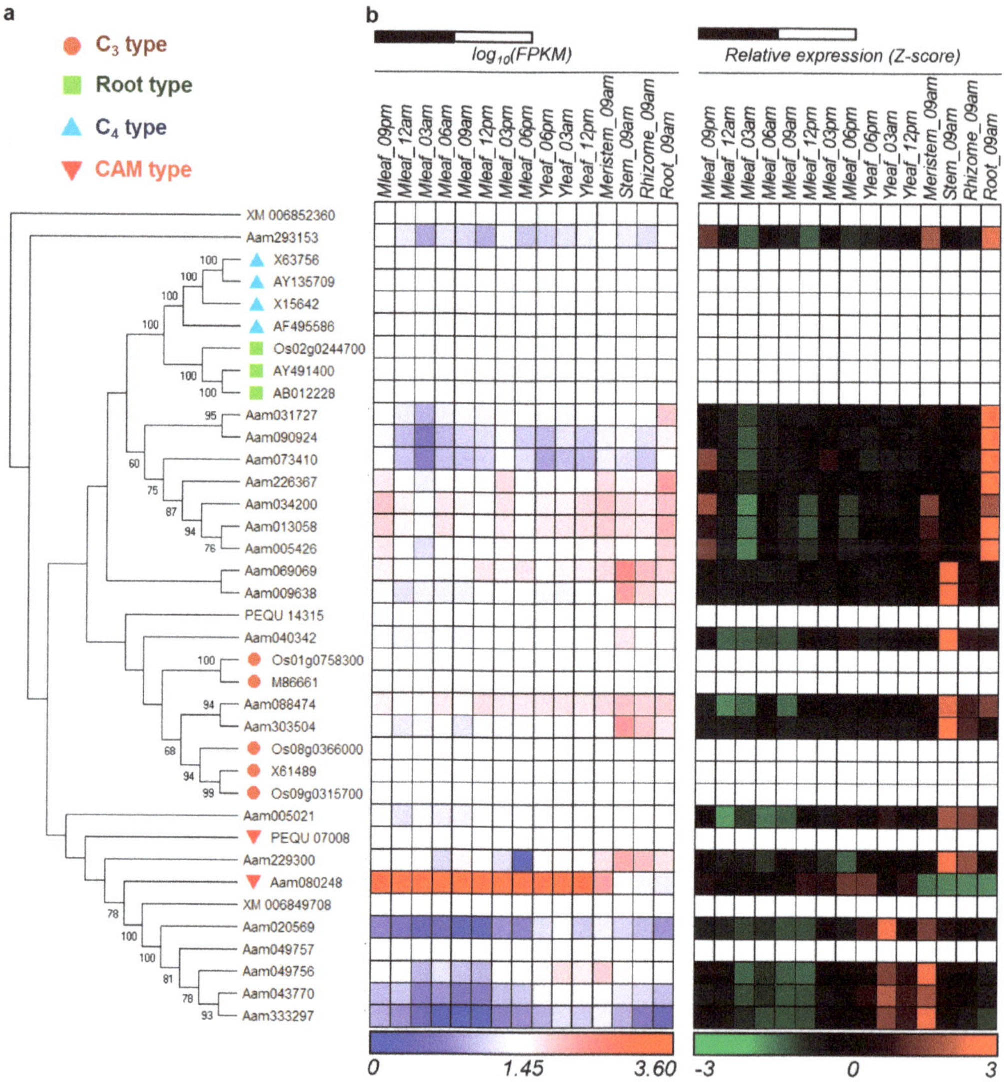

Figure 1. Identification of the crassulacean acid metabolism (CAM)-isoform of phospho*enol*pyruvate carboxylases (PEPC) in *Agave americana*. (**a**) Phylogenetic relationships of PEPCs in *A. americana* and plant-type PEPCs from other monocot plants. Bootstrap values are shown at nodes. The sources for the PEPC sequences (as well as species abbreviations, etc.) are provided in 'Materials and Methods'. (**b**) Diel transcript expression of the *PEPC* genes in *A. americana*. The Z-score is defined as $(x_i - \mu_i)/\sigma_i$, where x_i is the FPKM of gene i, μ_i is the mean and σ_i is the standard deviation of all 15 columns. White and black bars indicate daytime (12 h) and nighttime (12 h), respectively. Mleaf: mature leaf; Yleaf: young leaf.

Using the quantitative gene transcript abundance data obtained from the transcriptome analysis of 15 tissues, including mature leaves sampled over a diel cycle (eight time-points) and, young leaves (three time-points), meristem, stem, root and rhizome in *A. americana* [28], we obtained the transcript pattern of the 21 *PEPC* transcripts (Figure 1b), Among which Aam080248 was the most abundant transcript in mature leaves, with a transcript abundance peak during the late afternoon just before the start of the dark period.

These results supported the above computational phylogenetic prediction (Figure 1a) that the Aam080248 gene encodes the CAM-specific PEPC in *A. americana*.

3.2. *AaPEPC1 Binds to Phosphoenolpyruvate*

Phospho*enol*pyruvate (PEP) is the substrate for PEPC enzymes [48,49]. To understand whether AaPEPC1 binds to PEP, we developed a protein structural model using I-TASSER (v.5.1) accompanied by a 200 ns MD simulation (Figure 2a–c). The last 50 ns of the MD trajectory were taken to refine the AaPEPC1 structure. The root mean square deviation (RMSD) of the α-carbon atoms of all residues was approximately 2.0 Å (Figure 2c), which is reasonable for fluctuations of the 964 amino acids (AA) protein. From the last 50 ns MD trajectory, the snapshot that was closest to the average structure (with RMSD = 0.8 Å) was selected as the final model (Figure 2a,b). Since the crystal structures of bacterial (*Escherichia coli*) PEPC (PDB entry 1JQN) and *Z. mays* C_4-type PEPC (PDB entry 5VYJ) are well-characterized [48–50], we compared the protein structure of AaPEPC1 with the *E. coli* PEPC and *Z. mays* C_4-type PEPC. In line with *E. coli* PEPC and *Z. mays* C_4-type PEPC, the AaPEPC1 model was found to be dominated by α-helix regions (546 AAs, 57.7%). The Ramachandran plot of the AaPEPC1 structure was found to have 93.4% (804 AAs) in the most favorable regions, compared to 92.6% for the *E. coli* PEPC and 90.5% for the *Z. mays* C_4-type PEPC (Figure 2d–f). In addition, 5.2% (45 AAs) of AaPEPC1 were found to be in the additional allowed regions (7.1% for *E. coli* PEPC and 8.7% for *Z. mays* C_4-type PEPC), 1.2% (10 AAs) were found to be in the generously allowed regions (0.1% for *E. coli* PEPC and 0.2% for *Z. mays* C_4-type PEPC) and 0.2% (2 AAs) were found to be in disallowed regions (0.1% for *E. coli* PEPC and 0.3% for *Z. mays* C_4-type PEPC). The N-terminus of AaPEPC1 was found to contain the plant-specific serine residue (S9) (Figure 2a), which is located in the middle of an α-helix flanked by a long loop region. The plant-specific Ser residue near the N-terminus of PEPC is well-known in C_4 plant species [48,51]. This serine residue, however, was absent in bacterial PEPC (e.g., *E. coli* PEPC, PDB entry 1JQN) (Figure 2a,b). The AaPEPC1 model suggested that the AaPEPC1 can efficiently bind to PEP, which is bound by H171 and R640, where R640 is located at a GRGGXXGR^{640}GG motif (Figure 2a) and is overlapped with R647 in the *Z. mays* C_4-type PEPC, whereas H171 is overlapped with H177 in the *Z. mays* C_4-type PEPC, hence both H171 and R640 in AaPEPC1 may directly participate in the carboxylation reaction as proposed in the C_4-type PEPC [48]. In addition, based on the PEP-AaPEPC1 complex model, R449 (R456 in *Z. mays* C_4-type PEPC) may also be involved in PEP binding and PEPC catalysis.

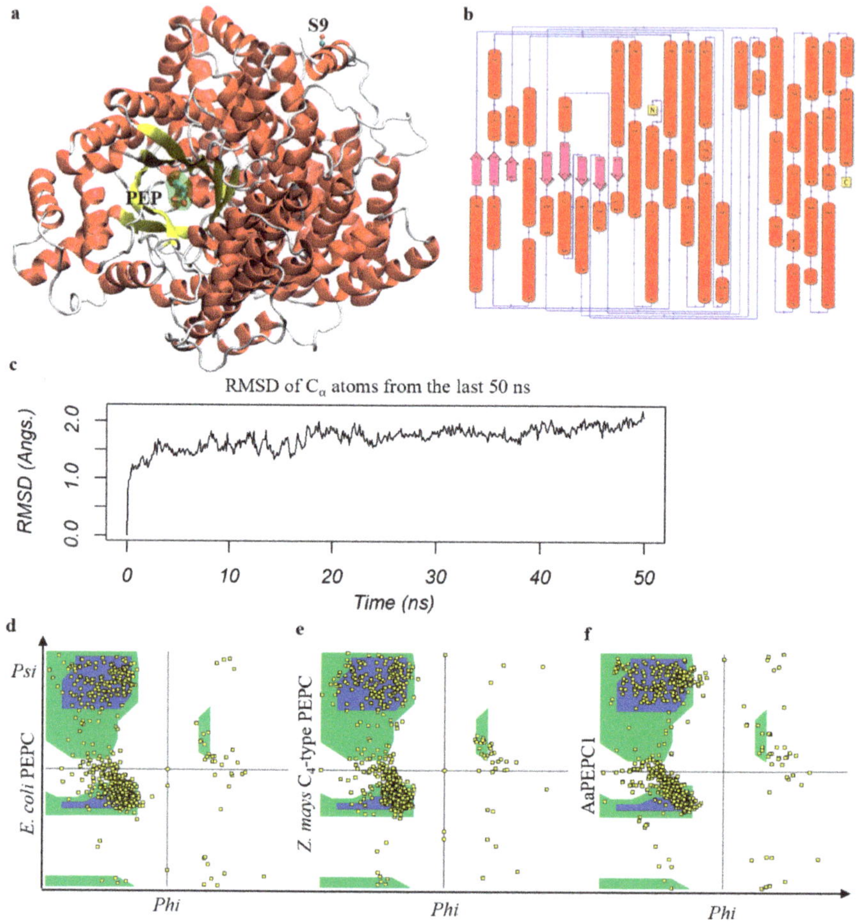

Figure 2. AaPEPC1 binds to phospho*enol*pyruvate (PEP) substrate. (**a**) Left: AaPEPC1 structure selected from 200 ns molecular dynamics (MD) simulation that is closest to the average structure of the whole trajectory (root mean square deviation RMSD = 0.8 Å). The protein structure is shown in cartoons with helices in red, strands in yellow and coils/turns in white. A PEP substrate that binds to the β-barrel (yellow) active site is shown in sticks and spheres. The plant-specific N-terminal serine residue (S9) is also shown. The binding position of the PEP substrate was obtained from the *Escherichia coli* PEPC–PEP complex (PDB entry 1JQN) [50]. (**b**) A cartoon representation of the topology of AaPEPC1 structure. The cylinders represent alpha-helices. (**c**) The RMSD (Å) profile of the final 50 ns of the MD simulation. The deviations of all snapshots to the average MD structure are 1.4 ± 0.2 Å (not shown). (**d–f**) Ramachandran plot of the *E. coli* (PDB entry 1JQN) PEPC, *Zea mays* C_4-type PEPC (PDB entry 5VYJ) [49], and AaPEPC1 (the final snapshot of the 200 ns MD trajectory), respectively.

3.3. Development of Transgenic Tobacco Lines Overexpressing AaPEPC1

After we identified the CAM-isoform of *PEPC* from *A. americana*, we wanted to determine the impact of overexpressing this *Agave* gene on the photosynthetic carbon metabolism and abiotic stress tolerance in *Nicotiana sylvestris* (tobacco), which is a C_3 plant species. The *AaPEPC1* coding sequence was cloned into binary vector pBI121, downstream of the cauliflower mosaic virus 35S (CaMV35S) promoter, to yield p35S::AaPEPC1 for transformation into *N. sylvestris* (Supplementary Figure S1a). The p35S::AaPEPC1 vector and empty vector (pBI121) were transferred into tobacco via *A. tumefaciens*-mediated genetic transformation [42]. Transformants harboring a single copy of transgene were identified from the segregation ratio for kanamycin resistance. Two T2 transgenic homozygous lines (p35S::AaPEPC1_OE1 and p35S::AaPEPC1_OE2) exhibiting different expression levels of the *AaPEPC1* (Supplementary Figure S1b) were selected as representative lines for subsequent phenotypic characterization.

3.4. Overexpression of AaPEPC1 Increases Photosynthetic Rate and Changes Stable Carbon Isotope Ratio

We determined whether the constitutive overexpression of *AaPEPC1* affected or influenced the photosynthetic rate in the transgenic tobacco plants, along with empty vector (EV) and WT control plants, grown under normal conditions (12 h light/12 h dark photoperiod; without drought- or salt-stress) at four time points (i.e., 3, 9, 15 and 21 h after the beginning of the light period). The transgenic plants overexpressing *AaPEPC1* showed significantly higher photosynthetic rates than the WT and EV controls in the light period (Figure 3a). A carbon isotope ratio $\delta^{13}C$ is a broadly accepted indicator of the extent to which the biomass is derived from PEPC-mediated CO_2 fixation in plants, because PEPC discriminates less against ^{13}C than Rubisco which is responsible for most CO_2 fixation in C_3 plants during the light period [52]. The positive correlation between the $\delta^{13}C$ values and CAM activity has been demonstrated to be a simple and reliable method for determining the type of photosynthesis, including that of C_3, C_3-CAM intermediate and CAM [52–54]. In this study, we found that the $\delta^{13}C$ values were significantly increased (i.e., became less negative) in the transgenic line (OE2), in which the expression level of *AaPEPC1* was 7.74-fold higher than that in the other transgenic line (OE1) (Supplementary Figure 1b), in comparison with the controls (Figure 3b), thus indicating that this transgenic line was using PEPC for photosynthetic carbon assimilation and production of biomass.

3.5. The Impact of AaPEPC1 Overexpression on the Accumulation of Malate and Glucose

PEPC functions in the production of OAA, thus leading to synthesis of malate/malic acid, which is a key intermediate of the tricarboxylic acid (TCA) cycle that links lipids and glucose metabolisms with photosynthesis in C_3 plants [47,55]. The diel fluctuation in malate content represents a central biochemical correlate of the CAM photosynthesis pathway [9]. Additionally, the glycolytic breakdown of glucose provides substrate for the primary carboxylation reaction at night whilst the day-time recovery of glucose via gluconeogenesis is an important sink for C released from malate decarboxylation [9,56]. To determine the impact of *AaPEPC1* overexpression on malate and glucose production in transgenic tobacco, the malate and glucose contents were measured at four time points (i.e., 3, 9, 15 and 21 h after the beginning of the light period). The transgenic plants overexpressing *AaPEPC1* showed higher malate and glucose contents than the WT and EV controls at all four time points during the day and night (Figure 3c,d). Furthermore, the transgenic plants expressing *AaPEPC1* showed a significant increase in malate content at 15 h compared to that at 9 h after the beginning of the light period, while no significant difference was found between the malate contents at 15 and 9 h in the control plants (Figure 3c). The rewired diel malate accumulation–depletion pattern suggested that *AaPEPC1* overexpression affects primary carboxylation in transgenic plants. Additionally, we found that higher amounts of glucose were broken down in the early night (at 15 h after the beginning of the dark period) in the transgenic plants overexpressing *AaPEPC1* compared with control plants (Figure 3d), suggesting that the overexpression of the *AaPEPC1* enhances glycolysis and consequently supplies more substrates for primary carboxylation.

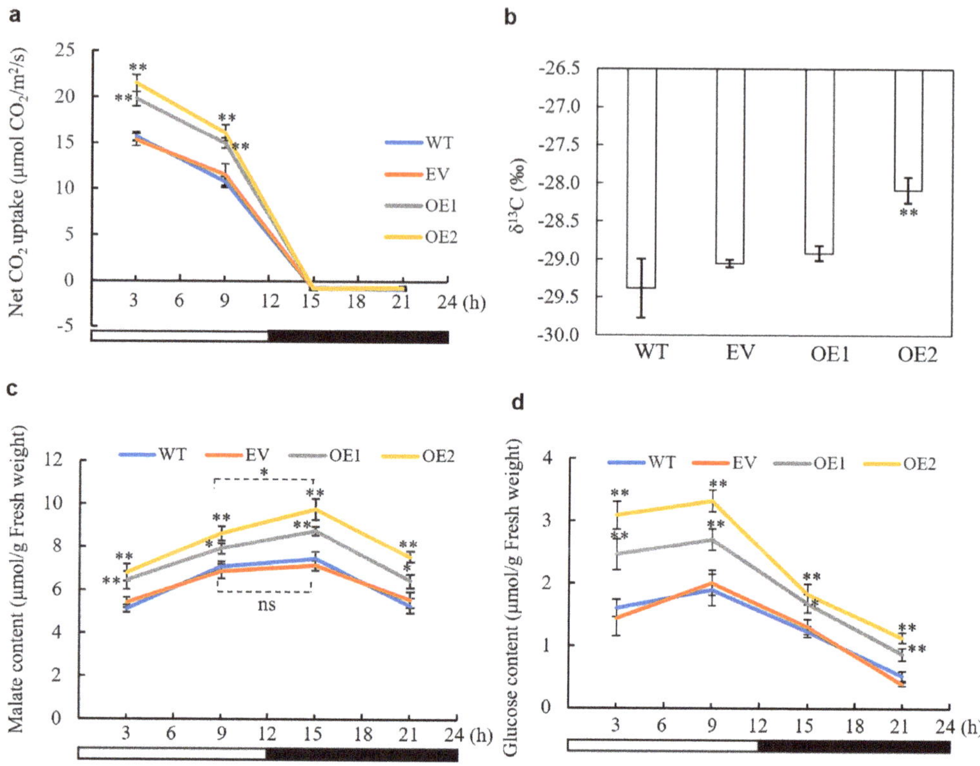

Figure 3. Impact of overexpressing *AaPEPC1* on photosynthetic rate (**a**), carbon isotope ratio $\delta^{13}C$ (**b**), malate content (**c**), and glucose content (**d**) in tobacco. OE1 and OE2: the transgenic plants expressing *AaPEPC1*. EV: empty vector control. WT: wild-type plants. White and black bars indicate the light period (12 h) and the dark period (12 h), respectively. *X*-axis represents the time after the beginning of the light period at 00:00 h. Values represent means ± SD (n = 3 individual replicate plants). * and ** indicated significant difference from that of WT at $p < 0.05$ and $p < 0.01$, respectively, by one-way ANOVA analysis with a post-hoc Tukey honestly significant difference (HSD) test. Ns = non-significant.

3.6. CAM-Related Genes Were Up-Regulated by AaPEPC1 Overexpression

To test whether the re-programed changes in diel malate and glucose content led to feedback regulation of CAM pathway genes, the transcript abundance of the orthologs of CAM genes in the transgenic and WT plants was analyzed using qRT-PCR (Figure 4a). The expression of carbonic anhydrase (NsyCA), an ortholog of CA that is responsible for rapid interconversion of CO_2 and HCO_3^- in CAM, was more increased in transgenic plants expressing *AaPEPC1* compared with that in the controls at 9, 15 and 21 h after the beginning of the light period (Figure 4b). The expression levels of malate dehydrogenase (*NsyMDH*) (an ortholog of MDH that is responsible for catalyzing the oxidation of malate to OAA in CAM) and tonoplast aluminum-activated malate transporter (*NsyALMT*) (an ortholog of ALMT that is responsible for transporting malate into vacuole in CAM) in the transgenic plants expressing *AaPEPC1* were higher than those in the controls at the early night period (15 h after the beginning of the light period)(Figure 4c,d). The expression of tonoplast dicarboxylate transporter (*NsyTDT*) (an ortholog of TDT that may be responsible for transporting malic acid out of vacuole in CAM) was up-regulated in transgenic plants expressing *AaPEPC1* at the early morning period (3 h after the beginning of the light period) (Figure 4e). The expression level of malic enzyme (*NsyME*) (an ortholog of ME that may be responsible for the decarboxylation of malate in CAM species) in transgenic *AaPEPC1* plants was increased during the light period (Figure 4f). These results demonstrated that the overexpression of the *AaPEPC1* up-regulates the expression of the orthologs of CAM pathway genes in transgenic plants.

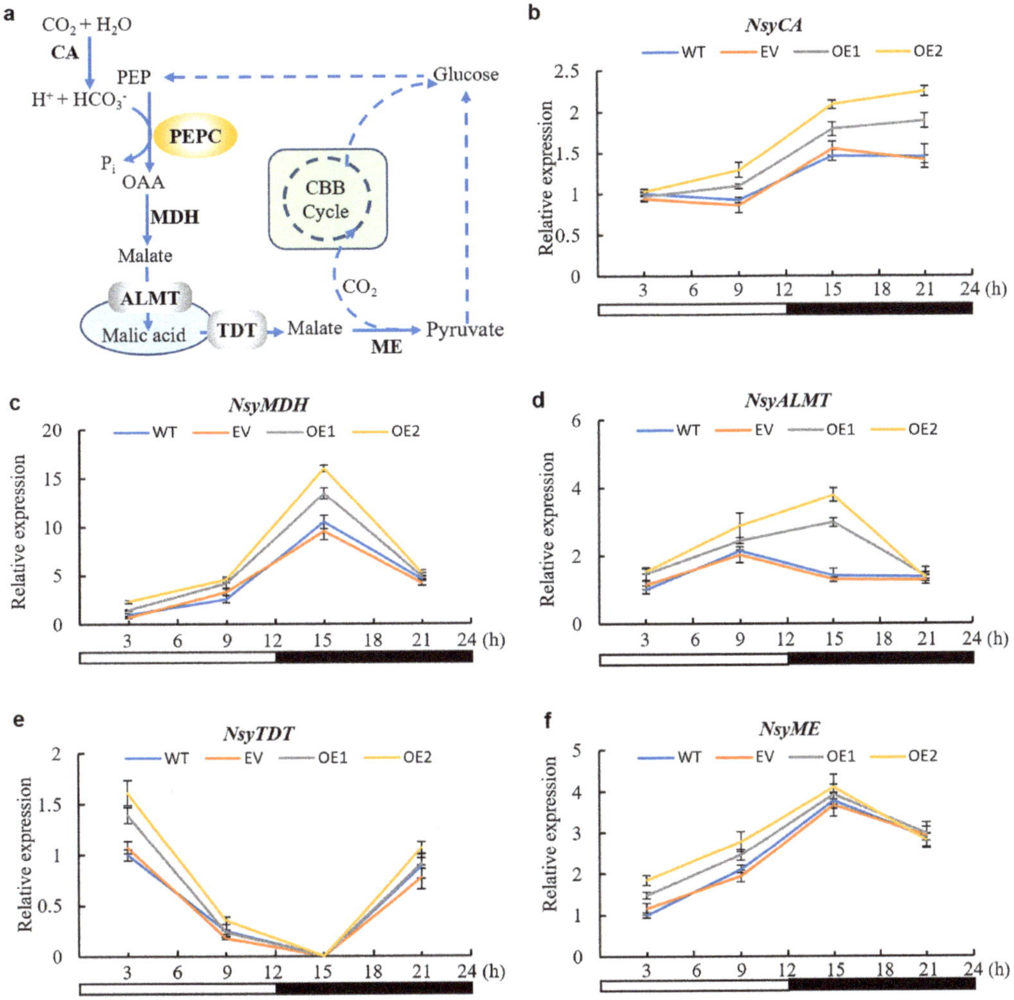

Figure 4. Relative expression level of the orthologs of CAM pathway genes in *AaPEPC1*-overexpressing tobacco plants. (**a**) Diagram of the CAM pathway in malic enzyme (ME) subtype [24]. CA: carbonic anhydrase; PEPC: phospho*enol*pyruvate carboxylase; OAA: oxaloacetate; MDH: malate dehydrogenase; ALMT: aluminum-activated malate transporter; TDT: tonoplast dicarboxylate transporter; CBB: Calvin-Benson-Bassham. (**b–f**) Relative expression level of *Nicotiana sylvestris* CA (*NsyCA*, XM_009805732.1), *NsyMDH* (XM_009784202.1), *NsyALMT* (XM_009797046.1), *NsyTDT*, (XM_009797970.1) and *NsyME* (XM_009781546.1), respectively. The *actin* gene (XM_009774717.1) was used as an internal control. The values were normalized to expression in the wild-type plants (WT) at the 3 h light time-point. White and black bars indicate daytime (12 h) and nighttime (12 h), respectively. *X*-axis represents the time after the beginning of the light period. Values represent means ± SD (n = 3).

3.7. Impact of AaPEPC1 Overexpression on Biomass Production

Based on the above, we speculated that the higher photosynthetic rates, glucose content and carbon isotope ratio $\delta^{13}C$ measured in transgenic plants overexpressing *AaPEPC1* may result in improved biomass production. To test this, we examined the growth of the transgenic plants overexpressing *AaPEPC1*, along with the EV and WT control plants, in growth chambers under well-water conditions. After six weeks of growth, the transgenic *AaPEPC1* plants exhibited larger physical size than the WT and EV controls (Figure 5a). The dry weight of the transgenic *AaPEPC1* plants was significantly increased in comparison with the control plants (Figure 5b), demonstrating that *AaPEPC1* overexpression increases biomass production in transgenic plants under normal conditions.

Figure 5. Growth characteristics of transgenic tobacco plants expressing *AaPEPC1*. (**a**) Phenotypes of transgenic plants expressing *AaPEPC1* (OE1 and OE2) as well as empty vector (EV), and wild-type plants (WT), grown in pots for 6 weeks under well-watered conditions. (**b**) Dry weight (shoot and root) of transgenic plants and WT. Values represent means ± SD (n = 3). * and ** indicate significant difference from that of WT at $p < 0.05$ and $p < 0.01$, respectively, by one-way ANOVA analysis with post-hoc Tukey HSD.

3.8. Impact of AaPEPC1 Overexpression on Salt and Drought Tolerance

Most crop plants are susceptible to salinity. The NaCl stress at concentrations of 100–200 mM can inhibit or even completely prevent plant growth, resulting in their death [57,58]. To investigate whether the overexpression of *AaPEPC1* enhanced salt tolerance in transgenic plants, the transgenic *AaPEPC1* plants as well as the EV and WT controls were grown in pots and irrigated with 200 mL of 200 mM NaCl solution once every two days for four weeks. The salt-stress treatment caused the death of the WT and EV control plants, while the transgenic plants overexpressing *AaPEPC1* maintained growth (Figure 6a). The dry weight of the transgenic plants overexpressing *AaPEPC1* was significantly increased compared to the controls (Figure 6b), establishing that the overexpression of *AaPEPC1* significantly enhanced tolerance to salt stress in transgenic tobacco plants.

To investigate whether overexpression of *AaPEPC1* enhanced drought tolerance in transgenic plants, the transgenic *AaPEPC1* plants, and EV and WT controls, were subjected to drought stress in growth chamber condition. After 15 days without watering, all WT and EV control plants displayed severe wilting (all leaves were severely curled and most leaves had turned yellow/or were dead), whereas the growth of transgenic tobacco plants expressing *AaPEPC1* was less affected and their youngest leaves were still green and expanded (Figure 7a). Three days after re-watering, all WT and EV controls were nearly dead, whereas all transgenic lines expressing *AaPEPC1* survived and started to regrow (Figure 7b). Dry weight of the transgenic plants expressing *AaPEPC1* was also significantly increased compared to the WT and EV controls (Figure 7c), providing evidence that overexpression of *AaPEPC1* improved the drought tolerance in the transgenic tobacco plants.

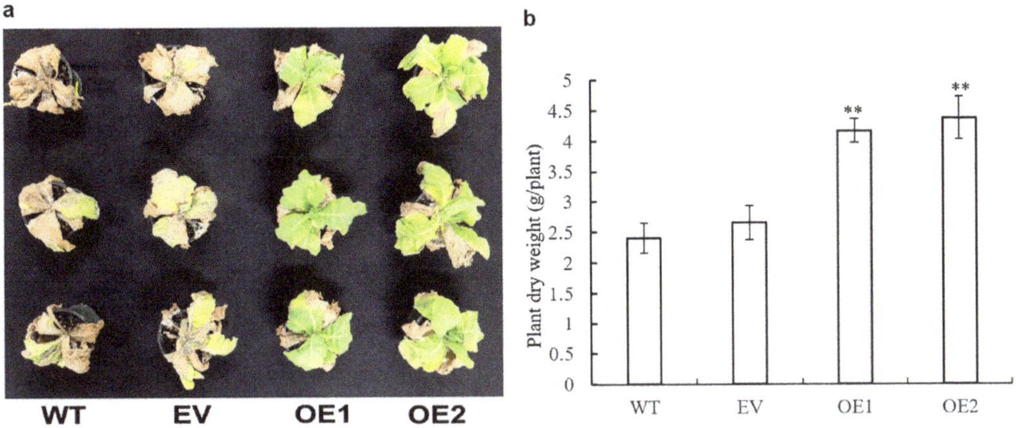

Figure 6. Responses of the *AaPEPC1*-overexpressing tobacco plants under salt stress. (**a**) Phenotypes of transgenic plants expressing *AaPEPC1* (OE1, OE2) or empty vector (EV), and wild-type plants (WT) grown in pots under 200 mM NaCl stress. The plants were irrigated with 200 mM NaCl solution once every 2 days for 4 weeks. (**b**) Dry weight of transgenic plants and WT. Values represent means ± SD (n = 3). ** significant difference from that of WT $p < 0.01$, by one-way ANOVA analysis with post-hoc Tukey HSD.

Figure 7. *Cont.*

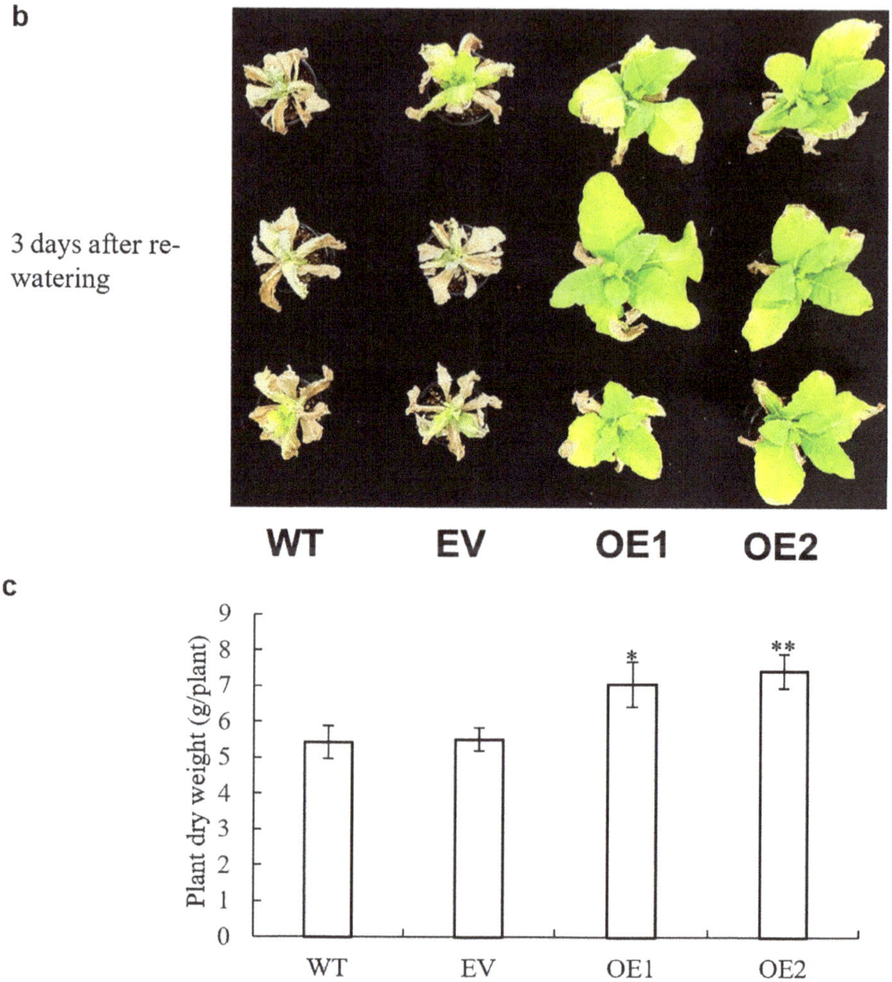

Figure 7. Responses of the *AaPEPC1*-overexpressing tobacco plants under drought stress. Phenotypes of transgenic plants expressing *AaPEPC1* (OE1 and OE2) or as well as empty vector (EV), and wild-type plants (WT) grown in pots under drought stress (**a**) and after re-watering (**b**). Transgenic and WT were grown in soil for 15 days without watering. (**c**) Dry weight of transgenic and WT plants. Values represent means ± SD (n = 3). * and ** indicate significant difference from that of WT at $p < 0.05$ and $p < 0.01$, respectively, by one-way ANOVA analysis with post-hoc Tukey HSD.

3.9. Proline Biosynthesis Is Enhanced by AaPEPC1 Overexpression

PEPC plays a crucial role in nitrogen metabolism in *Arabidopsis* [47] and loss-of-function of both *PEPC1* and *PEPC2* decreased the levels of glutamate in *Arabidopsis*. Glutamate can be converted into proline through pyrroline-5-carboxylate synthase (P5CS) and pyrroline-5-carboxylate reductase (P5CR) [11]. Proline plays important roles in stress tolerance, e.g., drought and salt stress tolerance, by regulating osmotic balance, activating the ROS scavenging system, protecting membrane integrity and photosynthesis [11,59]. We hypothesized that overexpression of *AaPEPC1* could enhance proline biosynthesis, and consequently increase the drought and salt stress tolerance in the transgenic plants. To test this hypothesis, proline content was analyzed in the transgenic *AaPEPC1* plants, along with the WT and EV controls. The proline contents in the transgenic plants expressing *AaPEPC1* were significantly higher than that in the WT and EV controls (Figure 8a). Furthermore, the expression levels of two key proline biosynthesis genes *NsyP5CS* and *NsyP5CR* were significantly higher in the transgenic *AaPEPC1* plants in comparison with the WT and EV controls (Figure 8b,c), supporting our hypothesis that the overexpression of *AaPEPC1* increased drought and salt stress tolerance through the enhancement of proline biosynthesis.

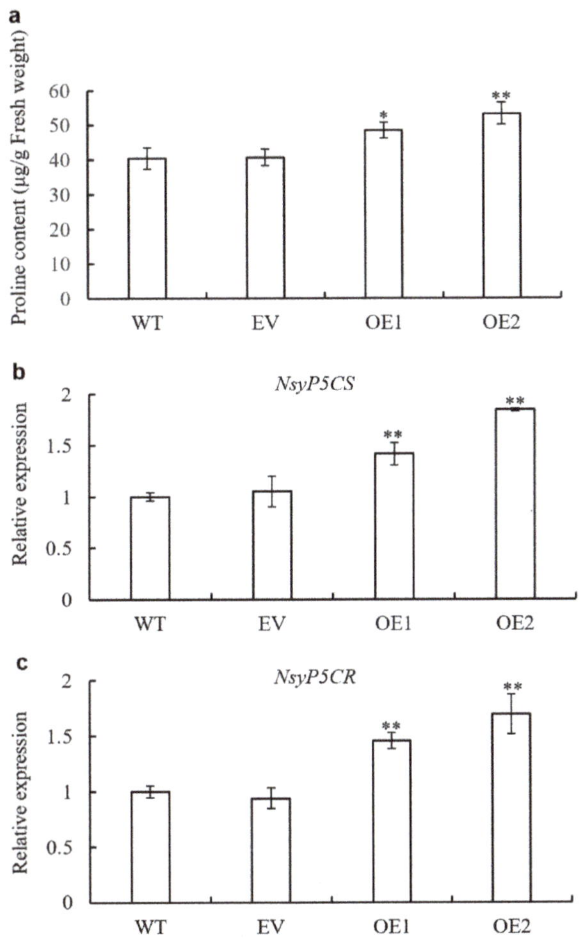

Figure 8. Changes in proline biosynthesis in *AaPEPC1*-overexpressing tobacco plants. (**a**) Proline content in the leaves of transgenic plants expressing *AaPEPC1* (OE1 and OE2) as well as empty vector (EV), and wild-type plants (WT). (**b**) and (**c**) Relative expression level of the proline biosynthesis genes pyrroline-5-carboxylate synthase (P5CS) and pyrroline-5-carboxylate reductase (P5CR) in the *AaPEPC1*-overexpressing tobacco plants. Values represent means ± SD (n = 3). * and ** indicate significant difference from that of WT at $p < 0.05$ and $p < 0.01$, respectively, by one-way ANOVA analysis with post-hoc Tukey HSD.

4. Discussion

The engineering of the water-conserving CO_2-concentrating mechanism of CAM has been proposed as a potential strategy for improving photosynthetic CO_2 fixation and abiotic stress tolerance in C_3 plants [1,2,9]. In this study, we found that the overexpression of one single CAM gene *AaPEPC1* enhanced the performance of transgenic plants in multiple ways, including photosynthesis, biomass production, drought tolerance and salt tolerance, which yet not be reported. We determined the impacts of the ectopic expression of the *AaPEPC1* on C_3-toward-CAM progression. Specifically, we identified a CAM-type *PEPC* (*AaPEPC1*) in *A. americana* and transformed the *AaPEPC1* gene into C_3 plant tobacco. Compared with the WT and EV controls, the transgenic plants expressing *AaPEPC1* showed several interesting traits: (1) a higher malate content at the onset of the dark period, (2) higher leaf carbon isotope ratios ($\delta^{13}C$ values), and (3) upregulated expression levels of the orthologs of several putative key CAM pathway genes (Figures 3 and 4). These changes in photosynthetic carbon metabolism driven by *AaPEPC1* improved photosynthetic CO_2-fixation, biomass production and tolerance to drought and salt stresses (Figure 9). These traits are not necessarily associated with CAM. Recently, Boxall, et al. [60] silenced *PEPC1* in a CAM plant species *Kalanchoë laxiflora* and found that the very low level of PEPC activity in the RNAi transgenic line was associated with the refixation of respiratory CO_2 and malate accumulation. Daloso, et al. [61] reported that tobacco guard cells fix CO_2 by both Rubisco and PEPC. The photosynthetic carbon metabolism changes in our study could be explained by PEPC-mediated CO_2 fixation during the light period or the refixation of respiratory CO_2, or an enhanced anaplerotic role for PEPC as in C_3 plants. A CO_2 labeled assay and metabolites analysis could be used to test this hypothesis and to identify the possible secondary products arising from *PEPC* overexpression in the future. Metabolite (e.g., glucose) sensing and signaling play important roles in regulating gene expression and controlling plant development [62,63]. We hypothesize that *AaPEPC1* overexpression perturbated the expression levels of metabolite signaling genes which subsequently caused the phenotypic changes (e.g., drought stress tolerance) in the transgenic tobacco plants. We found that the glucose, malate and proline contents were increased, and a few CAM-related genes were upregulated in the transgenic plants overexpressing *AaPEPC1*. Glucose, as a reducing sugar, is only ever present in modest concentrations, thus representing the partitioning of primary photosynthate or trafficking of reserves. A full starch–glucan–sucrose digestion could be performed to comprehensively evaluate the carbohydrate status in transgenic plants, and an integrative analysis of metabolomics, transcriptomics and proteomics could help to test this hypothesis in the future. This initial success provides a solid foundation for future effort to achieve a complete switch from C_3 to CAM photosynthesis by engineering additional CAM-related genes involved in carboxylation, decarboxylation, stomatal movement, the glycolytic-gluconeogenic pathway and carbohydrate turnover modules [1,2,9].

It is interesting that *AaPEPC1* overexpression increased the transcript abundance of several other CAM-related genes, including *CA*, *MDH*, *ALMT*, *TDT* and *ME* (Figure 4). As a CO_2-fixation enzyme, AaPEPC1 is not able to directly regulate the expression of other genes at the transcription level. We hypothesize that the change in cellular metabolic status caused by the *AaPEPC1* overexpression resulted in a rewiring of the regulatory network. This hypothesis can be tested through the transcriptomic and metabolomic analysis of gain-of-function and loss-of-function *AaPEPC1* mutants in the future. It was recently reported that phosphorylation of PEPC is essential for core circadian clock operation in the obligate CAM species *Kalanchoë fedtschenkoi* [20]. Therefore, it would be interesting to investigate the impact of *AaPEPC1* overexpression on the circadian rhythm in the transgenic plants in the future. In addition, on the premise of the coordinated regulation of multiple CAM-related genes by overexpressing one key gene such as *AaPEPC1*, we can argue that there is no need to transfer all the CAM pathway genes into C_3 species; consequently, we just need to focus on the engineering of a small number of "master" genes like *AaPEPC1*, which can upregulate the expression of multiple other CAM-related genes.

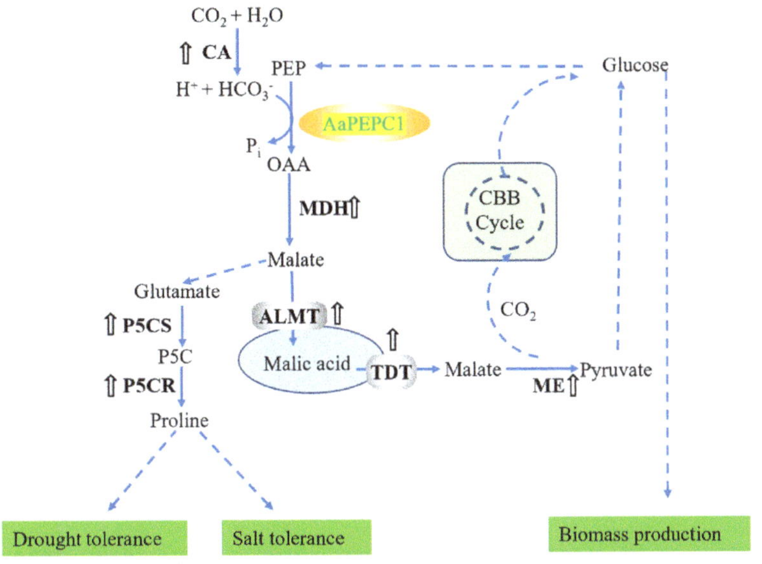

Figure 9. Regulation of photosynthetic CO_2-fixation and abiotic stress tolerance in the 'turbocharging' plants with the *AaPEPC1*. Overexpression of the *AaPEPC1*, a highly abundant enzyme catalyzing the primary fixation of CO_2 in CAM plants, increases photosynthetic CO_2-fixation, and rewires the diel accumulation-depletion pattern of malate and glucose. The re-programed malate-dependent carboxylation leads to the feedback up-regulation of the orthologs of key CAM pathway genes, i.e., nocturnal carboxylation and diurnal decarboxylation modules. The increased malate content up-regulates pyrroline-5-carboxylate synthase (P5CS) and pyrroline-5-carboxylate reductase (P5CR), which results in higher proline accumulation. Proline accumulation enhances the salt and drought tolerance of the transgenic plants expressing *AaPEPC1*. Additionally, a higher glucose content produced from the photosynthetic source is transported as sucrose or glucose to sink tissues and organs to promote cell proliferation, elongation, and expansion, as well as to maintain energy and metabolic homeostasis, resulting in improved biomass production [64,65]. The up arrows indicate up-regulation of expression of genes coding these enzymes or content. CA: carbonic anhydrase; OAA: oxaloacetate; MDH: malate dehydrogenase; ALMT: aluminum-activated malate transporter; TDT: tonoplast dicarboxylate transporter; ME: malic enzyme; PEP: phospho*enol*pyruvate; and CBB: Calvin–Benson–Bassham.

Although PEPC is well-known as a key enzyme for CO_2 fixation, its role in conferring resistance to salt stress in plants is not well defined [66]. In this study, we demonstrated that the overexpression of *AaPEPC1* significantly increased the salt tolerance in transgenic tobacco plants (Figure 6), likely a result of enhanced biosynthesis of proline (Figure 8), which plays important roles in regulating osmotic balance, activating ROS scavenging system, protecting membrane integrity and photosynthesis [11]. C_4-type PEPC and C_3-type PEPC have also been reported to be involved in drought tolerance improvement [67,68], but the mechanism is unclear. Thus far, the role of CAM-type PEPC in drought tolerance has not been reported yet. In this study, we found that overexpression of the *AaPEPC1* significantly increased the drought tolerance in the transgenic tobacco plants (Figure 7). Our results suggested that the improvement of photosynthetic carbon metabolism in the transgenic *AaPEPC1* plants enhanced proline metabolism pathway, which results in improved drought tolerance (Figure 9). In addition, compared to the previously published work for C_4-type PEPC or C_3-type PEPC overexpression in plants [67–70], this is the first time it has been shown that overexpression of one single *PEPC* gene enables the simultaneous improvement of photosynthesis and stress tolerance in plants.

Recently, photosynthesis and plant growth were significantly improved in tobacco plants by introducing a faster Rubisco of cyanobacterial origin [71], accelerating recovery from photoprotection [72], or engineering synthetic glycolate metabolism pathways [5]. However, none of these approaches enhanced tolerance to drought or salt stresses. In addition, previous genetic engineering efforts have reported progress in creating genetically-modified plants with enhanced tolerance to either drought stress [73,74] or salt stress [75–77], with limited success in conferring tolerance to both drought stress and salt stress in a single transgenic line. In this study, we created genetically modified tobacco plants that had enhanced performance in multiple aspects: photosynthesis, plant growth, drought tolerance, and salt tolerance. As such, we have provided new insights into the concept that the coordinated regulation of photosynthesis can increase biomass and stress tolerance [78]. These pleiotropic effects of *AaPEPC1*-overexpression open a new door to genetic improvement of crops for sustainable bioenergy and food production on marginal lands to alleviate the challenge caused by human population growth, urbanization, and global climate change.

In conclusion, we report the first successful effort of the engineering of a CAM pathway gene to improve photosynthetic CO_2 fixation and abiotic stress tolerance in tobacco, the model C_3 plant species. These findings have important implications for ultimate aspirations to engineer CAM into non-CAM crops as a means of improving productivity, and abiotic stress tolerance.

Supplementary Materials: The following are available online at https://www.mdpi.com/2073-4409/10/3/582/s1, Figure S1: Development of transgenic lines of tobacco overexpressing *Agave americana* PHOSPHO*ENOL*PYRUVATE CARBOXYLASE (AaPEPC1); Figure S2: Chlorophyll relative content in the leaves of transgenic plants expressing *Agave americana* PHOSPHO*ENOL*PYRUVATE CARBOXYLASE (*AaPEPC1*) (OE1, OE2) or empty vector (EV), and wild-types (WT) grown in pots under normal growth conditions. Table S1: Primers used in this study.

Author Contributions: X.Y. conceived, designed and supervised the project, as well as edited the manuscript. D.L. conceived, designed and performed the experiments, as well as wrote the manuscript. R.H. designed and constructed the *AaPEPC1* binary vector, performed molecular identification of transgenic plants and screened homozygous lines. J.Z. performed genome-wide analysis of the *PEPC* gene family and phylogenetic analysis. H.-B.G. and H.Q. performed structural modeling and molecular dynamics simulation analysis. H.C. performed the experiments. L.L. interpreted the data. G.A.T., A.M.B., J.-G.C. and W.M. reviewed and edited the manuscript. All authors have read and agreed to the published version of the manuscript.

Funding: This research is supported by the U.S. Department of Energy (DOE), Office of Science, Genomic Science Program under Award Number DE-SC0008834, and the Center for Bioenergy Innovation (CBI), which is a DOE Bioenergy Research Center supported by the Biological and Environmental Research program in the DOE Office of Science. Oak Ridge National Laboratory is managed by UT-Battelle, LLC for the DOE under Contract Number DE-AC05–00OR22725. HG and HQ also acknowledge the support of NSF Career award #1453078 (transferred to #1720215), BD Spoke #1761839, and internal funding of University of Tennessee at Chattanooga.

Informed Consent Statement: Not applicable.

Data Availability Statement: Data presented in this study are contained in this article, or available upon request to the corresponding author.

Conflicts of Interest: The authors declare no conflict of interest.

References

1. Yang, X.; Cushman, J.C.; Borland, A.M.; Edwards, E.J.; Wullschleger, S.D.; Tuskan, G.A.; Owen, N.A.; Griffiths, H.; Smith, J.A.C.; De Paoli, H.C. A roadmap for research on crassulacean acid metabolism (CAM) to enhance sustainable food and bioenergy production in a hotter, drier world. *New Phytol.* **2015**, *207*, 491–504. [CrossRef]
2. Liu, D.; Palla, K.J.; Hu, R.; Moseley, R.C.; Mendoza, C.; Chen, M.; Abraham, P.E.; Labbé, J.L.; Kalluri, U.C.; Tschaplinski, T.J. Perspectives on the basic and applied aspects of crassulacean acid metabolism (CAM) research. *Plant Sci.* **2018**, *274*, 394–401. [CrossRef] [PubMed]
3. Hanjra, M.A.; Qureshi, M.E. Global water crisis and future food security in an era of climate change. *Food Policy* **2010**, *35*, 365–377. [CrossRef]
4. Godfray, H.C.J.; Beddington, J.R.; Crute, I.R.; Haddad, L.; Lawrence, D.; Muir, J.F.; Pretty, J.; Robinson, S.; Thomas, S.M.; Toulmin, C. Food security: The challenge of feeding 9 billion people. *Science* **2010**, *327*, 812–818. [CrossRef]
5. South, P.F.; Cavanagh, A.P.; Liu, H.W.; Ort, D.R. Synthetic glycolate metabolism pathways stimulate crop growth and productivity in the field. *Science* **2019**, *363*, eaat9077. [CrossRef] [PubMed]
6. Salesse-Smith, C.E.; Sharwood, R.E.; Busch, F.A.; Kromdijk, J.; Bardal, V.; Stern, D.B. Overexpression of Rubisco subunits with RAF1 increases Rubisco content in maize. *Nat. Plants* **2018**, *4*, 802. [CrossRef]
7. Hibberd, J.M.; Sheehy, J.E.; Langdale, J.A. Using C_4 photosynthesis to increase the yield of rice—Rationale and feasibility. *Curr. Opin. Plant Biol.* **2008**, *11*, 228–231. [CrossRef]
8. Shameer, S.; Baghalian, K.; Cheung, C.M.; Ratcliffe, R.G.; Sweetlove, L.J. Computational analysis of the productivity potential of CAM. *Nat. Plants* **2018**, *4*, 165. [CrossRef] [PubMed]
9. Borland, A.M.; Hartwell, J.; Weston, D.J.; Schlauch, K.A.; Tschaplinski, T.J.; Tuskan, G.A.; Yang, X.; Cushman, J.C. Engineering crassulacean acid metabolism to improve water-use efficiency. *Trends Plant Sci.* **2014**, *19*, 327–338. [CrossRef] [PubMed]
10. Mahmood, T.; Khalid, S.; Abdullah, M.; Ahmed, Z.; Shah, M.K.N.; Ghafoor, A.; Du, X. Insights into drought stress signaling in plants and the molecular genetic basis of cotton drought tolerance. *Cells* **2019**, *9*, 105. [CrossRef] [PubMed]
11. Liu, D.; He, S.; Zhai, H.; Wang, L.; Zhao, Y.; Wang, B.; Li, R.; Liu, Q. Overexpression of *IbP5CR* enhances salt tolerance in transgenic sweetpotato. *Plant Cell. Tissue Organ Cult.* **2013**, *117*, 1–16. [CrossRef]
12. Zhai, H.; Wang, F.; Si, Z.; Huo, J.; Xing, L.; An, Y.; He, S.; Liu, Q. A myo-inositol-1-phosphate synthase gene, *IbMIPS1*, enhances salt and drought tolerance and stem nematode resistance in transgenic sweet potato. *Plant Biotechnol. J.* **2016**, *14*, 592–602. [CrossRef] [PubMed]
13. Ahmed, I.M.; Nadira, U.A.; Qiu, C.-W.; Cao, F.; Chen, Z.-H.; Vincze, E.; Wu, F. The barley S-adenosylmethionine synthetase 3 gene HvSAMS3 positively regulates the tolerance to combined drought and salinity stress in Tibetan wild barley. *Cells* **2020**, *9*, 1530. [CrossRef]
14. Movafeghi, A.; Khataee, A.; Abedi, M.; Tarrahi, R.; Dadpour, M.; Vafaei, F. Effects of TiO_2 nanoparticles on the aquatic plant *Spirodela polyrrhiza*: Evaluation of growth parameters, pigment contents and antioxidant enzyme activities. *J. Environ. Sci.* **2018**, *64*, 130–138. [CrossRef] [PubMed]
15. Tarrahi, R.; Khataee, A.; Movafeghi, A.; Rezanejad, F.; Gohari, G. Toxicological implications of selenium nanoparticles with different coatings along with Se^{4+} on *Lemna minor*. *Chemosphere* **2017**, *181*, 655–665. [CrossRef] [PubMed]
16. Turbat, A.; Rakk, D.; Vigneshwari, A.; Kocsubé, S.; Thu, H.; Szepesi, Á.; Bakacsy, L.; D Škrbić, B.; Jigjiddorj, E.-A.; Vágvölgyi, C. Characterization of the plant growth-promoting activities of endophytic fungi isolated from Sophora flavescens. *Microorganisms* **2020**, *8*, 683. [CrossRef]
17. Ehleringer, J.R.; Monson, R.K. Evolutionary and ecological aspects of photosynthetic pathway variation. *Annu. Rev. Ecol. Syst.* **1993**, *24*, 411–439. [CrossRef]
18. Lüttge, U. CO_2-concentrating: consequences in crassulacean acid metabolism. *J. Exp. Bot.* **2002**, *53*, 2131–2142. [CrossRef]
19. Owen, N.A.; Griffiths, H. A system dynamics model integrating physiology and biochemical regulation predicts extent of crassulacean acid metabolism (CAM) phases. *New Phytol.* **2013**, *200*, 1116–1131. [CrossRef]
20. Boxall, S.F.; Dever, L.V.; Knerova, J.; Gould, P.D.; Hartwell, J. Phosphorylation of phosphoenolpyruvate carboxylase is essential for maximal and sustained dark CO_2 fixation and core circadian clock operation in the obligate crassulacean acid metabolism species Kalanchoë fedtschenkoi. *Plant Cell* **2017**, *29*, 2519–2536. [CrossRef]
21. Eisenhut, M.; Weber, A.P. Improving crop yield. *Science* **2019**, *363*, 32–33. [CrossRef]
22. Miyamoto, S. *Salt Tolerance of Landscape Plants Common to the Southwest*; Texas Water Resources Institute: El Paso, TX, USA, 2008.
23. Davis, S.C.; Kuzmick, E.R.; Niechayev, N.; Hunsaker, D.J. Productivity and water use efficiency of *Agave americana* in the first field trial as bioenergy feedstock on arid lands. *GCB Bioenergy* **2016**, *9*, 314–325. [CrossRef]
24. Yang, X.; Hu, R.; Yin, H.; Jenkins, J.; Shu, S.; Tang, H.; Liu, D.; Weighill, D.A.; Yim, W.C.; Ha, J.; et al. The *Kalanchoë* genome provides insights into convergent evolution and building blocks of crassulacean acid metabolism. *Nat. Commun.* **2017**, *8*, 1899. [CrossRef]
25. Black, C.C.; Osmond, C.B. Crassulacean acid metabolism photosynthesis: 'Working the night shift'. *Photosynth. Res.* **2003**, *76*, 329–341. [CrossRef]
26. Klavsen, S.K.; Madsen, T.V.; Maberly, S.C. Crassulacean acid metabolism in the context of other carbon-concentrating mechanisms in freshwater plants: A review. *Photosynth. Res.* **2011**, *109*, 269–279. [CrossRef] [PubMed]

27. Silvera, K.; Neubig, K.M.; Whitten, W.M.; Williams, N.H.; Winter, K.; Cushman, J.C. Evolution along the crassulacean acid metabolism continuum. *Funct. Plant Biol.* **2010**, *37*, 995–1010. [CrossRef]
28. Abraham, P.E.; Yin, H.; Borland, A.M.; Weighill, D.; Lim, S.D.; De Paoli, H.C.; Engle, N.; Jones, P.C.; Agh, R.; Weston, D.J.; et al. Transcript, protein and metabolite temporal dynamics in the CAM plant *Agave*. *Nat. Plants* **2016**, *2*, 16178. [CrossRef] [PubMed]
29. Katoh, K.; Rozewicki, J.; Yamada, K.D. MAFFT online service: Multiple sequence alignment, interactive sequence choice and visualization. *Brief. Bioinform* **2019**, *20*, 1160–1166. [CrossRef] [PubMed]
30. Trifinopoulos, J.; Nguyen, L.-T.; von Haeseler, A.; Minh, B.Q. W-IQ-TREE: A fast online phylogenetic tool for maximum likelihood analysis. *Nucleic Acids Res.* **2016**, *44*, W232–W235. [CrossRef] [PubMed]
31. Masumoto, C.; Miyazawa, S.-I.; Ohkawa, H.; Fukuda, T.; Taniguchi, Y.; Murayama, S.; Kusano, M.; Saito, K.; Fukayama, H.; Miyao, M. Phosphoenolpyruvate carboxylase intrinsically located in the chloroplast of rice plays a crucial role in ammonium assimilation. *Proc. Natl. Acad. Sci. USA* **2010**, *107*, 5226–5231. [CrossRef] [PubMed]
32. Roy, A.; Kucukural, A.; Zhang, Y. I-TASSER: A unified platform for automated protein structure and function prediction. *Nat. Protoc.* **2010**, *5*, 725–738. [CrossRef] [PubMed]
33. Mirjalili, V.; Noyes, K.; Feig, M. Physics-based protein structure refinement through multiple molecular dynamics trajectories and structure averaging. *Proteins* **2014**, *82*, 196–207. [CrossRef] [PubMed]
34. Chen, V.B.; Arendall, W.B.; Headd, J.J.; Keedy, D.A.; Immormino, R.M.; Kapral, G.J.; Murray, L.W.; Richardson, J.S.; Richardson, D.C. MolProbity: All-atom structure validation for macromolecular crystallography. *Acta Crystallogr. D* **2010**, *66*, 12–21. [CrossRef] [PubMed]
35. Brooks, B.R.; Brooks, C.L.; Mackerell, A.D.; Nilsson, L.; Petrella, R.J.; Roux, B.; Won, Y.; Archontis, G.; Bartels, C.; Boresch, S.; et al. CHARMM: The biomolecular simulation program. *J. Comput. Chem.* **2009**, *30*, 1545–1614. [CrossRef] [PubMed]
36. Phillips, J.C.; Braun, R.; Wang, W.; Gumbart, J.; Tajkhorshid, E.; Villa, E.; Chipot, C.; Skeel, R.D.; Kale, L.; Schulten, K. Scalable molecular dynamics with NAMD. *J. Comput. Chem.* **2005**, *26*, 1781–1802. [CrossRef]
37. Best, R.B.; Zhu, X.; Shim, J.; Lopes, P.E.M.; Mittal, J.; Feig, M.; MacKerell, A.D. Optimization of the additive CHARMM all-atom protein force field targeting improved sampling of the backbone phi, psi and Side-Chain chi(1) and chi(2) dihedral angles. *J. Chem. Theory Comput.* **2012**, *8*, 3257–3273. [CrossRef] [PubMed]
38. Jorgensen, W.L.; Chandrasekhar, J.; Madura, J.D.; Impey, R.W.; Klein, M.L. Comparison of simple potential functions for simulating liquid water. *J. Chem. Phys.* **1983**, *79*, 926–935. [CrossRef]
39. Laskowski, R.A.; Jablonska, J.; Pravda, L.; Varekova, R.S.; Thornton, J.M. PDBsum: Structural summaries of PDB entries. *Protein Sci.* **2018**, *27*, 129–134. [CrossRef] [PubMed]
40. Terpe, K. Overview of tag protein fusions: from molecular and biochemical fundamentals to commercial systems. *Appl. Microbiol. Biot.* **2003**, *60*, 523–533. [CrossRef] [PubMed]
41. Höfgen, R.; Willmitzer, L. Storage of competent cells for Agrobacterium transformation. *Nucleic Acids Res.* **1988**, *16*, 9877. [CrossRef]
42. Zhang, L.; Jia, J.; Xu, Y.; Wang, Y.; Hao, J.; Li, T. Production of transgenic *Nicotiana* sylvestris plants expressing melatonin synthetase genes and their effect on UV-B-induced DNA damage. *Vitr. Cell. Dev. Biol. Plant* **2011**, *48*, 275–282. [CrossRef]
43. He, S.; Han, Y.; Wang, Y.; Zhai, H.; Liu, Q. In vitro selection and identification of sweetpotato *(Ipomoea batatas* (L.) Lam.) plants tolerant to NaCl. *Plant Cell. Tissue Organ Cult.* **2008**, *96*, 69–74. [CrossRef]
44. Xia, Z.; Wei, Y.; Sun, K.; Wu, J.; Wang, Y.; Wu, K. The maize AAA-type protein SKD1 confers enhanced salt and drought stress tolerance in transgenic tobacco by interacting with Lyst-interacting protein 5. *PLoS ONE* **2013**, *8*, e69787. [CrossRef]
45. Schmittgen, T.D.; Livak, K.J. Analyzing real-time PCR data by the comparative CT method. *Nat. Protoc.* **2008**, *3*, 1101–1108. [CrossRef]
46. O'Leary, B.; Park, J.; Plaxton, W.C. The remarkable diversity of plant PEPC (phosphoenolpyruvate carboxylase): Recent insights into the physiological functions and post-translational controls of non-photosynthetic PEPCs. *Biochem. J.* **2011**, *436*, 15–34. [CrossRef]
47. Shi, J.; Yi, K.; Liu, Y.; Xie, L.; Zhou, Z.; Chen, Y.; Hu, Z.; Zheng, T.; Liu, R.; Chen, Y.; et al. Phosphoenolpyruvate carboxylase in *Arabidopsis* leaves plays a crucial role in carbon and nitrogen metabolism. *Plant Physiol.* **2015**, *167*, 671–681. [CrossRef] [PubMed]
48. Kai, Y.; Matsumura, H.; Izui, K. Phosphoenolpyruvate carboxylase: Three-dimensional structure and molecular mechanisms. *Arch. Biochem. Biophys.* **2003**, *414*, 170–179. [CrossRef]
49. González-Segura, L.; Mújica-Jiménez, C.; Juárez-Díaz, J.A.; Güemez-Toro, R.; Martínez-Castilla, L.P.; Muñoz-Clares, R.A. Identification of the allosteric site for neutral amino acids in the maize C_4-isozyme of phospho*enol*pyruvate carboxylase: The critical role of Ser-100. *J. Biol. Chem.* **2018**, *293*, 9945–9957. [CrossRef] [PubMed]
50. Matsumura, H.; Xie, Y.; Shirakata, S.; Inoue, T.; Yoshinaga, T.; Ueno, Y.; Izui, K.; Kai, Y. Crystal structures of C_4 form maize and quaternary complex of *E. coli* phosphoenolpyruvate carboxylases. *Structure* **2002**, *10*, 1721–1730. [CrossRef]
51. Kai, Y.; Matsumura, H.; Inoue, T.; Terada, K.; Nagara, Y.; Yoshinaga, T.; Kihara, A.; Tsumura, K.; Izui, K. Three-dimensional structure of phosphoenolpyruvate carboxylase: A proposed mechanism for allosteric inhibition. 1999, 96, 823–828. *Proc. Natl. Acad. Sci. USA* **1999**, *96*, 823–828. [CrossRef]
52. Winter, K.; Holtum, J.A. How closely do the δ13C values of crassulacean acid metabolism plants reflect the proportion of CO2 fixed during fay and night? *Plant Physiol.* **2002**, *129*, 1843–1851. [CrossRef]

53. Zhang, L.; Chen, F.; Zhang, G.Q.; Zhang, Y.Q.; Niu, S.; Xiong, J.-S.; Lin, Z.; Cheng, Z.M.; Liu, Z.J. Origin and mechanism of crassulacean acid metabolism in orchids as implied by comparative transcriptomics and genomics of the carbon fixation pathway. *Plant J.* **2016**, *86*, 175–185. [CrossRef] [PubMed]
54. Holtum, J.A.; Aranda, J.; Virgo, A.; Gehrig, H.H.; Winter, K. δ^{13}C values and crassulacean acid metabolism in *Clusia* species from Panama. *Trees* **2004**, *18*, 658–668. [CrossRef]
55. Nunes-Nesi, A.; Araújo, W.L.; Fernie, A.R. Targeting mitochondrial metabolism and machinery as a means to enhance photosynthesis. *Plant Physiol.* **2010**, *155*, 101–107. [CrossRef]
56. Bräutigam, A.; Schlüter, U.; Eisenhut, M.; Gowik, U. On the evolutionary origin of CAM photosynthesis. *Plant Physiol.* **2017**, *174*, 473–477. [CrossRef]
57. Munns, R.; Termaat, A. Whole-plant responses to salinity. *Funct. Plant Biol.* **1986**, *13*, 143–160. [CrossRef]
58. Acosta-Motos, J.R.; Ortuño, M.F.; Bernal-Vicente, A.; Diaz-Vivancos, P.; Sanchez-Blanco, M.J.; Hernandez, J.A. Plant responses to salt stress: Adaptive mechanisms. *Agronomy* **2017**, *7*, 18. [CrossRef]
59. Patel, M.K.; Kumar, M.; Li, W.; Luo, Y.; Burritt, D.J.; Alkan, N.; Tran, L.-S.P. Enhancing salt tolerance of plants: From metabolic reprogramming to exogenous chemical treatments and molecular approaches. *Cells* **2020**, *9*, 2492. [CrossRef]
60. Boxall, S.F.; Kadu, N.; Dever, L.V.; Knerova, J.; Waller, J.L.; Gould, P.D.; Hartwell, J. Kalanchoë PPC1 is essential for crassulacean acid metabolism and the regulation of core circadian clock and guard cell signaling genes. *Plant Cell* **2020**, *32*, 1136–1160. [CrossRef] [PubMed]
61. Daloso, D.M.; Antunes, W.C.; Pinheiro, D.P.; Waquim, J.P.; AraÚJo, W.L.; Loureiro, M.E.; Fernie, A.R.; Williams, T. Tobacco guard cells fix CO 2 by both Rubisco and PEP case while sucrose acts as a substrate during light-induced stomatal opening. *Plant Cell Environ.* **2015**, *38*, 2353–2371. [CrossRef]
62. Li, L.; Sheen, J.J. Dynamic and diverse sugar signaling. *Curr. Opin. Plant Biol.* **2016**, *33*, 116–125. [CrossRef] [PubMed]
63. Wang, Y.-P.; Lei, Q.-Y. Metabolite sensing and signaling in cell metabolism. *Signal Transduct. Target. Ther.* **2018**, *3*, 1–9. [CrossRef] [PubMed]
64. Xiong, Y.; McCormack, M.P.; Li, L.; Hall, Q.; Xiang, C.; Sheen, J. Glucose–TOR signalling reprograms the transcriptome and activates meristems. *Nat. Cell Biol.* **2013**, *496*, 181–186. [CrossRef]
65. Sheen, J. Master regulators in plant glucose signaling networks. *J. Plant Biol.* **2014**, *57*, 67–79. [CrossRef]
66. Kandoi, D.; Mohanty, S.; Tripathy, B.C. Towards efficient photosynthesis: Overexpression of *Zea mays* phosphoenolpyruvate carboxylase in *Arabidopsis thaliana*. *Photosynth. Res.* **2016**, *130*, 47–72. [CrossRef]
67. Zhang, C.; Li, X.; He, Y.; Zhang, J.; Yan, T.; Liu, X. Physiological investigation of C_4-phosphoenolpyruvate-carboxylase-introduced rice line shows that sucrose metabolism is involved in the improved drought tolerance. *Plant Physiol. Biochem.* **2017**, *115*, 328–342. [CrossRef] [PubMed]
68. Qian, B.; Li, X.; Liu, X.; Chen, P.; Ren, C.; Dai, C. Enhanced drought tolerance in transgenic rice over-expressing of maize C_4 phosphoenolpyruvate carboxylase gene via NO and Ca^{2+}. *J. Plant Physiol.* **2015**, *175*, 9–20. [CrossRef]
69. Ku, M.S.; Agarie, S.; Nomura, M.; Fukayama, H.; Tsuchida, H.; Ono, K.; Hirose, S.; Toki, S.; Miyao, M.; Matsuoka, M. High-level expression of maize phosphoenolpyruvate carboxylase in transgenic rice plants. *Nat. Biotechnol.* **1999**, *17*, 76–80. [CrossRef]
70. Ding, Z.-S.; Huang, S.-H.; Zhou, B.-Y.; Sun, X.-F.; Zhao, M. Over-expression of phospho*enol*pyruvate carboxylase cDNA from C_4 millet (*Seteria italica*) increase rice photosynthesis and yield under upland condition but not in wetland fields. *Plant Biotechnol. Rep.* **2013**, *7*, 155–163. [CrossRef]
71. Lin, M.T.; Occhialini, A.; Andralojc, P.J.; Parry, M.A.J.; Hanson, M.R. A faster Rubisco with potential to increase photosynthesis in crops. *Nat. Cell Biol.* **2014**, *513*, 547–550. [CrossRef] [PubMed]
72. Kromdijk, J.; Głowacka, K.; Leonelli, L.; Gabilly, S.T.; Iwai, M.; Niyogi, K.K.; Long, S.P. Improving photosynthesis and crop productivity by accelerating recovery from photoprotection. *Science* **2016**, *354*, 857–861. [CrossRef]
73. Singha, D.L.; Tuteja, N.; Boro, D.; Hazarika, G.N.; Singh, S. Heterologous expression of *PDH47* confers drought tolerance in indica rice. *Plant Cell. Tissue Organ Cult.* **2017**, *130*, 577–589. [CrossRef]
74. Wang, L.; Zhao, R.; Li, R.; Yu, W.; Yang, M.; Sheng, J.; Shen, L. Enhanced drought tolerance in tomato plants by overexpression of SlMAPK1. *Plant Cell. Tissue Organ Cult.* **2018**, *133*, 27–38. [CrossRef]
75. Roy, S.J.; Negrão, S.; Tester, M. Salt resistant crop plants. *Curr. Opin. Biotechnol.* **2014**, *26*, 115–124. [CrossRef] [PubMed]
76. Li, R.; Kang, C.; Song, X.; Yu, L.; Liu, D.; He, S.; Zhai, H.; Liu, Q. A ζ-carotene desaturase gene, *IbZDS*, increases β-carotene and lutein contents and enhances salt tolerance in transgenic sweetpotato. *Plant Sci.* **2017**, *262*, 39–51. [CrossRef]
77. Liu, D.; He, S.; Song, X.; Zhai, H.; Liu, N.; Zhang, D.; Ren, Z.; Liu, Q. *IbSIMT1*, a novel salt-induced methyltransferase gene from *Ipomoea batatas*, is involved in salt tolerance. *Plant Cell. Tissue Organ Cult.* **2015**, *120*, 701–715. [CrossRef]
78. Ambavaram, M.M.; Basu, S.; Krishnan, A.; Ramegowda, V.; Batlang, U.; Rahman, L.; Baisakh, N.; Pereira, A. Coordinated regulation of photosynthesis in rice increases yield and tolerance to environmental stress. *Nat. Commun.* **2014**, *5*, 1–14. [CrossRef] [PubMed]

cells

MDPI

Article

Overproduction of Human Zip (SLC39) Zinc Transporters in *Saccharomyces cerevisiae* for Biophysical Characterization

Eva Ramos Becares [1], Per Amstrup Pedersen [2], Pontus Gourdon [1,3,*] and Kamil Gotfryd [1,*]

[1] Membrane Protein Structural Biology Group, Department of Biomedical Sciences, Faculty of Health and Medical Sciences, University of Copenhagen, Maersk Tower 7-9, DK-2200 Copenhagen N, Denmark; evabecares@sund.ku.dk

[2] Department of Biology, Faculty of Science, University of Copenhagen, Universitetsparken 13, DK-2100 Copenhagen OE, Denmark; papedersen@bio.ku.dk

[3] Department of Experimental Medical Science, Lund University, Sölvegatan 19, SE-221 84 Lund, Sweden

* Correspondence: pontus@sund.ku.dk (P.G.); kamil@sund.ku.dk (K.G.); Tel.: +45-503-39990 (P.G.); +45-414-02869 (K.G.)

Abstract: Zinc constitutes the second most abundant transition metal in the human body, and it is implicated in numerous cellular processes, including cell division, DNA and protein synthesis as well as for the catalytic activity of many enzymes. Two major membrane protein families facilitate zinc homeostasis in the animal kingdom, i.e., Zrt/Irt-like proteins (ZIPs aka solute carrier 39, SLC39, family) and Zn transporters (ZnTs), essentially conducting zinc flux in the opposite directions. Human ZIPs (hZIPs) regulate import of extracellular zinc to the cytosol, being critical in preventing overaccumulation of this potentially toxic metal, and crucial for diverse physiological and pathological processes, including development of neurodegenerative disorders and several cancers. To date, our understanding of structure–function relationships governing hZIP-mediated zinc transport mechanism is scarce, mainly due to the notorious difficulty in overproduction of these proteins for biophysical characterization. Here we describe employment of a *Saccharomyces cerevisiae*-based platform for heterologous expression of hZIPs. We demonstrate that yeast is able to produce four full-length hZIP members belonging to three different subfamilies. One target (hZIP1) is purified in the high quantity and homogeneity required for the downstream biochemical analysis. Our work demonstrates the potential of the described production system for future structural and functional studies of hZIP transporters.

Keywords: membrane proteins; overproduction; production platform; protein purification; *Saccharomyces cerevisiae*; solute carrier 39; SLC39; family; yeast; zinc; zinc transporters; ZIPs

Citation: Becares, E.R.; Pedersen, P.A.; Gourdon, P.; Gotfryd, K. Overproduction of Human Zip (SLC39) Zinc Transporters in *Saccharomyces cerevisiae* for Biophysical Characterization. *Cells* **2021**, *10*, 213. https://doi.org/10.3390/cells10020213

Received: 29 December 2020
Accepted: 19 January 2021
Published: 21 January 2021

1. Introduction

Zinc is one of the most crucial minerals for both plants and animals, and constitutes the second most abundant transition metal in humans. Up to 10% of the human proteome associates with zinc, and this metal exerts essential structural and/or functional roles, enabling correct activity of many proteins [1]. Notably, zinc is critical for many transcription factors and serves as a cofactor in all enzyme classes. In addition, it acts as a signaling molecule, and abnormalities of intracellular levels of zinc affect pathways controlling cell development, growth, differentiation, and death [2].

Since zinc is essential and yet toxic in excess, both body and cellular concentrations of this trace element have to be tightly regulated. In the animal kingdom, zinc transport is primarily achieved by membrane proteins (MPs) belonging to two solute carrier (SLC) families, i.e., SLC30 and SLC39. ZnT transporters (SLC30 family) reduce cytoplasmic zinc levels by providing cellular efflux or uptake into intracellular compartments. Conversely, Zrt-, Irt-like proteins (ZIPs) of the SLC39 family increase the intracellular concentration of the metal by mediating cellular uptake or efflux from intracellular compartments [3–7].

In humans, the ZIP family encompasses 14 members that, based on sequence similarity, are divided into four subfamilies, i.e., hZIP I, hZIP II, GufA, and LIV-1 (Figure 1A) [8,9]. Human ZIPs (hZIPs) have been proposed to regulate zinc homeostasis in a tissue- and cell type-specific manner, as reflected by their subcellular localization [10–13]. As examples, hZIP1 mediates microglial activation by autocrine and paracrine signaling [14]; hZIP2 is specifically expressed in the epidermis, responsible for differentiation of keratinocytes and skin turnover [15]; hZIP4 is located to the apical membrane of enterocytes and regulates dietary zinc absorption [16]; whereas hZIP10 plays an important role in maintenance of the immune system, as it is key for early β-cell survival [17]. In addition, altered protein expression or malfunction of hZIPs has been linked to a plethora of conditions, including several cancers, dermatitis, and neurodegenerative autoimmune diseases [6]. For instance, reduced levels of hZIP1, 2, and 3 were reported in the two types of prostate cancers [18,19], and upregulation of hZIP4 and 6 was shown to stimulate proliferation of pancreas and breast tumors, respectively [20,21]. Moreover, hZIP11 single nucleotide polymorphisms correlate with the development of bladder cancer, having either protective or increased risk of cancer progression [22]. Furthermore, loss-of-function mutations in hZIP4 lead to acrodermatitis enteropathica, a rare genetic disorder leading to Zn^{2+} deficiency manifested by skin lesions, growth retardation, immune system dysfunction, and neurological complications [23]. In addition, hZIP13 has been demonstrated to play a key role in development of the connective tissue and its malfunction can result in spondylocheiro dysplastic form of Ehlers-Danlos syndrome, a rare disorder of the connective tissue [24,25]. Consequently, hZIPs represent attractive targets to discover novel compounds for therapeutic purposes.

Despite the crucial role of hZIPs in health and disease, the understanding of this family of transporters remains sparse, as reflected by lack of structural and mechanistic information. Based on sequence analyses, hZIPs are predicted to consist of eight transmembrane segments (TMs), with N- and C-termini facing the noncytosolic space (Figure 1B) [13], of which the N-termini vary considerably in length (Figure 2A). Moreover, ZIPs usually form homodimers, but hZIP6/hZIP13 heterodimers have also been reported [10]. The only available structures are the inward-open metal-bound state of a full-length ZIP originate from a prokaryotic homologue, i.e., *Bordetella bronchiseptica* (BbZIP). These data indicate suggested that TM2, 4, 5, and 7 form a central helical core containing a binuclear metal binding center, while the remaining TMs (i.e., TM1, 3, 6, and 8) are located more peripherally (Figure 1C) [26,27]. In addition, the BbZIP structure indicated that the eight TMs are grouped into two units, one including TM1, 4, 5, and 6, while the second consists of the other four TMs (Figure 1C), forming the binuclear metal binding center in-between [27]. For hZIPs the available structural information is limited to the structure of the extracellular N-terminal domain of hZIP4 [28], a stretch that has been shown to play a critical role in dimerization, based on a highly conserved PAL motif [28]. Nevertheless, the function and transport mechanism of hZIPs remain largely elusive, and in addition, functional data suggest that ZIP members can bind different metals with varying specificities, affinities, and kinetics [29], implicating hitherto poorly understood protein-specific adaptations of the zinc transport. Thus, complementary structural information of other ZIP members and conformations are required to understand the molecular determinants governing the zinc transfer.

As for other MPs, structural investigation of ZIPs is obstructed by the difficulty in obtaining protein samples of sufficient quantity and quality. This is highlighted by the fact that ZIP transporters have been exceptionally difficult to overproduce and purify, perhaps related to their N_{out}-C_{out} topology [30,31]. In fact, BbZIP is the only ZIP member that has been successfully isolated to homogeneity in solution to date, although over 50 ZIP homologs have been attempted [31,32]. Moreover, even BbZIP was reported to be stable only in the presence of Cd^{2+}, resulting in tightly bound ions in the binuclear metal binding center [27]. Among the 14 hZIPs, only hZIP13 has been successfully overexpressed in sufficient quantities to perform biochemical characterization, through implementation of an insect Sf9 cell-based expression system, but no information on obtained protein yields is available [24].

Traditionally, due to the availability of expression plasmids, its easy genetic manipulation, high growth rate, and low cost, *Escherichia coli* has been the primary choice as an expression system for delivery of MPs [33,34]. However, *E. coli*-based platforms display major limitations when expressing proteins from higher organisms, where often expression systems of eukaryotic origin, such as yeast, insect, or mammalian cell-derived, are required [35]. Compared to *E. coli*, establishment of insect- or mammalian-based heterologous expression platforms can be cumbersome, tedious, and expensive. In contrast, yeast systems offer a cheap, easy, and robust large-scale production, combined with the ability to perform some of the post-translational modifications required for proper protein folding [36–38]. Hence, yeast represents an attractive complementary host for high-quality MP overproduction, permitting downstream structure–function characterization [39,40].

The *Saccharomyces cerevisiae* strain PAP1500 combined with a high copy number expression vector containing the CYC-GAL promoter (CG-P) has been successfully exploited for production of diverse classes of human MPs [39,41–43]. In this tandem system, the Galactose 4 (Gal4) transcription activator is overexpressed in the host upon galactose supplementation to the media, and the activator is employed for induction of the CYC-GAL (CG-P) promoter present in the expression plasmid [44]. The *URA3* and *leu2-d* selection markers are also included in the vector to enable easy selection of the transformants, and to ensure high amounts of plasmid prior to induction in a uracil- and leucine-deficient media [45].

Here we report development of a cost-effective and efficient method to express hZIPs, employing the PAP1500-based MP production system. Briefly, we attempted expression of four selected hZIPs, i.e., hZIP1, 2, 11, and 13, belonging to three different subfamilies, and produced them as full-length versions with C-terminal green fluorescent protein (GFP) fusions. Subsequently, utilizing GFP florescence, we screened to select the most promising target for large-scale production, both by analyzing expression levels and by identifying suitable detergent(s) for membrane extraction using florescence-detection size-exclusion chromatography (F-SEC). For the most promising candidate, i.e., hZIP1, we proceeded with affinity chromatography-based protein purification. We also refined the procedure through investigation of how different tags and their localization affect hZIP1 production and solubilization. Collectively, our results demonstrate that *S. cerevisiae* is capable of producing hZIPs in large scale for downstream biophysical characterization.

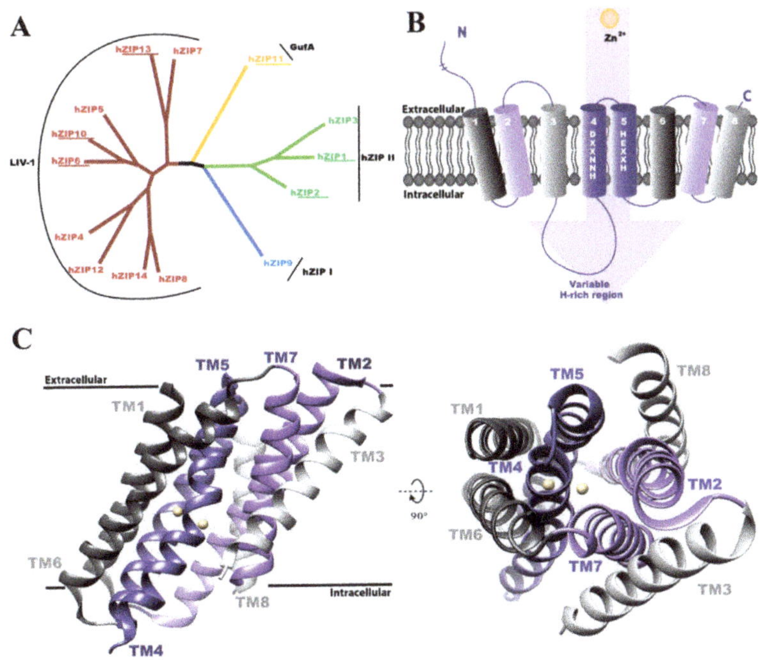

Figure 1. Overview of the Zrt-, Irt-like (ZIP) family of zinc transporters. (**A**) Phylogenetic distribution of the 14 human ZIP transporters (hZIP1-14) classified into four subfamilies, i.e., hZIP I, hZIP II, GufA, and LIV-1. The members investigated in this study are underlined. The phylogenetic tree was constructed using the MEGAX software (https://www.megasoftware.net/) [46]. (**B**) The predicted topology of hZIPs with selected characteristic features indicated. hZIPs likely form eight transmembrane segments (TMs), and the conserved motifs in TM4 and 5 (i.e., HNNXXD and HEXXH, respectively) establish a binuclear metal center. TM3 and 4 are linked by a variable loop containing a histidine-(H)-rich domain. The human ZIPs harbor N-terminal tails of varying length (from ≈6 aa in hZIP2 to ≈407 aa in hZIP10) and a short C-terminus, both oriented towards the extracellular side. (**C**) The only available structure of full-length ZIP from *Bordetella bronchiseptica* (BbZIP) (PDB ID:5TSB) [26,27]. The structure reveals that the TMs arrange pseudo-symmetrically into two domains: TM1, 4, 5, and 6 (dark shades) and TM2, 3, 7, and 8 (light shades). A central binuclear center has been suggested to be involved in Zn^{2+} transfer (yellow spheres) [27]. Side and top views are shown (left and right panels, respectively).

2. Materials and Methods

2.1. Engineering of Expression Plasmids

cDNAs encoding full-length hZIPs were commercially synthetized by GenScript (New Jersey, NJ, USA). All genes were codon-optimized for *Saccharomyces cerevisiae*-based expression using OptimumGene^{TM} algorithm, including adjustment of parameters crucial for transcription, translation, and protein folding. A panel of constructs was designed to screen for the levels of expression and to enable protein purification (Figure 2A). For the initial screening, engineered constructs included a C-terminally fused tag containing tobacco etch virus (TEV) protease cleavage site (ENLYFQ↓SQF), GFP, and a His_8 stretch (denoted hZIP-TEV-GFP-His throughout). Subsequently, optimization of production of the selected target, i.e., hZIP1, was performed with GFP-free constructs, including only either N- or C-terminal TEV protease-detachable purification tag enabling affinity chromatography step (His_8- or StrepII-tag). These comprised the following two variants, i.e., His-TEV-hZIP1 and StrepII-TEV-hZIP1, respectively. All expression constructs were generated by PCR amplification of the respective hZIP cDNAs with or without GFP fragment using AccuPol DNA polymerase (Amplicon, Odense, Denmark) and the primers listed in Table S1. Subsequently, each expression plasmid was assembled employing homologous recombination directly in the *S. cerevisiae* strain PAP1500 (α *ura3-52 trp1::GAL10-GAL4 lys2-801 leu2Δ1 his3Δ200 pep4::HIS3 prb1Δ1.6R can1 GAL*; Figure 2C) [47] with the corresponding hZIP/GFP PCR fragments and a *Bam*HI-, *Hin*dIII- and *Sal*I-digested pEMBLyex4 vector [44]. Transformants were selected on minimal medium agar plates supplemented with glucose (2 g L^{-1}), leucine (60 mg L^{-1}), and lysine (30 mg L^{-1}). DNA sequencing was performed on isolated plasmids to verify their identity.

2.2. Overproduction of hZIPs, Live-Cell Bioimaging and In-Gel GFP Fluorescence

All hZIPs were produced using the *S. cerevisiae* expression strain PAP1500 (Figure 2C) essentially as previously reported [47]. For screening purposes, expression of hZIP-TEV-GFP-His, His-TEV-hZIP1, and StrepII-TEV-hZIP1 constructs was performed in 2-L cell culture scale using shaker flasks. Briefly, a single colony of each transformant was used to inoculate 5 mL of minimal media containing glucose (2 g L^{-1}), leucine (60 mg L^{-1}), and lysine (30 mg L^{-1}), and grown overnight at 30 °C until OD_{450} ranged from 0.5 to 1.0. Subsequently, 0.5 mL of the preculture was transferred to 5 mL of the minimal medium w/o leucine and grown for another 24 h at 30 °C to increase plasmid copy number. The following day, the culture was scaled up to 100 mL in the same media for an additional 24 h and used to inoculate 2 L of media containing amino acids (alanine (20 mg L^{-1}), arginine (20 mg L^{-1}), aspartic acid (100 mg L^{-1}), cysteine (20 mg L^{-1}), glutamic acid (100 mg L^{-1}), histidine (20 mg L^{-1}), lysine (30 mg L^{-1}), methionine (20 mg L^{-1}), phenylalanine (50 mg L^{-1}), proline (20 mg L^{-1}), serine (375 mg L^{-1}), threonine (200 mg L^{-1}), tryptophan (20 mg L^{-1}), tyrosine (30 mg L^{-1}) and valine (150 mg L^{-1})), glucose (10 g L^{-1}) and glycerol (3%, *v/v*). Upon consumption of sugar, protein expression was induced by supplementing the culture with galactose (final concentration of 2%) and allowed for 48–72 h at 15 °C. Subsequently, cells were harvested (5800× *g*, 15 min, 4 °C) and stored at −80 °C. Obtained material typically yielded ≈3–4 g of wet cell pellet per L of cell culture. For initial purification of hZIP1-TEV-GFP-His, cell material obtained from 2-L yeast culture induced for expression for 72 h was used. Overproduction of His-TEV-hZIP1 and StrepII-TEV-hZIP1 constructs for large-scale purification was performed identically, but in 12-L scale with 68-h induction.

Localization of TEV-GFP-His hZIP fusions was performed by live-cell bioimaging where GFP fluorescence was analyzed using the Nikon Eclipse E600 microscope (Nikon, Tokyo, Japan) equipped with an Optronics MagnaFire S99802 camera (Optronics, Muskogee, OK, USA).

In-gel GFP fluorescence was detected on samples resolved by 4–20% SDS-PAGE (Thermo Fisher Scientific, Waltham, MA, USA), where the signal was immediately visualized with ImageQuant LAS 4000 imaging system and the corresponding control software (both from GE Healthcare, Copenhagen, Denmark).

2.3. Preparation of Crude Membranes

For initial screening purposes of hZIP-TEV-GFP-His expression and purification, wet cell pellets were resuspended in ice-cold lysis buffer (50 mM Tris-HCl pH 8.0, 200 mM NaCl and 30% glycerol) supplemented with 5 mM β-mercaptoethanol (BME), 1 mM PMSF, and SigmaFAST™ protease inhibitor cocktail (Sigma-Aldrich, St. Louis Missouri, MO, USA). Subsequently, cell suspension was mixed with glass beads and mechanically ruptured using a bead beater (BioSpec, Bartlesville, OK, USA). Following cell disruption, the supernatant was collected, and glass beads were washed in ice-cold lysis buffer. Unbroken cells and cell debris were removed from the homogenate by centrifugation (1000 or 3000× *g*, 10 min, 4 °C). Crude membranes were then pelleted by ultracentrifugation (138,000× *g*, 3 h, 4 °C), resuspended in the above-mentioned buffer (see Table S2) using a Potter–Elvehjem homogenizer and stored at −80 °C.

In contrast, cells overexpressing both N-terminal hZIP1 fusions, i.e., His-TEV-hZIP1 and StrepII-TEV-hZIP1, were disrupted by manual bead beating. Briefly, for His-TEV-hZIP1, the wet cell pellet obtained from 12-L culture (typically yielding ≈80 g) was resuspended in ice-cold lysis buffer (25 mM Tris-HCl pH 7.0, 500 mM NaCl, and 20% glycerol) supplemented with 250 μM EDTA, 250 μM EGTA, 5 mM BME, 1 mM PMSF, 1 μg mL^{-1} chymostatin, 1 μg mL^{-1} leupeptin, and 1 μg mL^{-1} pepstatin. Cells expressing StrepII-TEV-hZIP1 were lysed identically, but in the buffer with pH of 8.0 (25 mM Tris-HCl). Subsequently, 12.5 mL of the cell suspension was mixed with 15 mL of glass beads directly in 50-mL conical tubes and the mixture was homogenized by high-speed vortexing (eight cycles of 1 min with 1-min cooling in between). The resulting supernatant was collected and treated identically as described above for the C-terminally fused constructs. Following ultracentrifugation, crude membranes were resuspended in the respective solubilization buffers containing all the supplements present in the lysis buffer mentioned above (see Table S2).

2.4. Detergent Screening, Immunoblotting, and F-SEC Analysis

Following estimation of the total protein concentration using Bradford assay [48] (Sigma-Aldrich), crude membranes were diluted in the corresponding solubilization buffers to a final concentration of 2–5 mg L^{-1}. Subsequently, small-scale solubilization screens were performed using 0.5 mL of the respective membranes vigorously rotated (90–120 min, 4 °C) in the presence of the below-mentioned detergents. hZIP1-, hZIP2-, and hZIP13-TEV-GFP-His-expressing membranes were solubilized with a final concentration of 1% (w/v) of n-dodecylphosphocholine (FC-12), n-decyl-D-maltoside (DM), or n-dodecyl-D-maltoside (DDM) with or without 0.34% (w/v) of cholesteryl hemisuccinate Tris salt (CHS); all from Anatrace, Maumee, OH, USA. Solubilization efficacy of hZIP11-TEV-GFP-His was tested with a double final concentration of the above-mentioned surfactants, i.e., 2% of detergent ±0.68% of CHS. Extraction of His-TEV-hZIP1 was performed only with DDM and CHS tested at three ratios (1% + 0.1%, 2% + 0.2% and 5% + 0.5%, w/v, respectively). Subsequently, insoluble material was removed by ultracentrifugation (50,000× g, 20 min, 4 °C). Supernatant originating from the respective TEV-GFP-His hZIP fusions was used directly to measure GFP fluorescence (excitation 485 nm, emission 520 nm) to assess detergent extraction efficacy and/or for SDS-PAGE analysis. Extracted fractions of His-TEV- and StrepII-TEV-hZIP1 were evaluated by immunoblotting. Briefly, the respective samples were resolved by 4–20% SDS-PAGE and transferred to PVDF membranes (0.45 μm; both from Thermo Fisher Scientific). Following blocking, membranes were probed with 6×His mAb-HRP conjugate (Takara Bio, Mountain View, CA, USA) or Strep-Tactin® HRP conjugate (IBA GmbH, Göttingen, Germany) according to recommendations of the manufacturers. Immunoblots were visualized by chemiluminescence with SuperSignal™ West Femto Maximum Sensitivity substrate (Thermo Fisher Scientific) using AlphaImager™ (Alpha Innotech, San Leandro, CA, USA). Relative signal intensities of the bands were quantified using ImageJ software (https://imagej.nih.gov/ij/).

In addition, extracted TEV-GFP-His hZIP fusions were analyzed using florescence-detection size-exclusion chromatography (F-SEC). Briefly, solubilized material was analyzed on a Superdex 200 Increase 10/300 GL column (GE Healthcare) equilibrated with F-SEC buffer (20 mM Tris-HCl pH 8, 100 mM NaCl, 2 mM BME, 10% glycerol and 0.03% DDM; Anatrace) attached to ÄKTA Pure system (GE Healthcare) equipped with a Prominence RF-20A fluorescence detector (Shimadzu, Kyoto, Japan).

2.5. Purification of hZIP1, TEV Protease Cleavage, and SEC Analysis

Purification of the two His-tagged hZIP1 variants, i.e., hZIP1-TEV-GFP-His and His-TEV-hZIP1 was performed using immobilized metal ion affinity chromatography (IMAC), whereas StrepII-TEV-hZIP1 was purified employing StrepII-tag affinity chromatography. In all cases, the starting material included crude membranes isolated from flask-grown yeast cells (for the corresponding culture conditions see Section 2.3). Details about solubilization conditions and chromatography buffers are presented in Table S2. Following solubilization at 4 °C, insoluble material was removed by ultracentrifugation (80,000× g, 3 h, 4 °C) and subjected to IMAC-based purification. Briefly, solubilized His-tagged hZIP1 variants were diluted in the corresponding solubilization buffers (Table S2) to reduce detergent concentration and loaded onto a 5-mL HisTrap HP columns attached to an Äkta Pure system (both from GE Healthcare). Following column wash, bound proteins were eluted in IMAC buffers using a linear imidazole gradient (25/50–500 mM; Table S2). Subsequently, top IMAC fractions were pooled, concentrated on Vivaspin concentrators (MWCO 30 or 50 kDa; Sartorius, Göttingen, Germany) and obtained His-tagged hZIP1 samples were cleaved with TEV-His_{10}-tagged protease (home source; 16 h, 4 °C) used in TEV-to-hZIP1 ratio of 1:10 (*w/w*) in MWCO 10 kDa dialysis bags (Thermo Fisher Scientific) with concomitant dialysis against IMAC buffer containing 20 mM imidazole. Subsequently, reverse (R)-IMAC was performed to separate cleaved, i.e., TEV-GFP-His- or His-TEV-free hZIP1 from uncleaved fraction, TEV protease and released fusion tags, all containing His-tag and rebinding to the IMAC resin. TEV protease cleavage products were analyzed by SDS-PAGE and visualized by in-gel GFP fluorescence or immunoblotting.

Solubilized StrepII-TEV-hZIP1 membranes were also diluted to reduce detergent concentration and the pH of the solubilizate was adjusted to reach a final value of 8.0, prior to loading onto a 5-mL Strep-Tactin®XT Superflow® high capacity column (IBA GmbH) attached to an Äkta Pure system (GE Healthcare). Subsequently, the column was washed until reaching stable A_{280} signal and the protein was eluted in StrepII buffer supplemented with 50 mM biotin (Table S2).

R-IMAC- and StrepII affinity-pure hZIP1 samples were loaded onto a 24-mL Superdex 200 Increase 10/300 GL column (GE Healthcare) equilibrated with SEC buffer (20 mM Tris-NaOH pH 6, 100 mM NaCl and 10% glycerol) supplemented with 2 mM BME, 0.03% DDM, and 0.003% CHS. Fractions corresponding to the main elution peak were analyzed using SDS-PAGE.

3. Results

3.1. Overproduction of hZIP-TEV-GFP-His Fusions in S. cerevisiae-Based System

ZIP transporters have proven to be extremely challenging to overproduce and purify in a functional form [30,31]. As the selection of an adequate expression host is a bottleneck in protein production, we decided to investigate whether our established *S. cerevisiae*-based platform will be applicable in manufacturing hZIPs. Based on sequence analysis of the 14 hZIPs, we selected four members possessing the shortest N-termini (Figure 2A), i.e., hZIP1, 2, 11, and 13 with the assumption that the long tails may restrict overall protein rigidity and introduce an additional level of disorder that could be detrimental for the production, stability, and subsequent crystallization-based structure determination approaches. Following the same strategy, we also included the two longest hZIPs, i.e., hZIP6 and 10, as, upon successful production in a stable form, they may still represent the most suitable ZIP family members for structure determination efforts using cryo-electron microscopy (cryo-EM). Thus, this selection of targets widely covers the hZIP family, as these members represent three out of four known subfamilies. Moreover, the selected proteins are related to various pathologies, highlighting their importance as possible drug targets.

The applied *S. cerevisiae* system exploits modified high-copy vectors encoding codon-optimized full-length hZIPs C-terminally fused to tobacco etch virus (TEV) protease-detachable GFP-His-tag, enabling localization, quantification, quality control, and affinity purification (Figure 2B).

To be able to proceed with downstream studies, high protein yields are crucial. Thus, a satisfactory overproduction level is one of the first milestones to achieve when synthesizing proteins using heterologous systems. In order to assess the initial protein yields, we used the GFP attached to our targets to easily localize and quantify the obtained amounts of recombinant protein. As illustrated by the results from live-cell bioimaging fluorescent micrographs of cells derived from 2-L shaker cultures induced for expression for 48–72 h (Figure 2D), all four short constructs were expressed, and we were able to evaluate the cellular localization of the different targets. However, only minimum traces of protein were obtained for hZIP6 and 10 (data not shown), and hence further work with these members was discontinued. While hZIP1-, hZIP2-, and hZIP11-TEV-GFP-His accumulated mainly in the plasma membrane, hZIP13-TEV-GFP-His localized to intracellular membrane enclosed compartments. This space-varying accumulation in yeast correlates with the natural ability of hZIPs to localize in both plasma and intracellular membranes in human cells. In addition, as reported by the fluorescence signal originating from C-terminally fused GFP, the hZIPs are likely correctly folded and inserted into the yeast membranes. Moreover, by taking advantage of the GFP fusion, we also estimated protein yields per L of small-scale cell cultures (i.e., 2-L following 48–72 h induction). As determined from the whole-cell GFP fluorescence [49] compared with a signal from the GFP standard [41], the anticipated protein yields ranged from 3.3 to 5.7 mg protein per L of cell culture (Figure 2E), as obtained for hZIP1- and hZIP13-TEV-GFP-His, respectively. Such yields are compatible with biophysical, functional, and structural analysis, and thus, they set a promising starting point for further evaluation of the constructs.

Exploiting the ability of correctly folded GFP to exhibit resistance to SDS, we visualized expression of all targets by in-gel GFP fluorescence of whole-cell lysates resolved using SDS-PAGE (Figure 2F). The results demonstrate that the short hZIPs expressed under the above-mentioned conditions accumulated in the yeast membranes in a stable, full-length form, as no apparent fluorescent degradation products were visible. Moreover, the four targets migrated in SDS-PAGE according to their predicted molecular weight, with hZIP13-TEV-GFP-His being the heaviest target (MW of the fusion of 68 kDa, with TEV-GFP-His representing 28 kDa) and hZIP2-TEV-GFP-His being the lightest (62 kDa), respectively. In addition, for all evaluated targets additional weak fluorescent bands of higher molecular mass were observed, indicating presence of possible higher oligomeric states. Based on both the promising production levels and sample quality, we decided to proceed with solubilization screening of all the four short targets.

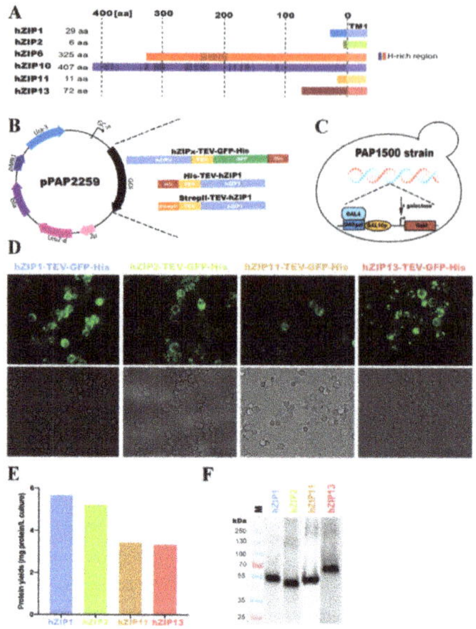

Figure 2. Overproduction of hZIP-TEV-GFP-His fusions in *Saccharomyces cerevisiae*-based system. (**A**) Generalized domain structure of the N-termini of human ZIP targets approached in this study, including the four shortest and the two longest human ZIPs. N-terminal segments of hZIP1 (Q9NY26), hZIP2 (Q9NP94), hZIP6 (Q13433), hZIP10 (Q9ULF5), hZIP11 (Q8N1S5), and hZIP13 (Q96H72) were assigned using TOPCONS topology prediction tool (http://topcons.cbr.su.se) [50]. UniProt (https://www.uniprot.org/) [51] accession numbers of full-length hZIPs protein sequences are shown in brackets. The length of the N-termini, and the positions of histidine-(H)-rich regions (dark boxes) and transmembrane segments TM1 are indicated. (**B**) Map of the *S. cerevisiae* expression plasmid pPAP2259 encoding the different hZIP genes of interest (GOIs) used in the study shown in order of their appearance. Expression constructs were designed to include the full-length forms of the respective hZIPs and different combinations of the following elements: tobacco etch virus (TEV) protease cleavage site, green fluorescent protein (GFP), an octa-histidine- (His) or a StrepII-tag (StrepII). The arrangement of regions of the respective constructs are shown. The expression plasmid also contains a hybrid promotor with the GAL10 upstream activation sequence in the 5′ nontranslated leader of the cytochrome-1 gene (CG-P), a yeast origin of replication (2μ), a β-isopropylmalate dehydrogenase gene with truncated promotor sequence resulting in poor expression of this gene (*leu2-d*), a β-lactamase gene (*bla*), an origin of replication (pMB1), and a yeast orothidine-5-phosphate decarboxylase gene (*URA3*). (**C**) The *S. cerevisiae* protein production strain PAP1500 used in the study. Upon induction with galactose, the strain overexpresses the Gal4 transcriptional activator that is the limiting factor for expression from galactose regulated promoter. GAL10p and UASgal constitute a specific DNA binding site for GAL4 activator. (**D**) Live-cell bioimaging of *S. cerevisiae* cells (derived from 2-L cultures) expressing hZIP1-, hZIP2-, and hZIP13-TEV-GFP-His (72-h induction at 15 °C) or hZIP11-TEV-GFP-His (48-h induction at 15 °C). For each construct GFP fluorescence (top) and differential interference contrast (bottom) micrographs are shown. Magnification: 1000×. (**E**) Estimates of protein production levels of four selected hZIPs expressed as in (**D**). The intensity of the whole-cell GFP fluorescence signal was transformed to the protein amount, by comparison against a free-GFP standard curve fluorescence signal, and the predicted yield is expressed as mg per L of cell culture. (**F**) In-gel GFP fluorescence of SDS-PAGE-resolved whole-cell lysates (i.e., supernatant from centrifuged homogenates) expressing different hZIP-TEV-GFP-His fusions as in (**D**). The predicted MWs of the respective fusions are 63.3 kDa (hZIP1), 61.8 kDa (hZIP2), 63.8 kDa (hZIP11), and 68 kDa (hZIP13). M: marker.

3.2. Detergent Screening and F-SEC Analysis of Solubilized hZIP-TEV-GFP-His Fusions

Prior to most biochemical and biophysical studies, MPs need to be extracted in their native form from the surrounding membranes. Thus, successful purification relies on the capacity of surfactant(s) to efficiently solubilize the membranes and maintain isolated protein stable in solution [52]. To systematically assess and identify detergents that can effectively extract each of the remaining hZIPs, we performed initial screening using a panel of surfactants tested in the presence or absence of cholesteryl hemisuccinate Tris salt (CHS) at a fixed pH value and ionic content (data not shown). In general, the four tested hZIP-TEV-GFP-His fusions appeared difficult to extract in milder surfactants. High solubilization efficacies were achieved only in harsh anionic or zwitterionic detergents (e.g., N-lauroylsarcosine sodium salt or foscholines, respectively), compounds typically not applicable for the downstream investigation of MPs. Three of the examined detergents, i.e., FC-12, a harsh zwitterionic compound, and two milder nonionic surfactants, i.e., DM and DDM, representing the two most commonly used detergents in MP studies [53], displayed favorable extraction efficiencies (Figure 3A). However, both DM and DDM performed only moderately when extracting hZIP2-, hZIP11-, and hZIP13-TEV-GFP-His, yielding solubilization efficacies >10%. Interestingly, for these three detergents, we observed an increase in solubilization efficacy upon supplementation with CHS. However, target-specific variations were observed, with hZIP1-TEV-GFP-His being the most prone and hZIP2-TEV-GFP-His the most refractory to membrane solubilization, respectively. Combining both estimated production levels and extraction efficacies, we decided to proceed with the three most promising candidates, i.e., hZIP1, 2, and 13. In contrast, hZIP11 was excluded due to its low expression levels and inefficient solubilization. Importantly, for the three selected targets, SDS-PAGE analysis of membranes solubilized in all three applied detergents exhibited the expected migration pattern and stability of TEV-GFP-His fusions (Figure 3B), without any visible signs of protein degradation.

In addition to high extraction efficiency, an optimal detergent should preserve the isolated MP in a stable, homogenous form. To assess stability of the extracted hZIP1, 2, and 13, we again took advantage of the TEV-GFP-His-tag and employed F-SEC to evaluate the elution profiles of the detergent-exposed samples (Figure 3C). Thus, for each target, six F-SEC runs were performed and the resulting F-SEC profiles of solubilized fusion proteins were monitored using fluorescence spectroscopy [54]. Overall, the F-SEC data indicate that although all tested hZIPs display some degree of aggregation (as reflected by the peaks eluting at the void volume of the column), the presence of CHS during solubilization significantly improved protein stability. In the case of FC-12, the resulting F-SEC profiles for all three targets were sharp and symmetrical, suggesting monodispersity of all hZIPs in this detergent. However, considering the harshness of this surfactant, as indicated by the highest extraction efficiencies observed in FC-12 (Figure 3A) and large fractions of aggregated protein present in the void, the use of this zwitterionic detergent entails the risk of protein denaturation due to disruption of the protein fold [54,55]. Therefore, we only considered DM and DDM when selecting a primary detergent to obtain hZIPs for downstream applications. Out of three tested targets, hZIP1-TEV-GFP-His clearly displayed the highest stability in both DM and DDM (Figure 3A, top panel), with only marginal fractions of aggregated protein, greatly reduced by CHS. hZIP1 solubilized in DM/DDM + CHS eluted mainly as a sharp, symmetrical peak preceded by a peripheral shoulder (Figure 3C, top panel), suggesting monodispersity of this target. F-SEC peak profiles obtained for the two remaining proteins, i.e., hZIP2 and 13, solubilized in DM/DDM + CHS were enriched in a significant fraction eluting in the void volume (Figure 3C, middle and bottom panels), indicating that a large population of these targets tend to aggregate in these detergents.

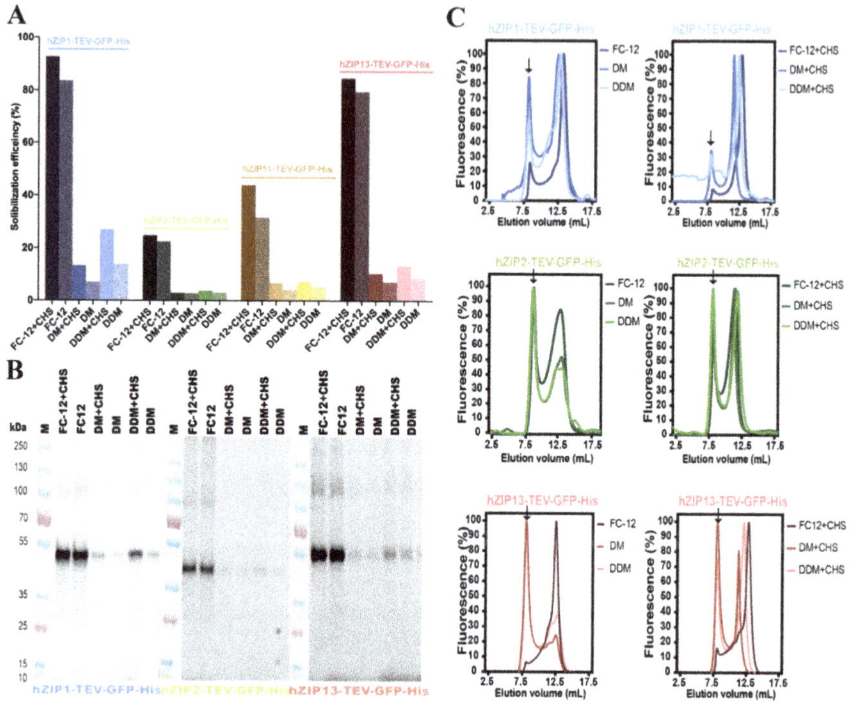

Figure 3. Detergent screening and fluorescence-detection size-exclusion chromatography (F-SEC) analysis of solubilized hZIP-TEV-GFP-His fusions. (**A**) Solubilization efficiency of crude membranes isolated from *Saccharomyces cerevisiae* cells (derived from 2-L cultures) expressing hZIP1-, hZIP2-, and hZIP13-TEV-GFP-His (72-h induction at 15 °C) or hZIP11-TEV-GFP-His (48-h induction at 15 °C). For hZIP1, 2, and 13 analyzed material was solubilized for 2 h at 4 °C with FC-12 (at final concentration of 1%, *w*/*v*), DM (1%, *w*/*v*), and DDM (1%, *w*/*v*) with and without CHS (0.34%, *w*/*v*). For hZIP11, the final concentrations were doubled. Following treatment with the respective detergents, GFP fluorescence was measured in the supernatant after ultracentrifugation and percentage of solubilization efficiency was calculated. CHS: cholesteryl hemisuccinate Tris salt. (**B**) In-gel GFP fluorescence of SDS-PAGE-separated detergent-solubilized crude *S. cerevisiae* membranes overexpressing hZIP1, 2, and 13 C-terminally fused to TEV-GFP-His-tag. Extraction conditions are identical to (**A**) and the analyzed material represents ultracentrifuged supernatant from the respective solubilization. The predicted MWs of the respective fusions are 63.3 kDa (hZIP1), 61.8 kDa (hZIP2), 63.8 kDa (hZIP11), and 68 kDa (hZIP13). M: marker. (**C**) F-SEC analysis of detergent-solubilized crude *S. cerevisiae* membranes overexpressing hZIP1, 2, and 13 C-terminally fused to TEV-GFP-His-tag. Extraction conditions are identical to (**A**) and analyzed material includes ultracentrifuged supernatant from the respective solubilization. Normalized F-SEC chromatograms were obtained after separation on Superdex 200 Increase 10/300 GL column where GFP fluorescence of the eluate was monitored. Arrows indicate the estimated elution positions of the void volume.

All-in-all, hZIP1-TEV-GFP-His exhibited the highest expression levels, extraction efficacy in mild nonionic detergents and post-solubilization stability. Hence, we selected this target for subsequent large-scale purifications in DDM + CHS, representing the most promising surfactant condition for the overproduction of hZIP1.

3.3. Purification of the hZIP1-TEV-GFP-His Fusion

hZIP1-TEV-GFP-His was purified from ≈7 g of flask-grown *S. cerevisiae* cells (material obtained from ≈2 L of shaker culture induced for expression for 72 h). Following isolation, crude yeast membranes were solubilized in DDM + CHS (for 2 h at 4 °C; Table S2) and subjected to immobilized metal ion affinity chromatography (IMAC; Figure 4A). The affinity purification yielded ≈1 mg of protein per L of the cell culture, overall, with relatively high sample purity (Figure 4B). The significant proportion of monomeric form visualized by the SDS-PAGE suggests that, in the applied conditions, hZIP1-TEV-GFP-His was purified mainly as the monomer, but both dimer and high-order oligomers are also visible. Importantly, the produced sample displayed high stability as no fluorescent signal originating from degraded TEV-GFP-His-tag was detected (Figure 4B, left panel). Subsequently, we attempted TEV-GFP-His-tag removal by treatment with TEV protease (Figure S1). However, hZIP1-TEV-GFP-His fusion remained refractory to the enzymatic cleavage, likely due to the single-amino acid length of the C-terminus that results in poor accessibility to the TEV recognition sequence, preventing proteolytic activity.

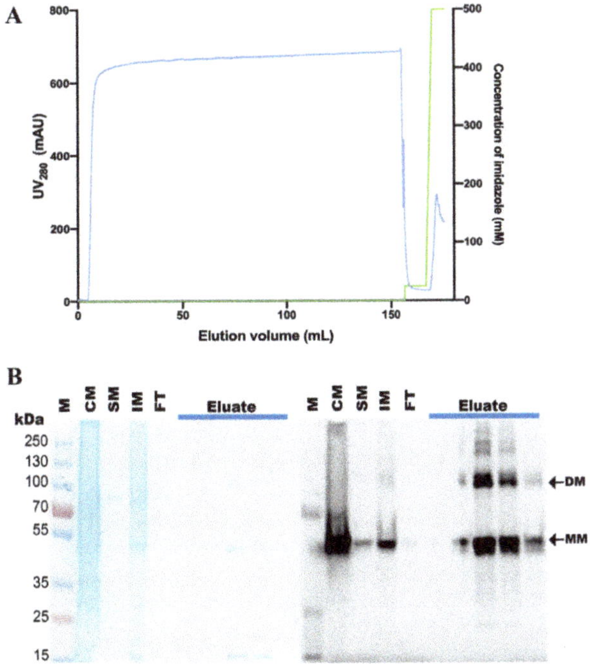

Figure 4. Purification of the hZIP1-TEV-GFP-His fusion. (**A**) Representative purification of hZIP1 C-terminally fused to TEV-GFP-His-tag. Protein was purified from crude *Saccharomyces cerevisiae* membranes isolated from ≈7 g of flask-grown cells and solubilized in 2% (*w*/*v*) DDM + 0.68% (*w*/*v*) CHS for 2 h at 4 °C. Profile from immobilized metal affinity chromatography (IMAC) indicates UV_{280} signal for protein (blue) and concentration of imidazole applied at each step of the affinity purification (green). CHS: cholesteryl hemisuccinate Tris salt. (**B**) Coomassie staining (left) and in-gel GFP fluorescence (right) of the corresponding SDS-PAGE-separated samples collected during IMAC purification. M: marker; CM: crude membranes; SM: solubilized membranes; IM: insoluble material; FT: IMAC flow-through; Eluate: IMAC peak nonconcentrated elution fractions eluted with 500 mM imidazole. Arrows indicate the monomeric (MM) and dimeric (DM) forms of purified hZIP-TEV-GFP-His fusion, with the predicted MWs of the respective fusions of 63.3 kDa (MM) and 126.6 kDa (DM).

3.4. Design and Purification of N-Terminal hZIP1 Fusions

Following fruitless TEV-GFP-His tag removal from the C-terminal fusion, we attempted design of alternative hZIP1 constructs, omitting GFP and placing the His-tag in either the N- or C-terminus. Initially, we engineered a hZIP1-TEV-His variant to resemble the first studied TEV-GFP-His fusion. However, both low expression levels and extraction efficiencies in nonionic detergents discouraged us from exploiting this construct (data not shown). Therefore, we extended our production strategy with N-terminal hZIP1 fusions. Two selected tags were applied, i.e., His-tag (delivering good expression levels and moderate solubilization properties, as already demonstrated for hZIP1-TEV-GFP-His), and StrepII-tag that has proven useful and convenient in production of other MPs [56], yielding His-TEV-hZIP1 and StrepII-TEV-hZIP1, respectively. After confirming the expression of the two new hZIP1 variants, we performed solubilization screening with increasing concentrations of DDM supplemented with CHS (Figure S2). For His-TEV-hZIP1, the highest solubilization efficacy (≈56%) was obtained for 2% DDM + 0.2% CHS (Figure S2A). In contrast, the StrepII-TEV-hZIP1 construct could be completely extracted by all tested concentrations, with the indication that 2% DDM + 0.2% CHS would perform the most efficiently in the large-scale purification (Figure S2B). Overall, encouraged by results obtained from this screening, yielding solubilization levels far beyond the C-terminal fusions, we purified both N-terminal hZIP1 variants using single-step affinity chromatography.

Purification of both constructs was performed on crude membranes isolated from ≈40 g of flask-grown *S. cerevisiae* cells derived from ≈12 L of cultures induced for expression for 68 h. Following solubilization of the respective membranes in 2% DDM + 0.2% CHS (Table S2), either IMAC or StrepII-tag affinity chromatography steps were run (Figure 5). Single-step affinity purification (Figure 5A) yielded ≈0.5 mg of protein per L of the cell culture, however, with moderate sample purity (Figure 5B). Moreover, the analysis of IMAC-pure fractions revealed that His-TEV-hZIP1 was also purified in different oligomeric forms, as reflected by the presence of both monomer and dimer (detectable mainly by immunoblotting; Figure 5B, left panel). Importantly, in contrast to the hZIP1-TEV-GFP-His construct, moving the TEV protease recognition sequence to the N-terminus resulted in marked increase of cleavage efficiency of the His-TEV-hZIP1 variant (Figure S3). Thus, following TEV protease treatment of this N-terminal fusion, tag-free sample of high purity was obtained after reverse-IMAC (R-IMAC) purification step.

Affinity purification of the StrepII-TEV-hZIP1 construct yielded protein samples of high purity (Figure 5C,D), however, obtained in low amounts (i.e., ≈0.3 mg of protein per L of the cell culture). Thus, in case of both tested N-terminal hZIP1 variants, protein yields were lower than for the hZIP1-TEV-GFP-His construct, perhaps indicative of a stabilizing role of the GFP fusion. Moreover, the presence of an additional band of lower molecular weight in the immunoblot analysis of the samples collected during StrepII-tag affinity purification (Figure 5D, left panel) may suggest either the presence of a contaminant recognized by the antibody or partial protein degradation, the latter supporting a stability enhancing effect of the GFP. Nevertheless, the final purity of the sample can be increased by an additional affinity chromatography step after TEV protease treatment, but this was not attempted here. To assess the stability of produced N-terminal hZIP1 fusions, we proceeded with SEC analysis.

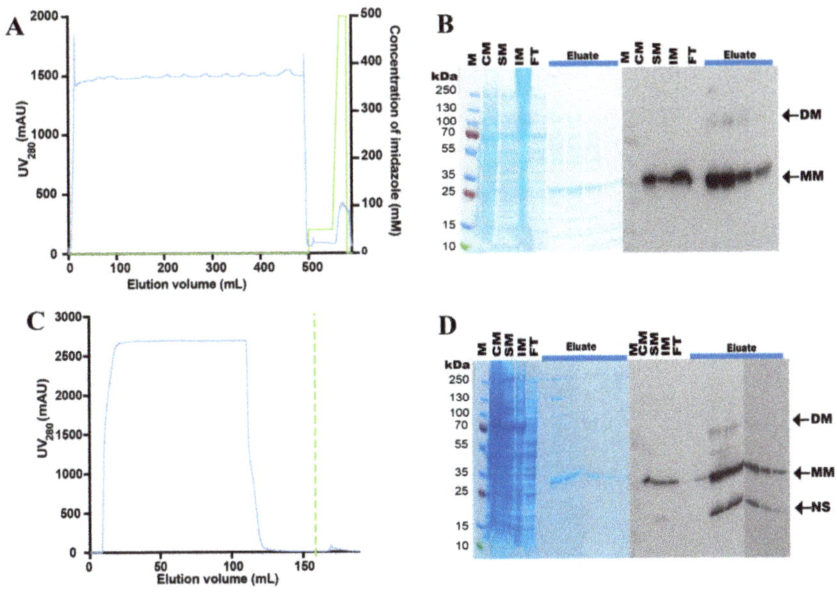

Figure 5. Purification of N-terminal hZIP1 fusions. Crude membranes isolated from *Saccharomyces cerevisiae* cells (derived from 12-L cultures) expressing (**A**,**B**) His-TEV-hZIP1 and (**C**,**D**) StrepII-TEV-hZIP1 (68-h induction at 15 °C) were solubilized in 2% (w/v) DDM + 0.2% (w/v) CHS for 2 h at 4 °C. Following solubilization, affinity-based purification was performed. CHS: cholesteryl hemisuccinate Tris salt. (**A**) Immobilized metal affinity chromatography (IMAC) of His-TEV-hZIP1. IMAC profile indicates UV_{280} signal for protein (blue) and concentration of imidazole applied at each step of the affinity purification (green). (**B**) Coomassie staining (left) and immunoblot probed with 6×His mAb-HRP conjugate (right) of SDS-PAGE-separated samples collected during IMAC purification. M: marker; CM: crude membranes; SM: solubilized membranes; IM: insoluble material; FT: IMAC flow-through; Eluate: IMAC peak fractions eluted with imidazole gradient. Arrows indicate the monomeric (MM) and dimeric (DM) forms of purified His-TEV-hZIP fusion, with the predicted MWs of the respective forms of 32.5 kDa (MM) and 65 kDa (DM). (**C**) StrepII-tag affinity chromatography of StrepII-TEV-hZIP1. The profile indicates UV_{280} signal for protein (blue). Green dashed line indicates the starting point of the elution phase with 50 mM biotin. (**D**) Coomassie staining (left) and immunoblot probed with Strep-Tactin® HRP conjugate (right) of SDS-PAGE-separated samples collected during StrepII-tag affinity purification. Letter code of the analyzed samples is to identical to (**B**). Arrows indicate the monomeric (MM) and dimeric (DM) forms of purified StrepII-TEV-hZIP fusion, with the predicted MWs of the respective fusions of 36.5 kDa (MM) and 73.0 kDa (DM). NS: nonspecific band/degradation product.

3.5. Stability of Affinity-Pure N-Terminal hZIP1 Fusions

In order to assess the quality of the two N-terminal hZIP1 fusions, we analyzed the samples originating from affinity purification by SEC performed in the primary detergent environment, i.e., in the buffer systems containing DDM and CHS (Table S2). Initially, protein homogeneity was monitored at different pH to enable selection of the most optimal conditions for the subsequent SEC analysis (data not shown). Interestingly, pH scouting revealed that higher sample stability was achieved at lower pH, similarly to what was already reported for hZIP4 [23]. The two hZIP1 variants displayed maximal stability at slightly different pH, i.e., highest homogeneity of His-TEV-free hZIP1 was achieved at pH 6.0 (Figure 6A), whereas the most symmetric SEC profile for StrepII-TEV-hZIP1 was obtained for pH 5.0 (Figure 6B). At the respective optimal pH both hZIP1 variants displayed only marginal aggregation. Moreover, in elution profiles of both hZIP1 samples, the main peak was preceded by a shoulder, hinting at the presence of higher oligomers or partial aggregation in the sample. In addition, as apparent from the SDS-PAGE analysis, SEC enabled the removal of the possible contaminant/degradation product seen in affinity-pure StrepII-TEV-hZIP1 sample. Overall, the generated SEC profiles for both detergent-solubilized N-terminal hZIP1 fusions were symmetric, suggesting high homogeneity and stability of the purified samples. Nevertheless, we observed a gradual decrease of protein stability after repeated rounds of freeze-thawing, as reflected by reduction of the height of the main SEC peak with concomitant increase of aggregated fraction eluting in the void volume (data not shown). Thus, utilization of freshly prepared protein samples should be considered for downstream applications. Finally, the SEC-pure samples exhibited a tendency to precipitate when concentrated above 9 mg mL^{-1}, indicating that further optimization of buffer conditions may be required.

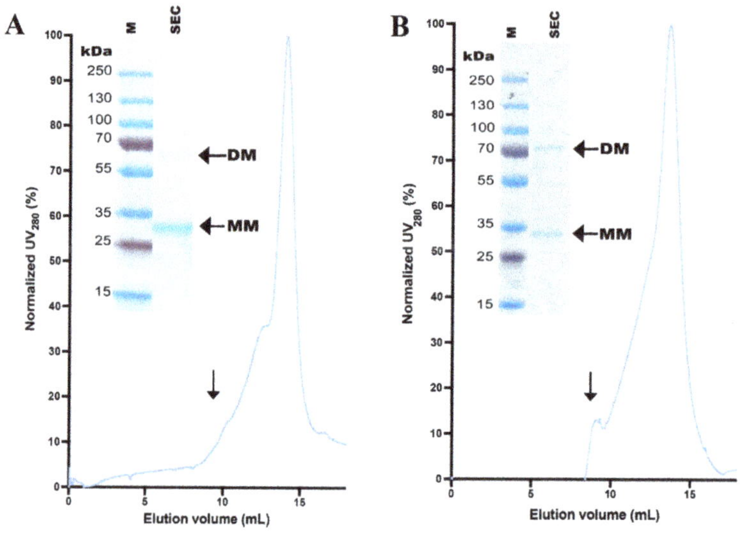

Figure 6. Size-exclusion chromatography (SEC) analysis of N-terminal hZIP1 fusions. (**A**) SEC profile of reverse-immobilized metal ion affinity chromatography-pure hZIP1 (i.e., TEV protease-digested His-TEV-hZIP1). (**B**) SEC profile of StrepII-tag affinity-pure StrepII-TEV-hZIP1. Normalized SEC chromatograms were obtained after separation on Superdex 200 Increase 10/300 GL column where A_{280} was monitored. Arrows indicate the estimated elution positions of the void volume. Insets: Coomassie staining of the resulting SEC-pure SDS-PAGE-separated samples (SEC). M: marker. Arrows indicate the monomeric (MM) and dimeric (DM) forms of the respective N-terminal hZIP1 fusions, with the predicted MWs of His-TEV-free hZIP1 of 32.5 kDa (MM) and 65 kDa (DM), and StrepII-TEV-hZIP1 of 36.5 kDa (MM) and 73 kDa (DM).

4. Discussion

The present paper addresses one of the bottlenecks that significantly restricts MP characterization, i.e., the production of prime-quality samples required for their biophysical characterization. Although MPs represent approximately 30% of the human proteome and constitute prominent drug targets [57–59], this class of proteins remains poorly understood relative to their soluble counterparts and is highly underrepresented in protein structure databases [60]. Therefore, there is a constant need for fine-tuning of existing heterologous expression systems to deliver recombinant MPs in an economic, efficient, and reliable manner. This dogma applies also to the ZIP family of transporters investigated here, for which no structure of a eukaryotic member has been reported to date. Thus, the overarching aim of this study was to develop an effective expression and purification platform to produce hZIPs.

Prior to this work, the closest reported attempt of hZIP purification was that of hZIP13 expressed using insect Sf9 cells to investigate the oligomerization state of this target, however, no data on obtained protein yield are presented [24]. In the same publication, the authors also utilized a human 293T cell-based platform, but protein purification was not attempted, and expression was visualized by immunoblotting only. In agreement with this notion, the usage of human cell-derived systems often results in low protein yields and high costs associated with the production [34,61], although strategies to improve expression levels have been proposed [62,63]. Conversely, the *S. cerevisiae* PAP1500 strain-based overproduction system has been employed for isolation of diverse classes of human MPs, and has delivered milligram quantities of samples in a cheap and easy-to-handle manner [39,41–43,64]. Importantly, human targets overproduced using this platform recently delivered both X-ray (human aquaporin 10) [39] and cryo-EM (human chloride channel ClC-1) [40] structures. Thus, in this study we took advantage of this protein production pipeline to attempt heterologous overproduction of hZIPs.

To ease downstream screening, all constructs were initially engineered to possess C-terminal TEV protease-detachable GFP-His-tags, enabling detection and affinity purification of the resulting fusions [49,65]. The first overproduction tests of these C-terminally-tagged variants immediately indicated the power of our production strategy, as we were able to recover the four short hZIP transporters selected for this study, all with promising expected expression levels, exceeding 5 mg of protein per L of cell culture in two of the cases, i.e., for hZIP1 and 2 (Figure 2E). On the contrary, both of the longest hZIPs attempted here, i.e., hZIP6 and 10, were refractory to expression using applied strategy. As these targets possess long N-termini (of more than 300 amino acids), it can be speculated that these additional segments may reduce overall protein stability that cannot be rescued by the yeast host. This is consistent with the previous study reporting that even production of an isolated extracellular domain of hZIP4 (with length of 327 amino acids) attempted in *E. coli* was fruitless, as the protein was severely aggregated [28]. Thus, assuming that the full-length hZIP6 and 10 may be more prone to protein degradation than the short hZIP counterparts, and combining it with the C-terminal location of the fusion tag, the detection of any partially translated products would not be possible here. Hence, optimization of the construct design and/or change of the expression host may be necessary for these two targets, but none of these were further explored here.

All four short, C-terminally-tagged hZIPs were synthesized as membrane-embedded full-length proteins. Interestingly, we observed differences in localization of the respective targets in the yeast cells. While hZIP1, 2, and 11 localized largely to the plasma membrane, hZIP13 accumulated intracellularly (Figure 2D), correlating with the hZIP-member specific naturally occurring localization in human cells [10].

As the next step, we sought to identify suitable detergents enabling efficient extraction and stabilizing solubilized hZIPs in their native form. We showed that, in general, the hZIPs investigated here were difficult to extract from the yeast membranes (Figure 3A), regardless of their cellular localization pattern. Based on data from the extensive detergent screening, we can conclude that only harsh zwitterionic detergents of the foscholine family displayed strong solubilization properties, whereas milder nonionic maltosides were rather ineffective, providing extraction efficacy higher than 20% only in the case of hZIP1. This finding agrees largely with the previously observed trend for other classes of MPs expressed in the same host [42,43]. However, the use of zwitterionic detergents entails the risk of protein denaturation leading to the loss of its structure [54,55]. Therefore, even though foscholine-solubilized material exhibited higher symmetry and less void in the F-SEC profiles, DDM was selected as a "safer option" solubilization detergent. Nevertheless, the homogeneity of the DDM-extracted samples was also relatively high, indicating that milder detergents are promising for isolation of hZIPs. Importantly, the presence of CHS increased both the efficiency of solubilization and protein stability. This suggests that this additional presence of cholesterol during solubilization may increase fluidity of the lipid environment of the yeast membranes or the CHS even directly interacts with expressed hZIPs to enhance their stability. The latter, i.e., stabilizing effect of cholesterol, is a well-known phenomenon in the MP field [66].

As the selected construct for initial characterization, i.e., hZIP1-TEV-GFP-His, was refractory to the TEV treatment, likely reflecting poor accessibility of the C-terminus preventing proteolytic activity, design of new constructs was attempted. As the His-tag position is known to contribute to the protein expression levels and stability [67,68], in our strategy we moved the tag to the N-terminus and investigated whether such modification could have beneficial or deleterious effects on the production. Remarkably, the construct comprising an N-terminal His-tag, but devoid of GFP, proved to be TEV-cleavable (Figure S3), but displayed significantly lower purification yields compared with the original C-terminal GFP fusion, although we did not observe major differences in the solubilization efficacy (Figure S2A). This observed decrease may suggest lowered levels of expression either as a direct consequence of different location of the tag and/or it may reflect an auxiliary role of the GFP fusion. Indeed, the latter has been reported to preserve the protein fold of other MPs, e.g., G protein-coupled receptors [69]. Thus, extending the N-terminally-tagged hZIP1 construct with GFP may provide more direct possibility of comparing such variant with the C-terminal TEV-GFP-His fusion, but this was not attempted here. Moreover, it cannot be excluded that production of hZIPs can be especially sensitive to the placement of charged tags, such as the His-tag, near the transmembrane domain. Although still not completely understood, it has been shown that cytoplasmic segments of MPs are typically enriched in positively charged residues (mostly lysine and arginine), the so-called "positive-inside rule" [70]. Thus, introduction of an additional positively charged stretch, here in a form of octa-histidine-tag, preceding the noncytoplasmic N-terminal segments of hZIPs may violate the rule, resulting in a lower protein stability. To corroborate this possible effect of the charged tag on production, we included a construct where the His-tag was exchanged to StrepII-tag. StrepII-tag, an octa-amino acid polypeptide, typically does not alter bioactivity of its fusion partner and can be used for affinity purification under mild conditions, maintaining the functionality of produced protein [71]. Interestingly, the production levels of the StrepII-TEV-hZIP1 variant were marginally decreased compared to the His-TEV-hZIP1 construct, again without any detectable changes in detergent extraction efficacy (Figure S2B). Collectively, this indicates that the expression of both the N-terminal hZIP1 variants described here may require further optimization of the fusion tag, possibly due to the predicted N_{out}-C_{out} topology. Similar conclusions were drawn in a work describing development of purification strategies of bacterial ZIPs sharing the same location of the N- and C-termini, where the insertion of a maltose-binding protein and a signal sequence between the His-tag and the scare N-terminal domain was shown to rescue protein production [32].

hZIPs have been shown to form homodimers in a native state, but hZIP6 and 10 can also heterodimerize [13]. Analysis of the in-gel GFP fluorescence of the SDS-PAGE-separated IMAC-pure hZIP1-TEV-GFP-His samples implies that this construct preserves homodimerization to some degree (Figure 4B). The proportion of dimeric versus monomeric form decreases in the N-terminally-tagged GFP-free constructs (Figures 4 and 5), which may correlate with lower stability and/or yields observed for these variants. Hence, while for the StrepII-TEV-hZIP1 construct a small population of the dimeric form was detectable after SEC, hZIP1 (TEV protease-digested His-TEV-hZIP1) eluted mostly as monomer (Figure 6). This may indicate that the His-tag is directly assisting in dimerization, similarly to what was observed for BbZIP, however with the opposing effect [31].

Affinity-based protein purification yielded milligram amounts of both N-terminal hZIP1 fusions, although the relative purity of the samples marginally differed (Figure 5). However, while a single chromatography round was sufficient to produce a highly pure sample of StrepII-TEV-hZIP1, two additional steps were included for His-TEV-hZIP1, i.e., TEV protease cleavage and R-IMAC (Figure S3). Although both samples exhibited high monodispersity, as assessed by SEC (Figure 6), the purified variants displayed limited stability, tending to precipitate after freeze–thawing cycles. Moreover, freshly prepared, nonfrozen samples precipitated also at concentrations exceeding 9 mg mL^{-1}. However, such concentrations are compatible with requirements for most of the biophysical experiments, including X-ray crystallography [72] and cryo-EM, where even low amounts of the sample can be utilized [73].

In conclusion, we outlined a *S. cerevisiae*-based platform to approach milligram production of hZIPs, with hZIP1 being an example target that was purified to monodispersity. We believe that our strategy for protein expression and purification described here provides a new encouraging option for sample preparation of hZIPs for more detailed biochemical and structural analysis.

Supplementary Materials: The following are available online at https://www.mdpi.com/2073-4409/10/2/213/s1, Table S1: Sequences of cloning primers used to engineer expression constructs of human Zrt-, Irt-like proteins (hZIPs) described in the study, Table S2: Solubilization conditions and buffers used during affinity purification and size-exclusion chromatography of the respective hZIP1 fusions, Figure S1: Cleavage of hZIP1-TEV-GFP-His fusion with tobacco etch virus (TEV) protease, Figure S2: Detergent screening of N-terminal hZIP1 fusions, Figure S3: Cleavage of His-TEV-hZIP1 fusion with tobacco etch virus (TEV) protease.

Author Contributions: E.R.B. and K.G. cloned the expression constructs, and performed protein expression, purification, and biochemical characterization from cell material provided by P.A.P. Furthermore, P.A.P. supplied the expression vector and yeast strain, and assisted in cloning of the expression constructs. E.R.B., P.G. and K.G. designed the project. E.R.B. and K.G. wrote the paper with contribution from all authors. All authors have read and agreed to the published version of the manuscript.

Funding: E.R.B. was funded by The Independent Research Fund Denmark through PhD scholarship. P.G. was supported by the following Foundations: Lundbeck, Knut and Alice Wallenberg, Carlsberg, Novo-Nordisk, Brødrene Hartmann, Agnes og Poul Friis, Augustinus, Crafoord, and The Per-Eric and Ulla Schyberg. Funding was also obtained from The Independent Research Fund Denmark, The Swedish Research Council and through a Michaelsen scholarship. K.G. was supported by The Independent Research Fund Denmark and The Lundbeck Foundation.

Institutional Review Board Statement: Not applicable.

Informed Consent Statement: Not applicable.

Data Availability Statement: All data presented in this study are contained within this article (Overproduction of Human Zip (SLC39) Zinc Transporters in Saccharomyces Cerevisiae for Biophysical Characterization) and the Supplementary Materials.

Acknowledgments: We are thankful to David Sørensen for his excellent technical assistance.

Conflicts of Interest: The authors declare no conflict of interest.

References

1. Andreini, C.; Banci, L.; Bertini, I.; Rosato, A. Counting the Zinc-Proteins Encoded in the Human Genome. *J. Proteome Res.* **2006**, *5*, 196–201. [CrossRef]
2. Chasapis, C.T.; Loutsidou, A.C.; Spiliopoulou, C.A.; Stefanidou, M. Zinc and human health: An update. *Arch. Toxicol.* **2012**, *86*, 521–534. [CrossRef] [PubMed]
3. Thingholm, T.E.; Rönnstrand, L.; Rosenberg, P.A. Why and how to investigate the role of protein phosphorylation in ZIP and ZnT zinc transporter activity and regulation. *Cell. Mol. Life Sci.* **2020**, *77*, 3085–3102. [CrossRef] [PubMed]
4. Guerinot, M.L. The ZIP Family of Metal Transporters. *Biochim. Biophys. Acta Biomembr.* **2000**, *1465*, 190–198. [CrossRef]
5. Bafaro, E.; Liu, Y.; Xu, Y.; Dempski, R.E. The emerging role of zinc transporters in cellular homeostasis and cancer. *Signal Transduct. Target. Ther.* **2017**, *2*, 17029. [CrossRef] [PubMed]
6. Kambe, T.; Hashimoto, A.; Fujimoto, S. Current understanding of ZIP and ZnT zinc transporters in human health and diseases. *Cell. Mol. Life Sci.* **2014**, *71*, 3281–3295. [CrossRef] [PubMed]
7. Kambe, T.; Tsuji, T.; Hashimoto, A.; Itsumura, N. The Physiological, Biochemical, and Molecular Roles of Zinc Transporters in Zinc Homeostasis and Metabolism. *Physiol. Rev.* **2015**, *95*, 749–784. [CrossRef]
8. Dufner-Beattie, J.; Langmade, S.J.; Wang, F.; Eide, D.; Andrews, G.K. Structure, Function, and Regulation of a Subfamily of Mouse Zinc Transporter Genes. *J. Biol. Chem.* **2003**, *278*, 50142–50150. [CrossRef]
9. Gaither, L.A.; Eide, D.J. Eukaryotic zinc transporters and their regulation. *BioMetals* **2001**, *14*, 251–270. [CrossRef]
10. Bowers, K.; Srai, S.K.S. The trafficking of metal ion transporters of the Zrt- and Irt-like protein family. *Traffic* **2018**, *19*, 813–822. [CrossRef]

11. Kimura, T.; Kambe, T. The Functions of Metallothionein and ZIP and ZnT Transporters: An Overview and Perspective. *Int. J. Mol. Sci.* **2016**, *17*, 336. [CrossRef] [PubMed]
12. Hennigar, S.R.; McClung, J.P. Zinc Transport in the Mammalian Intestine. *Compr. Physiol.* **2018**, *9*, 59–74. [CrossRef]
13. Bin, B.-H.; Seo, J.; Kim, S.T. Function, Structure, and Transport Aspects of ZIP and ZnT Zinc Transporters in Immune Cells. *J. Immunol. Res.* **2018**, *2018*, 1–9. [CrossRef] [PubMed]
14. Higashi, Y.; Segawa, S.; Matsuo, T.; Nakamura, S.; Kikkawa, Y.; Nishida, K.; Nagasawa, K. Microglial zinc uptake via zinc transporters induces ATP release and the activation of microglia. *Glia* **2011**, *59*, 1933–1945. [CrossRef]
15. Inoue, Y.; Hasegawa, S.; Ban, S.; Yamada, T.; Date, Y.; Mizutani, H.; Nakata, S.; Tanaka, M.; Hirashima, N. ZIP2 Protein, a Zinc Transporter, Is Associated with Keratinocyte Differentiation. *J. Biol. Chem.* **2014**, *289*, 21451–21462. [CrossRef] [PubMed]
16. Wang, X.; Zhou, B. Dietary zinc absorption: A play of Zips and ZnTs in the gut. *IUBMB Life* **2010**, *62*, 176–182. [CrossRef]
17. Miyai, T.; Hojyo, S.; Ikawa, T.; Kawamura, M.; Irié, T.; Ogura, H.; Hijikata, A.; Bin, B.-H.; Yasuda, T.; Kitamura, H.; et al. Zinc transporter SLC39A10/ZIP10 facilitates antiapoptotic signaling during early B-cell development. *Proc. Natl. Acad. Sci. USA* **2014**, *111*, 11780–11785. [CrossRef]
18. Franz, M.C.; Pujol-Giménez, J.; Montalbetti, N.; Fernandez-Tenorio, M.; DeGrado, T.R.; Niggli, E.; Romero, M.F.; Hediger, M.A. Reassessment of the Transport Mechanism of the Human Zinc Transporter SLC39A2. *Biochemistry* **2018**, *57*, 3976–3986. [CrossRef]
19. Franklin, L.C.C.R.B. Evidence that Human Prostate Cancer is a ZIP1-Deficient Malignancy that could be Effectively Treated with a Zinc Ionophore (Clioquinol) Approach. *Chemother. Open Access* **2014**, *4*, 1–10. [CrossRef]
20. Li, M.; Zhang, Y.; Liu, Z.; Bharadwaj, U.; Wang, H.; Wang, X.; Zhang, S.; Liuzzi, J.P.; Chang, S.-M.; Cousins, R.J.; et al. Aberrant expression of zinc transporter ZIP4 (SLC39A4) significantly contributes to human pancreatic cancer pathogenesis and progression. *Proc. Natl. Acad. Sci. USA* **2007**, *104*, 18636–18641. [CrossRef]
21. Taylor, K.M.; Morgan, H.E.; Smart, K.; Zahari, N.M.; Pumford, S.; Ellis, I.O.; Robertson, J.F.R.; Nicholson, R.I. The Emerging Role of the LIV-1 Subfamily of Zinc Transporters in Breast Cancer. *Mol. Med.* **2007**, *13*, 396–406. [CrossRef] [PubMed]
22. Wu, L.; Chaffee, K.G.; Parker, A.S.; Sicotte, H.; Petersen, G.M. Zinc transporter genes and urological cancers: Integrated analysis suggests a role for ZIP11 in bladder cancer. *Tumor Biol.* **2015**, *36*, 7431–7437. [CrossRef] [PubMed]
23. Hoch, E.; Levy, M.; Hershfinkel, M.; Sekler, I. Elucidating the H+ Coupled Zn2+ Transport Mechanism of ZIP4; Implications in Acrodermatitis Enteropathica. *Int. J. Mol. Sci.* **2020**, *21*, 734. [CrossRef] [PubMed]
24. Bin, B.-H.; Fukada, T.; Hosaka, T.; Yamasaki, S.; Ohashi, W.; Hojyo, S.; Miyai, T.; Nishida, K.; Yokoyama, S.; Hirano, T. Biochemical characterization of human ZIP13 protein a homo-dimerized zinc transporter involved in the spondylocheiro dysplastic ehlers-danlos syndrome. *J. Biol. Chem.* **2011**, *286*, 40255–40265. [CrossRef]
25. Dusanic, M.; Dekomien, G.; Lucke, T.; Vorgerd, M.; Weiß, J.; Epplen, J.T.; Köhler, C.; Hoffjan, S. Novel Nonsense Mutation in SLC39A13 Initially Presenting as Myopathy: Case Report and Review of the Literature. *Mol. Syndr.* **2018**, *9*, 100–109. [CrossRef]
26. Zhang, T.; Sui, D.; Zhang, C.; Cole, L.; Hu, J. Asymmetric functions of a binuclear metal center within the transport pathway of a human zinc transporter ZIP4. *FASEB J.* **2020**, *34*, 237–247. [CrossRef]
27. Zhang, T.; Liu, J.; Fellner, M.; Zhang, C.; Sui, D.; Hu, J. Crystal structures of a ZIP zinc transporter reveal a binuclear metal center in the transport pathway. *Sci. Adv.* **2017**, *3*, e1700344. [CrossRef]
28. Zhang, T.; Sui, D.; Hu, J. Structural insights of ZIP4 extracellular domain critical for optimal zinc transport. *Nat. Commun.* **2016**, *7*, 11979. [CrossRef]
29. Dempski, R.E. The Cation Selectivity of the ZIP Transporters. In *Na Channels from Phyla to Function*; Elsevier: Amsterdam, The Netherlands, 2012; Volume 69, pp. 221–245.
30. Blindauer, C.A. Advances in the molecular understanding of biological zinc transport. *Chem. Commun.* **2015**, *51*, 4544–4563. [CrossRef]
31. Lin, W.; Chai, J.; Love, J.; Fu, D. Selective Electrodiffusion of Zinc Ions in a Zrt-, Irt-like Protein, ZIPB. *J. Biol. Chem.* **2010**, *285*, 39013–39020. [CrossRef]
32. Ma, C.; Hao, Z.; Huysmans, G.; Lesiuk, A.; Bullough, P.; Wang, Y.; Bartlam, M.; Phillips, S.E.; Young, J.D.; Goldman, A.; et al. A Versatile Strategy for Production of Membrane Proteins with Diverse Topologies: Application to Investigation of Bacterial Homologues of Human Divalent Metal Ion and Nucleoside Transporters. *PLoS ONE* **2015**, *10*, e0143010. [CrossRef] [PubMed]
33. Jia, B.; Jeon, C.O. High-Throughput Recombinant Protein Expression in Escherichia Coli: Current Status and Future Perspec-tives. *Open Biol.* **2016**, *6*, 2046–2441. [CrossRef] [PubMed]
34. Pandey, A.; Shin, K.; Patterson, R.E.; Liu, X.Q.; Rainey, J.K. Current Strategies for Protein Production and Purification Ena-bling Membrane Protein Structural Biology1. *Biochem. Cell Biol.* **2016**, *94*, 507–527. [CrossRef] [PubMed]
35. Terpe, K. Overview of bacterial expression systems for heterologous protein production: From molecular and biochemical fundamentals to commercial systems. *Appl. Microbiol. Biotechnol.* **2006**, *72*, 211–222. [CrossRef] [PubMed]
36. Gomes, A.R.; Byregowda, S.M.; Veeregowda, B.M.; Balamurugan, V. An Overview of Heterologous Expression Host Systems for the Production of Recombinant Proteins. *Adv. Anim. Veter- Sci.* **2016**, *4*, 346–356. [CrossRef]
37. Liu, Z.; Tyo, K.E.; Martínez, J.L.; Petranovic, D.; Nielsen, J. Different expression systems for production of recombinant proteins in Saccharomyces cerevisiae. *Biotechnol. Bioeng.* **2012**, *109*, 1259–1268. [CrossRef]
38. Hou, J.; Tyo, K.E.J.; Liu, Z.; Petranovic, D.; Nielsen, J. Metabolic Engineering of Recombinant Protein Secretion by Saccharo-myces Cerevisiae. *FEMS Yeast Res.* **2012**, *12*, 491–510. [CrossRef]

39. Gotfryd, K.; Mósca, A.F.; Missel, J.W.; Truelsen, S.F.; Wang, K.; Spulber, M.; Krabbe, S.L.; Nielsen, C.; Laforenza, U.; Soveral, G.; et al. Human adipose glycerol flux is regulated by a pH gate in AQP10. *Nat. Commun.* **2018**, *9*, 4749. [CrossRef]
40. Wang, K.; Preisler, S.S.; Zhang, L.; Cui, Y.; Missel, J.W.; Grønberg, C.; Gotfryd, K.; Lindahl, E.; Andersson, M.; Calloe, K.; et al. Structure of the human ClC-1 chloride channel. *PLoS Biol.* **2019**, *17*, e3000218. [CrossRef]
41. Molbaek, K.; Scharff-Poulsen, P.; Nielsen, C.; Klaerke, D.A.; Pedersen, P.A. High yield purification of full-length functional hERG K+ channels produced in Saccharomyces cerevisiae. *Microb. Cell Factories* **2015**, *14*, 15. [CrossRef]
42. Bjørkskov, F.B.; Krabbe, S.L.; Nurup, C.N.; Missel, J.W.; Spulber, M.; Bomholt, J.; Molbaek, K.; Helix-Nielsen, C.; Gotfryd, K.; Gourdon, P.E.; et al. Purification and functional comparison of nine human Aquaporins produced in Saccharomyces cerevisiae for the purpose of biophysical characterization. *Sci. Rep.* **2017**, *7*, 1–21. [CrossRef] [PubMed]
43. Zhang, L.; Wang, K.; Klaerke, D.A.; Calloe, K.; Lowrey, L.; Pedersen, P.A.; Gourdon, P.; Gotfryd, K. Purification of Functional Human TRP Channels Recombinantly Produced in Yeast. *Cells* **2019**, *8*, 148. [CrossRef] [PubMed]
44. Cesareni, G.; Murray, J.A.H. Plasmid Vectors Carrying the Replication Origin of Filamentous Single-Stranded Phages. In *Genetic Engineering*; Springer: Berlin, Germany, 1987; pp. 135–154.
45. Erhart, E.; Hollenberg, C.P. The presence of a defective LEU2 gene on 2 mu DNA recombinant plasmids of Saccharomyces cerevisiae is responsible for curing and high copy number. *J. Bacteriol.* **1983**, *156*, 625–635. [CrossRef] [PubMed]
46. Stecher, G.; Tamura, K.; Kumar, S. Molecular Evolutionary Genetics Analysis (MEGA) for macOS. *Mol. Biol. Evol.* **2020**, *37*, 1237–1239. [CrossRef] [PubMed]
47. Pedersen, P.A.; Rasmussen, J.H.; Jørgensen, P.L. Expression in High Yield of Pig $\alpha 1\beta 1$ Na,K-ATPase and Inactive Mutants D369N and D807N in Saccharomyces cerevisiae. *J. Biol. Chem.* **1996**, *271*, 2514–2522. [CrossRef] [PubMed]
48. Bradford, M.M. A rapid and sensitive method for the quantitation of microgram quantities of protein utilizing the principle of protein-Dye binding. *Anal. Biochem.* **1976**, *72*, 248–254. [CrossRef]
49. Drew, D.; Newstead, S.; Sonoda, Y.; Kim, H.; Von Heijne, G.; Iwata, S. GFP-based optimization scheme for the overexpression and purification of eukaryotic membrane proteins in Saccharomyces cerevisiae. *Nat. Protoc.* **2008**, *3*, 784–798. [CrossRef]
50. Tsirigos, K.D.; Peters, C.; Shu, N.; Käll, L.; Elofsson, A. The TOPCONS web server for consensus prediction of membrane protein topology and signal peptides. *Nucleic Acids Res.* **2015**, *43*, W401–W407. [CrossRef]
51. The UniProt Consortium UniProt: A worldwide hub of protein knowledge. *Nucleic Acids Res.* **2019**, *47*, D506–D515. [CrossRef]
52. Arachea, B.T.; Sun, Z.; Potente, N.; Malik, R.; Isailovic, D.; Viola, R.E. Detergent selection for enhanced extraction of membrane proteins. *Protein Expr. Purif.* **2012**, *86*, 12–20. [CrossRef]
53. Stetsenko, A.; Guskov, A. An Overview of the Top Ten Detergents Used for Membrane Protein Crystallization. *Crystals* **2017**, *7*, 197. [CrossRef]
54. Kawate, T.; Gouaux, E. Fluorescence-Detection Size-Exclusion Chromatography for Precrystallization Screening of Integral Membrane Proteins. *Structure* **2006**, *14*, 673–681. [CrossRef] [PubMed]
55. Geertsma, E.R.; Groeneveld, M.; Slotboom, D.-J.; Poolman, B. Quality control of overexpressed membrane proteins. *Proc. Natl. Acad. Sci. USA* **2008**, *105*, 5722–5727. [CrossRef] [PubMed]
56. Schmidt, T.G.M.; Skerra, A. The Strep-tag system for one-step purification and high-affinity detection or capturing of proteins. *Nat. Protoc.* **2007**, *2*, 1528–1535. [CrossRef] [PubMed]
57. Arinaminpathy, Y.; Khurana, E.; Engelman, D.M.; Gerstein, M.B. Computational analysis of membrane proteins: The largest class of drug targets. *Drug Discov. Today* **2009**, *14*, 1130–1135. [CrossRef]
58. Edwards, A.M.; Berman, J.; Sundström, M. Structural Genomics and Drug Discovery. *Annu. Rep. Med. Chem.* **2005**, *40*, 349–369. [CrossRef]
59. Kotov, V.; Bartels, K.; Veith, K.; Josts, I.; Subhramanyam, U.K.T.; Günther, C.; Labahn, J.; Marlovits, T.C.; Moraes, I.; Tidow, H.; et al. High-throughput stability screening for detergent-solubilized membrane proteins. *Sci. Rep.* **2019**, *9*, 1–19. [CrossRef]
60. Almeida, J.G.; Preto, A.J.; Koukos, P.I.; Bonvin, A.M.J.J.; Moreira, I.S. Membrane Proteins Structures: A Review on Computa-tional Modeling Tools. *Biochim. Biophys. Acta Biomembr.* **2017**, *1859*, 2021–2039. [CrossRef]
61. Andréll, J.; Tate, C.G. Overexpression of membrane proteins in mammalian cells for structural studies. *Mol. Membr. Biol.* **2012**, *30*, 52–63. [CrossRef]
62. Lin, C.-Y.; Huang, Z.; Wen, W.; Wu, A.; Wang, C.; Niu, L. Enhancing Protein Expression in HEK-293 Cells by Lowering Culture Temperature. *PLoS ONE* **2015**, *10*, e0123562. [CrossRef]
63. Rana, M.S.; Wang, X.; Banerjee, A. An Improved Strategy for Fluorescent Tagging of Membrane Proteins for Overexpression and Purification in Mammalian Cells. *Biochemistry* **2018**, *57*, 6741–6751. [CrossRef] [PubMed]
64. Molbaek, K.; Tejada, M.; Ricke, C.H.; Scharff-Poulsen, P.; Ellekvist, P.; Helix-Nielsen, C.; Kumar, N.; Klaerke, D.A.; Pedersen, P.A. Purification and initial characterization of Plasmodium falciparum K+ channels, PfKch1 and PfKch2 produced in Saccharomyces cerevisiae. *Microb. Cell Factories* **2020**, *19*, 1–16. [CrossRef] [PubMed]
65. Drew, D.; Lerch, M.; Kunji, E.R.S.; Slotboom, D.-J.; De Gier, J.-W. Optimization of membrane protein overexpression and purification using GFP fusions. *Nat. Chem. Biol.* **2006**, *3*, 303–313. [CrossRef] [PubMed]
66. Fantini, J.; Barrantes, F.J. How cholesterol interacts with membrane proteins: An exploration of cholesterol-binding sites including CRAC, CARC, and tilted domains. *Front. Physiol.* **2013**, *4*, 31. [CrossRef] [PubMed]
67. Mohanty, A.K.; Wiener, M.C. Membrane protein expression and production: Effects of polyhistidine tag length and position. *Protein Expr. Purif.* **2004**, *33*, 311–325. [CrossRef] [PubMed]

68. Booth, W.T.; Schlachter, C.R.; Pote, S.; Ussin, N.; Mank, N.J.; Klapper, V.; Offermann, L.R.; Tang, C.; Hurlburt, B.K.; Chruszcz, M. Impact of an N-terminal Polyhistidine Tag on Protein Thermal Stability. *ACS Omega* **2018**, *3*, 760–768. [CrossRef]
69. Chun, E.; Thompson, A.A.; Liu, W.; Roth, C.B.; Griffith, M.T.; Katritch, V.; Kunken, J.; Xu, F.; Cherezov, V.; Hanson, M.A.; et al. Fusion Partner Toolchest for the Stabilization and Crystallization of G Protein-Coupled Receptors. *Structure* **2012**, *20*, 967–976. [CrossRef]
70. Wallin, E.; Von Heijne, G. Genome-wide analysis of integral membrane proteins from eubacterial, archaean, and eukaryotic organisms. *Protein Sci.* **1998**, *7*, 1029–1038. [CrossRef]
71. Locatelli-Hoops, S.C.; Yeliseev, A.A. Use of tandem affinity chromatography for purification of cannabinoid receptor CB_2. *Breast Cancer* **2014**, *1177*, 107–120. [CrossRef]
72. Newby, Z.E.R.; O'Connell, J.D.; Gruswitz, F.; Hays, F.A.; Harries, W.E.C.; Harwood, I.M.; Ho, J.D.; Lee, J.K.; Savage, D.F.; Miercke, L.J.W.; et al. A general protocol for the crystallization of membrane proteins for X-ray structural investigation. *Nat. Protoc.* **2009**, *4*, 619–637. [CrossRef]
73. Tonggu, L.; Wang, L. Cryo-EM sample preparation method for extremely low concentration liposomes. *Ultramicroscopy* **2020**, *208*, 112849. [CrossRef] [PubMed]

cells

MDPI

Review

Legume Lectins with Different Specificities as Potential Glycan Probes for Pathogenic Enveloped Viruses

Annick Barre [1], Els J. M. Van Damme [2], Bernard Klonjkowski [3], Mathias Simplicien [1], Jan Sudor [1], Hervé Benoist [1] and Pierre Rougé [1,*]

1 UMR 152 PharmaDev, Institut de Recherche et Développement, Faculté de Pharmacie, Université Paul Sabatier, 35 Chemin des Maraîchers, F-31062 Toulouse, France; annick.barre@univ-tlse3.fr (A.B.); simplicien.mathias@gmail.com (M.S.); jan.sudor1@univ-tlse3.fr (J.S.); herve.benoist@ird.fr (H.B.)

2 Department of Biotechnology, Faculty of Bioscience Engineering, Ghent University, Coupure Links 653, B-9000 Ghent, Belgium; elsjm.vandamme@ugent.be

3 UMR Virologie, INRA, ANSES, Ecole Nationale Vétérinaire d'Alfort, F-94700 Maisons-Alfort, France; bernard.klonjkowski@vet-alfort.fr

* Correspondence: pierre.rouge.perso@gmail.com; Tel.: +33-069-552-0851

Abstract: Pathogenic enveloped viruses are covered with a glycan shield that provides a dual function: the glycan structures contribute to virus protection as well as host cell recognition. The three classical types of *N*-glycans, in particular complex glycans, high-mannose glycans, and hybrid glycans, together with some *O*-glycans, participate in the glycan shield of the Ebola virus, influenza virus, human cytomegalovirus, herpes virus, human immunodeficiency virus, Lassa virus, and MERS-CoV, SARS-CoV, and SARS-CoV-2, which are responsible for respiratory syndromes. The glycans are linked to glycoproteins that occur as metastable prefusion glycoproteins on the surface of infectious virions such as gp120 of HIV, hemagglutinin of influenza, or spike proteins of beta-coronaviruses. Plant lectins with different carbohydrate-binding specificities and, especially, mannose-specific lectins from the Vicieae tribe, such as pea lectin and lentil lectin, can be used as glycan probes for targeting the glycan shield because of their specific interaction with the α1,6-fucosylated core $Man_3GlcNAc_2$, which predominantly occurs in complex and hybrid glycans. Other plant lectins with Neu5Ac specificity or GalNAc/T/Tn specificity can also serve as potential glycan probes for the often sialylated complex glycans and truncated *O*-glycans, respectively, which are abundantly distributed in the glycan shield of enveloped viruses. The biomedical and therapeutical potential of plant lectins as antiviral drugs is discussed.

Keywords: enveloped virus; Ebola virus; HIV; herpes simplex virus; human cytomegalovirus; influenza virus; MERS-CoV; SARS-CoV-2; *N*-glycosite; *O*-glycosite; high-mannose glycan; complex *N*-glycans; Vicieae man-specific lectin; T/Tn-specific lectin; specific interaction

Citation: Barre, A.; Van Damme, E.J.M.; Klonjkowski, B.; Simplicien, M.; Sudor, J.; Benoist, H.; Rougé, P. Legume Lectins with Different Specificities as Potential Glycan Probes for Pathogenic Enveloped Viruses. *Cells* **2022**, *11*, 339. https://doi.org/10.3390/cells11030339

Academic Editor: Suleyman Allakhverdiev

Received: 17 December 2021
Accepted: 18 January 2022
Published: 20 January 2022

1. Introduction

Many pathogenic viruses for humans are so-called enveloped viruses with a lipid bilayer that allows the infectious virions to fuse with the cell membrane, followed by the entry and replication of the viral genetic material into the host cells. Ebola virus (EBOV), influenza virus (IV), herpes simplex virus (HSV), human immunodeficiency virus (HIV), human cytomegalovirus (HCMV), Lassa virus (LASV), and the beta-coronaviruses responsible for the Middle East respiratory syndrome (MERS-CoV) and the severe acute respiratory syndrome (i.e., SARS-CoV and SARS-CoV-2) belong to this group of pathogenic enveloped viruses [1]. Among the surface glycoproteins that are embedded in the lipid bilayer of enveloped viruses, so-called fusion glycoproteins play a key role in mediating the recognition of infectious virions by the host cell membrane receptors and their subsequent anchorage to the host cells [2]. Energetically driven conformational changes occurring in the metastable fusion proteins, which usually occur in a prefusion state, are responsible for an enhanced exposure of the receptor-binding domain (RBD) of the fusion proteins that favors their recognition by the host cell receptors [1]. Fusion proteins of enveloped viruses usually consist of the non-covalent association of three monomers to build a homotrimeric structure exposed on the surface of the virions. However, depending on the enveloped viruses, the structure, shape, and size of the monomers building the homotrimers are highly variable from one virus to another (Table 1). Other surface glycoproteins, such as the so-called B glycoprotein from HSV [3], and E proteins from flaviviruses responsible for some severe diseases, including chikungunya virus (CHIV), dengue virus (DENV), and Zika virus (ZIV), also contribute to the glycan shield covering the infectious virions [4–6].

Table 1. Structural properties of fusion protein and E glycoprotein associations of enveloped viruses as parts of the glycan shield covering infectious virions.

Enveloped Virus	Homotrimer	Monomer	PDB Entry *	Reference
Ebola virus (EBOV)	Homotrimer		7JPH	[7]
Influenza virus (IV)	Hemagglutinin A Homotrimer	hemagglutinin A	6Y5G	[8]
Human cytomegalovirus (HCMV)	Homotrimer	B glycoprotein	5CXF	[9]
Herpes simplex virus (HSV)	Homotrimer	B glycoprotein	2GUM	[3]
Human immunodeficiency virus (HIV)	Homotrimer	gp140	4TVP	[10]
Lassa virus (LASV)	Homotrimer	GPC glycoprotein	5VK2	[11]
Middle east respiratory syndrome coronavirus (MERS-CoV)	Spike	S protein	5W9H	[12]
Severe acute respiratory syndrome coronavirus-1 (SAR-CoV)	Spike	S protein	6ACD	[13]
Severe acute respiratory syndrome coronavirus-2 (SARS-CoV-2)	Spike	S protein	6VXX	[14]
Chikungunya virus (CHIV)	Homodimer	E protein	3N40	[4]
Dengue virus (DENV)	Homodimer	E protein	1UZG	[5]
Zika virus (ZIV)	Homodimer	E protein	57BUB	[6]

* A single PDB code is indicated but several PDB entries are available at the PDB.

Although *N*- and *O*-glycans decorate the fusion glycoproteins, the three classical types of *N*-glycans, including complex-type glycans, high-mannose-type glycans, and hybrid-type glycans, are predominantly distributed along the fusion proteins of pathogenic enveloped viruses. In addition, a high proportion of complex glycans are α1,6-fucosylated on the first GlcNAc linked to the Asn residue and often sialylated on their terminal Gal residues [15]. Moreover, the extreme diversity of complex glycans appears as a characteristic of enveloped viruses. In addition to the *N*-glycans, *O*-glycans have been identified on the envelope glycoproteins of infectious virions, especially in pathogenic coronaviruses such as SARS-CoV-2 [16]. In fact, most of the Ser and Thr residues of unoccupied NXT/S glycosylation sites of SARS-CoV-2 are *O*-glycosylated by short Gal/GalNAc/T/Tn-containing *O*-glycan chains [17]. However, the *O*-glycan content of SARS-CoV-2 is much lower than the level of *N*-glycans.

Lectins are known as a group of carbohydrate-binding proteins of non-immune origin that are widely distributed in plants. Many lectins have been studied for their role in the protection of plants against pathogens, aiming to resolve the function of the lectins inside different plant tissues. In addition, these carbohydrate-binding proteins have been proven to be important tools for glycobiology, allowing for the investigation of the importance of protein–carbohydrate interactions. Several plant lectins have been reported as potent molecules with anti-infectivity properties for RNA viruses including pathogenic enveloped viruses. Depending on their carbohydrate-binding specificity, lectins can recognize and bind particular types of glycan structures present in the glycan shield of viruses. Over the past decades, lectins from different legume species, referred to as legume lectins, have been studied in great detail. Despite the fact that legume lectins represent a large family of proteins with important similarity in their amino acid sequences, these lectins show remarkable variability in their carbohydrate-binding properties. Many legume lectins have been reported to recognize glycoconjugates on cells and viruses and can discriminate between diverse glycan structures, making them interesting research tools for glycomic research [18].

The primary purpose of this review was to give an overview of the types of glycans present in the glycan shield of different pathogenic enveloped viruses and how legume lectins with different specificities can act as carbohydrate-binding agents (CBAs) for these viruses. Finally, biomedical perspectives for plant lectins with antiviral properties are also discussed.

2. The Glycan Shield of Pathogenic Enveloped Viruses

Glycoproteins that are part of the glycan shield that covers the enveloped viruses are modified with three types of *N*-glycans including complex glycans, high-mannose glycans, and hybrid glycans (Figure 1):

- Complex-type *N*-glycans are most abundant on all the envelope proteins except for the gp120 and hemagglutinin from HIV and IV, respectively, which predominantly contain high-mannose-type *N*-glycans. Complex glycans exhibit a high diversity in their glycan structure, including bi-, tri-, and tetra-antennary glycans which are often sialylated on their terminal Gal residues and fucosylated on sub-terminal GlcNAc residues. Most of these complex *N*-glycans possess an α1,6-fucosylated $Man_3GlcNAc_2$ core;
- High-mannose *N*-glycans are less abundant and offer less diversity than complex *N*-glycans because they consist exclusively of Man residues. High-mannose *N*-glycans from enveloped viruses include oligomannosides containing 4–9 (Man_{4-9}) Man residues, and all of them possess a non-fucosylated $Man_3GlcNAc_2$ core;
- Hybrid *N*-glycans are least abundant on enveloped viruses.

A detailed study of the *N*-glycan structures occurring on the beta-coronaviruses, MERS-CoV, SARS-CoV, and SARS-CoV-2, confirmed the high heterogeneity of the complex glycans of the S glycoprotein forming the spikes and revealed important differences depending on the type of beta-coronavirus [15]. In addition, although some of the *N*-glycosylation sites, NXT/S, are often occupied with variable proportions of complex and high-mannose *N*-glycans, the complex *N*-glycans are largely predominant [16,27–29].

Furthermore, a few *O*-glycans also occur, especially on the S protein from beta-coronaviruses [16]. Interestingly, the Thr and Ser residues of *N*-glycosylation sites unoccupied by *N*-glycans are modified with short *O*-glycan chains [17]. Usually, these *O*-glycans are less exposed on the surface of S proteins, mainly due to the fact of their smaller size compared to the large and highly exposed *N*-glycans [27].

Both the homotrimeric organization of the fusion proteins and the homodimeric organization of E glycoproteins on the surface of pathogenic enveloped viruses favor the exposure of their glycan shield (Figures 2 and 3). However, the distribution of *N*-glycans, especially at the top of the fusion protein homotrimer, provides areas devoid of glycans allowing for the recognition of pathogenic viruses by the corresponding DPP4 and ACE2 receptors located on the host cells. These glycan-free areas, which correspond to the so-called RBDs of S proteins from MERS-CoV, SARS-CoV, and SARS-CoV-2, contribute to the infectious potential developed by the pathogenic beta-coronaviruses [28–30].

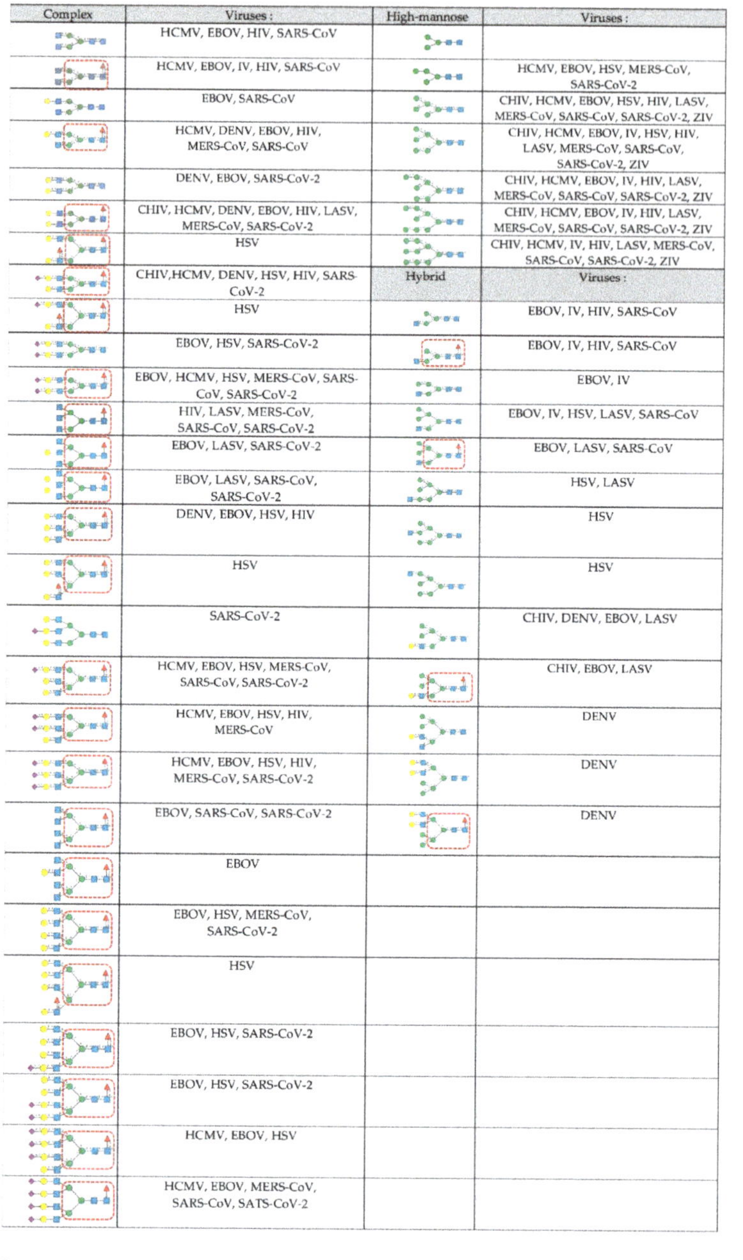

Figure 1. Diversity of the different types of *N*-glycans forming the glycan shield covering the pathogenic enveloped viruses: Ebola virus (EBOV) [19], herpes simplex virus (HSV) [20], human cytomegalovirus (HCMV) [21], human immunodeficiency virus (HIV) [22], influenza virus (IV) [23], chikungunya virus (CHIV) [24], Lassa virus (LASV) [25], MERS-CoV (MERS-CoV) [15], SARS-CoV (SARS-CoV) [15], SARS-CoV-2 (SARS-CoV-2) [15], and Zika virus (ZIV) [26]. Symbols representing the glycan structures are as follows: GlcNAc (blue square), Gal (yellow circle), Man (green circle), Fuc (red triangle), and Neu5Ac (purple diamond).

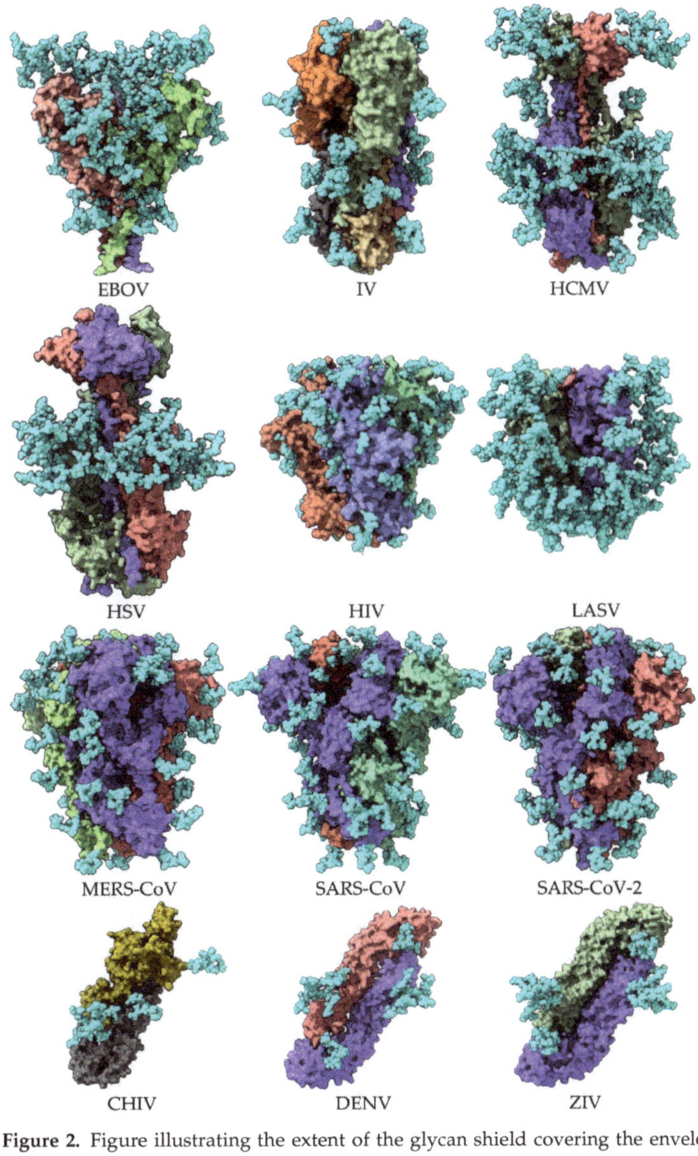

Figure 2. Figure illustrating the extent of the glycan shield covering the envelope glycoproteins of pathogenic enveloped viruses. The illustrations show the lateral face of the molecular surface of the homotrimeric organization of the envelope glycoprotein of Ebola virus (EBOV) (PDB code 7JPH), influenza virus (IV) (PDB code 6Y5G), human cytomegalovirus (HCMV) (PDB code 5CXF), herpes simplex virus (HSV) (PDB code 2GUM), human immunodeficiency virus HIV (PDB code 4TVP), Lassa virus (LASV) (PDB code 5VK2), Middle East respiratory syndrome virus (MERS-CoV) (PDB code 5W9H), severe acute respiratory syndrome (SARS-CoV) (PDB code 6ACD), and severe acute respiratory syndrome-2 (SARS-CoV-2) (PDB code 6VXX). Lateral face of the molecular surface of the dimeric organization of the E glycoprotein of chikungunya virus CHIV (PDB code 3N40), dengue virus DENV (PDB code 1UZG), and Zika virus ZIV (PDB code 7BUB). Monomers forming the homotrimeric and homodimeric associations of envelope glycoproteins are colored differently, and *N*-glycan chains forming the glycan shield are represented by cyan colored balls.

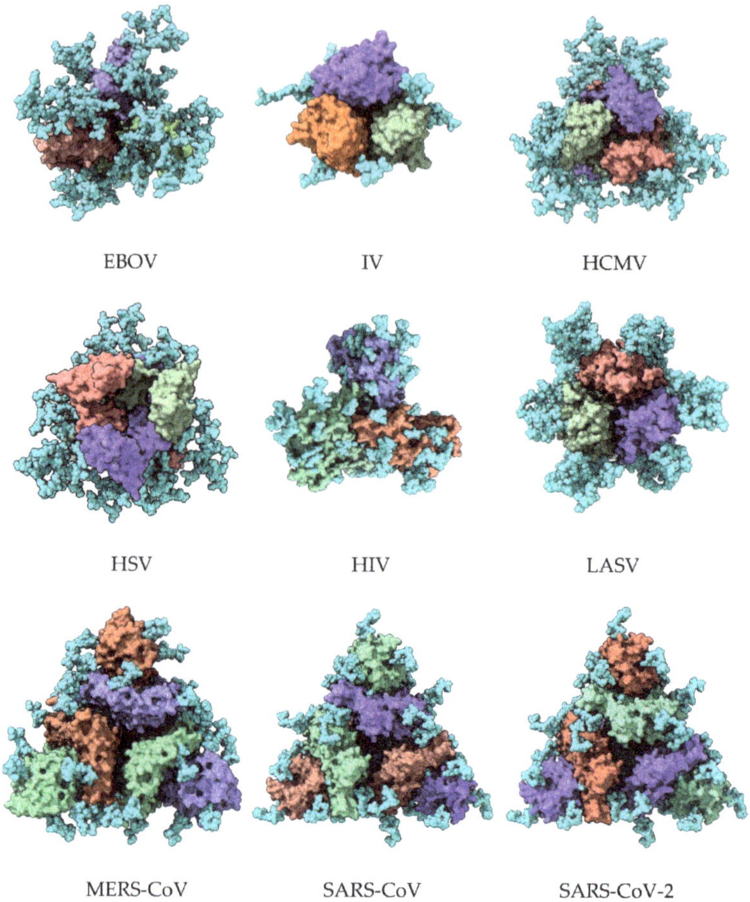

Figure 3. Figure illustrating the extent of the glycan shield covering the envelope glycoproteins of pathogenic enveloped viruses. Illustrations show the top face of the molecular surface of the homotrimeric organization of the envelope glycoprotein of Ebola virus (EBOV) (PDB code), influenza virus (IV) (PDB code), human cytomegalovirus (HCMV) (PDB code 5CXF), herpes simplex virus (HSV) (PDB code 2GUM), human immunodeficiency virus (HIV) (PDB code 4TVP), Lassa virus (LASV) (PDB code 5VK2), Middle East respiratory syndrome virus (MERS-CoV) (PDB code 5W9H), severe acute respiratory syndrome (SARS-CoV) (PDB code 6ACD), and severe acute respiratory syndrome-2 (SARS-CoV-2) (PDB code 6VXX). Monomers forming the homotrimeric associations of envelope glycoproteins are colored differently, and *N*-glycan chains forming the glycan shield are represented by cyan colored balls.

In spite of the glycan-free character of the RBDs, it should be noted that these areas are surrounded by glycan chains that should be accessible to CBAs, such as lectins, which could hamper the proper recognition of RBDs by their corresponding host cell receptors (Figure 3) [14].

3. Plant Lectins with Different Specificities Are Potential CBAs for Pathogenic Enveloped Viruses

Lectins from higher plants offer extreme diversity in terms of structural organization and recognition of simple and complex glycans [31]. Owing to the high diversity that characterizes the glycan shield of pathogenic enveloped viruses, the heterogeneous group of Man-specific lectins and, especially, the group of two-chain legume lectins, emerges as a potential tool for specific targeting of the *N*-glycan shield of enveloped viruses. Two-chain legume lectins form a particular group of Man-specific lectins that display an enhanced affinity for complex *N*-glycans possessing an α1,6-fucosylated trimannoside core $Man_3GlcNAc_2$ [32,33]. Seed lectins from pea (*Pisum sativum*) (PsA), lentil (*Lens culinaris*) (LcA), Cyprus-vetch (*Lathyrus ochrus*) (LoL-I/II), and faba bean (*Vicia faba*) (VfA) belong to this group of two-chain lectins [34–37]. They are built from the non-covalent association of two identical monomers built up from a heavy (β-chain) and a light (α-chain) subunit and possess an overall jelly roll structure similar to that of Con A, the single-chain Man-binding lectin from Jackbean (*Canavalia ensiformis*) [38]. A detailed crystallographic study of the *Lathyrus ochrus* isolectin-II (LoL-II) in complex with an octasaccharide derived from the human lactotransferrin (PDB code 1LGC) [39] revealed that the enhanced affinity of Vicieae lectins towards the α1,6-fucosylated Man3GlcNAc2 core depends on the direct interaction of the α1,6-linked Fuc residue with some of the amino acid residues forming the carbohydrate-binding site (CBS) of the lectin via a few hydrogen bonds (Figure 4).

In addition, complexes of pea lectin and *Lathyrus ochrus* lectin with non-fucosylated trisaccharides, indicated that Vicieae lectins also interact with one of the terminal Man of the trimannosyl core from the non-fucosylated *N*-glycans [40]. In this respect, LoL-I from *L. ochrus* seeds interacted with the trisaccharide Manα1,3Manβ1,4GlcNAc in such a way that the terminal Man residue occupies the monosaccharide-binding site of the lectin (Figure 5).

This binding pattern allows Vicieae lectins to readily interact with the trimannosyl core from the three types of complex, high-mannose, and hybrid *N*-glycans, irrespective of the possible α1,6-fucosylation on the first GlcNAc residue of the *N*-glycan chain.

A survey of the glycan array analyses performed by the Consortium for Functional Glycomics (CFG) (http://www.functionalglycomics.org (accessed on 15 December 2021)) for PsA, LcA, and VfA, all members of the two-chain lectins from the Vicieae tribe, yielded the best results with glycans possessing the a1,6-fucosylated $Man_3GlcNAc_2$ core. As an example, most of the top five glycans displaying the best affinity for PsA, LcA, and VfA occur in the glycan shield covering the pathogenic enveloped viruses (Figure 6).

Other Man-specific lectins, such as the GNA-related lectins from different families of monocot plants, including Liliaceae, Amaryllidaceae, Polygonaceae, and Orchidaceae, preferentially interact with high-mannose glycans that contain a non-fucosylated $Man_3GlcNAc_2$ core. As an example, the top five glycans interacting with GNA in glycan array experiments mainly consist of high-mannose glycans (Figure 7). These high-mannose glycans were present in all the investigated enveloped viruses.

In addition, the *N*-glycans of pathogenic enveloped viruses are often sialylated on their terminal Gal antennae residues, which offers another potential recognition target for lectins that specifically recognize terminal sialylated Gal residues. The black elderberry (*Sambucus nigra*) bark lectin I (SNA-I) specifically interacts with these sialylated termini. In this respect, the top five glycans interacting with SNA-I in glycan array experiments contained sialylated Gal residues that occur in the glycan shield of enveloped viruses (Figure 8).

Finally, GalNAc/T/Tn lectins that recognize O-glycans, such as jacalin from *Artocarpus integer*, PNA from peanut (*Arachis hypogaea*), and Morniga-G from *Morus nigra*, should especially interact with the few *O*-glycans exposed at the surface of beta-coronaviruses [41]. The top five *O*-glycans interacting with Morniga-G in glycan array experiments are present in the glycan shield of the SARS-CoV-2 particles (Figure 9).

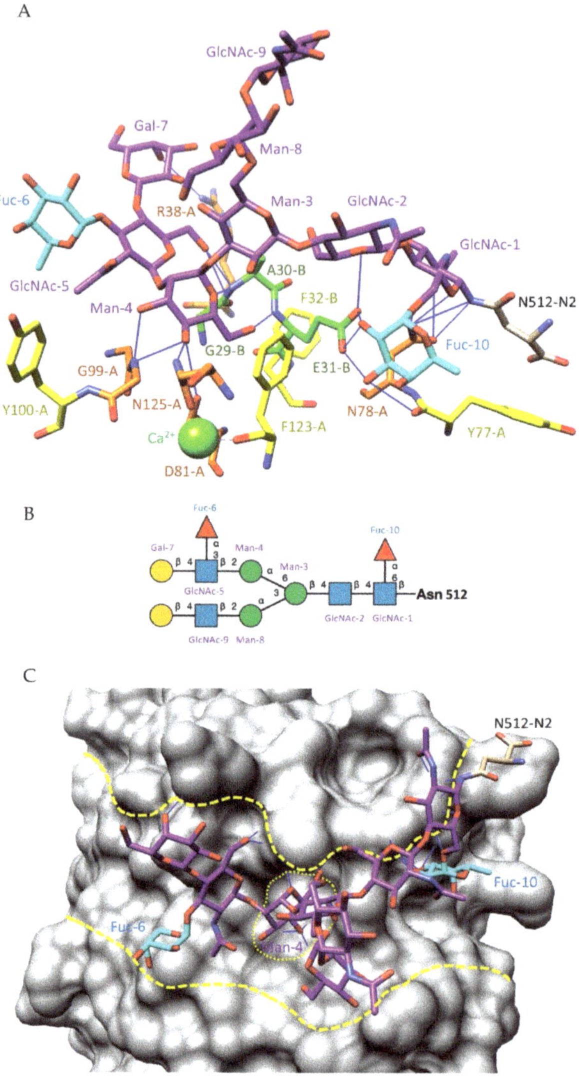

Figure 4. Figure illustrating the interaction of the Man-specific lectin LoL-II from *Lathyrus ochrus* with an oligosaccharide chain. (**A**) Network of hydrogen bonds (black lines) anchoring N2 oligosaccharide (colored purple) to LoL-II isolectin from *Lathyrus ochrus* (PDB code 1LGC). Hydrophilic residues R38, N78, D81, G99, and N125 of the α-chain and E31 of the β-chain, which participate in hydrogen bonds, are colored orange and green, respectively. Aromatic residues Y77, Y100, F123, Y124, and W128 of the α-chain and F32 of the β-chain, involved in stacking interactions with the pyranose rings of the oligosaccharide, are colored yellow. The α1,6-linked fucose (Fuc), which participates in the H-bond network, is colored cyan. (**B**) Depiction of the N2 oligosaccharide using the symbol nomenclature for glycans: Fuc (red triangle), Gal (yellow circle), GalNAc (yellow square), GlcNAc (blue square), Man (green circle), and sialic acid/Neu5Ac (purple diamond). (**C**) Molecular surface of the N2 oligosaccharide–Lo-LII complex, showing how the isolectin accommodates the oligosaccharide via a network of hydrogen bonds and stacking interactions. The groove harboring the N2 oligosaccharide and the central monosaccharide-binding site of the lectin are delineated with yellow, dashed lines.

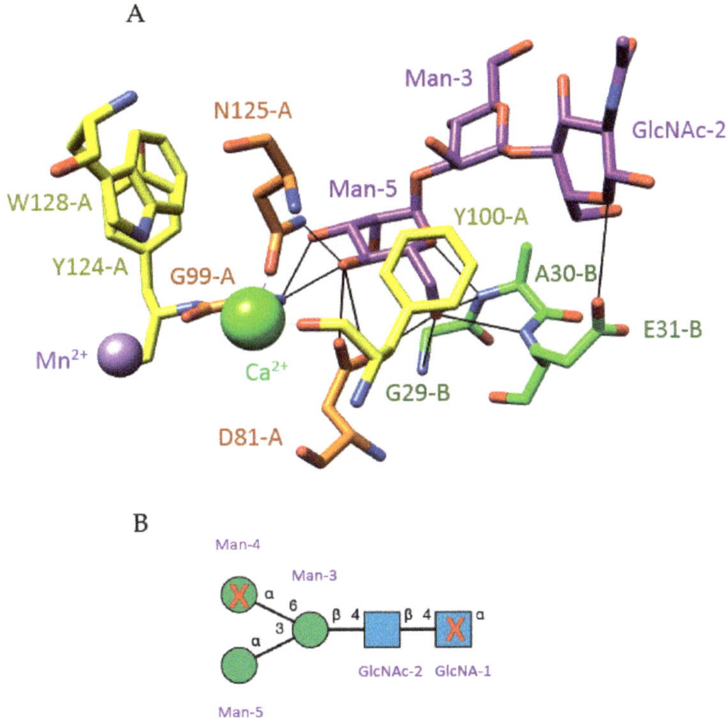

Figure 5. Figure illustrating the interaction of the Man-specific lectin LoL-I from *Lathyrus ochrus* with a trimannoside. (**A**) Network of hydrogen bonds (black lines) anchoring the trisaccharide Manα1,3Manβ1,4GlcNAc (colored purple) to LoL-I isolectin from *Lathyrus ochrus* (PDB code 1LOG). Hydrophilic residues D81, G99, and N125 of the α-chain and G29, A30, and E31 of the β-chain, which participate in hydrogen bonds, are colored orange and green, respectively. Aromatic residues Y100, Y124, and W128 of the α-chain, involved in stacking interactions with the pyranose rings of the oligosaccharide, are colored yellow. (**B**) Illustration of the $Man_3GlcNAc_2$ oligosaccharide using the symbol nomenclature for glycans: Fuc (red triangle), Gal (yellow circle), GalNAc (yellow square), GlcNAc (blue square), Man (green circle), and sialic acid/Neu5Ac (purple diamond), showing the sugar units that participate in the trisaccharide–LoL-I complex.

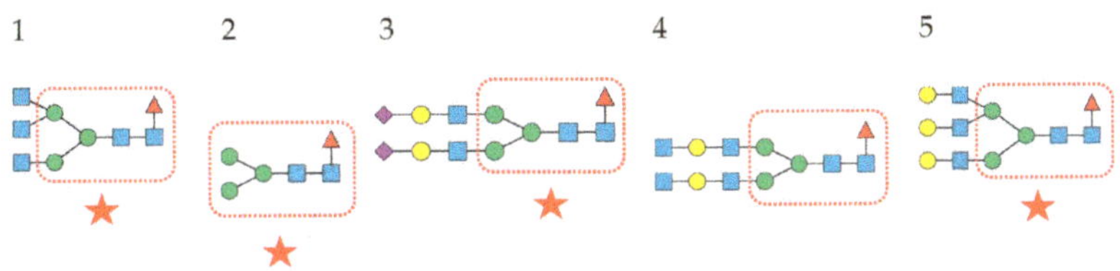

Figure 6. Structures of *N*-glycans recognized by PsA. Top 5 *N*-glycans are arranged in decreasing order of affinity for PsA. The α1,6-fucosylated $Man_3GlcNAc_2$ core is delineated with a red square. Glycans occurring in the glycan shield of pathogenic enveloped viruses are indicated by a red star. Fuc (red triangle), Gal (yellow circle), GalNAc (yellow circle), GlcNAc (blue square), Man (green circle), and sialic acid/Neu5Ac (purple diamond).

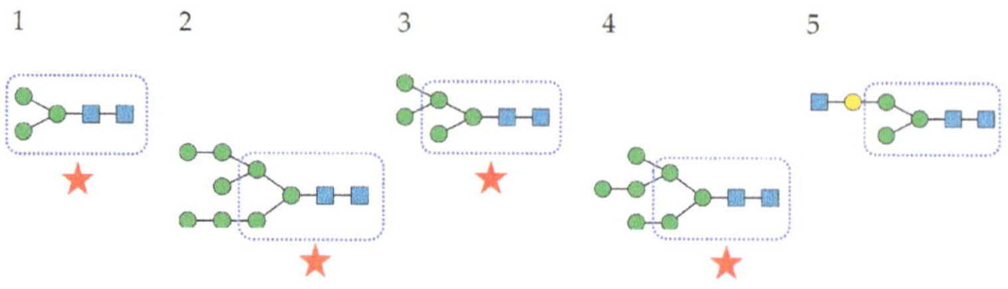

Figure 7. Structures of *N*-glycans recognized by GNA. The top 5 *N*-glycans are arranged in decreasing order of affinity for GNA. The $Man_3GlcNAc_2$ core is delineated by a blue square. High-mannose glycans occurring in the glycan shield of pathogenic enveloped viruses are indicated by a red star. Gal (yellow circle), GalNAc (yellow circle), GlcNAc (blue square), Man (green circle).

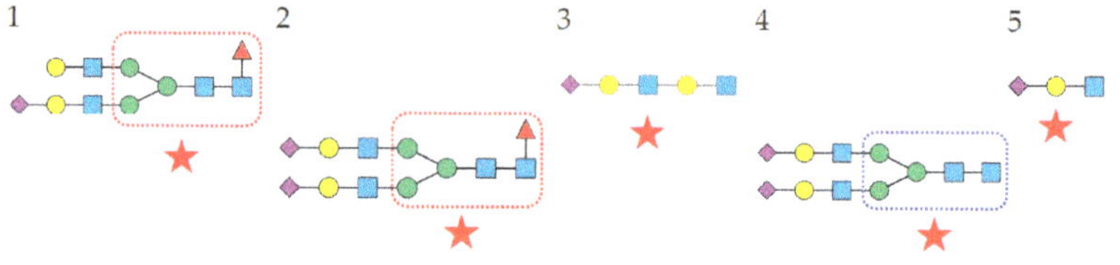

Figure 8. Structures of *N*-glycans recognized by SNA-I. The top five sialylated *N*-glycans are arranged in decreasing order of affinity for SNA-I. The $Man_3GlcNAc_2$ core is delineated by a blue square. The fucosylated $Man_3GlcNAc_2$ core is indicated with a red circled. Sialylated glycans occurring in the glycan shield of pathogenic enveloped viruses are indicated by a red star. Fuc (red triangle), Gal (yellow circle), GalNAc (yellow circle, GlcNAc (blue square), Man (green circle), and sialic acid/Neu5Ac (purple diamond).

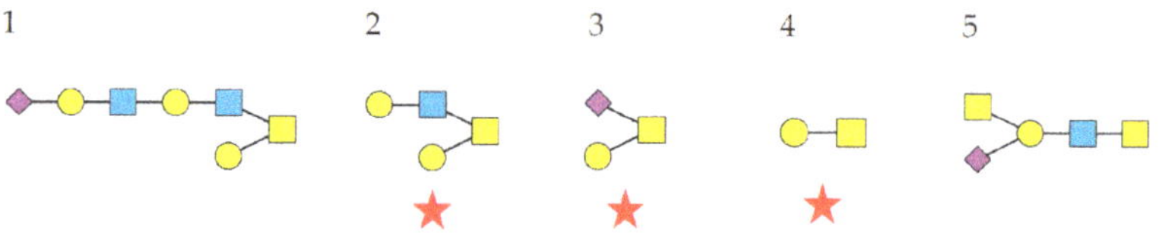

Figure 9. Structures of *O*-glycans recognized by Morniga-G. The top 5 *O*-glycans are arranged in decreasing order of affinity for PNA. The *O*-glycans occurring in the glycan shield of SARS-CoV-2 spike protein are indicated by a red star. Gal (yellow circle), GalNAc (yellow square), GlcNAc (blue square), and sialic acid/Neu5Ac (purple diamond).

4. Man-Specific and Neu5Ac-Specific Lectins as Potential CBAs for Pathogenic Enveloped Viruses

Plant lectins with different carbohydrate-binding specificities have been identified as CBAs for pathogenic enveloped virus including HIV, HCMV, the hepatitis C virus HCV, HSV, IV, and the coronaviruses MERS-CoV, SARS-CoV, and SARS-CoV-2. Targets for these CBAs are the glycan structures present on the envelope proteins of pathogenic enveloped viruses (Table 2).

In addition to CBAs from higher plants, it should be noted that other lectins isolated from algae and Cyanobacteria (formerly classified as blue algae), which essentially recognize high-mannose glycans, have been identified as potential CBAs for pathogenic enveloped viruses (Table 3). In this respect, griffithsin, the Man-specific lectin purified from the red alga *Griffithsia* sp. [67], was investigated in detail as a relevant CBA for targeting the envelope protein from pathogenic enveloped viruses [68] because of its high affinity for oligomannosides [69] (Figure 10).

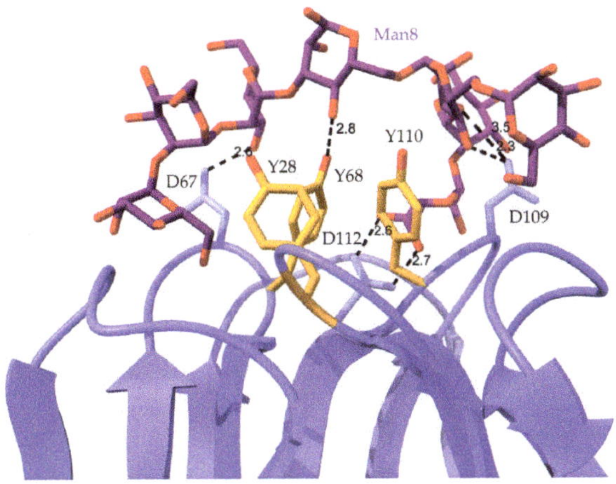

Figure 10. Figure illustrating how the Man-specific lectin griffithsin interacts with an oligomannoside. Network of hydrogen bonds (black, dashed lines) connecting the griffithsin monomer (colored violet) to a linear Man8 chain (colored purple) (PDB code 3LL2). Amino acid D residues participating in hydrogen bonds are labeled (i.e., D67, D109, and D112). Aromatic Y residues involved in stacking interactions with the sugar rings are labeled and colored orange (i.e., Y28, Y68, and Y110).

Table 2. List of plant lectins identified as carbohydrate-binding agents (CBAs) for envelope proteins from pathogenic enveloped viruses.

Lectin	Plant Species	Carbohydrate-Binding Specificity	Targeted Envelope Protein	Virus	Ref.
APA	*Allium porum*	Man	S-protein	SARS-CoV	[41]
AUA	*Allium ursinum*	Man	S-protein	SARS-CoV	[41]
BanLec	*Musa acuminata*	Man	gp120	HIV	[42–44]
			hemagglutinin	IV	[45]
			E-glycoprotein	HCMV	[46]
			E-glycoprotein	EBOV	[46,47]
			E-glycoprotein	LASV	[46]
Con A	*Canavalia ensiformis*	Man	gp120	HIV	[48]
			S-protein	SARS-CoV-2	[49]
Succinyl-Con A	*Canavalia ensiformis*	Man	S-protein	Mers-CoV	[50]
			S-protein	SARS-CoV	[50]
			S-protein	SARS-CoV-2	[50]
ConBr	*Canavalia brasiliensis*	Man	S-protein	SARS-CoV-2	[51]
ConM	*Canavalia maritima*	Man	S-protein	SARS-CoV-2	[51]
CLA	*Cladastris lutea*	Man	S-protein	SARS-CoV	[41]
CHA	*Cymbidium* hybrid	Man	β-glycoprotein	HCMV	[52]
			E-glycoprotein	HCV	[52]
			gp120	HIV	[52]
			S-protein	SARS-CoV	[53]
DSL	*Datura stramonium*	Neu5Ac-Gal/GalNAc	S-protein	MERS-CoV	[50]
			S-protein	SARS-CoV	[50]
			S-protein	SARS-CoV-2	[50]
DLasL	*Dioclea lasiocarpa*	Man	S-protein	SARS-CoV-2	[51]
DSclerL	*Dioclea sclerocarpa*	Man	S-protein	SARS-CoV-2	[51]
EHA	*Epipactis helleborine*	Man	β-glycoprotein	HCMV	[52]

Table 2. *Cont.*

Lectin	Plant Species	Carbohydrate-Binding Specificity	Targeted Envelope Protein	Virus	Ref.
			gp120	HIV	[52]
			hemagglutinin	IV	[52]
GNA	*Galanthus nivalis*	Man	β-glycoprotein	HCMV	[53]
			E-glycoprotein	HCV	[54,55]
			gp120	HIV	[53]
			S-protein	SARS-CoV	[41]
			S-protein	SARS-CoV-2	[56]
			hemagglutinin	IV	[57]
HHA	*Hyppeastrum* hybrid	Man	β-glycoprotein	HCMV	[53]
			E-glycoprotein	HCV	[52]
			gp120	HIV	[52]
			S-protein	SARS-CoV	[41]
			hemagglutinin	IV	[57]
Horcolin	*Hordeum vulgare*	Man	gp120	HIV	[58]
IRA	*Iris* hybrid	GalNAc/Gal	S-protein	SARS-CoV	[41]
FRIL	*Lablab purpureus*	Man	S-protein	SARS-CoV	[59]
			hemagglutinin	IV	[59]
LcA	*Lens culinaris*	Man	S-protein	MERS-CoV	[50]
			S-protein	SARS-CoV	[50]
			S-protein	SARS-CoV-2	[50]
LOA	*Listera ovata*	Man	β-glycoprotein	HCMV	[53]
			gp120	HIV	[53]
MAL	*Maackia amurensis*	Neu5Ac	S-protein	SARS-CoV-2	[60]
Morniga-G	*Morus nigra*	Gal	S-protein	SARS-CoV	[41]
Morniga-M	*Morus nigra*	Man	S-protein	SARS-CoV	[41]
NPA	*Narcissus pseudonarcissus*	Man	β-glycoprotein	HCMV	[53]
			gp120	HIV	[53]
Nictaba	*Nicotiana tabacum*	(GlcNAc)n	S-protein	SARS-CoV	[41,61]
Orysata	*Oryza sativa*	Man	gp120	HIV	[62]
			S-protein	SARS-CoV	[62]
PHA	*Phaseolus vulgaris*	Complex glycans	S-protein	MERS-CoV	[50]
			S-protein	SARS-CoV	[50]
			S-protein	SARS-CoV-2	[50]
PCA	*Polygonatum cyrtonema*	Man	gp120	HIV	[63]
SSL	*Sambucus sieboldiana*	Neu5Ac-Gal/GalNAc	S-protein	SARS-CoV	[50]
			S-protein	SARS-CoV-2	[50]
TLC II	*Tulipa* hybrid	Man	S-protein	SARS-CoV	[41]
TDL	*Typhonium divaricatum*	Man	E-glycoprotein	HSV	[64]
UDA	*Urtica dioica*	(GlcNAc)n	β-glycoprotein	HCMV	[52]
			E-glycoprotein	HCV	[51]
			gp120	HIV	[52]
			S-protein	SARS-CoV	[41]
			hemagglutinin	IV	[57]
ML II	*Vicum album*	Gal/GalNAc	S-protein	SARS-CoV	[41]
ML III	*Viscum album*	Gal/GalNAc	S-protein	SARS-CoV	[41]
WGA	*Triticum aestivum*	GlcNAc/Neu5Ac	S-protein	MERS-CoV	[50]
			S-protein	SARS-CoV	[41,50]
			S-protein	SARS-CoV-2	[50,65]
GNAmaize	*Zea mays*	Man	S-protein	SARS-CoV	[66]

Table 3. List of algal and cyanobacterial lectins identified as carbohydrate-binding agents (CBAs) for envelope proteins from pathogenic enveloped viruses.

Lectin	Algal Species	Carbohydrate-Binding Specificity	Targeted Envelope Protein	Virus	Ref.
AML	*Amantia multifida*	Fetuin, mannan	hemagglutinin	IV	[52]
			E-glycoprotein	HSV	[52]
			gp120	HIV	[52]
BSL	*Bryothamnion seaforthii*	Fetuin, mucin	gp120	HIV	[52]
			E-glycoprotein	HSV	[52]
			hemagglutinin	IV	[52]
ESA-2	*Eucheuma serra*	Man	hemagglutinin	IV	[70]
GCL	*Grateloupia chiangii*	Man	hemagglutinin	IV	[71]
			E-glycoprotein	HSV	[71]
Griffithsin	*Griffithsia* sp.	Man	gp120	HIV	[72]
			E-glycoprotein	HCV	[73]
			S-protein	MERS-CoV	[74]
			S-protein	SARS-CoV	[68]
			S-protein	SARS-CoV-2	[75]
HML	*Hypnea musciformis*	Thyroglobulin, mucin	gp120	HIV	[52]
			hemagglutinin	IV	[52]
			E-glycoprotein	HSV	[51]
HTL-40	*Halimeda renschii*	Man	hemagglutinin	IV	[76]
KAA-2	*Kappaphycus alvarezii*	Man	hemagglutinin	IV	[77]
			gp120	HIV	[78]
MEL	*Meristiella echinocarpa*	Mannan	hemagglutinin	IV	[52]
			hemagglutinin	IV	[52]
SfL	*Solieria filiformis*	Mannan	E-glycoprotein	HSV	[52]
			gp120	HIV	[52]
			hemagglutinin	IV	[52]
BCA	*Boodlea coacta*	Man	E-glycoprotein	HSV	[79]
			hemagglutinin	IV	[79]
Lectin	**Cyanobacterial Species**	**Carbohydrate-Binding Specificity**	**Targeted Envelope Protein**	**Virus**	**Ref.**
MVN	*Microcystis aeruginosa*	Man	gp120	HIV	[80,81]
			E-glycoprotein	HSV	[81]
MVL	*Microcystis viridis*	Man	gp120	HIV	[82]
			E-glycoprotein	HCV	[83]
Cyanovirin-N	*Nostoc ellipsosporum*	Man	gp120	HIV	[49,84]
(CV-N)			E-glycoprotein 1,2	EBOV	[85,86]
			hemagglutinin	IV	[86]
			E-glycoprotein	HCV	[87]
			E-glycoprotein	HSV	[88]
			S-protein	SARS-CoV-2	[89]
	Oscillatoria agardhii	Man	gp120	HIV	[90]
OAA	*Scytonema varium*	Man	gp120	HIV	[91]
SVN	*Scytonema varium*		E-glycoprotein	DENV	[92]
			E-glycoprotein	EBOV	[93]

Although cyanobacterial lectins exhibit similar antiviral activities against pathogenic enveloped viruses, compared to other lectins from higher plants, they readily differ by the different fold and the smaller size of their structural scaffolds [93,94]. Moreover, cyanobacteria contain other small metabolites that could be used as valuable tools for combating enveloped viruses and, especially SARS-CoV-2 responsible for the COVID-19 pandemic [95].

5. How Can the Infectivity of Pathogenic Enveloped Viruses Be Affected by Lectins?

The glycan-mediated interaction of lectins with pathogenic enveloped viruses has a direct effect on virus infectivity, essentially by interfering with the recognition of their corresponding host cell receptors via different mechanisms. However, a dichotomy must be introduced between experiments performed in vitro on cultured cells and experiments achieved in vivo on animals.

Experimental studies performed in vitro, especially on HIV-infected cultured cells, have focused on higher plant lectins and cyanobacterial lectins [96]. Mannose-specific lectins have been recognized as the most efficient inhibitors of the HIV entry into the target cells by interacting with the glycan shield of gp120 and gp41, preventing their recognition by the CD4 receptors present on CD4+ T cells. Mannose-specific lectins of monocot plant species, such as *Cymbidium* hybrid (CHA), *Epipactis helleborine* (EHA), *Hippeastrum* hybrid (HHA), *Galanthus nivalis* (GNA), *Listera ovata* (LOA), and *Narcissus pseudonarcissus* (NPA), have been widely investigated in this respect by Balzarini and co-workers [53,54]. Other lectins with similar Man-binding specificity like BanLec from *Musa acuminata* [43] and Con A from *Canavalia ensiformis* [97] or different carbohydrate-binding specificities, such as the $(GlcNAc)_n$-specific lectins Nictaba from *Nicotiana tabacum* [62], UDA from *Urtica dioica* [53], and WGA from *Triticum aestivum* [98], were also identified as potential inhibitors of the HIV entry into target cells in vitro and the syncytium formation resulting from the fusion of HIV-infected and HIV-uninfected $CD4^+$ T lymphocytes [99,100].

A rather different situation can occur under in vivo conditions due to the multiplicity of cells susceptible to interacting with the virus. In addition, to block the entry of HIV particles into the cells and syncytium formation, plant lectins interfere with other mechanisms of virus infection and transmission, for example, by preventing the recognition of high-mannose glycans of gp120 by the DC-SIGN receptor of dendritic cells [101–103] or by blocking the transmission of DC-SIGN-captured virions to the $CD4^+$ T lymphocytes [102]. In addition, as reported in [96], interaction with lectins of different carbohydrate-binding specificities can result in cytotoxic side effects on host cells, e.g., caspase-dependent apoptotic responses, due to the activation of different signaling pathways leading to apoptotic and necrotic responses that result from the recognition of surface-exposed *N*- and *O*-glycans by lectins.

The effects of plant lectins on other pathogenic enveloped viruses have been more scarcely investigated. Plant lectins were identified as blocking agents for the entry of IV [45,52,57], HSV [57], HCMV [47,53,54], EBOV [47,48], LASV [47], and the coronaviruses MERS-CoV [51], SARS-CoV [42,51,54,62,63,104], and SARS-CoV-2 [50–52,58,60,61,66] in their corresponding host cells. However, depending on the viruses, the envelope glycoprotein(s) targeted by plant lectins are extremely diverse as mentioned in Table 2. In this respect, an engineered banana lectin, BanLec, which has lost its mitogenic potential but retained its mannose-binding property, interacted with the envelope E-glycoprotein and inhibited both the entry and replication of Ebola virus in cell cultures [47,48]. The cyanobacterial lectin, cyanovirin-N (CV-N), also displayed similar inhibition towards Ebola virus [85]. Similarly, lentil lectin, LcA, inhibited the early steps of the host cell infection by SARS-CoV-2 and variants B.1.1.7 (α variant), B.1.351 (β variant), and P1 (γ variant), by blocking the recognition of their spike S protein by the ACE2 receptor [51].

The effects resulting from the binding of plant lectins on the different enveloped viruses also depend on the mechanisms of infection and transmission of these viruses, which may differ from those developed by HIV. In addition to preventing the entry and replication of Ebola viruses into the host cells in cell cultures, the engineered BanLec lectin, pre-administered to virus-infected mice, were highly protective against a lethal EBOV infection in vivo (~80% of mice protected) [48]. The engineered BanLec lectin was similarly efficient for protecting influenza virus-infected mice, by inhibiting the virus-endosome fusion occurring after the exogenous lectin has been internalized in the late endosomal/lysosomal compartment of the host cells [46]. Both plant lectins and griffithsin, the Man-specific lectin from the red alga *Griffithsia* sp., also inhibited the entry of SARS-CoV and MERS-CoV, respectively, in virus-infected cultured cells [42,74]. The Neu5Ac-specific *Maackia amurensis* lectin, MAL, inhibited the interaction of the SARS-CoV-2 S protein with the ACE2 receptor in cell cultures and decreased the expression of inflammatory mediators associated with COVID-19 disease progression [61].

The effects of plant lectins on coronaviruses were also investigated in virus-infected mice models. The $(GlcNAc)_n$-specific lectin from the stinging nettle (*Urtica dioica*) (UDA), was shown to prevent virus entry and replication in a dose-dependent manner and reduced the virus infectivity significantly in a lethal SARS-CoV BALB/c mouse model [105]. Under in vivo conditions, the lectin from the hyacinth bean (*Lallab purpureus*), FRIL, neutralized H1N1 influenza by aggregating and trapping virions in the late endosomes of the host cells, thus preventing their nuclear internalization [60]. The lectin similarly neutralized SARS-CoV-2 by preventing viral protein production and cytotoxic effects on the host cells.

Algal and cyanobacterial Man-specific lectins, such as griffithsin and cyanovirin-N, also inhibited the entry of HIV and other enveloped viruses in the host cells in vitro and exerted in vivo cytotoxic effects very similar to those of plant lectins [74–103,105–107].

6. Biomedical Perspectives for Antiviral Lectins

Depending on their affinity towards surface-exposed glycans of enveloped viruses, plant lectins are considered as potential CBAs useful for combating viral infections, even though little evidence exists to date for their efficacy as relevant therapeutic tools [97,108–112]. Beyond their possible use as well-adapted tools for the diagnosis of viral infection, the therapeutic use of plant lectins as virus blockers faces practical and functional challenges which mainly concern (1) their large-scale production and (2) their unwanted immunomodulatory properties.

In most higher plants, lectins of different specificities that could be used as virus blockers occur as storage proteins in seeds and other vegetative organs such as tubers and rhizomes [113]. Man-specific two-chain (LcA, PsA, VfA, and LoL-I/II) and single-chain lectins (Con A, PHA, and SBA) from the Fabaceae are sequestered in the protein bodies of the cotyledonary cells in rather low amounts [114]. Accordingly, the extraction yield of legume seed lectins is rather low, in the range 50–80 mg/100 g (dry weight) seed [115]. However, the degree of purity of the extracted lectins is excellent since the introduction of affinity chromatography techniques using carbohydrate-immobilized columns. Different strategies have been developed recently to improve the extraction yield of griffithsin, the Man-specific lectin from the red alga *Griffithsia* sp., for the purpose of obtaining a large-scale production of the lectin able to supply the quantities of lectins needed for therapeutic applications [116–121]. These strategies are based on the continuous improvement of yields obtained from the high-level expression and extraction of griffithsin from transformed tobacco (*Nicotiana benthamiana*) leaves. In addition, griffithsin is easily purified and recovered from ensiled dried tobacco leaves, which allows for a low-cost production of lectin quickly adaptable to demand.

Most plant lectins consist of oligomeric structures built up from the non-covalent association of 15–20 kDa monomers in dimers and tetramers, more rarely in hexamers or octamers [31]. Depending on their structural organization, plant lectins usually exhibit a high degree of resistance to the degradation by trypsin-like proteases together with a pronounced capacity to trigger the synthesis of specific anti-lectin IgG. In this respect, IgG-binding epitopes have been identified on the molecular surface of Man-specific two-chain lectins from the Vicieae tribe [122,123], and monoclonal antibodies that specifically recognize lentil and *Lathyrus ochrus* lectins, were easily prepared [124,125]. Even Man-specific dietary lectins, such as BanLec from banana and ASA from garlic, have been reported to induce an immune response since specific anti-lectin IgG have been identified in the serum of banana and garlic consumers [126,127]. Through a specific interaction with the plant lectins associated to the enveloped viruses, these IgG could eventually neutralize the effects of lectins in the host cells. The Man-specific algal griffithsin and cyanobacterial lectins, such as cyanovirin-N and scytovirin, could overcome this challenge because of the small size of their composing monomers [128,129]. In addition to griffithsin, grifonin-1 (GRFN-1), an even smaller peptide of eighteen amino acids derived from griffithsin, has proven its efficacy as a blocking agent against HIV [130].

Initially recognized as potent mitogenic proteins [131–133], plant lectins have been known for a long time as non-specific immune-modulatory proteins that are susceptible to interaction with various cell surface glycoproteins/glycolipids and for interfering with various signaling pathways triggering cytotopathologic effects on the targeted cells. Although most plant lectins with antiviral activity activate different sets of T lymphocytes and, more scarcely, B lymphocytes, they also activate both the apoptotic and necrotic pathways in many other types of healthy and cancer cells [134–143]. The cellular activation mediated by plant lectins on healthy and transformed cells elicits the release of various chemokines and/or cytokines that are, in turn, susceptible to interfere with the cytokine stimulation associated with the viral infection, e.g., HIV infection [59]. However, plant lectins readily differ from each other by their capacity to elicit a cytokine production, some of them, such as PHA, Con A, and cyanovirin-N, being more active to induce the synthesis and release of activation markers [144,145], while Man-specific GNA-like lectins, such as GNA and HHA, were virtually incapable of triggering a relevant cytokine production [145,146]. Recently, a single-point mutation performed on an engineered banana lectin, BanLec, and an engineered Malaysian banana lectin, Malay BanLec, allowed to produce an active Man-binding lectin significantly devoid of mitogenic/cytotoxic activity [46,47]. If applicable for other Man-specific lectins, this point mutation approach would be an elegant way to attenuate or suppress the unwanted mitogenic/cytotoxic effects of lectins on target cells.

In spite of these limitations hampering the use of plant lectins as CBAs for combating pathogenic enveloped viruses, some ex vivo applications of plant lectins have been successfully developed. An important decrease in the plasma load with Ebola virus was achieved by extracorporeal affinity plasmapheresis of the contaminated blood through a GNA-immobilized matrix [147]. Recently, *ex-vivo* plasmapheresis on a Man-specific lectin-immobilized column of blood spoiled by the MERS-CoV and the Marburg viruses has proven its efficacity to purge the blood samples from virus particles [148]. Although essentially theoretical, the risk of a possible transfusion transmission of SARS-CoV-2 with spoiled blood samples should be avoided by a simple lectin plasmapheresis step of the suspected blood samples [149,150]. Another ex vivo application of plant lectins has been proposed on the web (Pittsburgh University, 2020) in the form of a nasal spray of griffithsin that could be used to prevent the infection by SARS-CoV-2 and other pathogenic enveloped viruses, e.g., in immune-compromised people. A lectin spray could also be used to detect the enveloped viruses on various domestic surfaces, such as doorknobs, handrails, computers, and cooking tools, under UV illumination after labeling with specific anti-lectin antibodies coupled to a fluorochrome.

7. Bioinformatics

Atomic coordinates of fusion proteins, B glycoproteins, and E glycoproteins were taken from the Protein Data Bank (PDB): 7JPH (EBOV) [5], 6Y5G (IV) [6], 5CXF (HCMV) [7], 2GUM (HSV) [8], 4TVP (HIV) [9], 5VK2 (LASV) [10], 5W9H (MERS-CoV) [11], 6ACD (SARS-CoV) [12], 6VXX (SARS-CoV-2) [13], 3N40 (CHIV) [14], 1UZG (DENV) [15], and 7BUB (ZIV) [16].

The molecular surface of the lectins and envelope glycoproteins from pathogenic enveloped viruses were calculated and displayed with Chimera [151] and Chimera-X [152]. Assuming that putative *N*-glycosylation sites, NXT/S, of envelope glycoproteins are actually glycosylated, a classic *N*-glycan chain corresponding to the trimannoside core $Man_3GlcNAc_2$, was modeled using the GlyProt server (http://www.glycosciences.de/

modeling/glyprot/php/main.php) (accessed on 22 December 2021) [153] and represented in CPK on the molecular surface of the envelope glycoproteins.

The illustrations of the high-mannose *N*-glycans, complex *N*-glycans, hybrid *N*-glycans, and *O*-glycans were built and represented with the DrawGlycan SNFG package for Mac [154]. Colored symbols were used to represent Fuc (red triangle), Gal (yellow circle), Glc (blue circle), GalNAc (yellow square), GlcNAc (blue square), Man (green circle), and sialic acid/Neu5Ac (purple diamond).

8. Discussion

The glycan shield covering pathogenic enveloped viruses plays a role not only in the protection of viruses but also in various important mechanisms insuring the entry and replication of viruses in the host cells [30]. Thus, the recognition of the glycan shield by plant lectins provides a way to fight viral infection and, especially, the SARS-CoV-2 infection, by competing with the spike-mediated attachment of viral particles to the host cell virus receptors. However, due to the extreme diversity of *N*-glycan types covering enveloped viruses, especially beta-coronaviruses [15], plant lectins with different carbohydrate-binding specificities should be tested for this purpose. From experiments performed under in vitro and in vivo conditions, it follows that plant, algal, and cyanobacterial lectins with different carbohydrate-binding specificities represent well-adapted CBAs for blocking the entry of pathogenic enveloped viruses into the host cells. In this respect, Man-specific lectins of the Vicieae tribe, which specifically recognize the α1,6-fucosylated $Man_3GlcNAc_2$ core of *N*-glycans of the complex- and hybrid-type, are particularly relevant as glycan probes for the beta-coronaviruses MERS-CoV, SARS-CoV, and SARS-CoV-2 [51,106]. However, Man-specific lectins are not considered as replication blockers for coronaviruses, since they do not interfere with the coronavirus replication within the cell.

Despite the accumulating evidences that plant lectins and, especially, Man-specific plant lectins, could be used as tools for preventing infection by pathogenic enveloped virus, in particular SARS-CoV-2 responsible for the COVID-19 pandemic, some unwanted characteristics of plant lectins make these molecules difficult to use for a therapeutic purpose. Due to the fact of their high molecular size, which favors the synthesis of anti-lectin antibodies, and their mitogenic/cytotoxic properties, which interfere with the cytokine response of infected individuals, their use is limited to external treatments. However, the promising results obtained with a single-mutated engineered banana lectin, BanLec, which retains its carbohydrate-binding ability but loses its mitogenic property [46,47], could pave the way for the forthcoming production of innocuous mutated plant lectins available for a therapeutic use.

In addition to lectins, other small molecules could be used as blockers for the SARS-CoV-2/ACE2 interaction. Recently, some small molecular drugs, including dyes, glycosides, tannins, and immunosuppressors, were characterized as either spike or ACE2 binders susceptible to blocking the attachment of the SARS-CoV-2 spikes to ACE2 by interfering with the ligand and/or the receptor surface [155]. Depending on the *N*-glycan types that are linked to the receptor DPP4 for MERS-CoV and SARS-CoV viruses [156] and ACE2 for SARS-CoV-2 virus [157], plant lectins with Man-binding activity could interfere with the capture of the beta-coronavirus spikes by their corresponding host cell receptors. This dual activity of plant lectins towards the glycans of spikes and their receptors is of paramount importance for reinforcing the antiviral properties of plant lectins against pathogenic beta-coronaviruses.

Author Contributions: Conceptualization, P.R.; software, P.R.; validation, E.J.M.V.D. and P.R.; resources, A.B., B.K., M.S., J.S. and H.B.; writing—original draft preparation, P.R.; writing—review and editing, E.J.M.V.D. and P.R.; supervision, E.J.M.V.D. and P.R.; project administration, A.B. All authors have read and agreed to the published version of the manuscript.

Funding: This research received no external funding.

Institutional Review Board Statement: Not applicable.

Informed Consent Statement: Not applicable.

Data Availability Statement: Not applicable.

Conflicts of Interest: The authors declare no conflict of interest.

References

1. Rey, F.A.; Lok, S.-M. Common features of enveloped viruses and implications for immunogen design for next-generation vaccines. *Cell* **2018**, *172*, 1319–1334. [CrossRef] [PubMed]
2. Thorley, J.A.; McKeating, J.A.; Rappoport, J.Z. Mechanisms of viral entry: Seaking in the front door. *Protoplasma* **2010**, *244*, 15–24. [CrossRef] [PubMed]
3. Heldwein, E.E.; Lou, H.; Bender, F.C.; Cohen, G.H.; Eisenberg, R.J.; Harrison, S.C. Crystal structure of glycoprotein B from herpes simplex virus 1. *Science* **2006**, *313*, 217–220. [CrossRef] [PubMed]
4. Voss, J.E.; Vaney, M.-C.; Duquerroy, S.; Vonrhein, C.; Girard-Blanc, C.; Crublet, E.; Thompson, A.; Bricogne, G.; Rey, F.A. Glycoprotein organization of Chikungunya virus particles revealed by X-ray crystallography. *Nature* **2010**, *468*, 707–712. [CrossRef] [PubMed]
5. Modis, Y.; Ogata, S.; Clements, D.; Harrison, S.C. Variable surface epitopes in the crystal structure of dengue virus type 3 envelope glycoprotein. *J. Virol.* **2005**, *79*, 1223–1231. [CrossRef]
6. Zhang, S.; Loy, T.; Ng, T.S.; Lim, X.-N.; Chew, S.V.; Tan, T.Y.; Xu, M.; Kostyuchenko, V.A.; Tukijan, F.; Shi, J.; et al. A human antibody neutralizes different flaviviruses by using different mechanisms. *Cell Rep.* **2020**, *31*, 107584. [CrossRef] [PubMed]
7. He, L.; Chaudhary, A.; Lin, X.; Sou, C.; Alkutkar, T.; Kumar, S.; Ngo, T.; Kosviner, E.; Ozorowski, G.; Stanfield, R.L.; et al. Single-component multilayer self-assembling nanoparticles presenting rationally designed glycoprotein trimers as Ebola virus vaccines. *Nat. Commun.* **2021**, *12*, 2633. [CrossRef] [PubMed]
8. Benton, D.J.; Gamblin, S.; Rosenthal, P.B.; Skehel, J. Structural transitions in influenza haemagglutinin at membrane fusion pH. *Nature* **2020**, *583*, 150–153. [CrossRef] [PubMed]
9. Burke, H.G.; Heldwein, E.E. Crystal structure of the Human cytomegalovirus glycoprotein B. *PLoS Pathog.* **2015**, *11*, e1005227. [CrossRef]
10. Pancera, M.; Zhou, T.; Druz, A.; Georgiev, I.S.; Soto, C.; Gorman, J.; Huang, J.; Acharya, P.; Chuang, G.-Y.; Ofek, G.; et al. Structure and immune recognition of trimeric prefusion HIV-1 Env. *Nature* **2014**, *514*, 455–461. [CrossRef]
11. Hastie, K.; Zandonatti, M.A.; Kleinfelter, L.M.; Heinrich, M.L.; Rowland, M.M.; Chandran, K.; Branco, L.M.; Robinson, J.E.; Garry, R.F.; Saphire, E.O. Structural basis for antibody-mediated neutralization of Lassa virus. *Science* **2017**, *356*, 923–928. [CrossRef]
12. Pallesen, J.; Wang, N.; Corbett, K.S.; Wrapp, D.; Kirchdoerfer, R.N.; Turner, H.L.; Cottrell, C.A.; Becker, M.M.; Wang, L.; Shi, W.; et al. Immunogenicity and structures of a rationally designed prefusion MERS-CoV spike antigen. *Proc. Natl. Acad. Sci. USA* **2017**, *114*, E7348–E7357. [CrossRef]
13. Song, W.; Gui, M.; Wang, X.; Xiang, Y. Cryo-EM structure of the SARS coronavirus spike glycoprotein in complex with its host cell receptor ACE2. *PLoS Pathog.* **2018**, *14*, e1007236. [CrossRef]
14. Walls, A.C.; Park, Y.-J.; Tortorici, M.A.; Wall, A.; McGuire, A.T.; Veesler, D. Structure, function, and antigenicity of the SARS-CoV-2 spike glycoprotein. *Cell* **2020**, *181*, 281–292. [CrossRef] [PubMed]
15. Chon, B.G.; Gautam, S.; Peng, W.; Huang, Y.; Goli, M.; Mechref, Y. Direct comparison of N-glycans and their isomers derived from spike glycoprotein 1 of MERS-CoV, SARS-CoV, and SARS-CoV-2. *J. Proteom. Res.* **2021**, *20*, 4357–4365.
16. Shajahan, A.; Supekar, N.T.; Gleinich, A.S.; Azadi, P. Deducing the N- and O-glycosylation profile of the spike protein of novel coronavirus SARS-CoV-2. *Glycobiology* **2020**, *30*, 981–988. [CrossRef] [PubMed]
17. Bagdonaite, I.; Thompson, A.J.; Wang, X.; Søgaard, M.; Fougeroux, C.; Frank, M.; Diedrich, J.K.; Yates III, J.R.; Salanti, A.; Vakhrushev, S.Y.; et al. Site-specific O-glycosylation analysis of SARS-CoV-2 spike protein produced in insect and Human cells. *Viruses* **2021**, *13*, 551. [CrossRef]
18. Van Damme, E.J.M. 35 years in plant lectin research: A journey from basic science to applications in agriculture and medicine. *Glycoconj. J.* **2021**, in press. [CrossRef] [PubMed]
19. Collar, A.L.; Clarke, E.C.; Anaya, E.; Merrill, D.; Yarborough, S.; Anthony, S.M.; Kuhn, J.H.; Merle, C.; Theisen, M.; Bradfute, S.B. Comparison of N- and O-linked glycosylation patterns of ebolavirus glycoproteins. *Virology* **2017**, *502*, 39–47. [CrossRef]
20. Luo, S.; Hu, K.; He, S.; Wang, P.; Zhang, M.; Huang, X.; Du, T.; Zheng, C.; Liu, Y.; Hu, Q. Contribution of N-linked glycans on HSV-2 gG to cell-cell fusion and viral entry. *Virology* **2015**, *483*, 72–82. [CrossRef]
21. Smargiasso, N.; Nader, J.; Rioux, S.; Mazzucchelli, G.; Boutry, M.; De Pauw, E.; Chaumont, F.; Navarre, C. Exploring the N-glycosylation profile of glycoprotein B from Human cytomegalovirus expressed in CHO and *Nicotiana tabacum* BY-2 cells. *Int. J. Mol. Sci.* **2019**, *20*, 3741. [CrossRef] [PubMed]
22. Behrens, A.-J.; Vasiljevic, S.; Pritchard, L.K.; Harvey, D.J.; Andev, R.S.; Krumm, S.A.; Struwe, W.B.; Cupo, A.; Kumar, A.; Zitzmann, N.; et al. Composition and antigenic effects of individual glycan sites of a trimeric HIV-1 envelope glycoprotein. *Cell Resp.* **2016**, *14*, 2695–2706. [CrossRef]
23. Li, J.; Liu, S.; Gao, Y.; Tian, S.; Yang, Y.; Ma, N. Comparison of N-linked glycosylation on hemagglutinins derived from chicken embryos and MDCK cells: A case of the production of a trivalent seasonal influenza vaccine. *Appl. Microbiol. Biotechnol.* **2021**, *105*, 3559–3572. [CrossRef]

24. Lancaster, C.; Pristatsky, P.; Van Hoang, M.; Casimiro, D.R.; Schwartz, R.M.; Rustandi, R.; Ha, S. Characterization of *N*-glycosylation profiles from mammalian and insect cell derived chikungunya VLP. *J. Chromatogr. B Analyt. Technol. Life Sci.* **2016**, *1032*, 218–223. [CrossRef] [PubMed]
25. Watanabe, Y.; Raghwani, J.; Allen, J.D.; Seabright, G.E.; Li, S.; Moser, F.; Huiskonen, J.T.; Strecker, T.; Bowden, T.A.; Crispin, M. Structure of the Lassa virus glycan shield provides a model for immunological resistance. *Proc. Natl. Acad. Sci. USA* **2018**, *115*, 7320–7325. [CrossRef]
26. Pralow, A.; Nikolay, A.; Leon, A.; Genzel, Y.; Rapp, E.; Reich, U. Site-specific N-glycosylation analysis of animal cell culture-derived Zika virus proteins. *Sci. Rep.* **2021**, *11*, 5147. [CrossRef] [PubMed]
27. Watanabe, Y.; Allen, J.D.; Wrapp, D.; McLellan, J.S.; Crispin, M. Site-specific glycan analysis of the SARS-CoV-2 spike. *Science* **2020**, *369*, 330–333. [CrossRef] [PubMed]
28. Hatmal, M.M.; Alshaer, W.; Al-Hatamleh, M.A.I.; Hatmal, M.; Smadi, O.; Taha, M.O.; Oweida, A.J.; Boer, J.C.; Mohamud, R.; Plebanski, M. Comprehensive structural and molecular comparison of spike proteins of SARS-CoV-2, SARS-CoV and MERS-CoV, and their interactions with ACE2. *Cells* **2020**, *9*, 2638. [CrossRef] [PubMed]
29. Zhao, P.; Praissman, J.L.; Grant, O.C.; Xiao, T.; Rosenbalm, K.E.; Aoki, K.; Kellman, B.P.; Bridger, R.; Barouch, D.H.; Brindley, M.A.; et al. Virus-receptor interactions of glycosylated SARS-CoV-2 spike and Human ACE2 receptor. *Cell Host Microbe* **2020**, *28*, 586–601. [CrossRef] [PubMed]
30. Li, Y.; Liu, D.; Wang, Y.; Su, W.; Liu, G.; Dong, W. The importance of glycans of viral and host proteins in enveloped virus infection. *Front. Immunol.* **2021**, *12*, 638573. [CrossRef]
31. Peumans, W.J.; Van Damme, E.J.M.; Barre, A.; Rougé, P. Classification of plant lectins in families of structurally and evolu- tionary related proteins. *Adv. Exp. Med. Biol.* **2001**, *491*, 27–54. [PubMed]
32. Debray, H.; Decout, D.; Strecker, G.; Spik, G.; Montreuil, J. Specificity of twelve lectins towards oligosaccharides and glyco peptides related to *N*-glycosylproteins. *Eur. J. Biochem.* **1981**, *117*, 41–55. [CrossRef]
33. Debray, H.; Rougé, P. The fine sugar specificity of the *Lathyrus ochrus* seed lectin and isolectins. *FEBS Lett.* **1984**, *176*, 120–124. [CrossRef]
34. Einspahr, H.; Pareks, E.H.; Suguna, K.; Subramanian, E.; Suddath, F.L. The crystal structure of pea lectin at 3.0-Å resolution. *J. Biol. Chem.* **1986**, *261*, 16518–16527. [CrossRef]
35. Foriers, A.; Van Driessche, E.; De Neve, R.; Kanarek, L.; Strosberg, A.D. The subunit structure and N-terminal sequences of theα- and β-subunits of the lentil lectin (*Lens culinaris*). *FEBS Lett.* **1977**, *75*, 237–240. [CrossRef]
36. Bourne, Y.; Abergel, C.; Cambillau, C.; Frey, M.; Rougé, P.; Fontecilla-Camps, J.C. X-ray crystal structure determination and refinement at 1.9 Å resolution of isolectin I from the seeds of *Lathyrus ochrus*. *J. Mol. Biol.* **1990**, *214*, 571–584. [CrossRef]
37. Reeke, G.N., Jr.; Becker, J.W. Three-dimensional structure of favin: Saccharide binding-cyclic permutation in leguminous lectins. *Science* **1986**, *234*, 1108–1111. [CrossRef] [PubMed]
38. Hardman, K.D.; Ainsworth, C.F. Structure of concanavalin A at 2.4-Å resolution. *Biochemistry* **1972**, *11*, 4910–4919. [CrossRef]
39. Bourne, Y.; Mazurier, J.; Legrand, D.; Rougé, P.; Montreuil, J.; Spik, G.; Cambillau, C. Structures of a legume lectin complexed with the human lactotransferrin N2 fragment, and with an isolated biantennary glycopeptide: Role of the fucose moiety. *Structure* **1994**, *15*, 209–219. [CrossRef]
40. Bourne, Y.; Rougé, P.; Cambillau, C. X-ray structure of (α-Man(1-3)β-Man(1-4)GlcNAc)-lectin complex at 2.1-Å resolution. The role of water in sugar-lectin interaction. *J. Biol. Chem.* **1990**, *265*, 18161–18165. [CrossRef]
41. Rougé, P.; Peumans, W.J.; Van Damme, E.J.M.; Barre, A.; Singh, T.; Wu, J.H.; Wu, A.M. Glycotope structures and intramole-cular affinity factors of plant lectins for Tn/T antigens. *Adv. Exp. Med. Biol.* **2011**, *705*, 143–154. [PubMed]
42. Keyaerts, E.; Vijgen, L.; Pannecouque, C.; Van Damme, E.; Peumans, W.; Egberink, H.; Balzarini, J.; Van Ranst, M. Plant lectins are potent inhibitors of coronaviruses by interfering with two targets in the viral replication cycle. *Antivir. Res.* **2007**, *75*, 179–187. [CrossRef]
43. Swanson, M.D.; Winter, H.C.; Goldstein, I.J.; Markovitz, D.M. A lectin isolated from banana is a potent inhibitor of HIV replication. *J. Biol. Chem.* **2010**, *285*, 8646–8655. [CrossRef]
44. Swanson, M.D.; Boudreaux, D.M.; Salmon, L.; Clugh, J.; Winter, H.C.; Meagher, J.L.; André, S.; Murphy, P.V.; Oscarson, S.; Roy, R.; et al. Engineering a therapeutic lectin by uncoupling mitogenicity from antiviral activity. *Cell* **2015**, *163*, 746–758. [CrossRef]
45. Mazalovska, M.; Kouokam, J.C. Lectins as promising therapeutics for the prevention and treatment of HIV and other potential coinfections. *BioMed. Res. Int.* **2018**, *8*, 3750646. [CrossRef]
46. Covés-Datson, E.M.; King, S.R.; Legendre, M.; Gupta, A.; Chan, S.M.; Gitlin, E.; Kulkarni, V.V.; Pantaleón García, J.; Smee, D.F.; Lipka, E.; et al. A molecularly engineered antiviral banana lectin inhibits fusion and is efficacious against influenza virus infection in vivo. *Proc. Natl. Acad. Sci. USA* **2020**, *117*, 2122–2132. [CrossRef] [PubMed]
47. Covés-Datson, E.M.; King, S.R.; Legendre, M.; Swanson, M.D.; Gupta, A.; Claes, S.; Meagher, J.L.; Boonen, A.; Zhang, L.; Kalveram, B.; et al. Targeted disruption of pi-pi stacking in Malaysian banana lectin reduces mitogenicity while preserving antiviral activity. *Sci. Rep.* **2021**, *11*, 656. [CrossRef]
48. Covés-Datson, E.M.; Dyall, J.; DeWald, L.E.; King, S.R.; Dube, D.; Legendre, M.; Nelson, E.; Drews, K.C.; Gross, R.; Gerhardt, D.M.; et al. Inhibition of Ebola virus by a molecularly engineered banana lectin. *PLoS Negl. Trop. Dis.* **2019**, *13*, e0007595. [CrossRef] [PubMed]

49. Witvrouw, M.; Fikkert, V.; Hantson, A.; Pannecouque, C.; O'Keefe, B.R.; McMahon, J.; Stamatatos, L.; de Clercq, E.; Bolmstedt, A. Resistance to human immunodeficiency virus type 1 to the high-mannose binding agents cyanovirin N and concanavalin A. *J. Virol.* **2005**, *79*, 7777–7784. [CrossRef]
50. Jang, H.; Lee, D.-H.; Kang, H.G.; Lee, S.J. Concanavalin A targeting N-linked glycans in spike proteins influence viral interactions. *Dalton Trans.* **2020**, *49*, 13538–13543. [CrossRef]
51. Wang, W.; Li, Q.; Wu, J.; Hu, Y.; Wu, G.; Yu, C.; Xu, K.; Liu, X.; Wang, Q.; Huang, W.; et al. Lentil lectin from *Lens culinaris* exhibit broad antiviral activities against SARS-CoV-2 variants. *Emerg. Microbes Infect.* **2021**, *10*, 1519–1529. [CrossRef] [PubMed]
52. Gondim, A.C.S.; da Silva, S.R.; Mathys, L.; Noppen, S.; Liekens, S.; Sampaio, A.H.; Nagano, C.S.; Costa Rocha, C.R.; Nasci Mento, K.S.; Cavada, B.S.; et al. Potent antiviral activity of carbohydrate-specific algal and leguminous lectins from the Brazilian biodiversity. *Med. Chem. Commun.* **2019**, *10*, 390–398. [CrossRef] [PubMed]
53. Balzarini, J.; Neyts, J.; Schols, D.; Hosoya, M.; Van Damme, E.; Peumans, W.; De Clercq, E. The mannose-specific plant lectins from *Cymbidium* hybrid and *Epipactis helleborine* and the (*N*-acetylglucosamine)n-specific plant lectin from *Urtica dioica* are potent and selective inhibitors of human immunodeficiency virus and cytomegalovirus replication in vitro. *Antivir. Res.* **1992**, *18*, 191–207.
54. Balzarini, J.; Schols, D.; Neyts, J.; Van Damme, E.; Peumans, W.; De Clercq, E. Alpha-(1-3)- and alpha-(1-6)-D-mannose-specific plant lectins are markedly inhibitory to human immunodeficiency virus and cytomegalovirus infections in vitro. *Antimicrob. Agents Chemother.* **1991**, *35*, 410–416. [CrossRef] [PubMed]
55. Bertaux, C.; Daelemans, D.; Meertens, L.; Cormier, E.G.; Reinus, J.F.; Peumans, W.J.; Van Damme, E.J.M.; Igarashi, T.; Oki, T.; Schols, D.; et al. Entry of hepatitis C virus and human immunodeficiency virus is selectively inhibited by carbohydrate-binding agents but not by polyanions. *Virology* **2007**, *366*, 40–50. [CrossRef]
56. Ashfaq, U.; Masoud, M.; Khaliq, S.; Nawaz, Z.; Riazuddin, S. Inhibition of hepatitis C virus 3a genotype entry through *Galanthus nivalis* agglutinin. *Virol. J.* **2011**, *8*, 248. [CrossRef] [PubMed]
57. Luo, Y.; Xu, X.; Liu, J.; Li, J.; Sun, Y.; Liu, Z.; Liu, J.; Van Damme, E.; Balzarini, J.; Bao, J. A novel mannose-binding tuber lectin from *Typhonium divaricatum* (L.) Decne (family Araceae) with antiviral activity against HSV-II and anti-proliferative effect on human cancer cell lines. *J. Biochem. Mol. Biol.* **2007**, *40*, 358–367. [CrossRef]
58. Amundson, D.E.; Shah, U.; de Necochea-Campion, R.; Jacobs, M.; LaRosa, S.P.; Fisher, C.J., Jr. Removal of COVID-19 spike protein, whole virus, exosomes, and exosomal microRNAs by the Hemopurifier®lectin-affinity cartridge in critically Ill patients with COVID-19 infection. *Front. Med.* **2021**, *8*, 744141. [CrossRef] [PubMed]
59. Jayaprakash, N.G.; Singh, A.; Vivek, R.; Yadav, S.; Pathak, S.; Trivedi, J.; Jayaraman, N.; Nandi, D.; Mitra, D.; Surolia, A. The barley lectin, horcolin, binds high-mannose glycans in a multivalent fashion, enabling high-affinity, specific inhibition of cellular HIV infection. *J. Biol. Chem.* **2020**, *295*, 12111–12129. [CrossRef]
60. Liu, Y.-M.; Shahed-Al-Mahmud, M.D.; Chen, X.; Chen, T.-H.; Liao, K.-S.; Lo, J.M.; Wu, Y.-M.; Ho, M.-C.; Wu, C.-Y.; Wong, C.-H.; et al. A carbohydrate-binding protein from the edible lallab beans effectively blocks the infections of influenza viruses and SARS-CoV-2. *Cell Rep.* **2020**, *32*, 108016. [CrossRef]
61. Sheehan, S.A.; Hamilton, K.L.; Retzbach, E.P.; Balachandran, P.; Krishnan, H.; Leone, P.; Lopez-Gonzalez, M.; Suryavanshi, S.; Kumar, P.; Russo, R.; et al. Evidence that *Maackia amurensis* seed lectin (MASL) exerts pleiotropic actions on oral squamous cells with potential to inhibit SARS-CoV-2 infection and COVID-19 disease progression. *Exp. Cell Res.* **2021**, *403*, 112594. [CrossRef] [PubMed]
62. Gordts, S.C.; Renders, M.; Férir, G.; Huskens, D.; Van Damme, E.J.M.; Peumans, W.; Balzarini, J.; Schols, D. NICTABA and UDA, two GlcNAc-binding lectins with unique antiviral activity profiles. *J. Antimicrob. Chemother.* **2015**, *70*, 1674–1685. [CrossRef]
63. Al Atalah, B.; Fouquaert, E.; Vanderschaeghe, D.; Proost, P.; Balzarini, J.; Smith, D.F.; Rougé, P.; Lasanajak, Y.; Callewaert, N.; Van Damme, E.J.M. Expression analysis of the nucleocytoplasmic lectin 'Orysata' from rice in *Pichia pastoris*. *FEBS J.* **2011**, *278*, 2064–2079. [CrossRef] [PubMed]
64. An, J.; Liu, J.-Z.; Wu, C.-F.; Li, J.; Dai, L.; Van Damme, E.; Balzarini, J.; De Clercq, E.; Chen, F.; Bao, J.-K. Anti-HIV I/II activity and molecular cloning of a novel mannose/sialic acid-binding lectin from rhizome of *Polygonatum cyrtonema* Hua. *Acta Biochim. Biophys. Sin.* **2006**, *38*, 70–78. [CrossRef]
65. Vanderlinden, E.; Van Winkel, N.; Naesens, L.; Van Damme, E.J.M.; Persoons, L.; Schols, D. In vitro characterization of the carbohydrate-binding agents HHA, GNA, and UDA as inhibitors of influenza A and B virus replication. *Antimicrob. Agents Chemother.* **2021**, *65*, e01732-20. [CrossRef] [PubMed]
66. Auth, J.; Fröba, M.; Grobe, M.; Rauch, P.; Ruetalo, N.; Schindler, M.; Morokutti-Kurz, M.; Fraf, P.; Dolischka, A.; Prieschl-Grassauer, E.; et al. Lectin from *Triticum vulgaris* (WGA) inhibits infection with SARS-CoV-2 and its variants of concern alpha and beta. *Int. J. Mol. Sci.* **2021**, *22*, 10205. [CrossRef]
67. Mori, T.; O'Keefe, B.R.; Sowder, R.C., II; Bringans, S.; Gardella, R.; Berg, S.; Cochran, P.; Turpin, J.A.; Buckheit, R.W., Jr.; McMahon, J.B.; et al. Isolation and characterization of griffithsin, a novel HIV-inactivating protein, from the red alga *Griffithsia* sp. *J. Biol. Chem.* **2005**, *280*, 9345–9353. [CrossRef]
68. O'Keefe, B.R.; Giomarelli, B.; Barnard, D.L.; Shenoy, S.R.; Chan, P.K.S.; McMahon, J.B.; Palmer, K.E.; Barnett, B.W.; Mey Erholz, D.K.; Wohlford-Lenane, C.L.; et al. Broad-spectrum in vitro activity and in vivo efficacy of the antiviral protein griffithsin against emerging viruses of the family Coronaviridae. *J. Virol.* **2010**, *84*, 2511–2521. [CrossRef]

69. Moulaei, T.; Shenoy, S.R.; Giomarelli, B.; Thomas, C.; McMahon, J.B.; Dauter, Z.; O'Keefe, B.R.; Wlodawer, A. Monomerizationof viral entry inhibitor griffithsin elucidates the relationship between multivalent binding to carbohydrates and anti-HIV activity. *Structure* **2010**, *18*, 1104–1115. [CrossRef]
70. Sato, Y.; Morimoto, K.; Kubo, T.; Sakaguchi, T.; Nishizono, A.; Hirayama, M.; Hori, K. Entry inhibition of influenza viruses with high mannose binding lectin ESA-2 from the red alga *Eucheuma serra* through the recognition of viral hemagglutinin. *Mar. Drugs* **2015**, *13*, 3454–3465. [CrossRef]
71. Hwang, H.-J.; Han, J.-W.; Jeon, H.; Cho, K.; Kim, J.-H.; Lee, D.-S.; Han, J.W. Characterization of a novel mannose binding lectin with antiviral activities from red alga, *Grateloupia chiangii*. *Biomolecules* **2020**, *10*, 333. [CrossRef] [PubMed]
72. Xue, J.; Hoorelbeke, B.; Kagiampakis, I.; Demeler, B.; Balzarini, J.; Liwang, P.J. The griffithsin dimer is required for high-potency inhibition of HIV-1: Evidence for manipulation of the structure of gp120 as part of the griffithsin dimer mechanism. *Antimicrob. Agents Chemother.* **2013**, *57*, 3976–3989. [CrossRef] [PubMed]
73. Meuleman, P.; Albecka, A.; Belouzard, S.; Vercauteren, K.; Verhoye, L.; Wychowski, C.; Leroux-Roels, G.; Palmer, K.E.; Dubuisson, J. Griffithsin has antiviral activity against hepatitis C virus. *Antimicrob. Agents Chemother.* **2011**, *55*, 5159–5167. [CrossRef]
74. Millet, J.K.; Séron, K.; Labitt, R.N.; Danneels, A.; Palmer, K.E.; Whittaker, G.R.; Dubuisson, J.; Belouzard, S. Middle East respiratory syndrome coronavirus infection is inhibited by griffithsin. *Antivir. Res.* **2016**, *133*, 1–8. [CrossRef] [PubMed]
75. Alsaidi, S.; Cornejal, N.; Mahoney, O.; Melo, C.; Verma, N.; Bonnaire, T.; Chang, T.; O'Keefe, B.R.; Sailer, J.; Zydowsky, T.M.; et al. Griffithsin and carrageenan combination results in antiviral synergy against SARS- CoV-1 and 2 in a pseudoviral model. *Mar. Drugs* **2021**, *19*, 418. [CrossRef]
76. Mu, J.; Hirayama, M.; Sato, Y.; Morimoto, K.; Hori, K. A novel high-mannose specific lectin from the green alga *Halimeda renschii* exhibits a potent anti-influenza virus activity through high-affinity binding to the viral hemagglutinin. *Mar. Drugs* **2017**, *15*, 255. [CrossRef]
77. Sato, Y.; Morimoto, K.; Hirayama, M.; Hori, K. High-mannose-specific lectin (KAA-2) from the red alga *Kappaphycus alvarezii* potently inhibits influenza virus infection in a strain-independent manner. *Biochem. Bipophys. Res. Commun.* **2011**, *405*, 291–296. [CrossRef]
78. Hirayama, M.; Shibata, H.; Imamura, K.; Sakaguchi, T.; Hori, K. High-mannose specific lectin and its recombinants from a Carrageenophyta *Kappaphycus alvarezii* represent a potent anti-HIV activity through high-affinity binding to the viral envelope glycoprotein gp120. *Mar Biotechnol.* **2016**, *18*, 144–160. [CrossRef]
79. Sato, Y.; Hirayama, M.; Morimoto, K.; Yamamoto, N.; Okuyama, S.; Hori, K. High mannose-binding lectin with preference for the cluster of α1-2-mannose from the green alga *Boodlea coacta* is a potent entry inhibitor of HIV-1 and influenza viruses. *J. Biol. Chem.* **2011**, *286*, 19446–19458. [CrossRef] [PubMed]
80. Huskens, D.; Férir, G.; Vermeire, K.; Kehr, J.-C.; Balzarini, J.; Dittmann, E.; Schols, D. Microvirin, a novel α(1,2)-mannose-specific lectin isolated from *Microcystis aeruginosa*, has anti-HIV-1 activity comparable with that of cyanovirin-N but a much higher safety profile. *J. Biol. Chem.* **2010**, *285*, 24845–24854. [CrossRef]
81. Shahid, M.; Qadir, A.; Yang, J.; Ahmad, I.; Zahid, H.; Mirza, S.; Windisch, M.P.; Shahzad-Ul-Hussan, S. An engineered microvirin with identical structural domains potently inhibits human immunodeficiency virus and hepatitis C virus cellular entry. *Viruses* **2020**, *12*, 199. [CrossRef]
82. Ziólkowska, N.E.; Wlodawer, A. Structural studies of algal lectins with anti-HIV activity. *Acta Biochim. Pol.* **2006**, *53*, 617–626. [CrossRef]
83. Kachko, A.; Loesgen, S.; Shahzad-Ul-Hussan, S.; Tan, W.; Zubkova, I.; Takeda, K.; Wells, F.; Rubin, S.; Bewley, C.A.; Major, M.E. Inhibition of hepatitis C virus by the cyanobacterial protein *Microcystis viridis* lectin: Mechanistic differences between the high-mannose specific lectins MVL, CV-N, and GNA. *Mol. Pharm.* **2013**, *10*, 4590–4602. [CrossRef] [PubMed]
84. Shenoy, S.R.; O'Keefe, B.R.; Bolmstedt, A.J.; Cartner, L.K.; Boyd, M.R. Selective interactions of the human immunodeficiency virus-inactivating protein cyanovirin-N with high-mannose oligosaccharides on gp120 and other glycoproteins. *J. Pharmacol. Exp. Ther.* **2001**, *297*, 704–710. [PubMed]
85. Barrientos, L.G.; O'Keefe, B.R.; Bray, M.; Sanchez, A.; Gronenborn, A.M.; Boyd, M.R. Cyanovirin-N binds to the viral surface glycoprotein, GP1,2 and inhibits infectivity of Ebola virus. *Antivir. Res.* **2003**, *58*, 47–56. [CrossRef]
86. Maier, I.; Schiestl, R.H.; Kontaxis, G. Cyanovirin-N binds viral envelope proteins at the low-affinity carbohydrate binding site without direct virus neutralization ability. *Molecules* **2021**, *26*, 3621. [CrossRef]
87. Helle, F.; Wychowski, C.; Vu-Dac, N.; Gustafson, K.R.; Voisset, C.; Dubuisson, J. Cyanovirin-N inhibits hepatitis C virus entry by binding to envelope protein glycans. *J. Biol. Chem.* **2006**, *281*, 15177–25183. [CrossRef]
88. Tiwari, V.; Shukla, S.; Shukla, D. A sugar binding protein cyanovirin-N blocks herpes simplex virus type-1 entry and cell fusion. *Antivir. Res.* **2009**, *84*, 67–75. [CrossRef]
89. Naidoo, D.; Kar, P.; Roy, A.; Mutanda, T.; Bwapwa, J.; Sen, A.; Anandraj, A. Structural insight into the binding of cyanovirin- N with the spike glycoprotein, Mpro and PLpro of SARS-CoV-2: Protein-protein interactions, dynamics simulations and free energy calculations. *Molecules* **2021**, *26*, 5114. [CrossRef]
90. Sato, Y.; Okuyama, S.; Hori, K. Primary structure and carbohydrate binding specificity of a potent anti-HIV lectin isolated from the filamentous cyanobacterium *Oscillatoria agardhii*. *J. Biol. Chem.* **2007**, *282*, 11021–11029. [CrossRef]

91. Bokesch, H.R.; O'Keefe, B.R.; McKee, T.C.; Pannell, L.K.; Patterson, G.M.L.; Gardella, R.S.; Sowder, R.C., II; Turpin, J.; Watson, K.; Buckheit, R.W., Jr.; et al. A potent novel anti-HIV protein from the cultured cyanobacterium *Scytonema varium*. *Biochemistry* **2003**, *42*, 2578–2584. [CrossRef]
92. Siqueira, A.S.; Lima, A.R.J.; de Sousa, R.C.; Santos, A.S.; da Silva Gonçalves Vianez Júnior, J.L.; Costa Gonçalves, E. Anti-dengue virus activity of scytovirin and evaluation of point mutation effects by molecular dynamics and binding free energy calculations. *Biochem. Biophys. Res. Commun.* **2017**, *490*, 1033–1038. [CrossRef]
93. Garrison, A.R.; Giomarelli, B.G.; Lear-Rooney, C.M.; Saucedo, C.J.; Yellayi, S.; Krumpe, L.R.H.; Rose, M.; Paragas, J.; Bray, M.; Olinger, G.G.; et al. The cyanobacterial lectin scytovirin displays potent in vitro and in vivo activity against Zaire Ebola virus. *Antivir. Res.* **2014**, *112*, 1–7. [CrossRef]
94. Koharudin, L.M.; Furey, W.; Gronenborn, A.M. Novel fold and carbohydrate specificity of the potent anti-HIV cyanobacterial lectin from *Oscillatoria agardhii*. *J. Biol. Chem.* **2011**, *286*, 1588–1597. [CrossRef] [PubMed]
95. Mazur-Marzec, H.; Ceglowska, M.; Konkel, R.; Pyrc, K. Antiviral cyanometabolites—A review. *Biomolecules* **2021**, *11*, 474. [CrossRef]
96. François, K.O.; Balzarini, J. Potential of carbohydrate-binding agents as therapeutics against enveloped viruses. *Med. Res. Rev.* **2012**, *32*, 349–387. [CrossRef] [PubMed]
97. Matsui, T.; Kobayashi, S.; Yoshida, O.; Ishii, S.; Abe, Y.; Yamamoto, N. Effects of succinylated concanavalin A on infectivity and syncytial formation of human immunodeficiency virus. *Med. Microbiol. Immunol.* **1990**, *179*, 225–235. [CrossRef]
98. Banks, W.A.; Ibrahimi, F.; Farr, S.A.; Flood, S.A.; Morley, J.E. Effects of wheat germ agglutinin and aging on the regional brain uptake of HIV-1 gp120. *Life Sci.* **1999**, *65*, 81–89. [CrossRef]
99. Balzarini, J.; Van Laethem, K.; Hatse, S.; Vermeire, K.; De Clercq, E.; Peumans, W.; Van Damme, E.; Vandamme, A.M.; Bolmstedt, A.; Schols, D. Profile of resistance of human immunodeficiency virus to mannose-specific plant lectins. *J. Virol.* **2004**, *78*, 10617–10627. [CrossRef]
100. Balzarini, J. Targeting the glycans of gp120: A novel approach aimed at the Achilles heel of HIV. *Lancet Infect. Dis.* **2005**, *5*, 727–731. [CrossRef]
101. Turville, S.G.; Vermeire, K.; Balzarini, J.; Schols, D. Sugar-binding proteins potently inhibit dendritic cell human immunodeficiency virus type 1 (HIV-1) infection and dendritic-cell-directed HIV-1 transfer. *J. Virol.* **2005**, *79*, 13519–13527. [CrossRef]
102. Balzarini, J.; Van Herrewege, Y.; Vermeire, K.; Vanham, G.; Schols, D. Carbohydrate-binding agents efficiently prevent dendritic cell-specific intercellular adhesion molecule-3-grabbing nonintegrin (DC-SIGN)-directed HIV-1 transmission to T lymphocytes. *Mol. Pharmacol.* **2007**, *71*, 3–11. [CrossRef]
103. Auwerx, J.; François, K.O.; Vanstreels, E.; Van Laethem, K.; Daelemans, D.; Schols, D.; Balzarini, J. Capture and transmission of HIV-1 by the C-type lectin L-SIGN (DC-SIGNR) is inhibited by carbohydrate-binding agents and polyanions. *Antivir. Res.* **2009**, *83*, 61–70. [CrossRef]
104. Hoorelbeke, B.; Van Damme, E.J.M.; Rougé, P.; Schols, D.; Van Laethem, K.; Fouquaert, E.; Balzarini, J. Differences in the mannose oligomer specificities of the closely related lectins from *Galanthus nivalis* and *Zea mays* strongly determine their eventual anti-HIV activity. *Retrovirology* **2011**, *8*, 10. [CrossRef]
105. Kumaki, Y.; Wandersee, M.K.; Smith, A.J.; Zhou, Y.; Simmons, G.; Nelson, N.M.; Bailey, K.W.; Vest, Z.G.; Li, J.K.-K.; Chan, P.K.-S.; et al. Inhibition of severe acute respiratory syndrome coronavirus replication in a lethal SARS-CoV BALB/c mouse model by stinging nettle lectin, *Urtica dioica* agglutinin. *Antivir. Res.* **2011**, *90*, 22–32. [CrossRef]
106. Barre, A.; Van Damme, E.J.M.; Simplicien, M.; Le Poder, S.; Konjklowski, B.; Benoist, H.; Peyrade, D.; Rougé, P. Man-specific lectins from plants, fungi, algae and cyanobacteria, as potential blockers for SARS-CoV, MERS-CoV and SARS-CoV-2 (COVID-19) coronaviruses: Biomedical perspectives. *Cells* **2021**, *10*, 1619. [CrossRef]
107. Ahmed, N.; Jahan, R.; Nissapatorn, V.; Wilairatana, P.; Rahmatullah, M. Plant lectins as prospective antiviral biomolecules in the search for COVID-19 eradication strategies. *Miomed. Pharmacother.* **2022**, *146*, 112507. [CrossRef]
108. Singh, R.S.; Walia, A.K.; Khattar, J.S.; Singh, D.P.; Kennedy, J.F. Cyanobacterial lectins characteristics and their role as antiviral agents. *Int. J. Biol. Macromol.* **2017**, *102*, 475–496. [CrossRef]
109. Singh, R.S.; Walia, A.K. Lectins from red algae and their biomedical potential. *J. Appl. Phycol.* **2018**, *30*, 1833–1858. [CrossRef]
110. Breitenbach Barroso Coelho, L.C.; Marcelino dos Santos Silva, P.; Felix de Oliveira, W.; De Moura, M.C.; Viana Pontual, E.; Soares Gomes, F.; Guedes Paiva, P.M.; Napoleão, T.H.; dos Santos Correia, M.T. Lectins as antimicrobial agents. *J. Appl. Microbiol.* **2018**, *125*, 1238–1252. [CrossRef]
111. Carneiro, D.C.; Fernandez, L.G.; Monteiro-Cunha, J.P.; Benevides, R.G.; Cunha Lima, S.T. A patent review of the antimicrobial applications of lectins: Perspectives on therapy of infectious diseases. *J. Appl. Microbiol.* 2021; *online ahead of print*. [CrossRef]
112. Martinez, D.; Amaral, D.; Markovitz, D.; Pinto, L. The use of lectins as tools to combat SARS-CoV-2. *Curr. Pharm. Des.* **2021**, *27*, 4212–4222. [CrossRef]
113. Peumans, W.J.; Van Damme, E.J.M. Lectins as plant defense proteins. *Plant Physiol.* **1995**, *109*, 347–352. [CrossRef]
114. Boisseau, C.; Moisand, A.; Père, D.; Rougé, P. Immunocytochemical localization of the *Lathyrus ochrus* (L.) DC. Seed lectin in seeds and seedlings. *Plant Sci.* **1985**, *42*, 25–34. [CrossRef]
115. Rougé, P.; Sousa-Cavada, B. Isolation and partial characterization of two isolectins from *Lathyrus ochrus* (L.) DC. Seeds. *Plant Sci. Lett.* **1984**, *37*, 21–27.

116. Fuqua, J.L.; Wanga, V.; Palmer, K.E. Improving the large-scale purification of the HIV microbicide, griffithsin. *BMC Biotechnol.* **2015**, *15*, 12. [CrossRef]
117. Fuqua, J.L.; Hamorsky, K.; Khalsa, G.; Matoba, N.; Palmer, K.E. Bulk production of the antiviral lectin griffithsin. *Plant Biotechnol.* **2015**, *13*, 1160–1168. [CrossRef]
118. Alam, A.; Jiang, L.; Kittleson, G.A.; Steadman, K.D.; Nandl, S.; Fuqua, J.L.; Palmer, K.E.; Tusé, D.; McDonald, K.A. Technoeconomic modeling of plant-based griffithsin manufacturing. *Front. Bieng. Biotechnol.* **2018**, *6*, 102. [CrossRef]
119. Hoelscher, M.; Tiller, N.; The, A.Y.-H.; Wu, G.-Z.; Ma, J.K.-C.; Bock, R. High-level expression of the HIV entry inhibitor griffithsin from the plastid genome and retention of biological activity in dried tobacco leaves. *Plant Mol. Biol.* **2018**, *97*, 357–370. [CrossRef]
120. Eapen, P.; Cates, J.; Mundell, R.; Palmer, K.E.; Fuqua, J.L. In preparation for outdoor pharming: Griffithsin can be expressed in *Nicotiana excelsiana* and retains activity after storage and silage. *Front. Bioeng. Biotechnol.* **2020**, *8*, 199. [CrossRef]
121. Decker, J.S.; Menacho-Melgar, R.; Lynch, M.D. Low-cost, large-scale production of the anti-viral lectin griffithsin. *Front. Bioeng. Biotechnol.* **2020**, *8*, 1020. [CrossRef]
122. Lueken, K.; Mazarguil, H.; Rougé, P. The identification of two peptide sequences of light subunits of the *Lathyrus ochrus* isolectins containing a sequential epitope. *Immunol. Lett.* **1988**, *19*, 309–312. [CrossRef]
123. Lueken, K.; Liboz, T.; Mazarguil, H.; Rougé, P. Localization of amino acid sequence stretches containing a continuous epitope on the surface of the two *Lathyrus ochrus* isolectins. *Immunol. Lett.* **1989**, *23*, 223–226. [CrossRef]
124. Kolberg, J.; Ayouba, A.; Rougé, P. Production and characterization of a mouse monoclonal antibody specific for lentil lectin. *Biol. Chem. Hoppe-Seyler* **1991**, *372*, 57–61.
125. Lueken, K.; Kolberg, J.; Cambillau, C.; Bourne, Y.; Rougé, P. Monoclonal antibody 117, C-11 recognizes three exposed regions on the surface of the *Lathyrus ochrus* isolectin I. *Immunol. Lett.* **1991**, *30*, 47–52. [CrossRef]
126. Koshte, V.L.; Aalbers, M.; Calkhoven, P.G.; Aalberse, R.C. The potent IgG4-inducing antigen in banana is a mannose-binding lectin, BanLec-I. *Int. Arch. Allergy Immunol.* **1992**, *97*, 17–24. [CrossRef]
127. Tchernychev, B.; Rabinkov, A.; Mirelman, D.; Wilchek, M. Natural antibodies to dietary proteins—The existence of natural antibodies to alliinase (alliin lyase) and mannose-specific lectin from garlic (*Allium sativum*) in human serum. *Immunol. Lett.* **1995**, *47*, 53–57. [CrossRef]
128. Bewley, C.A.; Gustafson, K.R.; Boyd, M.R.; Covell, D.G.; Bax, A.; Clore, G.M.; Gronenborn, A.M. Solution structure of cyanovirin-N, a potent HIV-inactivating protein. *Nat. Struct. Biol.* **1998**, *5*, 571–578. [CrossRef]
129. Moulaei, T.; Botos, I.; Ziólkowska, N.E.; Bokesch, H.; Krumpe, L.R.; McKee, T.C.; O'Keefe, B.R.; Dauter, Z.; Wlodawer, A. Atomic-resolution crystal structure of the antiviral lectin scytovirin. *Protein Sci.* **2007**, *16*, 2756–2760. [CrossRef]
130. Micevicz, E.D.; Cole, A.L.; Jung, C.L.; Phillips, M.L.; Pratikhya, P.; Sharma, S.; Waring, A.J.; Cole, A.M.; Ruchala, P. Grifonin-I: A small HIV-1 entry inhibitor derived from the algal lectin, griffithsin. *PLoS ONE* **2010**, *5*, e14360. [CrossRef]
131. Cooper, H.L.; Rubin, A.D. Synthesis of nonribosomal RNA by lymphocytes: A response to phytohemagglutinin treatment. *Science* **1966**, *152*, 516–518. [CrossRef]
132. Phillips, B.; Weisrose, E. The mitogenic response of human B lymphocytes to phytohaemagglutinin. *Clin. Exp. Immunol.* **1974**, *16*, 383–392.
133. Lis, H.; Sharon, N. Biological properties of lectins. In *The Lectins, Properties, Functions, and Applications in Biology and Medicine*; Liener, I.E., Sharon, N., Goldstein, I.J., Eds.; Academic Press Inc.: Orlando, FL, USA; San Diego, CA, USA; New York, NY, USA; Austin TX, USA, 1986; pp. 265–291.
134. Liu, T.; Wu, L.; Wang, H.; Chen, J.; Yang, C.; Bao, J.; Wu, C. Role of reactive oxygen species-mediated MAPK and NF-κB activation on *Polygonatum cyrtonema* lectin-induced apoptosis and autophagy in human lung adenocarcinoma A549 cells. *J. Biochem.* **2016**, *160*, 315–324. [CrossRef]
135. Chowdhury, S.R.; Ray, U.; Chatterjee, B.; Roy, S.S. Targeted apoptosis in ovarian cancer cells through mitochondrial dysfunction in response to *Sambucus nigra* agglutinin. *Cell Death Dis.* **2017**, *8*, e2762. [CrossRef] [PubMed]
136. Poiroux, G.; Barre, A.; Van Damme, E.J.M.; Benoist, H.; Rougé, P. Plant lectins targeting *O*-glycans at the cell surface as tools for cancer diagnosis, prognosis and therapy. *Int. J. Mol. Sci.* **2017**, *18*, 1232. [CrossRef] [PubMed]
137. Naik, S.; Rawat, R.S.; Khandai, S.; Kumar, M.; Jena, S.S.; Vijayalakshmi, M.A.; Kumar, S. Biochemical characterization of lectin from Indian hyacinth plant bulbs with potential inhibitory action against human cancer cells. *Int. J. Biol. Macromol.* **2017**, *105*, 1349–1356. [CrossRef] [PubMed]
138. Jang, S.; Yayeh, T.; Leem, Y.-H.; Park, E.-M.; Ito, Y.; Oh, S. Concanavalin A induces cortical neuron apoptosis by causing ROS accumulation and tyrosine kinase activation. *Neurochem. Res.* **2017**, *42*, 3504–3514. [CrossRef]
139. Islam, F.; Gopalan, V.; Lam, A.K.-Y.; Rashel Kabir, S. Pea lectin inhibits cell growth by inducing apoptosis in SW480 and SW48 cell lines. *Int. J. Biol. Macromol.* **2018**, *117*, 1050–1057. [CrossRef]
140. Poiroux, G.; Barre, A.; Simplicien, M.; Pelofy, S.; Segui, B.; Van Damme, E.J.M.; Rougé, P.; Benost, H. Morniga-G, a T/Tn-specific lectin, induces leukemic cell death via caspase and DR5 receptor-dependent pathways. *Int. J. Mol. Sci.* **2019**, *20*, 230. [CrossRef]
141. Bhutia, S.; Panda, P.K.; Sinha, N.; Praharaj, P.P.; Bhol, C.S.; Panigrahi, D.P.; Mahapatra, K.K.; Saha, S.; Patra, S.; Mishra, S.R.; et al. Plant lectins in cancer therapeutics: Targeting apoptosis and autophagy-dependent cell dath. *Pharmacol. Res.* **2019**, *144*, 8–18. [CrossRef]
142. Lalli, R.C.; Kaur, K.; Chakraborti, A.; Srinivasan, R.; Ghosh, S. *Maackia amurensis* agglutinin induces apoptosis in cultured resistant human non-small lung cancer cells. *Glycoconj. J.* **2019**, *36*, 473–485. [CrossRef]

143. Rashidbaghan, A.; Mostafaie, A.; Yazdani, Y.; Mansouri, K. *Urtica dioica* agglutinin (a plant lectin) has a caspase-dependent apoptosis induction effect on the acute lymphoblastic leukemia cell line. *Cell Mol. Biol.* **2020**, *66*, 121–126. [CrossRef]
144. Balzarini, J.; Van Laethem, K.; Peumans, W.J.; Van Damme, E.J.; Bolmstedt, A.; Cago, F.; Schols, D. Mutational pathways, resistance profile, and side effects of cyanovirin relative to human immunodeficiency virus type 1 strains with N-glycan deletions in their gp120 envelopes. *J. Virol.* **2006**, *80*, 8411–8421. [CrossRef]
145. Huskens, D.; Vermeire, K.; Vandermeulebroucke, E.; Balzarini, J. Safety concerns for the potential use of cyanovirin-N as a microbicidal anti-HIV agent. *Int. J. Biochem. Cell Biol.* **2008**, *40*, 2802–2814. [CrossRef]
146. Balzarini, J.; François, K.O.; Van Laethem, K.; Hoorelbeke, B.; Renders, M.; Auwers, J.; Liekens, S.; Oki, T.; Igarashi, Y.; Schold, D. Pradimicin S, a highly-soluble non-peptidic small-size carbohydrate-binding antibiotic, is an anti-HIV drug lead for both microbicidal and systemic use. *Antimicrob. Agents Chemother.* **2010**, *54*, 1425–1435. [CrossRef] [PubMed]
147. Büttner, S.; Koch, B.; Dolnik, O.; Eickmann, M.; Freiwald, T.; Rudolf, S.; Engel, J.; Becker, S.; Ronco, C.; Geiger, H. Extracorporeal virus elimination for the treatment of severe Ebola virus disease—First experience with lectin affinity plasmapheresis. *Blood Purif.* **2014**, *38*, 286–291. [CrossRef] [PubMed]
148. Koch, B.; Schult-Dietrich, P.; Büttner, S.; Dilmaghani, B.; Lohmann, D.; Baer, P.C.; Dietrich, U.; Geiger, H. Lectin affinity plasmapheresis for middle east respiratory syndrome-coronavirus and Marburg virus glycoprotein elimination. *Blood Purif.* **2018**, *46*, 126–133. [CrossRef]
149. Leblanc, J.-F.; Germain, M.; Delage, G.; O'Brien, S.; Drews, S.J.; Lewin, A. Risk of transmission of severe acute respiratory syndrome coronavirus 2 by transfusion: A literature review. *Transfusion* **2020**, *60*, 3046–3054. [CrossRef]
150. Gaussen, A.; Hornby, L.; Rockl, G.; O'Brien, S.; Delage, G.; Sapir-Pichhadze, R.; Drews, S.J.; Weiss, M.J.; Lewin, A. Evidence of SARS-CoV-2 infection in cells, tissues and organs and the risk of transmission through transplantation. *Transplantation* **2021**, *105*, 1405–1422. [CrossRef] [PubMed]
151. Pettersen, E.F.; Goddard, T.D.; Huang, C.C.; Couch, G.S.; Greenblatt, D.M.; Meng, E.C.; Ferrin, T.E. UCSF Chimera—A visualization system for exploratory research and analysis. *J. Comput. Chem.* **2004**, *25*, 1605–1612. [CrossRef]
152. Pettersen, E.F.; Goddard, T.D.; Huang, C.C.; Meng, E.C.; Couch, G.S.; Croll, T.I.; Morris, J.H.; Ferrin, T.E. UCSF ChimeraX: Structure visualization for researchers, educators, and developers. *Protein Sci.* **2021**, *30*, 70–82. [CrossRef]
153. Böhm, M.; Bohne-Lang, A.; Frank, M.; Loss, A.; Rojas-Macias, M.A.; Lütteke, T. Glycosciences.DB: An annotated data collec- tion linking glycomics and proteomics data (2018 update). *Nucleic Acids Res.* **2019**, *47*, D1195–D1201. [CrossRef] [PubMed]
154. Cheng, K.; Zhou, Y.; Neelamegham, S. DrawGlycan-SNFG: A robust tool to render glycans and glycopeptides with frag mentation information. *Glycobiology* **2017**, *27*, 200–205.
155. Day, C.J.; Bailly, B.; Guillon, P.; Dirr, L.; Jen, F.E.-C.; Spillings, B.L.; Mak, J.; Itzstein, M.; Haselhorst, T.; Jennings, M.P. Multidisciplinary approaches identify compounds that bind to human ACE2 or SARS-CoV-2 spike protein as candidates to block SARS-CoV-2-ACE2 receptor interactions. *mBio* **2021**, *12*, e03681-20. [CrossRef] [PubMed]
156. Kawasaki, N.; Lin, C.-W.; Inoue, R.; Khoo, K.-H.; Kawasaki, N.; Ma, B.Y.; Oka, S.; Ishiguro, M.; Sawada, T.; Ishida, H.; et al. Highly fucosylated N-glycan ligand for mannan-binding protein expressed specifically on CD26(DPPIV) isolated from a human colorectal carcinoma cell line, SW1116. *Glycobiology* **2009**, *19*, 437–450. [CrossRef] [PubMed]
157. Allen, J.D.; Watanabe, Y.; Chawla, H.; Newby, M.L.; Crispin, M. Subtle influence of ACE2 glycan processing on SARS-CoV-2 recognition. *J. Mol. Biol.* **2021**, *433*, 166762. [CrossRef] [PubMed]

cells MDPI

Review

Deciphering the Role of Ion Channels in Early Defense Signaling against Herbivorous Insects

Akanksha Gandhi [1], Rupesh Kariyat [1], Amaravadhi Harikishore [2], Marzieh Ayati [3], Anirban Bhunia [4] and Nirakar Sahoo [1,*]

1 Department of Biology, University of Texas Rio Grande Valley, Edinburg, TX 78539, USA; akankshagandhi037@gmail.com (A.G.); Rupesh.kariyat@utrgv.edu (R.K.)
2 School of Biological Sciences, Nanyang Technological University, 60 Nanyang Drive, Singapore 637551, Singapore; AMAR0002@e.ntu.edu.sg
3 Department of Computer Science, University of Texas Rio Grande Valley, Edinburg, TX 78539, USA; marzieh.ayati@utrgv.edu
4 Department of Biophysics, Bose Institute, Kolkata 700054, India; anirbanbhunia@gmail.com
* Correspondence: nirakar.sahoo@utrgv.edu; Tel.: +1-956-665-5361

Abstract: Plants and insect herbivores are in a relentless battle to outwit each other. Plants have evolved various strategies to detect herbivores and mount an effective defense system against them. These defenses include physical and structural barriers such as spines, trichomes, cuticle, or chemical compounds, including secondary metabolites such as phenolics and terpenes. Plants perceive herbivory by both mechanical and chemical means. Mechanical sensing can occur through the perception of insect biting, piercing, or chewing, while chemical signaling occurs through the perception of various herbivore-derived compounds such as oral secretions (OS) or regurgitant, insect excreta (frass), or oviposition fluids. Interestingly, ion channels or transporters are the first responders for the perception of these mechanical and chemical cues. These transmembrane pore proteins can play an important role in plant defense through the induction of early signaling components such as plasma transmembrane potential (V_m) fluctuation, intracellular calcium (Ca^{2+}), and reactive oxygen species (ROS) generation, followed by defense gene expression, and, ultimately, plant defense responses. In recent years, studies on early plant defense signaling in response to herbivory have been gaining momentum with the application of genetically encoded GFP-based sensors for real-time monitoring of early signaling events and genetic tools to manipulate ion channels involved in plant-herbivore interactions. In this review, we provide an update on recent developments and advances on early signaling events in plant-herbivore interactions, with an emphasis on the role of ion channels in early plant defense signaling.

Keywords: reactive oxygen species; herbivory; membrane potential; ion channel

Citation: Gandhi, A.; Kariyat, R.; Harikishore, A.; Ayati, M.; Bhunia, A.; Sahoo, N. Deciphering the Role of Ion Channels in Early Defense Signaling against Herbivorous Insects. *Cells* **2021**, *10*, 2219. https://doi.org/10.3390/cells10092219

Academic Editor: Suleyman Allakhverdiev

Received: 1 July 2021
Accepted: 20 August 2021
Published: 27 August 2021

1. Introduction

Plants regularly encounter a wide range of abiotic and biotic stresses in nature. Abiotic stress includes drought, salinity, extreme temperatures, radiation, floods, and heavy metals, whereas biotic stressors include insect, animal herbivores, and microbial pathogens. Plant and insect-herbivore interactions are among the most significant species interactions found in nature [1,2], and it is estimated that, annually, herbivory causes a 20% loss in the total productivity of agricultural crops [3]. However, plants are not totally defenseless against herbivory and are able to perceive and respond to this onslaught. They can perceive the insect attack through both mechanical and chemical cues. Mechanical signals are elicited through the damage caused by herbivores by piercing, chewing, or biting of plant tissues, and chemical signals are relayed via herbivore-associated elicitors (HAEs) such as oral secretions (OS) or regurgitant, insect excreta (frass), or oviposition fluids, to name a few [4,5]. Plants not only actively respond to herbivory, but also initiate a series of biochemical responses following the perception of herbivory. These biochemical cascades are initiated through ion channels that control the changes in the plasma membrane potential (V_m), generation of reactive oxygen species (ROS), cytosolic calcium fluxes, and ultimately induce plant defense genes to mount a multi-layered defense response that can act at both local and systemic levels [4,6–10]. In recent years, there have been several reviews on plant-herbivore interactions [4,5,7,11–26]. Here we complement these existing reviews with current research and recent discoveries on plant-herbivore interactions, focusing on early plant defense signaling, with a particular emphasis on ion channels involved in early plant defense signaling.

2. Long-Distance Communication in Plant Defense

During herbivory, the damaged areas of the plant need to inform the rest of the plant to keep them ready for the imminent herbivory threat. Therefore, plants need to alert their unaffected parts by sending long-distance signals from the site of damage to various parts of the plant to appraise the threat. Plants respond to diverse stimuli by communicating amongst cells from distinct tissues or organs, a process called systemic signaling [27]. Studies have revealed the existence of complex regulatory mechanisms that allow the plant to activate resistance in systemic tissues, commonly referred to as systemic acquired resistance (SAR) [28]. SAR is characterized by a more potent and faster response to future encounters with microbes, insects, or abiotic stress.

Considerable progress has been made in understanding this intricate relationship between plants and herbivores with a plethora of field and lab studies. These include studies that have dissected pairwise interactions between a specific herbivore and its host; interactions at species, genus, and community levels with multiple hosts and herbivores; and studies examining plant defense signaling networks through molecular genetics genomics, to name a few [29,30]. However, our knowledge of how plants perceive these cues and how that leads to specific and tightly regulated defense responses is still in its infancy. It has been proposed that following the insect attack, the foremost event is the recognition of the cue and its perception by specific membrane receptors and the transduction of these signals into the plant cell. These cues are termed as "early defense signaling molecules" such as the depolarization of plasma membrane along with the generation of secondary messengers such as cytosolic Ca^{2+} [31], reactive oxygen species (ROS), and reactive nitrogen species (RNS) [32–35] that contribute to plant defense signal transduction events.

Long-distance communication in plants has been linked with ion channels or membrane transporters. These are transmembrane pore proteins involved in the movement of ions across the cell membrane. In recent years, with electrophysiological tools, the research on ion channels in plants has been gaining momentum. Studies have reported that ion channels facilitate long-distance communication via V_m, Ca^{2+}, and ROS (Figure 1). Ion channels have been shown to mediate systemic signaling by modulating the influx of ions into different plant tissues [36]. They sense signals from the functional cells at the site of herbivory to activate other cells, which in turn relay this signal to induce defense responses. For example, a recent study [37] identified glutamate receptor-like channels (GLRs) in *Arabidopsis thaliana* that are related to mammalian ionotropic glutamate receptors, play a role in Ca^{2+} signaling during herbivory, nutrient transport, root gravitropism, and plant defense [38,39]. However, in mammals, these channels are involved in neurotransmission, and their openings are stimulated by glutamate binding to the postsynaptic neuron, resulting in Ca^{2+} and other cations influx. The signal is transmitted because of voltage changes caused by ion flux [40]. Remarkably, these GLRs are also responsible for long-distance Ca^{2+} transmission in plants in response to herbivory or mechanical injury, efficiently communicating herbivore attacks to surrounding cells.

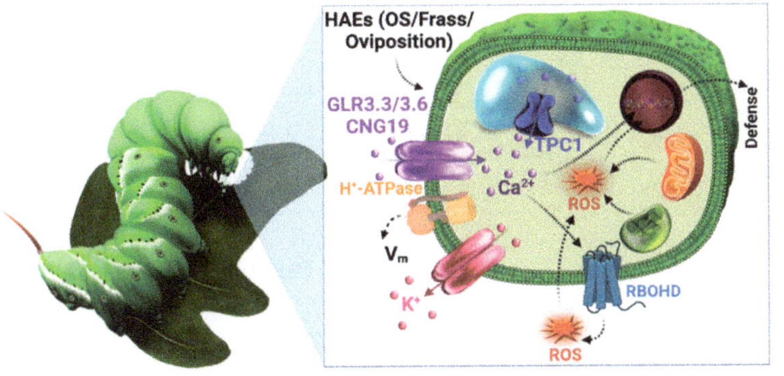

Figure 1. Initiation of early defense signaling mechanisms in response to insect herbivore attack. Schematic diagram showing herbivore *M. sexta* feeding induced signaling events, which include the perception of HAEs such as OS, frass, and oviposition by specialized receptors on the outer plasma membrane, which trigger modulation of V_m via H^+-ATPase and Ca^{2+} ion influx into the cell via Ca^{2+} channels, GLR3.3/3.6 and/or CNGC19. The increase in cytosolic Ca^{2+} may trigger the further release of vacuolar Ca^{2+} via the TPC1 channel. The subsequent release of Ca^{2+} may activate nicotinamide adenine dinucleotide phosphate (NADPH oxidase) and respiratory burst oxidase homologues (RBOHDs), leading to ROS generation, and induction of plant defense responses. Illustration by Annette Diaz.

There has been considerable research on identifying the factors that are involved in long-distance signaling. Plants can appraise their unaffected parts by extensive network of intracellular regulators, V_m, Ca^{2+}, and ROS [18,41]. The transmission rate of all these waves ranges from ~100 to >1000 µm/sec [41,42]. The process starts with the propagation of long-distance electrical signals as a result of variation in membrane potential due to potassium (K^+) and Ca^{2+} flux. Variation in V_m is critical for plant wounding responses [43]. Finally, Ca^{2+} and ROS, versatile secondary messenger, were generated that plants use to sense and transform environmental stimuli into an adaptive intracellular response [44]. Insect feeding and OS can lead to changes to the cytosolic Ca^{2+} concentration, and these spatiotemporal variations have been shown to yield Ca^{2+} signatures [45–48]. On the other hand, ROS are extremely reactive and hazardous chemicals formed from oxygen. Among them are O_2, H_2O_2, and OH^-. ROS which has been demonstrated to act as a self-propagating long-distance and fast wound signal [49]. Throughout this review, we will focus on the role of ion channels, V_m, Ca^{2+}, and ROS in plant response to herbivory and provide an overview of what is currently known about the role of ion channels in plant-herbivore interactions.

3. Membrane Potential (V_m)

The V_m is an electrical potential of the cell membrane that is maintained via the balance of ion fluxes across the plasma membrane. V_m indicates whether a cell is excited or not. It is responsible for generating action potentials in tissues, muscles, and nerves in animals and plays a crucial role in diverse biological functions such as biological sensing, hearing, cell cycle, proliferation, contractility, and circadian rhythm, to name a few [50]. Unlike animals, plants use V_m to regulate plant cellular functions such as maintaining turgor pressure, osmotic balance, and stomatal closure. There is no net flux of ions through the membrane when in equilibrium, called the resting membrane potential. Changes in the resting membrane potential will occur due to an unbalanced movement of ions, thus leading to V_m being more positive (depolarization) or more negative (hyperpolarization). In general, plants maintain a negative resting membrane potential in the order of −110 to −150 mV [51,52]. It has been reported that the signal transduction mechanism of plants to respond to minor changes in V_m leads to plant defense responses. The way plants sense insect cues and initiate defense responses has been a point of interest for many years. One hypothesis that has evolved by studying cellular responses following herbivory suggests that the first event following herbivory generates the fluctuation in V_m [53]. Maffei et al. [43] has also demonstrated that both mechanical wounding and OS of cotton leafworm (*Spodoptera littoralis*) alter V_m in lima bean (*Phaseolus lunatus* L.) at increasing distances of 5, 30, and 60 mm from the bite zone. V_m depolarization was observed within the first 15 min of feeding by *S. littoralis* in the palisade cells. The effect of *S. littoralis* regurgitant and its components were also tested on V_m in *P. lunatus* leaf and the results showed that V_m alterations were independent of regurgitate concentration. In addition, they also examined changes in V_m in response to the application of various H_2O_2 concentrations to mechanically damaged and herbivore-wounded *P. lunatus* leaves. H_2O_2 treatment induced a robust V_m that was significantly greater in herbivory-wounded plants than in mechanically injured leaves [54].

Bricchi et al. [55] studied V_m alterations in wild-type and plasmodesmata mutated *A. thaliana pdko3* lines; plasmodesmata are channels within the plant cell that allow chemicals to pass through, establishing a pathway for cell-to-cell communication. A strong V_m depolarization occurred in wild-type *A. thaliana* plants within 7 to 8 min after herbivory, but the *pdko3* mutant did not exhibit V_m depolarization in response to herbivory or application of OS from *S. littoralis*. However, Ca^{2+} elevation was observed in both wild types as well as in *pdko3* mutant. This observation ruled out the possibility of Ca^{2+} channels being involved in V_m depolarization. To dissect the dependence of V_m depolarization on potassium (K^+) channels, the K^+ channel activity was measured using fluorescent indicator FluxOR^{TM}. A significant increase in K^+ channel activity was observed in wild-type plants, whereas a complete loss of K^+ channel activity was observed in *pdko3* plants. This finding also suggests that K^+ channels are involved in V_m depolarization and supports the hypothesis that plant cells respond to OS by a V_m-mediated signal transduction pathway.

The fluctuation in V_m has been known to be induced by the binding of specific components from herbivore OS with the receptors present at the plasma membrane [56]. These components can alter ion channel activities, causing an imbalance in ion movement, which influences the membrane potential of the plasma membrane [43]. A study by Mohanta et al. [57] showed that Kew tree (*Ginkgo biloba*), a living fossil plant, responds to *S. littoralis* herbivory by inducing V_m depolarization, which was evident up to 6 h. Another study using *A. thaliana* also showed that the extent of V_m depolarization was the same for *S. littoralis*, green peach aphid (*Myzus persicae*), and the plant pathogenic bacteria *Pseudomonas syringe*, but the timing of the occurrence of V_m depolarization was different for each of these biotrophs. Moreover, the magnitude of early defense response depends upon the amount of tissue damage by the biotroph. V_m depolarization was rapid upon the attack of chewing herbivore, *S. littoralis* (30 min to 2 h), as it caused substantial tissue loss, since it consumed large amounts of leaf tissue. On the other hand, less damage was observed by a phloem feeder, *M. persicae* (4 to 6 h), that delayed the plant defense response since phloem feeders with sucking mouthparts feed on vascular tissues without visible tissue damage as observed with chewing herbivores [58]. It is apparent that Ca^{2+} and ROS generation are directly tied to V_m when herbivores interact with plants, and V_m is essential for plant defense responses.

4. Calcium (Ca^{2+})

Ca^{2+} is a ubiquitous signaling molecule in plants. It functions as a secondary messenger in cellular pathways that regulate plant growth and development, cell polarity, cytoskeleton organization, ion transport across membranes, stomatal regulation, root growth, fertilization, nutrient signaling, and plant immunity [59]. Consequently, each of these processes has its own "Ca^{2+} signature," linked with distinct fluctuations in Ca^{2+} concentration in the cytosol and sometimes in a particular intracellular compartment. Therefore, Ca^{2+} fluxes, especially oscillations between calcium stores and the cytosol, are important for cell signaling [60–62].

In plants, the cytosolic Ca^{2+} concentration is maintained at or below 100 nM; however, the majority of Ca^{2+} is stored in the apoplast, vacuole, endoplasmic reticulum (ER), and Golgi apparatus. The apoplast serves as the first Ca^{2+} reservoir of a cell that can store 0.33 mM free resting Ca^{2+} and the first area that responds to stimuli, while the vacuole serves as the largest Ca^{2+} pool of a cell that can store up to 0.2–5 mM free resting Ca^{2+} [60,63].

The Ca^{2+} signature plays an important role in long-distance signal transduction during herbivore attack through which HAEs such as OS, oviposition, and frass is sensed by the cell membrane, and then, a Ca^{2+} is rapidly propagated in the cytosol and travels throughout the plant to induce defense responses. The shaping of this "Ca^{2+} signature" during plant-herbivore interactions is achieved through the amplification and integration of Ca^{2+} signals. The amplification step is mediated via specific ion channels or transporter proteins and enhances Ca^{2+} fluxes at sites of herbivore attack, whereas the integration step is mediated via Ca^{2+} sensor proteins, which allow efficient transmission of Ca^{2+} signals from one cell to another in a tissue or organ. Herbivory induces Ca^{2+} entry from the apoplast to the cytosol via plasma membrane Ca^{2+} channels which stimulates Ca^{2+} signals in the cytosol leading to the amplification of Ca^{2+} signals. The localized Ca^{2+} signals from the cytosol are distributed throughout the whole plant. In this way, amplification, and integration of Ca^{2+} signals constitute two important ways by which "Ca^{2+} signature" contributes as a signaling molecule during plant-herbivore interactions [64].

The amplification of intracellular Ca^{2+} signal requires selective Ca^{2+} sensor proteins that respond to changes in cytosolic Ca^{2+} levels and encipher the frequency, amplitude, and signal localization of Ca^{2+} signatures. It is estimated that *A. thaliana* contains around 250 Ca^{2+} sensor proteins [65]. These can be classified into three main categories: (1) the calcineurin B-like proteins (CBLs) [66]; (2) the calmodulin (CaM), and calmodulin-like proteins (CMLs) [67]; and (3) the Ca^{2+} dependent protein kinases (CPKs) and the Ca^{2+} and calmodulin-dependent protein kinase (CCPK) [68]. All of these sensors contain EF-hand motifs, which enable Ca^{2+} binding and cause conformational changes in their structure [69].

CaM functions as a sensor relay protein since it lacks an enzymatic function. The *Arabidopsis* genome has seven calmodulin genes encoding four different isoforms (CaM1/4; CaM2/3/5; CaM6; and CaM7) [70]. CaM/CaM-like proteins (CML) regulate a variety of transcription factors, protein kinases, phosphatases, metabolic enzymes, ion pumps, and ion exchangers [71]. *A. thaliana* signal responsive (AtSR1) proteins [67], also known as CaM-binding transcription activators (AtCAMTAs) [72], have been shown to participate in wound-mediated defense responses. *Atsr1* mutants of *A. thaliana* were sensitive to attack by dark winged fungus gnats (*Bradysia impatiens*), suggesting the role of CaM as an important sensor in the early stages of the insect-plant attack [73]. Along with CaM, the plant has CML that undergo secondary structural changes in response to Ca^{2+} binding and act as Ca^{2+} relays/sensors [74]. CML and CAM share a 16% amino acid sequence similarity and include two to six EF-hand motif [70]. CML42 gene expression was shown to be increased in *A. thaliana* upon *S. littoralis* OS treatment, implying a function in early defense plant signaling [75]. CPKs have been classified as sensor responders because they combine a Ca^{2+} binding domain and a serine/threonine kinase domain into a single protein that performs the fundamental function of converting Ca^{2+} signals to phosphorylation events [76,77]. *A. thaliana* contains 34 CPK family genes that play a role in plant defense responses. CPK 3 and CPK 13 both participate in signaling after Ca^{2+} influx upon *S. littoralis* attack through regulation of plant defensin gene (PDF1.2) by phosphorylation of the transcription factor, HsfB2a [78]. The *cpk3* and *cpk13* mutants had much lower transcript levels of the plant defensin gene PDF1.2 in comparison to wild-type plants.

Tools Used to Monitor Ca^{2+} Signaling in Plant-Herbivore Interactions

In recent years, the research on Ca^{2+} signaling has gained momentum with the advance in Ca^{2+} imaging techniques. Therefore, it is important to discuss different plant Ca^{2+} imaging methods, which are widely used in the context of plant-herbivore interactions to observe and record cytosolic Ca^{2+} concentration in herbivore-infested plants. These techniques include the use of Ca^{2+} sensing fluorescent dyes and genetically encoded Ca^{2+} indicators. Various fluorescent Ca^{2+} sensing dyes, such as Fluo-3, Calcium Orange, etc., have been used to investigate the dynamics of cytosolic Ca^{2+} signals in plant- herbivore interaction [33,43,55,57,58,79–81]. For example, the Ca^{2+} indicator Ca^{2+} orange was utilized to identify changes in cytosolic Ca^{2+} concentrations in *P. lunatus* following *S. littoralis* herbivory. The changes in Ca^{2+} concentration were compared in response to a single wounding (MD) event, continual mechanical damage caused by a robotic worm (MecWorm, MW), and herbivory. After 30 min, a considerable increase in Ca^{2+} fluorescence was observed due to herbivory in the wounding zone, which persisted for 4 h, but in MD and MW plants, just a faint fluorescence was noticed [33]. Even though these dye-based markers have been demonstrated to be quite effective, these Ca^{2+} sensing dyes have some limitations, including toxicity, fragility, low fluorescence signals, and they cannot be imaged in living plants without permeabilization. To overcome these limitations, researchers have initiated research on the use of genetically encoded Ca^{2+} indicators. The most widely used Ca^{2+} imaging method includes genetically encoded Ca^{2+} indicators, such as GCaMP, Yellow Cameleon (YC) Ca^{2+}-sensors. The Ca^{2+} sensors were developed from GFP by combining them with calmodulin. These Ca^{2+} sensors can be expressed in the whole plant and are functional throughout the entire plant. Therefore, it can be used to monitor cytosolic Ca^{2+} in plants subjected to various herbivore attack conditions [37,42,82–84]. For example, Toyota et al. [37] showed that the *P. rapae* caterpillars induced cytosolic Ca^{2+} responses in the leaves of *A. thaliana* can be monitored with GCaMP3. This study reported that the increases in cytosolic Ca^{2+} concentration were associated with ion influx through plasma membrane Ca^{2+} channels such as GLR3.3/GLR3.6. Another example is Verrillo et al. [83], who showed that Ca^{2+} induction could be monitored with YC3.60, a YC-based Ca^{2+} sensor, following application of *S. littoralis* OS on mechanically damaged *A. thaliana* leaves. By using these tools, it is now possible to study the dynamics of Ca^{2+} signaling in plant-herbivore interactions at single-leaf, whole-plant, and whole-plant-insect herbivore attack conditions.

Intracellular Ca^{2+} level is controlled by the influx of Ca^{2+} ions from extracellular through apoplastic and vacuolar membranes. Therefore, plant ion channels play an important role in regulating plant development and the perception of many stimuli, including herbivory.

5. Plant Ion Channels

Ion channels are macromolecular pores in the membrane that regulate the influx and efflux of ions across the membrane at a rate of 10^6 ions per second. Ion channels can control ion fluxes in their target compartment and, thus, modify cellular homeostasis, and are vital in osmoregulation, development, signaling, mobility, and uptake of nutrients by the root and long-distance communication [85,86]. The first plant ion channel discovered, in 1984, is a K^+ channel, Stelar K^+ outward rectifier (SKOR) [87]. The last two to three decades have seen a dramatic increase in the number of ion channel subfamilies and their diverse functions. A large proportion of plant ion channel families have an analogous expression in animals. Ion channels are arranged into large families and are generally classified as cation, anion, or ligand-gated channels. Cation channels include voltage-gated K^+ channels such as the shaker family (AKT1, AKT2, AKT6, KAT1, KAT2, KAT3, GORK, and SKOR; K^+ transport), tandem pore, and two-pore K^+ channels (TPK1, TPK4; K^+ transport and TPC1; Ca^{2+} and other cation transport), are responsible for permeation of K^+ ion across the plasma membrane and tonoplast membrane. Anion channels include slowly activating anion channels (SLAC1, SLAH1, SLAH2, SLAH3; Cl^-/NO_3^- transport), aluminum-activated malate transporters (ALMT1, ALMT6, ALMT9, ALMT12; Malate, Cl^- transport), chloride channels/transporters (CLCc, CLCg, CLCe; Cl^- transport), and detoxification efflux carrier (DTX33, DTX35; Cl^- transport). Ligand-gated channels include cyclic nucleotide-gated channel (CNGC2, CNGC4, CNGC5, CNGC14, CNGC15, CNGC18, CNGC19, CNGC20; Ca^{2+}/Ba^{2+} transport) and glutamate receptor-like channels (GLR3.1,

GLR3.3, GLR3.4, GLR3.5, GLR3.6; Ca^{2+} and other cations transport) [88] (Figure 2). These channels are responsible for setting up membrane potential, signal transduction, water, and solute transport [89], stomatal opening and closure [90,91], pollination [92], salt tolerance [93], and plant defense [94], to name a few. However, four distinct families of Ca^{2+}-transporting ion channels have been shown to play a role in plant-herbivore interactions, including cyclic nucleotide-gated channels (CNGC19) [95,96], glutamate receptor-like channels (GLR3.3, GLR3.6) [37,42,97], two-pore channel 1 (TPC1) [59,84,98], and annexins (ANNEXIN 1) [99,100].

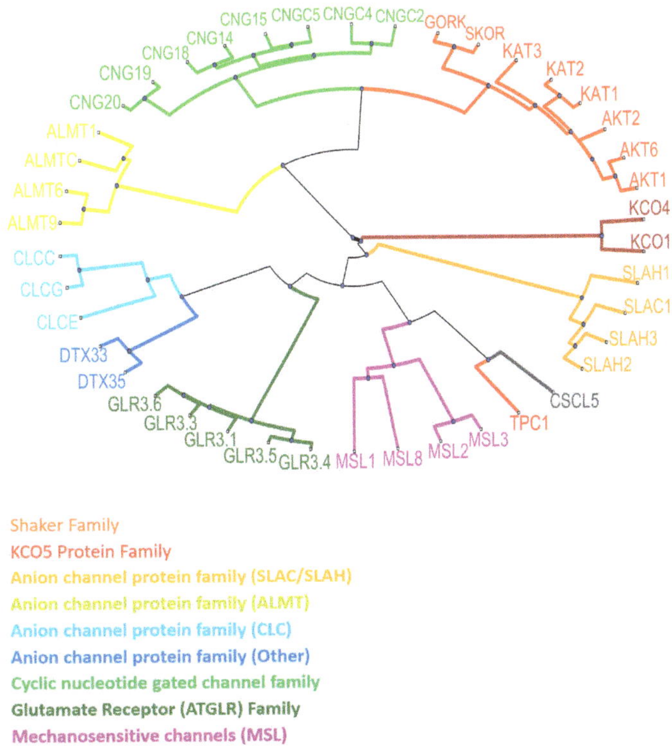

Figure 2. Phylogeny of plant ion channels. Representation of the phylogenetic tree of plant ion channels listed in Pantoja, 2020 [88], based on the analysis of protein homologs extracted from Uniprot.org. Progressive alignment and BLOSUM30 scoring method were used for multiple sequence alignment. The distance between the aligned sequences was calculated using Jukes-Cantor method. The phylogenetic tree was created by using the distance matrix. Unweighted pair group method average (UPGMA) was used to calculate group distance in the tree. Different colors represent different families of ion channels.

5.1. Cyclic Nucleotide Gated Channels (CNGC)

The cyclic nucleotide-gated channels (CNGCs) are ligand-gated Ca^{2+} channels, first discovered in retinal photoreceptors and olfactory neurons [101]. They play a role in signal transduction in animals and are also present in other non-neuronal tissues [102]. These ion transport proteins have also been identified in plants [74,103,104] and have been known to be involved in a variety of biological processes, ranging from plant development and stress tolerance, disease resistance [105,106], thermal tolerance [107], and salt stress [108]. These channels are typically localized at the plasma membrane and in the model plant *A. thaliana*, which consists of 20 family members [109].

CNGC channel is composed of four subunits, and each of these subunits consists of six membrane-spanning regions and a pore domain [110]. There is a cyclic-nucleotide binding (CNB) and a calmodulin-binding domain (CaMB) present at the C-termini of the channel (Figure 3) [111]. In contrast, the animal system has a CaMB domain at the *N*-termini [112,113]. The plant and the animal CNGC differ in their pore amino acid sequence as well as the selectivity for various cations [105,114]. The amino acids that form the CaM binding domain overlap with the polypeptide region that includes the CNBD [115]. This overlapping affects the channel activation as the binding of CaM at the C termini hinders cyclic nucleotide-binding, suggesting variability in plant and animal CNGC channel regulation [116,117]. These channels are activated by the binding of cyclic nucleotides such as cAMP (cyclic adenosine monophosphate) and cGMP (cyclic guanosine monophosphate) [118–120], and inhibited by calmodulin binding [121]. These channels also show similarity with shaker-like K^+ channels [105]. Patch-clamp recordings on plant cell protoplasts membrane directly show that CNGC activation can be achieved by the application of hyperpolarizing potentials (more negative than −120 MV), which allow Ca^{2+} entry into the cell [111,121].

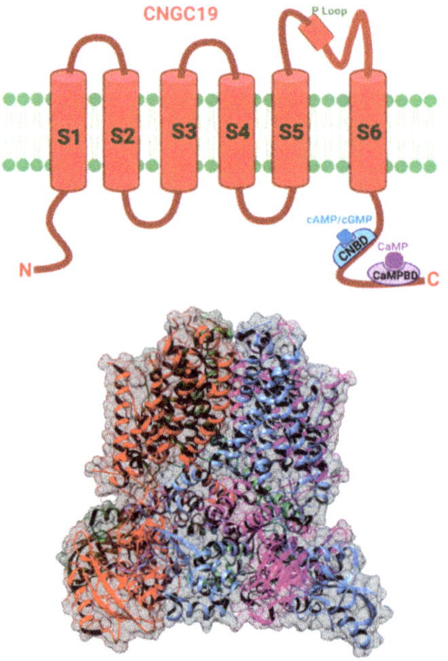

Figure 3. Putative structure of CNGC19 channel. (Top) Schematic cartoon representation of CNGC19 channel subunit showing six membrane-spanning regions (S1–S6) and a large pore domain (S5–S6). Functionally relevant sites in the *C*-terminus consist of a CNB, cyclic nucleotide-binding domain which can bind cAMP/cGMP, and a CaMBD, calmodulin-binding domain which can bind calmodulin. The functional channel is formed by four subunits. (Bottom) The structure of CNGC19 has not been solved to date but is likely to show similarities with the animal CNG family of channels. Therefore, the structure shown in the figure is an approximation based on homology to other channels. The predicted CNGC19 secondary 3D structure model, showing four subunits in transparent surface view, was developed from the closest homolog PDB structure, 5VA1 (human ether-a-go-go related K^+ channel) using PHYRE 2.0 program. The image was prepared using Chimera software [122]. Created with BioRender.com (accessed on 30 August 2021).

It has been demonstrated that CNGC channels are important in modulating biotic stress responses such as Ca^{2+} influx in plant responses mediated by insect herbivore feeding [95]. A recent study by Meena et al. [96] has shown that the *A. thaliana* CNGC19 is responsible for generating and transmitting Ca^{2+} signals in local and systemic leaves mediated by the herbivore *S. litura*. A loss-of-function CNGC19 mutant in which the Ca^{2+} signals were attenuated was found to be more susceptible to attack by *S. litura*. In addition, jasmonic acid, a key signaling molecule in plant defense, was also observed in lower amounts in the CNGC19 mutant. These results suggest that CNGCs are involved in modulating plant resistance to insect herbivores, thus playing a role in the modulation of plant-herbivore interactions.

5.2. Glutamate Receptor-Like Channels

Glutamate receptor-like (GLR) is a non-selective ion channel responsible for permeating Ca^{2+} ions across the plasma membrane of animals and plants. Plant glutamate receptor-like (GLR) channels are ionotropic glutamate receptor homologs in mammals (iGluRs). The iGluRs have been extensively studied for their central nervous system and have been known to play a vital role in synaptic transmission, learning, and memory [123,124]. It is intriguing that GLRs also exist in plants despite the absence of the central nervous system [125]. In plants, GLRs play a crucial role in carbon and nitrogen metabolism [126], gravitropism [127], pollen tube growth [128,129], immune defense reactions [38,130–133], and wound-induced intracellular signaling [97]. *Arabidopsis* consists of 20 GLR genes; each subunit hosts a *N*-terminal domain, two extracellular ligand-binding sites (L1, L2), and transmembrane domains (S1–S4), including a pore region (P) and the *C*-terminal domain [134] (Figure 4). In mammals, iGluRs are divided into three groups according to their sequence diversity and ligand specificities [124]. These include N-methyl-D-aspartate (NMDA), α-amino-3-hydroxy-5-methyl-4-isoxazole propionic acid (AMPA), and Kainate receptors. Plant GluRs share a high degree of similarity with the NMDA receptors that range from 16 to 63% in the ligand-binding domains and the transmembrane domains [135]. These channels are not only present at the plasma membrane but can also be found in chloroplasts, mitochondria [136], and vacuolar membranes [129]. Unlike their mammalian counterparts, the plant GLRs have much broader ligand selectivity. The major difference in plant and animal iGLR is the pore region. These non-selective cation channels are activated by amino acid glutamate, which acts as a metabolite, energy source, and neurotransmitter in animals [137,138].

Electrophysiological studies have shown the involvement of GLRs in inducing a Ca^{2+} influx in plants that leads to the modulation of plant defense signaling to insect herbivores [139,140]. A study by Vasta et al. [140] showed that the application of GLR agonists such as glutamate induced a strong and rapid cytosolic Ca^{2+} increase in tobacco (*Nicotiana tabacum)* var *xanthi* while the application of lanthanum and Ca^{2+} chelator, BAPTA, inhibited glutamate-induced Ca^{2+} increase. This observation suggests that the plant GLR has a role in the modulation of Ca^{2+} influx that ensures plant defense responses against insect herbivores.

GLR3.3 has been implicated in the transmission of signals in the form of Ca^{2+} waves from wounded to unwounded sections of the plant. When *S. littoralis* larvae were allowed to feed on *A. thaliana* wild-type plants, wound-induced surface potential alterations were detected. However, wounding reduced the surface potential alterations in the four GLR mutants GLR3.1, GLR3.2, GLR3.3, and GLR 3.6. [97]. This suggests that GLR3.3 plays an important role in the modulation of plant defense signaling to insect herbivores. Recently, Toyota et al. [37] showed that GLRs are activated by wounding and upon herbivory by cabbage butterfly (*Pieris rapae*) caterpillars in *A. thaliana* leaf expressing genetically encoded Ca^{2+} sensor GCaMP3. The cytosolic Ca^{2+} elevation and subsequent defense gene expression were observed after the application of glutamate and not with other amino acids such as sorbitol. Furthermore, the Ca^{2+} signals were completely abolished in the GLR3.3/GLR3.6 double mutant in *A. thaliana*, suggesting that GLR3.3 and GLR3.6 are essential for transmitting Ca^{2+} signals induced by wounding and herbivory. Another recent study by Shao et al. [42] demonstrated that wounding of the main root at a distance of 2 cm from the root-shoot junction increased the Ca^{2+} elevation and surface wave potential (SWP) in *A. thaliana* expressing calcium sensor GCaMP6. Additionally, the application of glutamate to the wound site in the root induced an increase in Ca^{2+} and SWP in all leaves. Interestingly, in the GLR3.3/GLR3.6 double mutant, this wound and glutamate-induced rise in root to shoot Ca^{2+} was attenuated. This finding suggests that GLR3.3 and GLR3.6 are involved in propagating systemic Ca^{2+} signaling from leaf to leaf and root to shoot. These results provide evidence for the role of plant GLRs in the modulation of Ca^{2+} signaling during plant defense responses against insect herbivores.

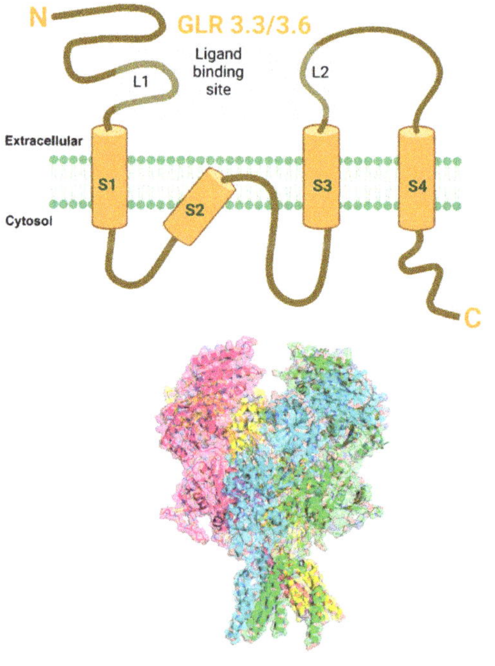

Figure 4. Putative structure of GLR3.3/3.6 channel. (Top) Schematic cartoon representation of GLR3.3/3.6 channel subunit showing extracellular *N*-terminus, four membrane-spanning regions (S1–S4), 2 extracellular ligand-binding sites (L1, L2), and intracellular *C*-terminus. (Bottom) The structure of GLR3.3/3.6 has not been solved to date but is likely to show similarities with the animal NMDA receptor family of channels. Therefore, the structure shown in the figure is an approximation based on homology to other channels. The predicted GLR3.3/3.6 secondary 3D structure model showing four subunits in transparent surface view was developed from closest homolog PDB structure 4TLL (*Xenopus laevis* GluN1/GluN2B NMDA receptor), using PHYRE 2.0 program. The image was prepared using PyMol software (PyMOL Molecular Graphics System, Version 2.4, Schrödinger, LLC, New York, NY, USA). Created with BioRender.com (accessed on 30 August 2021).

5.3. *ANNEXIN1*

Annexins are the phospholipid-binding proteins and are considered novel mechanosensitive Ca^{2+} channels [141,142]. In animal cells, annexins are present in the cytoplasm and cellular membranes [143]. They are involved in vital cellular processes such as membrane trafficking, ion flux, mitotic signaling, and cytoskeleton rearrangement [143,144]. Eight annexin genes have been identified in *A. thaliana* by genome sequencing [145]. Plant annexins are structurally different from their animal homologs but have a conserved primary sequence. These 32–42 kDa proteins have two major domains: a *N*-terminal head and a *C*-terminal annexin core [143] (Figure 5). The annexin core is composed of four annexin domains (I–IV), each of which is 70 amino acids in length and contains five short helices and a conserved endonexin fold (G-X-G-T-{38-40}-D/E). Ca^{2+} binding activity occurs in type II and III binding sites of annexin proteins [141,143]. Plant annexins have a shorter *N*-terminal region than their animal counterparts [146] and are crucial for actin binding, inhibition of callose synthase, and oxidative stress responses [147–150]. The functional diversity of annexins is due to the variable *N*-terminal region that interacts with other proteins.

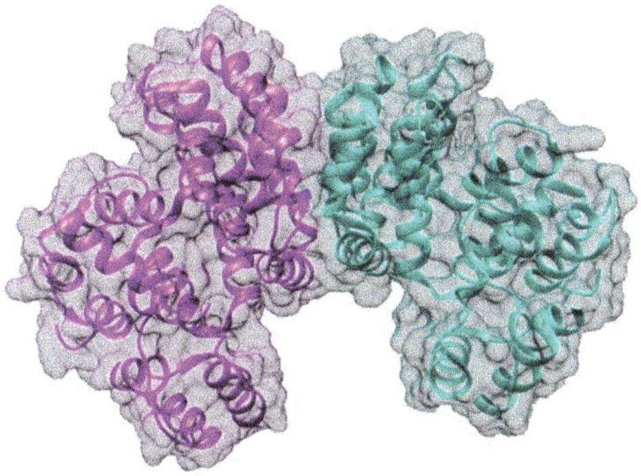

Figure 5. Putative structure of ANNEXIN1 channel. ANNEXIN1 secondary 3D structure model showing two subunits (homodimer) in transparent surface view was developed from PDB structure 1YCN (*Arabidopsis thaliana* ANNEXIN). The presence of Ca^{2+} or H_2O_2 appears to be required for homodimerization. The image was prepared using Chimera software [122].

A recent study by Malabarba et al. [100] reported the role of ANNEXIN1 (ANN-1) in initiating systemic defense in *A. thaliana* in response to Egyptian cotton leafworm (*S. littoralis*) herbivory. The study found that annexin 1 was responsible for inducing cytosolic free Ca^{2+} elevation upon wounding and simulated herbivory in *A. thaliana*. ANN-1 knock-out and ANN-1 overexpressing lines were employed in this work to evaluate their role in herbivory-mediated Ca^{2+} signaling. The result showed that in the ANN-1 deletion line, the increase in cytosolic Ca^{2+} upon herbivory by *S. littoralis* was impaired, and the larvae gained more weight than those fed on wild-type plants. On the other hand, weight increase was significantly lower in larvae that fed on the ANN-1 overexpressed line compared to the wild type. Additionally, jasmonate accumulation and defense responses were diminished in ANN-1 systemic leaves, demonstrating that ANN-1 is involved in systemic cytosolic Ca^{2+}-dependent jasmonate induction. This finding suggests that ANN-1 modulates plant defenses against herbivore damage through the Ca^{2+}-dependent jasmonate signaling pathway and is required for systemic rather than local defense activation in plants attacked by herbivorous insects.

5.4. Two Pore Channel 1 (TPC1)

Two pore channels (TPCs) are organellar cation channels that are widely expressed in animals and plants. In animals, they are localized in the endolysosomal membrane, while in plants they reside in the tonoplast of plant vacuoles [151–154]. They are members of the voltage-gated ion channel superfamily. The vacuolar TPC1 channel, also known as the slowly activating vacuole (SV) channel, has been implicated in a variety of processes in plants, including nutrient sensing, pH homeostasis, and modulation of the membrane potential. The first plant TPC1 gene was cloned in *A. thaliana* (AtTPC1), with 733 amino acids identical to the rat TPC1 sequence [152].

Plant and animal TPCs are similar in sequence to voltage-gated Ca^{2+} and Na^{+} channels and feature two shaker-like units with six transmembrane domains (S1–S6), each joined by a cytosolic linker containing two Ca^{2+}-binding EF-hands (EF1 and EF2). (Figure 6). Voltage and an increase in the cytosolic Ca^{2+} level both influence the activity of plant TPCs. Ca^{2+} binding to the cytosolic EF-hand domain induces conformational changes in the pair of pore-lining inner helices from the first 6-TM domains, whereas membrane potential activates the second voltage-sensing domain, which undergoes conformational changes and facilitates pore opening [155]. The SV channel transports Ca^{2+} in addition to Na^+ and K^+ and has a permeability ratio of 3:1 for Ca^{2+} to K^+ [156,157]. Ca^{2+} release is substantially dependent on the concentration of cytosolic free Ca^{2+}, indicating that this channel is involved in Ca^{2+}-induced Ca^{2+} release [156,158]. The plant TPC1 has been implicated in insect-plant interactions. A study by Kiep et al. [98] has shown that an increase in local cytosolic Ca^{2+} and systemic Ca^{2+} response was induced in response to *S. littoralis* feeding on *A. thaliana*. By using real-time imaging in *A. thaliana* expressing the Ca^{2+} reporter aequorin to monitor the induction of local and systemic cytosolic Ca^{2+} signals, this study showed that simulated herbivory by wounding inhibited the systemic Ca^{2+} signal in the *tpc1* knockout mutant. These results indicated that the TPC1 channel plays a key role in the systemic $[Ca^{2+}]$ cyt signal induced by insect herbivory in *A. thaliana*. Another study by Vincent et al. [84] employed *A. thaliana* plants expressing the GFP-based Ca^{2+} indicator GCaMP3 to visualize Ca^{2+} accumulation in response to aphid *M. persicae* feeding. Within 95 s of the aphids settling, a robust fluorescence burst was seen, indicating cytosolic Ca^{2+} elvation. The rise in cytosolic Ca^{2+} was strongly dependent on Brassinosteroid Insensitive Associated Kinase I (BAK1), the plasma membrane Ca^{2+} permeable ion channels glutamate receptor-like 3.3 and 3.6 (GLR3.3 and GLR3.6), which are critical regulators of extracellular Ca^{2+} import into the cytoplasm of plant cells. In addition, this study also revealed that the increase in cytosolic Ca^{2+} induced TPC1 mediated vacuolar Ca^{2+} release in response to aphid feeding, suggesting that the TPC1 channel operates in conjunction with the plasma membrane Ca^{2+} permeable ion channels GLR3.3 and GLR3.6 in mediating cytosolic Ca^{2+} increase during insect herbivory [84].

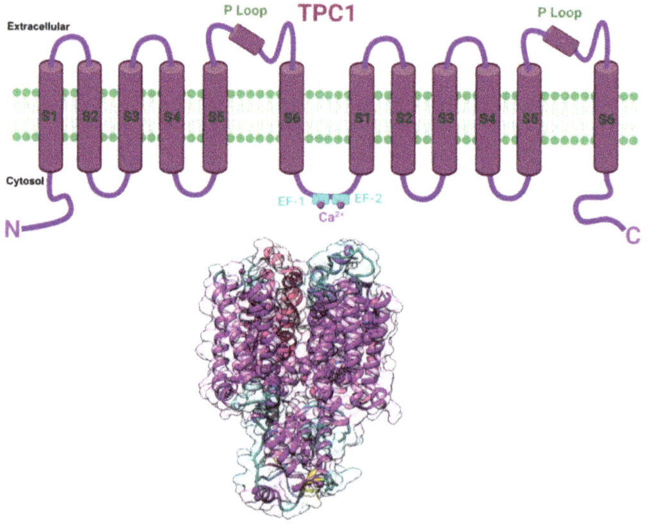

Figure 6. Putative structure of TPC1 channel. (Top) Schematic cartoon representation of individual plant TPC1 channel subunit comprising two repeated domains showing six membrane-spanning regions (S1–S6), two pore loops (P), and joined via a cytosolic linker containing two Ca^{2+} binding EF-hands (EF1 and EF2). (Bottom) TPC1 secondary 3D structure model showing two subunits in transparent surface view was developed from PDB structure 5DQQ (*Arabidopsis thaliana* TPC1). The image was prepared using Chimera software [122]. Created with BioRender.com (accessed on 30 August 2021).

5.5. H^+-ATPase

The proton-pumping ATPases (H^+-ATPases) are the primary pumps responsible for the generation of a proton gradient across cellular membranes. This electrogenic transporter uses energy from ATP hydrolysis to drive the translocation of protons against their concentration gradient from the cytosol to the external aqueous environment [159]. The H^+-ATPase is located in the plasma membrane (PM) of plant cells. It has been demonstrated that the activation and suppression of the H^+-ATPase activity in the plant plasma membrane modulate V_m, resulting in the alteration of PM ion channels and transporters functions [160]. The PM H^+-ATPase is a single 100 kDa polypeptide and a member of the large family of phosphorylation (P)-type ATPases. It is composed of six transmembrane helices (M1–M6) and a cytoplasmic domain containing phosphorylation (P), nucleotide-binding (N), and actuator (A) domains involved in ATP hydrolysis. The PM H^+-ATPase has been implicated in various physiological processes, including cell development, intracellular pH regulation, food uptake, stomatal opening, salt tolerance, and cellular expansion [161–165].

Plant PM H^+-ATPase has been shown to contribute in the propagation of the intracellular defense signaling cascade by modifying V_m in response to herbivore feeding [166]. A study by Camoni et al. [167] demonstrated that *S. littoralis* oral secretions effectively inhibited *Phaseolus lunatus* PM H^+-ATPase, resulting in decreased H^+ extrusion from the cytosol and modification of the V_m. This observation implied that H^+ extrusion by the plant H^+-ATPase was involved in V_m regulation and might initiate a plant defensive response to herbivory. Another recent study by Kumari et al. [168] has revealed that *Arabidopsis* H^+-ATPase 1 (AHA1) is involved in the formation of slow wave potentials (SWPs), which are required for long-distance electrical transmission during herbivore-induced plant defense. Fusicoccin, a PM H^+-ATPase activator, prolonged the SWP repolarization phase in leaves distal to wounds. The repolarization phase was significantly prolonged in reduced function *aha1* mutants, whereas the duration of SWP repolarization was reduced in the presence of a gain-of-function mutant *ost2-2D*. Additionally, *S littoralis* larvae performed better on *aha1-7* mutants than on wild-type plants. Overall, these observations suggest that the PM H^+-ATPase is required for the regulation of the V_m and electrical signal propagation between different parts of a plant during insect herbivory.

6. Reactive Oxygen Species (ROS)

Reactive oxygen species (ROS) are highly reactive molecules generated by redox reactions. They are part of several biological processes, such as photorespiration, oxidative phosphorylation, the electron transport chain (ETC), as well as a plant defense against pathogens and herbivores. ROS is produced in the mitochondria, chloroplast, and peroxisomes. There are several forms of ROS like superoxide anion ($O^{2-\bullet-}$), hydrogen peroxide ($H_2O_2{}^{\bullet}$), hydroxyl radical ($HO^{\bullet}$), peroxynitrite (ONOO), and singlet oxygen ($^1O_2$) [169]. ROS is typically produced by the nicotinamide adenine dinucleotide phosphate (NADPH) oxidase complex, which catalyzes the reduction of molecular oxygen to superoxide anion, which is then converted to H_2O_2. In plants, respiratory burst oxidase homologs (RBOHs) were found to be the key enzymes that catalyze the formation of ROS, which is a key step in plant protection against herbivores [170–172]. The respiratory burst oxidase homolog D (RBOHD) has been found to be essential for the propagation of ROS waves [173]. The significance of RBOHs in organizing responses against chewing insect herbivores was verified in *N. attenuate* where tobacco hornworm (*Manduca sexta*) OS enhanced NaRBOHD (*N. attenuata* NADPH oxidase homolog) on damaged leaves. ROS accumulation was diminished in *M. sexta* OS treated NaRBOHD-silenced *N. attenuata* plants without affecting OS-induced gene expression of defense-related genes [174].

The production of ROS is an inevitable by-product of metabolism in many cell types. Previously, it was assumed that ROS are toxic molecules that cause cellular damage to macromolecules [175]. Still, the role of ROS in plant defense has only recently emerged. It is well established that ROS can act as early defense signaling molecules that promote plant defense responses against a variety of pathogens and herbivores [54,176]. ROS act as secondary messengers that can penetrate up to 8.4 cm/min in *A. thaliana* [177]. Plants use ROS to alert the non-injured tissue about a plant attack by either releasing small quantities, which activates certain defense responses or prevent cell death by limiting the production of ROS [178]. ROS production has also been suggested to be involved in plant-microbe interactions as ROS can activate or repress the expression of defense-related genes [179,180]. The role of ROS in plant resistance to herbivores has been demonstrated in resistant and near-isogenic susceptible wheat after the attack of Russian wheat aphid (*Diuraphis noxia*). A strong burst of H_2O_2, as well as NADPH oxidase activity, was observed in resistant plants 3 h after infestation in comparison to susceptible plants. Treatments of plants with diphenyleneiodonium (DPI), an inhibitor of NADPH oxidase, suppressed the H_2O_2 production. Elevation in H_2O_2 levels (47%) was observed by treating resistant wheat plants with a mixture of glucose and glucose oxidase [181], suggesting that H_2O_2 plays a role in the defense response against *D. noxia* infestation.

Studies have shown that ROS serve as early defense signaling molecules in response to herbivore-induced wounding and secretions such as OS and oviposition. Imbiscuso et al. [182] investigated the effect of brake fern (*Pteris vittata*) response to herbivory by *S. littoralis*. The *P. vittata* plants responded to the attack of *S. littoralis* by activating peroxidases which produced H_2O_2. The concentration of H_2O_2 in leaves was lower in mechanically wounded young leaves than herbivory wounded leaves, suggesting that *P. vittata* can distinguish between mechanical and herbivory wounding by modulating the amount of ROS production. A study by Shinya et al. [183] demonstrated that the application of OS isolated from generalist herbivore, nightfeeding rice armyworm, (*Mythimna loreyi*), caused a strong intracellular ROS generation on rice cells, and a similar effect was obtained upon application of synthetically prepared N-linolenoyl-L-Glu, the most abundant FAC present in OS of *M. loreyi*, indicating that FAC from *M. loreyi* OS promoted ROS production in rice cells.

Recently, our group Gandhi et al. [184] demonstrated that *M. sexta* oral secretions (OS) induced ROS generation in isolated tomato protoplasts. Interestingly, our study showed that the application of tomato plant-fed (PF) *M. sexta* OS enhanced ROS generation while artificial diet-fed (DF) OS could not induce ROS in tomato protoplasts, suggesting that the oral secretions of *M. sexta* play an indispensable role in inducing ROS generation in tomato protoplasts. Our study also showed that the *M. sexta* PF-OS induced ROS increase was diminished in the presence of a Ca^{2+} chelator, BAPTA-AM, suggesting that there is a link between Ca^{2+} and ROS signaling. Several lines of evidence have indicated the existence of a positive feedback mechanism between ROS and Ca^{2+} production. In a heterologous expression system, treatment with ionomycin, an ionophore that leads to Ca^{2+} influx into cells, resulting in activation of RHD2 NADPH oxidase (root hair defective 2 reduced nicotinamide adenine dinucleotide phosphate) in root tips of *A. thaliana* confirming Ca^{2+} triggered RHD2 NADPH oxidase activity. These observations suggest that Ca^{2+} acts upstream of ROS production [185].

Compelling evidence indicates that ROS production by RBOHD is dependent on the Ca^{2+} binding [186,187]. RBOHD carries 2 EF-hands which are known to participate in Ca^{2+} dependent modulation [188]. Abscisic acid (ABA) signaling in guard cells involves both Ca^{2+} and ROS. *A. thaliana* mutants lacking certain NADPH oxidases (AtRBOHD and AtRBOHF) do not close their stomata and produce ROS, Ca^{2+}, and Ca^{2+} channel activation when they are exposed to ABA. Supplementation of H_2O_2 to guard cells rescues the mutant phenotype, implying that Ca^{2+} entry proceeds downstream of ROS generation in ABA signaling [189,190].

In *A. thaliana*, the production of H_2O_2 was observed in leaves 72 h after oviposition by cabbage moth (*Pieris brassicae*) and was recognized by the formation of a reddish-brown precipitate. This result indicates that oviposition can trigger a localized response that resembles the hypersensitive response induced by pathogens [191]. A recent study by Stahl et al. [192] showed that eggs of *P. brassicae* induced generation of H_2O_2, salicylic acid and defense gene expression in *A. thaliana*. This study also revealed phosphatidylcholines (PCs) released from eggs is the key signaling molecule that activates gene expression and triggers various defenses in the plants.

Tools Used to Monitor ROS Signaling in Plant-Herbivore Interactions

While ROS relevance in plant-herbivore interaction is gaining momentum, the detection and characterization of ROS are still a significant bottleneck in this field. The early detection and quantification of ROS can be carried out by either utilizing genetically encoded fluorescent ROS sensors such as redox-sensitive green fluorescence protein (Ro-GFP), or synthetic fluorescent probes, such as 3,3′-diaminobenzidine (DAB) and 2′,7′-dichlorofluorescein diacetate (H_2DCFDA). Genetically encoded ROS sensors "Ro-GFP" can monitor the cellular redox status at a high spatiotemporal resolution [193–199]. A recent study by Hipsch et al. [200] measured the whole plant ROS generation in response to high light, cold, and drought by using a chloroplast-targeted redox-sensitive green fluorescence protein 2 (RoGFP2). This finding suggests that whole-plant redox imaging using genetically encoded ROS sensors can be applied in a wide range of abiotic and biotic stress conditions, including plant-herbivore interaction. Despite the promising findings, the application of genetically encoded ROS sensors in plant-herbivore interactions is still limited due to the laborious and time-consuming method of its application. In contrast, synthetic fluorescent probes such as DAB and H_2DCFDA are easier to use and can measure ROS in real-time with high sensitivity [201]. DAB has been used in many studies on plants as a reliable biomarker for reactive oxygen species (ROS) production [202–204]. However, in recent years, H_2DCFDA has gained attention for its potential to measure the ROS levels in real-time on whole plants and as well as plant protoplasts [184,205,206]. Fichman et al. [205] measured the effect of light stress, injury, and pathogen, *P. syringae* pv. tomato DC 3000 on ROS signaling in H_2DCFDA dye sprayed *A. thaliana* by using whole plant-live imaging. This study suggests that the combination of live-cell imaging and the use of H_2DCFDA enables real-time monitoring of ROS in plants in response to various stress and pathogen treatments. This study also utilized an RBOHD (*rbohD*) knockout, and upon treatment with different stimuli, less ROS generation was observed. In contrast, another cytosolic ascorbate peroxidase 1 (*apx*) knockout produced more local as well as systemic ROS upon wounding or light stress treatments implying that this mutant had less ROS quenching capacity.

7. Conclusions

Recent years have witnessed immense progress in identifying the early defense signaling components in plant defense against herbivores, but studies on the molecular identification and characterization of these components are still a work in progress. However, with the advent of state-of-the-art imaging techniques, physiological and biochemical assays, and genomics may help us to understand the early defense signaling events by coordinating the plasma membrane potential changes, ion channels modulation, intracellular Ca^{2+} and ROS generation, gene expression, and, ultimately, the host plant defense response against herbivores. Transforming plants with biosensors such as GCaMP-Ca^{2+} and Ro-GFP-ROS sensors can help in the early identification of the plant defense responses. HAEs such as OS, frass, and oviposition could be used to develop strategies for early detection of the impending herbivory. So far, only a handful of Ca^{2+} permeable channels have been identified that plays a role in plant-herbivore interactions. Further studies are needed to unravel other ion channels that may be contributing to the modulation of V_m, Ca^{2+}, and ROS, the downstream signaling cascade, and, more importantly, the role of these ion channels in triggering a rapid defense response. A deeper understanding of these early signaling events will eventually help us to minimize herbivory by developing pest management strategies based on plant-herbivore monitoring systems. Such knowledge can be instrumental in the design of plants with improved resistance against herbivores. As such, in the future, it will be important to develop effective small-molecule modulators that can inhibit or enhance the early signaling events in plant-herbivore interactions. Such an approach would not only facilitate research on early plant signaling events but also help in developing novel strategies for the development of herbivore-resistant crops.

Author Contributions: N.S. conceptualized and wrote the manuscript with A.G.; N.S. designed the final figures and edited the final manuscript draft with A.G. and R.K.; A.G., A.H., M.A., A.B. and N.S. contributed to the preparation of figures. A.G., R.K. and N.S. proofread and revised the manuscript. All authors have read and agreed to the published version of the manuscript.

Funding: This research was funded by the College of Sciences, University of Rio Grande Valley startup fund and the University of Texas System Rising STARs Award to N.S. and College of Sciences Seed grant to R.K.

Institutional Review Board Statement: Not applicable.

Informed Consent Statement: Not applicable.

Data Availability Statement: Not applicable.

Conflicts of Interest: The authors declare no conflict of interest.

References

1. Johnson, M.T. Evolutionary ecology of plant defences against herbivores. *Funct. Ecol.* **2011**, *25*, 305–311. [CrossRef]
2. Schäfer, M.; Fischer, C.; Meldau, S.; Seebald, E.; Oelmüller, R.; Baldwin, I.T. Lipase activity in insect oral secretions mediates defense responses in *Arabidopsis*. *Plant Physiol.* **2011**, *156*, 1520–1534. [CrossRef] [PubMed]
3. Van der Meijden, E. *Herbivorous Insects—A Threat for Crop Production;* Springer: Cham, Switzerland, 2014; pp. 103–114.
4. Felton, G.W.; Tumlinson, J.H. Plant–insect dialogs: Complex interactions at the plant–insect interface. *Curr. Opin. Plant Biol.* **2008**, *11*, 457–463. [CrossRef] [PubMed]
5. Wu, J.; Baldwin, I.T. Herbivory-induced signalling in plants: Perception and action. *Plant Cell Environ.* **2009**, *32*, 1161–1174. [CrossRef] [PubMed]
6. Kessler, A. Defensive function of herbivore-induced plant volatile emissions in nature. *Science* **2001**, *291*, 2141–2144. [CrossRef] [PubMed]
7. Mithofer, A.; Boland, W. Recognition of herbivory associated molecular patterns. *Plant Physiol.* **2008**, *146*, 825–831. [CrossRef] [PubMed]
8. Bonaventure, G.; VanDoorn, A.; Baldwin, I.T. Herbivore-associated elicitors: FAC signaling and metabolism. *Trends Plant Sci.* **2011**, *16*, 294–299. [CrossRef] [PubMed]
9. Felton, G.W.; Chung, S.H.; Hernandez, M.G.; Louis, J.; Peiffer, M.; Tian, D. Herbivore oral secretions are the first line of protection against plant-induced defences. *Annu. Plant Rev.* **2014**, *47*, 37–76.
10. Acevedo, F.E.; Rivera-Vega, L.J.; Chung, S.H.; Ray, S.; Felton, G.W. Cues from chewing insects—The intersection of DAMPs, HAMPs, MAMPs and effectors. *Curr. Opin. Plant Biol.* **2015**, *26*, 80–86. [CrossRef]
11. Howe, G.A.; Jander, G. Plant Immunity to Insect Herbivores. *Annu. Rev. Plant Biol.* **2008**, *59*, 41–66. [CrossRef]
12. Walling, L.L. The myriad plant responses to herbivores. *J. Plant Growth Regul.* **2000**, *19*, 195–216. [CrossRef]
13. Gatehouse, J.A. Plant resistance towards insect herbivores: A dynamic interaction. *New Phytol.* **2002**, *156*, 145–169. [CrossRef]
14. Peng, J.Y.; Huang, Y.P. The signaling pathways of plant defense response and their interaction. *Zhi WU sheng Li Yu Fen Zi Sheng Wu Xue Xue Bao = J. Plant Physiol. Mol. Biol.* **2005**, *31*, 347–353.
15. Hogenhout, S.A.; Bos, J.I. Effector proteins that modulate plant–insect interactions. *Curr. Opin. Plant Biol.* **2011**, *14*, 422–428. [CrossRef]
16. Fürstenberg-Hägg, J.; Zagrobelny, M.; Bak, S. Plant defense against insect herbivores. *Int. J. Mol. Sci.* **2013**, *14*, 10242–10297. [CrossRef]
17. Zebelo, S.A.; Maffei, M.E. Role of early signaling events in plant-insect interactions. *J. Exp. Bot.* **2015**, *66*, 435–448. [CrossRef]
18. Gilroy, S.; Białasek, M.; Suzuki, N.; Górecka, M.; Devireddy, A.R.; Karpiński, S.; Mittler, R. ROS, calcium, and electric signals: Key mediators of rapid systemic signaling in plants. *Plant Physiol.* **2016**, *171*, 1606–1615. [CrossRef]
19. Luo, S.; Zhang, X.; Wang, J.; Jiao, C.; Chen, Y.; Shen, Y. Plant ion channels and transporters in herbivory-induced signalling. *Funct. Plant Biol.* **2017**, *45*, 111–131. [CrossRef]
20. Demidchik, V.; Maathuis, F.; Voitsekhovskaja, O. Unravelling the plant signalling machinery: An update on the cellular and genetic basis of plant signal transduction. *Funct. Plant Biol.* **2018**, *45*, 1–8. [CrossRef]
21. Farmer, E.E.; Gao, Y.Q.; Lenzoni, G.; Wolfender, J.L.; Wu, Q. Wound- and mechanostimulated electrical signals control hormone responses. *New Phytol.* **2020**, *227*, 1037–1050. [CrossRef]
22. Tian, W.; Wang, C.; Gao, Q.; Li, L.; Luan, S. Calcium spikes, waves and oscillations in plant development and biotic interactions. *Nat. Plants* **2020**, *6*, 750–759. [CrossRef] [PubMed]
23. Vega-Muñoz, I.; Duran-Flores, D.; Fernández-Fernández, Á.D.; Heyman, J.; Ritter, A.; Stael, S. Breaking bad news: Dynamic molecular mechanisms of wound response in plants. *Front. Plant Sci.* **2020**, *11*, 1959. [CrossRef]

24. Oelmüller, R. Threat at one end of the plant: What travels to inform the other parts? *Int. J. Mol. Sci.* **2021**, *22*, 3152. [CrossRef] [PubMed]
25. Johns, S.; Hagihara, T.; Toyota, M.; Gilroy, S. The fast and the furious: Rapid long-range signaling in plants. *Plant Physiol.* **2021**, *185*, 694–706. [CrossRef] [PubMed]
26. Singh, S.; Kaur, I.; Kariyat, R. The multifunctional roles of polyphenols in plant-herbivore interactions. *Int. J. Mol. Sci.* **2021**, *22*, 1442. [CrossRef] [PubMed]
27. Mittler, R.; Blumwald, E. The roles of ROS and ABA in systemic acquired acclimation. *Plant Cell* **2015**, *27*, 64–70. [CrossRef] [PubMed]
28. Spoel, S.H.; Dong, X. How do plants achieve immunity? Defence without specialized immune cells. *Nat. Rev. Immunol.* **2012**, *12*, 89–100. [CrossRef]
29. Agrawal, A.A. Induced responses to herbivory and increased plant performance. *Science* **1998**, *279*, 1201–1202. [CrossRef]
30. Kariyat, R.R.; Hardison, S.B.; Ryan, A.B.; Stephenson, A.G.; De Moraes, C.M.; Mescher, M.C. Leaf trichomes affect caterpillar feeding in an instar-specific manner. *Commun. Integr. Biol.* **2018**, *11*, 1–6. [CrossRef]
31. Reddy, A.S.N.; Ali, G.S.; Celesnik, H.; Day, I.S. Coping with stresses: Roles of calcium- and calcium/calmodulin-regulated gene expression. *Plant Cell* **2011**, *23*, 2010–2032. [CrossRef]
32. Miller, G.A.D.; Mittler, R.O.N. Could heat shock transcription factors function as hydrogen peroxide sensors in plants? *Ann. Bot.* **2006**, *98*, 279–288. [CrossRef]
33. Bricchi, I.; Leitner, M.; Foti, M.; Mithöfer, A.; Boland, W.; Maffei, M.E. Robotic mechanical wounding (MecWorm) versus herbivore-induced responses: Early signaling and volatile emission in Lima bean (*Phaseolus lunatus* L.). *Planta* **2010**, *232*, 719–729. [CrossRef]
34. Arimura, G.-I.; Ozawa, R.; Maffei, M.E. Recent advances in plant early signaling in response to herbivory. *Int. J. Mol. Sci.* **2011**, *12*, 3723–3739. [CrossRef]
35. Marino, D.; Dunand, C.; Puppo, A.; Pauly, N. A burst of plant NADPH oxidases. *Trends Plant Sci.* **2012**, *17*, 9–15. [CrossRef]
36. Danilova, M.N.; Kudryakova, N.V.; Andreeva, A.A.; Doroshenko, A.S.; Pojidaeva, E.S.; Kusnetsov, V.V. Differential impact of heat stress on the expression of chloroplast-encoded genes. *Plant Physiol. Biochem.* **2018**, *129*, 90–100. [CrossRef]
37. Toyota, M.; Spencer, D.; Sawai-Toyota, S.; Jiaqi, W.; Zhang, T.; Koo, A.J.; Howe, G.A.; Gilroy, S. Glutamate triggers long-distance, calcium-based plant defense signaling. *Science* **2018**, *361*, 1112–1115. [CrossRef]
38. Manzoor, H.; Kelloniemi, J.; Chiltz, A.; Wendehenne, D.; Pugin, A.; Poinssot, B.; Garcia-Brugger, A. Involvement of the glutamate receptor AtGLR3.3 in plant defense signaling and resistance to *Hyaloperonospora arabidopsis*. *Plant J.* **2013**, *76*, 466–480. [CrossRef]
39. Salvador-Recatalà, V. New roles for the glutamate receptor-like 3.3, 3.5, and 3.6 genes as on/off switches of wound-induced systemic electrical signals. *Plant Signal. Behav.* **2016**, *11*, e1161879. [CrossRef]
40. Muday, G.K.; Brown-Harding, H. Nervous system-like signaling in plant defense. *Science* **2018**, *361*, 1068–1069. [CrossRef]
41. Choi, W.-G.; Hilleary, R.; Swanson, S.J.; Kim, S.-H.; Gilroy, S. Rapid, long-distance electrical and calcium signaling in plants. *Annu. Rev. Plant Biol.* **2016**, *67*, 287–307. [CrossRef] [PubMed]
42. Shao, Q.; Gao, Q.; Lhamo, D.; Zhang, H.; Luan, S. Two glutamate- and pH-regulated Ca^{2+} channels are required for systemic wound signaling in *Arabidopsis*. *Sci. Signal.* **2020**, *13*, 1453. [CrossRef] [PubMed]
43. Maffei, M.; Bossi, S.; Spiteller, D.; Mithöfer, A.; Boland, W. Effects of feeding *Spodoptera littoralis* on lima bean leaves. I. membrane potentials, intracellular calcium variations, oral secretions, and regurgitate components. *Plant Physiol.* **2004**, *134*, 1752–1762. [CrossRef] [PubMed]
44. Jammes, F.; Hu, H.-C.; Villiers, F.; Bouten, R.; Kwak, J.M. Calcium-permeable channels in plant cells. *FEBS J.* **2011**, *278*, 4262–4276. [CrossRef] [PubMed]
45. McAinsh, M.R.; Hetherington, A.M. Encoding specificity in Ca^{2+} signalling systems. *Trends Plant Sci.* **1998**, *3*, 32–36. [CrossRef]
46. Reddy, A.S.N. Calcium: Silver bullet in signaling. *Plant Sci.* **2001**, *160*, 381–404. [CrossRef]
47. Moore, C.A.; Bowen, H.C.; Scrase-Field, S.; Knight, M.R.; White, P.J. The deposition of suberin lamellae determines the magnitude of cytosolic Ca^{2+} elevations in root endodermal cells subjected to cooling. *Plant J.* **2002**, *30*, 457–465. [CrossRef]
48. Hetherington, A.M.; Brownlee, C. The generation of signals in plants. *Annu. Rev. Plant Biol.* **2004**, *55*, 401–427. [CrossRef]
49. Miller, G.; Schlauch, K.; Tam, R.; Cortes, D.; Torres, M.A.; Shulaev, V.; Dangl, J.L.; Mittler, R. The plant NADPH oxidase RBOHD mediates rapid systemic signaling in response to diverse stimuli. *Sci. Signal.* **2009**, *2*, ra45. [CrossRef]
50. Abdul Kadir, L.; Stacey, M.; Barrett-Jolley, R. Emerging roles of the membrane potential: Action beyond the action potential. *Front. Physiol.* **2018**, *9*, 1661. [CrossRef]
51. Thiel, G.; MacRobbie, E.A.C.; Blatt, M.R. Membrane transport in stomatal guard cells: The importance of voltage control. *J. Membr. Biol.* **1992**, *126*, 1–18. [CrossRef]
52. Roelfsema, M.R.; Steinmeyer, R.; Staal, M.; Hedrich, R. Single guard cell recordings in intact plants: Light-induced hyperpolarization of the plasma membrane. *Plant J.* **2001**, *26*, 1–13. [CrossRef]
53. Maffei, M.E.; Arimura, G.-I.; Mithöfer, A. Natural elicitors, effectors and modulators of plant responses. *Nat. Prod. Rep.* **2012**, *29*, 1288. [CrossRef]
54. Maffei, M.E.; Mithöfer, A.; Arimura, G.-I.; Uchtenhagen, H.; Bossi, S.; Bertea, C.M.; Cucuzza, L.S.; Novero, M.; Volpe, V.; Quadro, S.; et al. Effects of feeding *Spodoptera littoralis* on lima bean leaves. III. Membrane depolarization and involvement of hydrogen peroxide. *Plant Physiol.* **2006**, *140*, 1022–1035. [CrossRef]

55. Bricchi, I.; Occhipinti, A.; Bertea, C.M.; Zebelo, S.A.; Brillada, C.; Verrillo, F.; De Castro, C.; Molinaro, A.; Faulkner, C.; Maule, A.J.; et al. Separation of early and late responses to herbivory in *Arabidopsis* by changing plasmodesmal function. *Plant J.* **2012**, *73*, 14–25. [CrossRef]
56. Zebelo, S.A.; Maffei, M.E. *Signal Transduction in Plant–Insect Interactions: From Membrane Potential Variations to Metabolomics*; Springer: Berlin, Germany, 2012; pp. 143–172.
57. Mohanta, T.K.; Occhipinti, A.; Atsbaha Zebelo, S.; Foti, M.; Fliegmann, J.; Bossi, S.; Maffei, M.E.; Bertea, C.M. *Ginkgo biloba* responds to herbivory by activating early signaling and direct defenses. *PLoS ONE* **2012**, *7*, e32822. [CrossRef]
58. Bricchi, I.; Bertea, C.M.; Occhipinti, A.; Paponov, I.A.; Maffei, M.E. Dynamics of membrane potential variation and gene expression induced by *Spodoptera littoralis*, *Myzus persicae*, and *Pseudomonas syringae* in *Arabidopsis*. *PLoS ONE* **2012**, *7*, e46673. [CrossRef]
59. Edel, K.H.; Marchadier, E.; Brownlee, C.; Kudla, J.; Hetherington, A.M. The evolution of calcium-based signalling in plants. *Curr. Biol.* **2017**, *27*, R667–R679. [CrossRef]
60. Costa, A.; Navazio, L.; Szabo, I. The contribution of organelles to plant intracellular calcium signalling. *J. Exp. Bot.* **2018**, *69*, 4175–4193. [CrossRef]
61. Monshausen, G.B. Visualizing Ca^{2+} signatures in plants. *Curr. Opin. Plant Biol.* **2012**, *15*, 677–682. [CrossRef]
62. Whalley, H.J.; Knight, M.R. Calcium signatures are decoded by plants to give specific gene responses. *New Phytol.* **2012**, *197*, 690–693. [CrossRef]
63. Stael, S.; Wurzinger, B.; Mair, A.; Mehlmer, N.; Vothknecht, U.C.; Teige, M. Plant organellar calcium signalling: An emerging field. *J. Exp. Bot.* **2011**, *63*, 1525–1542. [CrossRef]
64. Tuteja, N. *Integrated Calcium Signaling in Plants*; Springer: Berlin, Germany, 2009; pp. 29–49.
65. Day, I.S.; Reddy, V.S.; Shad Ali, G.; Reddy, A.S.N. Analysis of EF-hand-containing proteins in *Arabidopsis*. *Genome Biol.* **2002**, *3*, 1–24. [CrossRef]
66. Luan, S. The CBL–CIPK network in plant calcium signaling. *Trends Plant Sci.* **2009**, *14*, 37–42. [CrossRef]
67. Yang, T.; Poovaiah, B.W. A calmodulin-binding/CGCG Box DNA-binding protein family involved in multiple signaling pathways in plants. *J. Biol. Chem.* **2002**, *277*, 45049–45058. [CrossRef] [PubMed]
68. Cheng, S.-H.; Willmann, M.R.; Chen, H.-C.; Sheen, J. Calcium signaling through protein kinases. The *Arabidopsis* Calcium-dependent protein kinase gene family. *Plant Physiol.* **2002**, *129*, 469–485. [CrossRef] [PubMed]
69. Batistič, O.; Kudla, J. Analysis of calcium signaling pathways in plants. *Biochim. Biophys. Acta (BBA)—Gen. Subj.* **2012**, *1820*, 1283–1293. [CrossRef] [PubMed]
70. McCormack, E.; Braam, J. Calmodulins and related potential calcium sensors of *Arabidopsis*. *New Phytol.* **2003**, *159*, 585–598. [CrossRef] [PubMed]
71. Bouché, N.; Yellin, A.; Snedden, W.A.; Fromm, H. Plant-specific calmodulin-binding proteins. *Annu. Rev. Plant Biol.* **2005**, *56*, 435–466. [CrossRef]
72. Bouché, N.; Scharlat, A.; Snedden, W.; Bouchez, D.; Fromm, H. A novel family of calmodulin-binding transcription activators in multicellular organisms. *J. Biol. Chem.* **2002**, *277*, 21851–21861. [CrossRef]
73. Qiu, Y.; Xi, J.; Du, L.; Suttle, J.C.; Poovaiah, B.W. Coupling calcium/calmodulin-mediated signaling and herbivore-induced plant response through calmodulin-binding transcription factor AtSR1/CAMTA3. *Plant Mol. Biol.* **2012**, *79*, 89–99. [CrossRef]
74. Kohler, C.; Merkle, T.; Neuhaus, G. Characterisation of a novel gene family of putative cyclic nucleotide- and calmodulin-regulated ion channels in *Arabidopsis thaliana*. *Plant J.* **1999**, *18*, 97–104. [CrossRef]
75. Vadassery, J.; Reichelt, M.; Hause, B.; Gershenzon, J.; Boland, W.; Mithöfer, A. CML42-mediated calcium signaling coordinates responses to *Spodoptera* herbivory and abiotic stresses in *Arabidopsis*. *Plant Physiol.* **2012**, *159*, 1159–1175. [CrossRef]
76. Tena, G.; Boudsocq, M.; Sheen, J. Protein kinase signaling networks in plant innate immunity. *Curr. Opin. Plant Biol.* **2011**, *14*, 519–529. [CrossRef]
77. Boudsocq, M.; Sheen, J. CDPKs in immune and stress signaling. *Trends Plant Sci.* **2013**, *18*, 30–40. [CrossRef]
78. Kanchiswamy, C.; Takahashi, H.; Quadro, S.; Maffei, M.E.; Bossi, S.; Bertea, C.; Zebelo, S.; Muroi, A.; Ishihama, N.; Yoshioka, H.; et al. Regulation of *Arabidopsis* defense responses against *Spodoptera Littoralis* by CPK-mediated calcium signaling. *BMC Plant Biol.* **2010**, *10*, 97. [CrossRef]
79. Mithöfer, A.; Mazars, C.; Maffei, M.E. Probing spatio-temporal intracellular calcium variations in plants. In *Plant Signal Transduction*; Humana Press: Totowa, NJ, USA, 2009; pp. 79–92.
80. Kanchiswamy, C.N.; Mohanta, T.; Capuzzo, A.; Occhipinti, A.; Verrillo, F.; Maffei, M.E.; Malnoy, M. Differential expression of CPKs and cytosolic Ca^{2+} variation in resistant and susceptible apple cultivars (Malus x domestica) in response to the pathogen *Erwinia amylovora* and mechanical wounding. *BMC Genom.* **2013**, *14*, 760. [CrossRef]
81. Kanchiswamy, C.; Malnoy, M.; Occhipinti, A.; Maffei, M. Calcium Imaging Perspectives in Plants. *Int. J. Mol. Sci.* **2014**, *15*, 3842–3859. [CrossRef]
82. Russell, J.T. Imaging calcium signals in vivo: A powerful tool in physiology and pharmacology. *Br. J. Pharmacol.* **2011**, *163*, 1605–1625. [CrossRef]
83. Verrillo, F.; Occhipinti, A.; Kanchiswamy, C.N.; Maffei, M.E. Quantitative analysis of herbivore-induced cytosolic calcium by using a Cameleon (YC 3.6) calcium sensor in *Arabidopsis thaliana*. *J. Plant Physiol.* **2014**, *171*, 136–139. [CrossRef]

84. Vincent, T.R.; Avramova, M.; Canham, J.; Higgins, P.; Bilkey, N.; Mugford, S.T.; Pitino, M.; Toyota, M.; Gilroy, S.; Miller, A.J.; et al. Interplay of plasma membrane and vacuolar ion channels, together with BAK1, elicits rapid cytosolic calcium elevations in *Arabidopsis* during aphid feeding. *Plant Cell* **2017**, *29*, 1460–1479. [CrossRef]
85. Hedrich, R. Ion channels in plants. *Physiol. Rev.* **2012**, *92*, 1777–1811. [CrossRef] [PubMed]
86. Ward, J.M.; Mäser, P.; Schroeder, J.I. Plant ion channels: Gene families, physiology, and functional genomics analyses. *Annu. Rev. Physiol.* **2009**, *71*, 59–82. [CrossRef] [PubMed]
87. Gaymard, F.; Pilot, G.; Lacombe, B.; Bouchez, D.; Bruneau, D.; Boucherez, J.; Michaux-Ferrière, N.; Thibaud, J.-B.; Sentenac, H. Identification and disruption of a plant shaker-like outward channel involved in K^+ release into the xylem sap. *Cell* **1998**, *94*, 647–655. [CrossRef]
88. Pantoja, O. Recent advances in the physiology of ion channels in plants. *Annu. Rev. Plant Biol.* **2021**, *72*, 463–495. [CrossRef] [PubMed]
89. Johansson, I.; Larsson, C.; Ek, B.; Kjellbom, P. The major integral proteins of spinach leaf plasma membranes are putative aquaporins and are phosphorylated in response to Ca^{2+} and apoplastic water potential. *Plant Cell* **1996**, *8*, 1181–1191. [CrossRef]
90. Suh, S.J.; Park, J.G.; Lee, Y. Possible involvement of phospholipase A2 in light signal transduction of guard cells of *Commelina communis*. *Physiol. Plant.* **1998**, *104*, 306–310. [CrossRef]
91. Armstrong, F.; Leung, J.; Grabov, A.; Brearley, J.; Giraudat, J.; Blatt, M.R. Sensitivity to abscisic acid of guard-cell K+ channels is suppressed by abi1-1, a mutant *Arabidopsis* gene encoding a putative protein phosphatase. *Proc. Natl. Acad. Sci. USA* **1995**, *92*, 9520–9524. [CrossRef]
92. Holdaway-Clarke, T.L.; Feijo, J.A.; Hackett, G.R.; Kunkel, J.G.; Hepler, P.K. Pollen tube growth and the intracellular cytosolic calcium gradient oscillate in phase while extracellular calcium influx is delayed. *Plant Cell* **1997**, *9*, 1999–2010. [CrossRef]
93. Suzuki, K.; Costa, A.; Nakayama, H.; Katsuhara, M.; Shinmyo, A.; Horie, T. OsHKT2;2/1-mediated Na^+ influx over K^+ uptake in roots potentially increases toxic Na^+ accumulation in a salt-tolerant landrace of rice *Nona Bokra* upon salinity stress. *J. Plant Res.* **2015**, *129*, 67–77. [CrossRef]
94. Nguyen, C.T.; Kurenda, A.; Stolz, S.; Chételat, A.; Farmer, E.E. Identification of cell populations necessary for leaf-to-leaf electrical signaling in a wounded plant. *Proc. Natl. Acad. Sci. USA* **2018**, *115*, 10178–10183. [CrossRef]
95. Ma, Y.; Walker, R.K.; Zhao, Y.; Berkowitz, G.A. Linking ligand perception by PEPR pattern recognition receptors to cytosolic Ca^{2+} elevation and downstream immune signaling in plants. *Proc. Natl. Acad. Sci. USA* **2012**, *109*, 19852–19857. [CrossRef]
96. Meena, M.K.; Prajapati, R.; Krishna, D.; Divakaran, K.; Pandey, Y.; Reichelt, M.; Mathew, M.K.; Boland, W.; Mithöfer, A.; Vadassery, J. The Ca^{2+} channel CNGC19 regulates *Arabidopsis* defense against *spodoptera* herbivory. *Plant Cell* **2019**, *31*, 1539–1562. [CrossRef]
97. Mousavi, S.A.; Chauvin, A.; Pascaud, F.; Kellenberger, S.; Farmer, E.E. Glutamate-receptor like genes mediate leaf-to-leaf wound signalling. *Nature* **2013**, *500*, 422–426. [CrossRef]
98. Kiep, V.; Vadassery, J.; Lattke, J.; Maaß, J.P.; Boland, W.; Peiter, E.; Mithöfer, A. Systemic cytosolic Ca^{2+} elevation is activated upon wounding and herbivory in *Arabidopsis*. *New Phytol.* **2015**, *207*, 996–1004. [CrossRef]
99. Dodd, A.N.; Kudla, J.; Sanders, D. The language of calcium signaling. *Annu. Rev. Plant Biol.* **2010**, *61*, 593–620. [CrossRef]
100. Malabarba, J.; Meents, A.K.; Reichelt, M.; Scholz, S.S.; Peiter, E.; Rachowka, J.; Konopka-Postupolska, D.; Wilkins, K.A.; Davies, J.M.; Oelmüller, R.; et al. ANNEXIN1 mediates calcium-dependent defense in *Arabidopsis* plants upon herbivory and wounding. *New Phytol.* **2021**, *231*, 243–254. [CrossRef]
101. Zagotta, W.N.; Siegelbaum, S.A. Structure and function of cyclic nucleotide-gated channels. *Annu. Rev. Neurosci.* **1996**, *19*, 235–263. [CrossRef]
102. Kaupp, U.B.; Seifert, R. Cyclic nucleotide-gated ion channels. *Physiol. Rev.* **2002**, *82*, 769–824. [CrossRef]
103. Schuurink, R.C.; Shartzer, S.F.; Fath, A.; Jones, R.L. Characterization of a calmodulin-binding transporter from the plasma membrane of barley aleurone. *Proc. Natl. Acad. Sci. USA* **1998**, *95*, 1944–1949. [CrossRef]
104. Leng, Q.; Mercier, R.W.; Yao, W.; Berkowitz, G.A. Cloning and first functional characterization of a plant cyclic nucleotide-gated cation channel. *Plant Physiol.* **1999**, *121*, 753–761. [CrossRef]
105. Kaplan, B.; Sherman, T.; Fromm, H. Cyclic nucleotide-gated channels in plants. *FEBS Lett.* **2007**, *581*, 2237–2246. [CrossRef]
106. Abdel-Hamid, H.; Chin, K.; Shahinas, D.; Moeder, W.; Yoshioka, K. Calmodulin binding to *Arabidopsis* cyclic nucleotide gated ion channels. *Plant Signal. Behav.* **2010**, *5*, 1147–1149. [CrossRef]
107. Finka, A.; Cuendet, A.F.; Maathuis, F.J.M.; Saidi, Y.; Goloubinoff, P. Plasma membrane cyclic nucleotide gated calcium channels control land plant thermal sensing and acquired thermotolerance. *Plant Cell* **2012**, *24*, 3333–3348. [CrossRef]
108. Kugler, A.; Köhler, B.; Palme, K.; Wolff, P.; Dietrich, P. Salt-dependent regulation of a CNG channel subfamily in *Arabidopsis*. *BMC Plant Biol.* **2009**, *9*, 140. [CrossRef] [PubMed]
109. Moon, J.; Belloeil, C.; Ianna, M.; Shin, R. *Arabidopsis* CNGC family members contribute to heavy metal ion uptake in plants. *Int. J. Mol. Sci.* **2019**, *20*, 413. [CrossRef] [PubMed]
110. Duszyn, M.; Świeżawska, B.; Szmidt-Jaworska, A.; Jaworski, K. Cyclic nucleotide gated channels (CNGCs) in plant signalling—Current knowledge and perspectives. *J. Plant Physiol.* **2019**, *241*, 153035. [CrossRef] [PubMed]
111. Dietrich, P.; Anschütz, U.; Kugler, A.; Becker, D. Physiology and biophysics of plant ligand-gated ion channels. *Plant Biol.* **2010**, *12*, 80–93. [CrossRef] [PubMed]
112. Liu, M.; Chen, T.-Y.; Ahamed, B.; Li, J.; Yau, K.-W. Calcium-calmodulin modulation of the olfactory cyclic nucleotide-gated cation channel. *Science* **1994**, *266*, 1348–1354. [CrossRef]

113. Grunwald, M.E.; Yu, W.-P.; Yu, H.-H.; Yau, K.-W. Identification of a domain on the β-Subunit of the rod cGMP-gated cation channel that mediates inhibition by calcium-calmodulin. *J. Biol. Chem.* **1998**, *273*, 9148–9157. [CrossRef]
114. Chin, K.; Moeder, W.; Yoshioka, K. Biological roles of cyclic-nucleotide-gated ion channels in plants: What we know and don't know about this 20-member ion channel. *Botany* **2009**, *87*, 668–677. [CrossRef]
115. Jha, S.K.; Sharma, M.; Pandey, G.K. Role of cyclic nucleotide gated channels in stress management in plants. *Curr. Genom.* **2016**, *17*, 315–329. [CrossRef]
116. Varnum, M.D.; Zagotta, W.N. Interdomain interactions underlying activation of cyclic nucleotide-gated channels. *Science* **1997**, *278*, 110–113. [CrossRef]
117. Trudeau, M.C.; Zagotta, W.N. Mechanism of calcium/calmodulin inhibition of rod cyclic nucleotide-gated channels. *Proc. Natl. Acad. Sci. USA* **2002**, *99*, 8424–8429. [CrossRef]
118. Swarbreck, S.M.; Colaço, R.; Davies, J.M. Plant Calcium-Permeable Channels. *Plant Physiol.* **2013**, *163*, 514–522. [CrossRef]
119. Talke, I. CNGCs: Prime targets of plant cyclic nucleotide signalling? *Trends Plant Sci.* **2003**, *8*, 286–293. [CrossRef]
120. Kudla, J.; Batistič, O.; Hashimoto, K. Calcium signals: The lead currency of plant information processing. *Plant Cell* **2010**, *22*, 541–563. [CrossRef]
121. Wang, Y.-F.; Munemasa, S.; Nishimura, N.; Ren, H.-M.; Robert, N.; Han, M.; Puzõrjova, I.; Kollist, H.; Lee, S.; Mori, I.; et al. Identification of Cyclic Gmp-activated Nonselective Ca^{2+}-permeable cation channels and Associated cngc5 and cngc 6 genes in *arabidopsis* Guard Cells. *Plant Physiol.* **2013**, *163*, 578–590. [CrossRef]
122. Pettersen, E.F.; Goddard, T.D.; Huang, C.C.; Couch, G.S.; Greenblatt, D.M.; Meng, E.C.; Ferrin, T.E. UCSF chimera? A visualization system for exploratory research and analysis. *J. Comput. Chem.* **2004**, *25*, 1605–1612. [CrossRef]
123. Dingledine, R.; Borges, K.; Bowie, D.; Traynelis, S.F. The glutamate receptor ion channels. *Pharmacol. Rev.* **1999**, *51*, 7–61.
124. Traynelis, S.F.; Wollmuth, L.P.; McBain, C.J.; Menniti, F.S.; Vance, K.M.; Ogden, K.K.; Hansen, K.B.; Yuan, H.; Myers, S.J.; Dingledine, R. Glutamate receptor ion channels: Structure, regulation, and function. *Pharmacol. Rev.* **2010**, *62*, 405–496. [CrossRef]
125. Vezzani, A.; Fujinami, R.S.; White, H.S.; Preux, P.-M.; Blümcke, I.; Sander, J.W.; Löscher, W. Infections, inflammation and epilepsy. *Acta Neuropathol.* **2015**, *131*, 211–234. [CrossRef]
126. Kang, J.; Turano, F.J. The putative glutamate receptor 1.1 (AtGLR1.1) functions as a regulator of carbon and nitrogen metabolism in *Arabidopsis thaliana*. *Proc. Natl. Acad. Sci. USA* **2003**, *100*, 6872–6877. [CrossRef]
127. Miller, N.D.; Durham Brooks, T.L.; Assadi, A.H.; Spalding, E.P. Detection of a gravitropism phenotype in glutamate receptor-like 3.3 mutants of *Arabidopsis thaliana* using machine vision and computation. *Genetics* **2010**, *186*, 585–593. [CrossRef]
128. Michard, E.; Lima, P.T.; Borges, F.; Silva, A.C.; Portes, M.T.; Carvalho, J.E.; Gilliham, M.; Liu, L.-H.; Obermeyer, G.; Feijo, J.A. Glutamate receptor-like genes form Ca^{2+} channels in pollen tubes and are regulated by pistil D-serine. *Science* **2011**, *332*, 434–437. [CrossRef]
129. Wudick, M.M.; Portes, M.T.; Michard, E.; Rosas-Santiago, P.; Lizzio, M.A.; Nunes, C.O.; Campos, C.; Santa Cruz Damineli, D.; Carvalho, J.C.; Lima, P.T.; et al. Cornichon sorting and regulation of GLR channels underlie pollen tube Ca^{2+} homeostasis. *Science* **2018**, *360*, 533–536. [CrossRef]
130. Li, F.; Wang, J.; Ma, C.; Zhao, Y.; Wang, Y.; Hasi, A.; Qi, Z. Glutamate Receptor-Like channel3.3 is involved in Mediating glutathione-triggered cytosolic Calcium Transients, Transcriptional changes, and innate Immunity responses in *Arabidopsis*. *Plant Physiol.* **2013**, *162*, 1497–1509. [CrossRef]
131. Kwaaitaal, M.; Huisman, R.; Maintz, J.; Reinstädler, A.; Panstruga, R. Ionotropic glutamate receptor (iGluR)-like channels mediate MAMP-induced calcium influx in *Arabidopsis thaliana*. *Biochem. J.* **2011**, *440*, 355–373. [CrossRef]
132. Weiland, M.; Mancuso, S.; Baluska, F. Signalling via glutamate and GLRs in *Arabidopsis thaliana*. *Funct. Plant Biol.* **2016**, *43*, 1–25. [CrossRef]
133. Forde, B.G.; Roberts, M.R. Glutamate receptor-like channels in plants: A role as amino acid sensors in plants. *F1000Prime Rep.* **2014**, *6*, 37. [CrossRef]
134. Davenport, R. Glutamate receptors in plants. *Ann. Bot.* **2002**, *90*, 549–557. [CrossRef]
135. Lam, H.-M.; Chiu, J.; Hsieh, M.-H.; Meisel, L.; Oliveira, I.C.; Shin, M.; Coruzzi, G. Glutamate-receptor genes in plants. *Nature* **1998**, *396*, 125–126. [CrossRef] [PubMed]
136. Teardo, E.; Carraretto, L.; De Bortoli, S.; Costa, A.; Behera, S.; Wagner, R.; Lo Schiavo, F.; Formentin, E.; Szabo, I. Alternative splicing-mediated targeting of the *Arabidopsis* glutamate receptor 3.5 to mitochondria affects organelle morphology. *Plant Physiol.* **2014**, *167*, 216–227. [CrossRef] [PubMed]
137. Young, V.R.; Ajami, A.M. Glutamate: An amino acid of particular distinction. *J. Nutr.* **2000**, *130*, 892S–900S. [CrossRef] [PubMed]
138. Forde, B.G.; Lea, P.J. Glutamate in plants: Metabolism, regulation, and signalling. *J. Exp. Bot.* **2007**, *58*, 2339–2358. [CrossRef] [PubMed]
139. Dennison, K.L.; Spalding, E.P. Glutamate-gated calcium fluxes in *Arabidopsis*. *Plant Physiol.* **2000**, *124*, 1511–1514. [CrossRef]
140. Vatsa, P.; Chiltz, A.; Bourque, S.; Wendehenne, D.; Garcia-Brugger, A.; Pugin, A. Involvement of putative glutamate receptors in plant defence signaling and NO production. *Biochimie* **2011**, *93*, 2095–2101. [CrossRef]
141. Mortimer, J.C.; Laohavisit, A.; Macpherson, N.; Webb, A.; Brownlee, C.; Battey, N.H.; Davies, J.M. Annexins: Multifunctional components of growth and adaptation. *J. Exp. Bot.* **2008**, *59*, 533–544. [CrossRef]
142. Clark, G.B.; Morgan, R.O.; Fernandez, M.P.; Roux, S.J. Evolutionary adaptation of plant annexins has diversified their molecular structures, interactions and functional roles. *New Phytol.* **2012**, *196*, 695–712. [CrossRef]

143. Gerke, V.; Moss, S.E. Annexins: From structure to function. *Physiol. Rev.* **2002**, *82*, 331–371. [CrossRef]
144. Seaton, B.A.; Dedman, J.R. Annexins. *Biometals* **1998**, *11*, 399–404. [CrossRef]
145. Cantero, A.; Barthakur, S.; Bushart, T.J.; Chou, S.; Morgan, R.O.; Fernandez, M.P.; Clark, G.B.; Roux, S.J. Expression profiling of the *Arabidopsis* annexin gene family during germination, de-etiolation and abiotic stress. *Plant Physiol. Biochem.* **2006**, *44*, 13–24. [CrossRef]
146. Hofmann, A.; Proust, J.; Dorowski, A.; Schantz, R.; Huber, R. Annexin 24 from *capsicum annuum*. *J. Biol. Chem.* **2000**, *275*, 8072–8082. [CrossRef]
147. Calvert, C.M.; Gant, S.J.; Bowles, D.J. Tomato annexins p34 and p35 bind to F-actin and display nucleotide phosphodiesterase activity inhibited by phospholipid binding. *Plant Cell* **1996**, *8*, 333.
148. McClung, A.D.; Carroll, A.D.; Battey, N.H. Identification and characterization of ATPase activity associated with maize (*Zea mays*) annexins. *Biochem. J.* **1994**, *303*, 709–712. [CrossRef]
149. Andrawis, A.; Solomon, M.; Delmer, D.P. Cotton fiber annexins: A potential role in the regulation of callose synthase. *Plant J.* **1993**, *3*, 763–772. [CrossRef]
150. Gidrol, X.; Sabelli, P.A.; Fern, Y.S.; Kush, A.K. Annexin-like protein from *Arabidopsis thaliana* rescues delta oxyR mutant of *Escherichia coli* from H_2O_2 stress. *Proc. Natl. Acad. Sci. USA* **1996**, *93*, 11268–11273. [CrossRef]
151. Ishibashi, K.; Suzuki, M.; Imai, M. Molecular cloning of a novel form (two-repeat) protein related to voltage-gated sodium and calcium channels. *Biochem. Biophys. Res. Commun.* **2000**, *270*, 370–376. [CrossRef]
152. Furuichi, T.; Cunningham, K.W.; Muto, S. A putative two pore channel AtTPC1 mediates Ca^{2+} flux in *arabidopsis* leaf cells. *Plant Cell Physiol.* **2001**, *42*, 900–905. [CrossRef]
153. Dadacz-Narloch, B.; Beyhl, D.; Larisch, C.; López-Sanjurjo, E.J.; Reski, R.; Kuchitsu, K.; Müller, T.D.; Becker, D.; Schönknecht, G.; Hedrich, R. A novel calcium binding site in the slow vacuolar cation channel TPC1 senses luminal calcium levels. *Plant Cell* **2011**, *23*, 2696–2707. [CrossRef]
154. Cang, C.; Bekele, B.; Ren, D. The voltage-gated sodium channel TPC1 confers endolysosomal excitability. *Nat. Chem. Biol.* **2014**, *10*, 463–469. [CrossRef]
155. Guo, J.; Zeng, W.; Chen, Q.; Lee, C.; Chen, L.; Yang, Y.; Cang, C.; Ren, D.; Jiang, Y. Structure of the voltage-gated two-pore channel TPC1 from *Arabidopsis thaliana*. *Nature* **2015**, *531*, 196–201. [CrossRef]
156. Ward, J.M.; Schroeder, J.I. Calcium-activated K^+ channels and calcium-induced calcium release by slow vacuolar ion channels in guard cell vacuoles implicated in the control of stomatal closure. *Plant Cell* **1994**, 669–683.
157. Pitt, S.J.; Funnell, T.M.; Sitsapesan, M.; Venturi, E.; Rietdorf, K.; Ruas, M.; Ganesan, A.; Gosain, R.; Churchill, G.C.; Zhu, M.X.; et al. TPC2 is a novel NAADP-sensitive Ca^{2+} release channel, operating as a dual sensor of luminal pH and Ca^{2+}. *J. Biol. Chem.* **2010**, *285*, 35039–35046. [CrossRef]
158. Bewell, M.A.; Maathuis, F.J.M.; Allen, G.J.; Sanders, D. Calcium-induced calcium release mediated by a voltage-activated cation channel in vacuolar vesicles from red beet. *FEBS Lett.* **1999**, *458*, 41–44. [CrossRef]
159. Sondergaard, T.E.; Schulz, A.; Palmgren, M.G. Energization of transport processes in plants. roles of the plasma membrane H+-ATPase. *Plant Physiol.* **2004**, *136*, 2475–2482. [CrossRef]
160. Haruta, M.; Gray, W.M.; Sussman, M.R. Regulation of the plasma membrane proton pump $(H(^+)$-ATPase) by phosphorylation. *Curr. Opin. Plant Biol.* **2015**, *28*, 68–75. [CrossRef]
161. Briskin, D.P.; Hanson, J.B. How does the plant plasma membrane H^+-ATPase pump protons? *J. Exp. Bot.* **1992**, *43*, 269–289. [CrossRef]
162. Sussman, M.R. Molecular analysis of proteins in the plant plasma membrane. *Annu. Rev. Plant Physiol. Plant Mol. Biol.* **1994**, *45*, 211–234. [CrossRef]
163. Morsomme, P.; Boutry, M. The plant plasma membrane H^+-ATPase: Structure, function and regulation. *Biochim. Biophys. Acta (BBA)—Biomembr.* **2000**, *1465*, 1–16. [CrossRef]
164. Palmgren, M.G. Plant plasma membrane h+-atpases: Powerhouses for nutrient uptake. *Annu. Rev. Plant Physiol. Plant Mol. Biol.* **2001**, *52*, 817–845. [CrossRef]
165. Lutsenko, S.; Kaplan, J.H. Organization of p-type atpases: Significance of structural diversity. *Biochemistry* **1995**, *34*, 15607–15613. [CrossRef] [PubMed]
166. Zimmermann, M.R.; Maischak, H.; Mithöfer, A.; Boland, W.; Felle, H.H. System potentials, a novel electrical long-distance apoplastic signal in plants, induced by wounding. *Plant Physiol.* **2009**, *149*, 1593–1600. [CrossRef]
167. Camoni, L.; Barbero, F.; Aducci, P.; Maffei, M.E. *Spodoptera littoralis* oral secretions inhibit the activity Of *Phaseolus lunatus* plasma membrane H+-ATPase. *PLoS ONE* **2018**, *13*, e0202142. [CrossRef] [PubMed]
168. Kumari, A.; Chételat, A.; Nguyen, C.T.; Farmer, E.E. *Arabidopsis* h+-atpase aha1 controls slow wave potential duration and wound response Jasmonate pathway activation. *Proc. Natl. Acad. Sci. USA* **2019**, *116*, 20226–20231. [CrossRef] [PubMed]
169. Miller, G.; Shulaev, V.; Mittler, R. Reactive oxygen signaling and abiotic stress. *Physiol. Plant.* **2008**, *133*, 481–489. [CrossRef] [PubMed]
170. Torres, M.A.; Dangl, J.L. Functions of the respiratory burst oxidase in biotic interactions, abiotic stress and development. *Curr. Opin. Plant Biol.* **2005**, *8*, 397–403. [CrossRef] [PubMed]
171. Suzuki, N.; Miller, G.; Morales, J.; Shulaev, V.; Torres, M.A.; Mittler, R. Respiratory burst oxidases: The engines of ROS signaling. *Curr. Opin. Plant Biol.* **2011**, *14*, 691–699. [CrossRef]

172. Baxter, A.; Mittler, R.; Suzuki, N. ROS as key players in plant stress signalling. *J. Exp. Bot.* **2013**, *65*, 1229–1240. [CrossRef]
173. Mittler, R.; Vanderauwera, S.; Suzuki, N.; Miller, G.; Tognetti, V.B.; Vandepoele, K.; Gollery, M.; Shulaev, V.; Van Breusegem, F. ROS signaling: The new wave? *Trends Plant Sci.* **2011**, *16*, 300–309. [CrossRef]
174. Wu, J.; Wang, L.; Wünsche, H.; Baldwin, I.T. Narboh D, a Respiratory Burst Oxidase Homolog in *Nicotiana attenuata*, is required for Late Defense responses after Herbivore attack. *J. Integr. Plant Biol.* **2013**, *55*, 187–198. [CrossRef]
175. Asada, K.; Takahashi, M. Production and scavenging of active oxygen in photosynthesis. In *Photoinhibition*; Kyle, D.J., Osmond, B., Arntzen, C.J., Eds.; Elsevier: Amsterdam, The Netherlands, 1987; pp. 227–287.
176. Lamb, C.; Dixon, R.A. The oxidative burst in plant resistance. *Annu. Rev. Plant Physiol. Plant Mol. Biol.* **1997**, *48*, 251–275. [CrossRef]
177. Gilroy, S.; Suzuki, N.; Miller, G.; Choi, W.-G.; Toyota, M.; Devireddy, A.R.; Mittler, R. A tidal wave of signals: Calcium and ROS at the forefront of rapid systemic signaling. *Trends Plant Sci.* **2014**, *19*, 623–630. [CrossRef]
178. Sharma, P.; Jha, A.B.; Dubey, R.S.; Pessarakli, M. Reactive oxygen species, oxidative damage, and antioxidative defense mechanism in plants under stressful conditions. *J. Bot.* **2012**, *2012*, 217037. [CrossRef]
179. Thordal-Christensen, H.; Zhang, Z.; Wei, Y.; Collinge, D.B. Subcellular localization of H_2O_2 in plants. H_2O_2 accumulation in papillae and hypersensitive response during the barley-powdery mildew interaction. *Plant J.* **1997**, *11*, 1187–1194. [CrossRef]
180. Jacks, T.J.; Davidonis, G.H. Superoxide, hydrogen peroxide, and the respiratory burst of fungally infected plant cells. *Mol. Cell. Biochem.* **1979**, *158*, 77–79. [CrossRef]
181. Moloi, M.J.; van der Westhuizen, A.J. The reactive oxygen species are involved in resistance responses of wheat to the Russian wheat aphid. *J. Plant Physiol.* **2006**, *163*, 1118–1125. [CrossRef]
182. Imbiscuso, G.; Trotta, A.; Maffei, M.; Bossi, S. Herbivory induces a ROS burst and the release of volatile organic compounds in the fern *Pteris vittate* L. *J. Plant Interact.* **2009**, *4*, 15–22. [CrossRef]
183. Shinya, T.; Hojo, Y.; Desaki, Y.; Christeller, J.T.; Okada, K.; Shibuya, N.; Galis, I. Modulation of plant defense responses to herbivores by simultaneous recognition of different herbivore-associated elicitors in rice. *Sci. Rep.* **2016**, *6*, 32537. [CrossRef]
184. Gandhi, A.; Kariyat, R.R.; Chappa, C.; Tayal, M.; Sahoo, N. Tobacco Hornworm (*Manduca sexta*) Oral Secretion Elicits Reactive Oxygen Species in Isolated Tomato Protoplasts. *Int. J. Mol. Sci.* **2020**, *21*, 8297. [CrossRef]
185. Takeda, S.; Gapper, C.; Kaya, H.; Bell, E.; Kuchitsu, K.; Dolan, L. Local positive feedback regulation determines cell shape in root hair cells. *Science* **2008**, *319*, 1241–1244. [CrossRef]
186. Ogasawara, Y.; Kaya, H.; Hiraoka, G.; Yumoto, F.; Kimura, S.; Kadota, Y.; Hishinuma, H.; Senzaki, E.; Yamagoe, S.; Nagata, K.; et al. Synergistic Activation of the *Arabidopsis* NADPH oxidase AtrbohD by Ca^{2+} and phosphorylation. *J. Biol. Chem.* **2008**, *283*, 8885–8892. [CrossRef] [PubMed]
187. Kimura, S.; Kaya, H.; Kawarazaki, T.; Hiraoka, G.; Senzaki, E.; Michikawa, M.; Kuchitsu, K. Protein phosphorylation is a prerequisite for the Ca^{2+}-dependent activation of *Arabidopsis* NADPH oxidases and may function as a trigger for the positive feedback regulation of Ca^{2+} and reactive oxygen species. *Biochim. Biophys. Acta (BBA)—Mol. Cell Res.* **2012**, *1823*, 398–405. [CrossRef] [PubMed]
188. Kadota, Y.; Shirasu, K.; Zipfel, C. Regulation of the NADPH oxidase RBOHD during plant immunity. *Plant Cell Physiol.* **2015**, *56*, 1472–1480. [CrossRef] [PubMed]
189. Pei, Z.-M.; Murata, Y.; Benning, G.; Thomine, S.; Klüsener, B.; Allen, G.J.; Grill, E.; Schroeder, J.I. Calcium channels activated by hydrogen peroxide mediate abscisic acid signalling in guard cells. *Nature* **2000**, *406*, 731–734. [CrossRef] [PubMed]
190. Kwak, J.M. NADPH oxidase AtrbohD and AtrbohF genes function in ROS-dependent ABA signaling in *Arabidopsis*. *EMBO J.* **2003**, *22*, 2623–2633. [CrossRef]
191. Little, D.; Gouhier-Darimont, C.; Bruessow, F.; Reymond, P. Oviposition by pierid butterflies triggers defense responses in *Arabidopsis*. *Plant Physiol.* **2006**, *143*, 784–800. [CrossRef]
192. Stahl, E.; Brillatz, T.; Ferreira Queiroz, E.; Marcourt, L.; Schmiesing, A.; Hilfiker, O.; Riezman, I.; Riezman, H.; Wolfender, J.-L.; Reymond, P. Phosphatidylcholines from *Pieris brassicae* eggs activate an immune response in *Arabidopsis*. *eLife* **2020**, *9*, e60293. [CrossRef]
193. Dooley, C.T.; Dore, T.M.; Hanson, G.T.; Jackson, W.C.; Remington, S.J.; Tsien, R.Y. Imaging dynamic redox changes in mammalian cells with green fluorescent protein indicators. *J. Biol. Chem.* **2004**, *279*, 22284–22293. [CrossRef]
194. Hanson, G.T.; Aggeler, R.; Oglesbee, D.; Cannon, M.; Capaldi, R.A.; Tsien, R.Y.; Remington, S.J. Investigating mitochondrial redox potential with redox-sensitive green fluorescent protein indicators. *J. Biol. Chem.* **2004**, *279*, 13044–13053. [CrossRef]
195. Jiang, K.; Schwarzer, C.; Lally, E.; Zhang, S.; Ruzin, S.; Machen, T.; Remington, S.J.; Feldman, L. Expression and characterization of a redox-sensing green fluorescent protein (reduction-oxidation-sensitive green fluorescent protein) in *Arabidopsis*. *Plant Physiol.* **2006**, *141*, 397–403. [CrossRef]
196. Meyer, A.J.; Brach, T.; Marty, L.; Kreye, S.; Rouhier, N.; Jacquot, J.-P.; Hell, R. Redox-sensitive GFP in *Arabidopsis thaliana* is a quantitative biosensor for the redox potential of the cellular glutathione redox buffer. *Plant J.* **2007**, *52*, 973–986. [CrossRef]
197. Gutscher, M.; Sobotta, M.C.; Wabnitz, G.H.; Ballikaya, S.; Meyer, A.J.; Samstag, Y.; Dick, T.P. Proximity-based protein thiol oxidation by H_2O_2-scavenging peroxidases. *J. Biol. Chem.* **2009**, *284*, 31532–31540. [CrossRef]
198. Swanson, S.J.; Choi, W.-G.; Chanoca, A.; Gilroy, S. In vivo imaging of Ca^{2+}, pH, and reactive oxygen species using fluorescent probes in plants. *Annu. Rev. Plant Biol.* **2011**, *62*, 273–297. [CrossRef]

199. Nietzel, T.; Elsässer, M.; Ruberti, C.; Steinbeck, J.; Ugalde, J.M.; Fuchs, P.; Wagner, S.; Ostermann, L.; Moseler, A.; Lemke, P.; et al. The fluorescent protein sensor roGFP2-Orp1 monitors in vivo H_2O_2 and thiol redox integration and elucidates intracellular H_2O_2 dynamics during elicitor-induced oxidative burst in *Arabidopsis*. *New Phytol.* **2018**, *221*, 1649–1664. [CrossRef]
200. Hipsch, M.; Lampl, N.; Zelinger, E.; Barda, O.; Waiger, D.; Rosenwasser, S. Sensing stress responses in potato with whole-plant redox imaging. *bioRxiv* **2021**, *6*, kiab159.
201. Janků, M.; Luhová, L.; Petřivalský, M. On the origin and fate of reactive oxygen species in plant cell compartments. *Antioxidants* **2019**, *8*, 105. [CrossRef]
202. Li, X.; Liu, Y.; He, Q.; Li, S.; Liu, W.; Lin, C.; Miao, W. A candidate secreted effector protein of rubber tree powdery mildew fungus contributes to infection by regulating plant ABA biosynthesis. *Front. Microbiol.* **2020**, *11*, 2788. [CrossRef]
203. Chen, K.; Guo, Y.; Song, M.; Liu, L.; Xue, H.; Dai, H.; Zhang, Z. Dual role of MdSND1 in the biosynthesis of lignin and in signal transduction in response to salt and osmotic stress in apple. *Hortic. Res.* **2020**, *7*, 1–13. [CrossRef]
204. Fu, H.; Zhao, M.; Xu, J.; Tan, L.; Han, J.; Li, D.; Wang, M.; Xiao, S.; Ma, X.; Deng, Z. Citron C-05 inhibits both the penetration and colonization of *Xanthomonas citri* subsp. citri to achieve resistance to citrus canker disease. *Hortic. Res.* **2020**, *7I*, 1–12. [CrossRef]
205. Fichman, Y.; Mittler, R. Rapid systemic signaling during abiotic and biotic stresses: Is the ROS wave master of all trades? *Plant J.* **2020**, *102*, 887–896. [CrossRef]
206. Bissoli, G.; Muñoz-Bertomeu, J.; Bueso, E.; Sayas, E.; Vilcara, E.A.; Felipo, A.; Niñoles, R.; Rubio, L.; Fernández, J.A.; Serrano, R. An *Arabidopsis* mutant over-expressing subtilase SBT4.13 uncovers the role of oxidative stress in the inhibition of growth by intracellular acidification. *Int. J. Mol. Sci.* **2020**, *21*, 1173. [CrossRef] [PubMed]

cells

MDPI

Review

Effects of Salinity Stress on Chloroplast Structure and Function

Abdul Hameed [1], Muhammad Zaheer Ahmed [1], Tabassum Hussain [1], Irfan Aziz [1], Niaz Ahmad [2,3], Bilquees Gul [1] and Brent L. Nielsen [4,*]

[1] Dr. M. Ajmal Khan Institute for Sustainable Halophyte Utilization, University of Karachi, Sindh 75270, Pakistan; ahameed@uok.edu.pk (A.H.); mzahmed@uok.edu.pk (M.Z.A.); thussain@uok.edu.pk (T.H.); irfanaziz@uok.edu.pk (I.A.); bilqueesgul@uok.edu.pk (B.G.)

[2] Agricultural Biotechnology Division, National Institute for Biotechnology & Genetic Engineering (NIBGE), Faisalabad 44000, Pakistan; niazbloch@yahoo.com

[3] Department of Biotechnology, Pakistan Institute of Engineering and Applied Science (PIEAS), Islamabad 44000, Pakistan

[4] Department of Microbiology & Molecular Biology, Brigham Young University, Provo, UT 84602, USA

* Correspondence: brentnielsen@byu.edu

Abstract: Salinity is a growing problem affecting soils and agriculture in many parts of the world. The presence of salt in plant cells disrupts many basic metabolic processes, contributing to severe negative effects on plant development and growth. This review focuses on the effects of salinity on chloroplasts, including the structures and function of these organelles. Chloroplasts house various important biochemical reactions, including photosynthesis, most of which are considered essential for plant survival. Salinity can affect these reactions in a number of ways, for example, by changing the chloroplast size, number, lamellar organization, lipid and starch accumulation, and interfering with cross-membrane transportation. Research has shown that maintenance of the normal chloroplast physiology is necessary for the survival of the entire plant. Many plant species have evolved different mechanisms to withstand the harmful effects of salt-induced toxicity on their chloroplasts and its machinery. The differences depend on the plant species and growth stage and can be quite different between salt-sensitive (glycophyte) and salt-tolerant (halophyte) plants. Salt stress tolerance is a complex trait, and many aspects of salt tolerance in plants are not entirely clear yet. In this review, we discuss the different mechanisms of salt stress tolerance in plants with a special focus on chloroplast structure and its functions, including the underlying differences between glycophytes and halophytes.

Keywords: salinity stress; photosynthesis; chloroplast; plastid; osmolytes; osmotic adjustment

Citation: Hameed, A.; Ahmed, M.Z.; Hussain, T.; Aziz, I.; Ahmad, N.; Gul, B.; Nielsen, B.L. Effects of Salinity Stress on Chloroplast Structure and Function. *Cells* **2021**, *10*, 2023. https://doi.org/10.3390/cells10082023

Academic Editor: Suleyman Allakhverdiev

Received: 30 July 2021
Accepted: 5 August 2021
Published: 7 August 2021

1. Introduction

Soil quality in many parts of the U.S. and worldwide is susceptible to a variety of stresses, including drought, temperature, deterioration due to erosion and other factors, and increasing salinity due to evaporation and/or irrigation practices. At the same time the human population is growing and in many regions high-quality agricultural land is decreasing due to the expansion of urban areas [1].

Salinity is inhibitory to the growth and development of many plants, including most crops [2–5]. It affects all cellular processes, including disruption of cellular homeostasis, impairment of photosynthesis, mRNA processing, transcription, protein synthesis, disruption of energy metabolisms, amino acid biosynthesis as well as lipid metabolism [6–10]. In response to increasing salt, plant cells activate specific Na^+ and Cl^- ion transporters and adjust the accumulation of cytosolic K^+ [10–12]. Plant cells must also undergo osmotic adjustment, which is accomplished in many ways, including the production of organic osmolytes such as glycine betaine, proline, some sugars, and polyamines, of which most are synthesized in the chloroplast [3,10].

Chloroplasts belong to a family of cellular organelles commonly found in plant and algal cells known as plastids. Green plastids—chloroplasts—are the site where atmospheric CO_2 fixation occurs through a series of biochemical reactions called the Calvin–Benson cycle by utilizing the energy produced by the light reactions of photosynthesis [13]. Elevated salinity levels affect many cellular processes, including photosynthesis, the major function of chloroplasts. The presence of salt in the soil may cause both osmotic and ionic stresses [14], which may hinder photosynthesis through the diffusional (stomatal, mesophyll and boundary layer resistance to CO_2) and/or non-diffusional (photochemical and biochemical) limitations of carbon fixation [6,15–20]. Salinity exposure is also known to decrease the chlorophyll content in many plants [21,22]. However, salt-resistant plants, particularly those with a C_4 mechanism, may overcome the inhibitory effect of salinity on CO_2 fixation more effectively [6,23].

In general, when plants are exposed to salt stress, the very first response is osmotic shock followed by induction of stomatal closure. Stomatal closure, in turn, limits photosynthetic capacity by the restriction of CO_2 supply. However, research has shown that increasing the external CO_2 concentration under salt stress did not lead to an increase in photosynthesis rates in many cases. This observation suggests the involvement of some non-stomatal components in photosynthesis reduction under salinity, such as overproduction of reactive oxygen species (ROS) and the depletion of K^+ inside plant cells due to the accumulation of Na^+ [24,25]. This results in the disruption of ionic homeostasis in chloroplasts.

Besides CO_2 fixation, thylakoid reactions are also affected by salinity [6,18,26]. The most commonly studied parameters in this context are the maximum quantum efficiency of the PSII reaction centers (F_v/F_m), quantum efficiency of PSII (ΦPSII), non-photochemical quenching (NPQ), photochemical quenching (qP) and electron transport rate (ETR), which defines the overall performance of plants under different stresses [27]. Salt-resistant plants are known to possess resilient thylakoid reactions to overcome salinity effects such as photodamage [28] and protection of the reaction centers [29]. This may include protective mechanisms such as cyclic electron flow, photorespiration in C_3 plants and regulation of NPQ [18,30]. CO_2 fixation and thylakoid reactions of photosynthesis take place in thylakoids and the stroma of the chloroplast, providing the essential carbon skeleton for growth, energy for driving various metabolic reactions as well as the biosynthesis of different metabolites. Salt-induced toxicity negatively affects all these processes, resulting in poor plant growth and reduction in yield. Chloroplasts are also major reactive oxygen species (ROS) production sites at the reaction centers of PSII and PSI, where charge separation occurs, and the electron transport chain (ETC) from PSII to PSI are highly sensitive to salt-induced toxicity under which ROS production is further increased [31]. Higher concentrations of ROS cause oxidative damage to membranes, lipids, nucleic acids, proteins and some photosynthetic enzymes, resulting in reduced CO_2 fixation, slower plant growth and consequently low crop yields. The ROS-scavenging system includes both enzymatic and non-enzymatic antioxidants that prevent oxidative damage. Therefore, manipulation of the components of this system holds great implications for improving the photosynthetic rates under salt stress in crop plants. This has been tested by overexpression of Cu/Zn superoxide dismutase (SOD) in the chloroplasts of tobacco [32,33] and Chinese cabbage [34]. Since chloroplasts are largely under the control of nuclear gene expression for growth and metabolic activities, chloroplasts have evolved a sophisticated signaling network to coordinate with the nucleus to control gene expression and maintain the balanced expression of genes in the two compartments. Chloroplasts also act as global sensors relaying changes in their own developmental status as well as in the environmental conditions, including light intensity and stresses to the nucleus. As a result, the nucleus adjusts the expression of its genes to ensure optimal plant performance under changing environmental conditions [35]. Until recently, this chloroplast–nucleus communication has been largely viewed as bilateral, ignoring the pivotal role of chloroplasts in adjusting gene expression and metabolic processes that affect photosynthesis and ultimately crop yields.

In this review, we discuss the effect of salt stress on chloroplasts, their structures, and various biochemical reactions occurring in them. We also compare the differences in how chloroplasts of glycophytes and halophytes respond to salinity stress.

2. Effects of Salinity on Chloroplast Ultrastructure

2.1. Changes in Chloroplast Structure in Plants

Chloroplasts are roughly 1–2 μm thick and 5–7 μm in diameter. They are enclosed in a chloroplast envelope, which consists of a double membrane with outer and inner layers; the space in between is called the intermembrane space. A third, internal membrane, extensively folded and characterized by the presence of closed disks (or thylakoids), is known as the thylakoid membrane. In higher plants, the thylakoids are arranged in tight stacks called grana. Grana are connected by stromal lamellae extended from one granum through the stroma into a neighboring granum. The thylakoid membrane envelops a central aqueous region known as the thylakoid lumen. The space between the inner membrane and the thylakoid membrane is filled with stroma, a matrix containing dissolved enzymes, starch granules and copies of the chloroplast genome [36].

Several changes have been associated with chloroplast structure in response to environmental factors and the availability of water and minerals [37]. These include modifications in the lamellar organization, resulting in chloroplast shrinkage [37], swelling of chloroplast lamellae and an unrecognizable grana structure under highly saline conditions [38]. Some plants, such as *Atriplex* spp., may undergo lipid deposition to counter the harmful effects of salt-induced toxicity [39,40]. In some instances, starch accumulation under high salinity has also been reported, such as in chloroplasts of wheat cultivars, which was related to damage to the sucrose-phosphate synthase in the cytosol, triggering the triose-phosphate pathway towards starch synthesis [41]. Changes in the ionic composition of starch-degrading enzymes may also be linked with excessive starch deposition [42]. Under saline conditions, reactions involving starch and sucrose biosynthesis are also known to be regulated by changes in the orthophosphate concentration [43,44]. Stress-induced destruction of the chloroplast envelope and an increase in the numbers of plastoglobuli in thylakoid membranes have also been reported in cucumber leaves [45]. Accumulation of starch grains in the chloroplasts of *Thellungiella* and tobacco plants is known to play an important role as osmotica in maintaining the structural integrity of the chloroplasts [46].

2.2. Changes in Ultrastructure of Chloroplasts in Glycophytes and Halophytes

Salt stress-induced alterations in the structure of chloroplasts or thylakoid membranes have been extensively examined in various salt-sensitive plants [47,48] as well as in facultative halophytes [49]. Swelling of thylakoids under salt stress (~200 mM NaCl) was reported in rice [50]. However, recent 3D analysis confirmed that rice chloroplasts became spherical under salt stress without any changes in the overall chloroplast volume [51]. Contrasting observations regarding chloroplast volume have been reported among different species. For example, chloroplasts of salt-sensitive cultivars of wheat exhibited an increase in volume possibly due to changes in the ionic composition of the stroma [42]. Spinach chloroplasts showed a decrease in volume with concomitant changes in light-scattering during electron transport [47]. Arabidopsis seedlings grown in the presence of salt also exhibited swollen chloroplasts with less developed granum structures [41]. Changes in the thylakoid ultrastructure of potato [52] and maize [53] under salinity have been previously related to perturbed ion homeostasis in chloroplasts.

In the case of halophytes, salt entry into the chloroplast stroma may be critical for grana formation and photosystem II activity, as halophytes have been reported to accumulate more chloride (Cl^-) than glycophytes and use sodium (Na^+) in different functions [54]. Chloroplast swelling of *Atriplex* leaf cells at 345 mM NaCl appeared to be a likely result of the osmotic effect of salinity while few changes were reported in the chloroplasts of hair cells [40]. Similarly, distinct thylakoid swelling in *Thellungiella* under saline conditions (400 mM NaCl) was attributed to the disturbance in osmotic equilibrium [54]. Other notable changes in halophyte chloroplasts include the formation of 'slim spindle-shaped' grana with a clear stromal matrix in the halophyte *Kandelia candel* at 400 mM and increases in the plastoglobuli numbers at 600 mM NaCl with intact grana thylakoids [49]. In *Bruguiera parviflora*, no major alterations in the structural integrity or absorption characteristics of the thylakoid membranes were noted at 400 mM NaCl; however, a reduction in antenna size, as well as cytochrome (Cyt) *f* and Cyt b_6 were observed [55].

2.3. Changes in the Chloroplast Ultrastructure of C_4 Plants

Plants with C_4 photosynthesis possess chloroplasts both in the bundle sheath cell (BSC) and the mesophyll cell (MC). Both of these chloroplast types, BSC and MC, have been reported to respond differently to salt stress. C_4 plants are further divided into three subtypes, namely, NADP-malic enzyme (NADP-ME), NAD-malic enzyme (NAD-ME) and phosphoenolpyruvate carboxykinase (PCK) types, with peculiar leaf anatomical characteristics [56]. The NADP-ME type 4 species showed little damage to BSC chloroplasts compared to MC chloroplasts [57]. The BSC chloroplasts showed the development of grana when exposed to salt stress [53,58]. On the other hand, thylakoid swelling and disruption of envelopes in MC chloroplasts of both the NAD-ME and PCK types were observed under saline conditions [57]. It has also been reported that exposure to salinity enhanced granal development in BSC chloroplasts and appressed thylakoid density [57]. However, granal development in the NAD-ME and PCK type species is not as pronounced as in the NADP-ME type species. It is believed that granal development in BSC chloroplasts could compensate for the loss of PSII activity in MC chloroplasts under salt stress [53]. In glycophyte plants of the NADP-ME and NAD-ME subtypes, salt stress causes grana unstacking in MC chloroplasts but induces grana formation in BSC chloroplasts. Interestingly, in halophytes of the same subtypes, the grana are constitutively present in BSC chloroplasts and the unstacking of grana is absent in MC chloroplasts [53,58].

2.4. Effects of Salinity on Chloroplast Multiplication

Although the number of chloroplasts per leaf cell varies from a few to hundreds [59], they may occupy more than half of the cytoplasm volume in mesophyll cells [60]. Salinity may increase the number of chloroplasts per cell, e.g., in spinach, sugar beet [61], wheat [62], *Arabidopsis* [63] and *Thellungiella halophila* [46]. Bose et al. [29] proposed that halophytes have a greater capacity to increase chloroplast number than glycophytes under salinity, which may help in storing Na^+ and Cl^- without compromising photosynthesis [63]. Increases in succulence help in cell expansion and thus enable housing more chloroplasts [29]. Halophytes can accumulate over 20-fold higher Na^+ levels in chloroplasts compared to glycophytes [63–68]. In addition to compensating for reduced photosynthesis, increases in chloroplast number may also contribute to increased energy demands for osmotic adjustment and ion homeostasis under salinity [29]. Although information about the genes involved in binary fission of chloroplasts is plentiful [68–70], the detailed effects of salinity on the mechanism/regulation of chloroplast multiplication, particularly among halophytes and glycophytes, are limited.

3. Effects of Salinity on Transport across Chloroplast Membranes

Most of the nuclear-encoded proteins destined for chloroplasts are synthesized as 'cytosolic preproteins' and imported by a major pathway consisting of transmembrane protein complexes or channels, '*Toc*' (translocons on outer chloroplast) and '*Tic*' (translocons on inner chloroplast). The '*Toc*' transmembrane channels import larger molecules (including nuclear-encoded proteins) while the '*Tic*' complexes are more restrictive, with transport limited to targeted proteins [71]. Located at the interface between the stroma and the cytosol, the envelope is also the site for the transport and exchange of ions and metabolites required for the integration of the plastid metabolism within the plant cell. In general, chloroplasts harbor three types of membranes and each of them is equipped with a unique set of ion channels and transporters enabling the transport of nutrients, solutes, and metabolites in and out of it. This is achieved by coordinated regulation of a variety of transport systems located in chloroplast membranes, such as porins, solute channels, ion-specific cation/anion channels and various primary and secondary active transport systems [29].

3.1. Protein Transport across Chloroplast Membranes

The chloroplast proteome comprises 3000 different proteins, including components of the photosynthetic apparatus, which are highly abundant [72]. Most chloroplast proteins are nuclear-encoded, synthesized in the cytosol, and their import is mediated by multiprotein complexes in the envelope membranes that surround each organelle. The *Toc* complex mediates client protein recognition and early stages of the import. The *Toc* apparatus is regulated by the ubiquitin-proteasome system (UPS) in a process controlled by the envelope-localized ubiquitin E3 ligase SUPPRESSOR OF PPI1 LOCUS1 (*SP1*) [73]. Salinity stress depletes the *Toc* apparatus by enhancing the expression of *SP1*, which in turn may result in the suppression of photosynthesis activity [74].

3.2. Ion Transport across Chloroplast Membranes

The proper ionic (K^+, Na^+, Cl^-) balance is essential to control chloroplast volume [73]. For example, Cl^- influx from stroma to the lumen is required for thylakoid swelling, while lumen shrinkage is due to K^+ (or Na^+) efflux from the lumen to the stroma [75]. The outer membrane is not freely permeable to ions as some porins (OEP23, OEP37) are reported to have high cation selectivity [76], although information regarding their role in plant salt tolerance is lacking. The literature reports several nucleus-encoded candidate ion channels and transporters that regulate Na^+, K^+ and Cl^- transport through the chloroplast envelope and thylakoid membranes [75–79]. A several-fold increased Na^+ and Cl^- concentration in barley chloroplasts under salt stress has been reported [76]. Slabu et al. [79] reported that salt-induced damage in broad bean chloroplasts is due to the accumulation of Na^+ and not of Cl^- or K^+. In contrast, salt toxicity and inhibition of photosynthesis in soybean were found associated with the hyperaccumulation of Cl^- but not that of Na^+ in chloroplasts [80,81].

3.3. Chloroplast Trafficking of Ions in Glycophytes vs. Halophytes

Halophytes preferentially accumulate ~20-fold higher Na^+ levels than glycophytes [64,65,67]. This high ion level is known to have some effect on chloroplast functions [63,64,67], especially in the case of CAM and C_4 plants [81]. The Na^+ contribution in the transport of pyruvate [82,83], ascorbate [84] and phosphate [85] into chloroplasts has been reported but the effect of salt stress on transport requires further elaboration. Salt stress induces K^+ loss from chloroplasts in both glycophytes and halophytes. Chloroplasts isolated from halophytes revealed better tolerance to high Na^+ (100 mmol L^{-1} Na^+) and low K^+ (50 mmol L^{-1} K^+) in the cytosol than chloroplasts of glycophytes [86]. Likewise, halophytes accumulate more Cl^- than glycophytes under low salt conditions ($\leq$1 mmol L^{-1} Cl^-), while at higher salinities some halophytes maintain steady Cl^- concentrations, and others show a slight increase within the chloroplasts [63,65]. These findings indicate that halophytes have mechanisms to regulate the Cl^- concentrations; however, the candidate transporters for Cl^- regulation during salt stress remain uncharacterized.

3.3.1. Aquaporins and Non-Selective ion Channels

Aquaporins (PIP2;1, PIP2.3, PIP2;7, PIP1;3 and PIP1;2) are reported on the chloroplast membrane [77,78]. Expression of both PIP2;1 and PIP2;7 is altered by salinity [87]. Some aquaporins also have the ability to transport ions [88], but little is known about their function/regulation.

Non-selective ion channels include mechanosensitive channels (MSL2 and MSL3) that help reduce chloroplast swelling during hypo-osmotic conditions by releasing ions from the stroma [89]. In general, the ion selectivity of MSLs varies from non-selective to Cl^-, K^+, Na^+ or Ca^{2+} selective channels [75].

3.3.2. Na^+, K^+ and Cl^- Transporters

Sodium ions (Na^+) can be transported into chloroplasts through an inner envelope membrane-localized Na^+-dependent pyruvate transporter (*BASS2*) that is abundantly found in halophyte species compared to glycophytes [82]. Introduction of a halophyte BASS2 gene into glycophyte chloroplasts resulted in improved salt tolerance [83]. The inorganic phosphate transporters (thylakoid membrane-localized PHT4;1 and inner envelope localized PHT4;4 and PHT4;5) can use Na^+ or H^+ as a co-transporting ion [79], thereby changing the Na^+ concentration inside the chloroplasts. The existence of the Na^+/H^+ antiporter (NhaD; hereafter NHD)-type transporters at the chloroplast membrane mediating Na^+ efflux from the stroma was also reported in a halophytic tree, *Populus euphratica* [82]. In Arabidopsis, salt stress did not alter the expression of *NHD1* but silencing *NHD1* resulted in high chloroplast Na^+ and poor growth and photosynthetic performance [67]. In contrast, analysis of *Mesembryanthemum crystallinum* (a halophyte) under salt stress showed an increase in *NHD1* expression that resulted in higher Na^+ accumulation, indicating the involvement of *NHD1* in Na^+ import into the chloroplasts instead of Na^+ export [41]. Such opposite regulation of ion transport mechanisms requires further investigation for a more complete understanding of the salt tolerance mechanisms.

Two K^+ efflux antiporters (*KEA1 and KEA2*) located at the membrane of Arabidopsis have been suggested to function as K^+/H^+ exchangers mediating K^+ export out of the stroma [90]. The Arabidopsis double loss-of-function *kea1kea2* mutant showed better growth under salt stress as compared to the wild type, due to low K^+ efflux in the mutant resulting in increased K^+ retention as well as maintenance of pH in the stroma leading to improved photosynthetic performance and growth [91]. Arabidopsis *KEA3*, located in the thylakoid membrane, has been suggested to import K^+ into the lumen in exchange for H^+ [91,92] and support in PSII quantum efficiency and CO_2 assimilation under low light [93]; however, no information is available regarding *KEA3* function during salt stress.

Electrophysiological studies have shown the existence of Cl^- permeable channels in the chloroplast envelope and thylakoid membranes [75]. A bestrophin-like protein from Arabidopsis has been discovered and shown to alter PMF portioning by functioning as a voltage-dependent Cl^- channel (*AtVCCN1*) on the thylakoid membrane [93]. The effects of salinity on chloroplasts are summarized in the model in Figure 1.

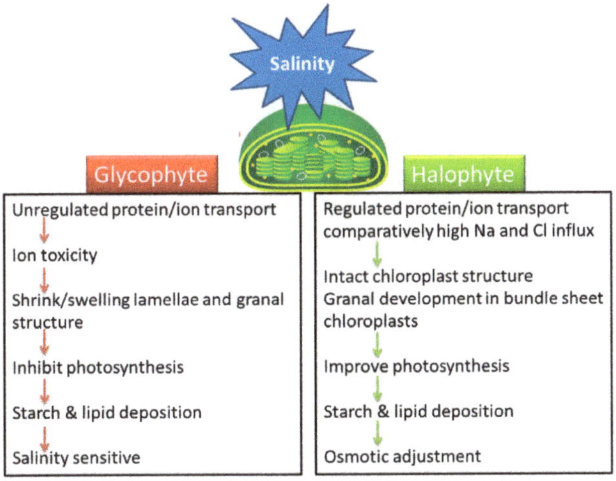

Figure 1. A model that summarizes the effects of salinity stress on chloroplasts in salt-sensitive (glycophyte) and salt-tolerant (halophyte) plants.

4. Effect of Salinity on Osmotic Adjustment in Chloroplasts

4.1. What Is Osmotic Adjustment and How Is It Achieved?

Hyper-osmotic stress due to salinity is well-known in plants and bacteria and may cause disrupted cell metabolism, turgor loss and growth arrest. However, an adaptive mechanism for water retention exists among organisms under stressed environments whereby they increase their osmolality, a phenomenon commonly termed as 'osmotic adjustment' [94]. Increases in osmolality are achieved by either of the following three mechanisms: (1) micro-organisms, such as bacteria or yeast, accumulate a range of osmolytes or compatible solutes available from the external medium; (2) plants activate genes for de novo synthesis of organic osmolytes (so-called 'compatible solutes'), such as glycine betaine, proline, sugars, polyols, etc.; and (3) plants regulate ion flux across cellular membranes [20,95].

4.2. Localization, Trafficking and Functions of Organic Osmolytes in Membrane-Bound Organelles

Among the organic osmolytes, glycine betaine (GB), sugars (mannitol, sorbitol and trehalose), polyamines and proline are the most important and are accumulated under abiotic stresses and confer tolerance to cells without interfering with the cellular machinery of the plant [96]. Of these osmolytes metabolism of proline (PRO) depends upon two important enzymes, catalyzing its synthesis from glutamate in the cytoplasm or chloroplast and two enzymes catalyzing proline catabolism back to glutamate in the mitochondria along with an alternative pathway of its synthesis via ornithine [97]. During water deficit or physiological drought under salinity PRO is known to protect the photosynthetic apparatus as well as in cytokinin-dependent photorespiration [98]. Studies on other osmolytes suggest that sugar alcohols, such as sorbitol and mannitol, and quaternary ammonium compounds, such as GB and their precursors, are highly localized in chloroplasts [99,100] and are somehow involved in protecting the photosystem (PSII) and membrane proteins against ROS under stress conditions [95,101,102]. The impairment of thylakoid membranes that results from salt stress may be alleviated by GB probably via protection and stabilization of the protein complexes as well as changes in lipid composition of the thylakoid membrane, thereby improving photosynthesis [102]. The accumulation of GB in higher concentrations in the chloroplasts of young leaves suggests that these are the main sites of its biosynthesis [98,103] from where it is translocated to other plant parts via phloem [104]. Sugar alcohols and polyols, such as mannitol, sorbitol, etc., regulate osmotic balance by sequestering Na^+ in the vacuole or apoplast, thus protecting membranes against drought [105] and salt stress [106]. These osmolytes also scavenge ROS, particularly hydroxyl radicals that do not require high concentrations of osmolytes as needed for osmotic adjustment [97]. This leads to the conclusion that such compounds may be more important in 'osmoprotection' rather than 'osmotic adjustment'.

4.3. Are Osmolytes Compatible for Osmotic Adjustment in Planta?

The classical concept of osmotic adjustment via accumulation of organic solutes in non-halophilic organisms still prevails [107,108] though it has been challenged by many physiologists [97,109,110]. A major shift in energy balance usually causes severe losses in growth yields of crop plants at the expense of other metabolic processes, raising the serious question of whether osmolytes are compatible in a real sense. Physiologists argue that conventional water retention under saline stress is not directly related to the contribution of organic solutes for many reasons. The first reason is the concentration of organic osmolytes, which seems to be too low compared to the inorganic solutes in cells. For instance, 3–10 mM in plants contributes less than 3% [111,112], while ~120–150 mM glycine betaine (GB) in plants contributes <50%, often ranging between 10 and 30% of the total cell solutes [112]. Even if it is assumed that most of the osmolytes are contained in the cytosol and chloroplasts (collectively constituting 10–15% of the cell volume) compared to a larger vacuolar fraction (~85%), this seems low given that 500–600 mM concentrations of Na^+ alone exist within the vacuole [99]. Osmolyte concentrations (GB in particular) between 200 and 300 mM may be sufficient to prevent cytoplasmic dehydration, thereby achieving osmotic adjustment. In some of the succulent halophytes (which accumulates up to 1000 mM Na^+ and Cl^-), ~200 mM plant water GB was reported in *Suaeda fruticosa* and about 600 mM in *Haloxylon stocksii* (sensu lato *recurvum*) under extreme saline conditions, which are exceptional as in other plants, including *Halopyrum mucronatum* and *Atriplex stocksii* (sensu lato *griffithii*), GB ranged between 100 and 150 mM [113]. The second reason for not considering organic osmolytes as 'compatible' is the cost of their synthesis, which is too high. For instance, 30–109 molecules of ATP may be required for osmolyte synthesis compared to one molecule of ATP for one K^+ and two Cl^- in bacteria [114], while plants require approximately 41 molecules of ATP for proline, 50 for glycine betaine and 52 for sucrose [115]. Thirdly, the synthesis of such organic solutes is very slow, often ranging from hours to many days while plants growing in water-stressed environments require rapid turgor recovery [102]. Moreover, salt-sensitive genotypes of many crop plants, e.g., rice, wheat, barley, etc., accumulate comparatively higher amounts of osmolytes than salt-tolerant varieties, which also creates ambiguity in the role of osmolytes in achieving osmotic adjustment [81,109,116].

4.4. Effects of Osmolytes on Organelles

Although the osmotic adjustment is based on the notion that osmolytes should not interfere with other metabolic processes, some studies on exogenous application of osmolytes suggest their toxic effects on plant growth as well as cell organelles [116,117]. Application of some L-amino acids (L-proline, L-alanine, etc.) in millimolar concentrations caused growth inhibition in suspension cultures of *Nicotiana silvestris* [116]. In another instance, a disruptive effect of PRO on the ultrastructure of chloroplasts in *Arabidopsis thaliana* was linked to feedback inhibition of its synthesis due to over-reduction of the photosynthetic electron acceptor pools [117]. In the same plant, exogenously supplied PRO was thought to have increased the rates of mitochondrial electron transport, resulting in elevated levels of ROS causing subcellular damage [117]. On the contrary, endogenous PRO did not seem to have a negative impact on the ultrastructure of chloroplasts and mitochondria in transgenic tobacco, indicating that this level of PRO had no toxic effects [118]. Though the assumption of osmolyte toxicity is largely inconclusive, it seems that plants treated with exogenous application of osmotica may suffer from an 'overdose' compared to their endogenous levels. In fact, in certain cases, exogenous application (both foliar as well as through the rooting medium) of osmolytes such as GB, PRO, inositol, and mannitol have indicated stress alleviation in many plants [29,119]. Exogenous application of osmotica, such as GB, may also enhance the membrane integrity of chloroplasts and also increase PS II efficiency [97,98], suggesting an osmoprotective role. Experiments on exogenous application of osmolytes have intrigued geneticists to manipulate the biosynthetic pathway of compatible solutes to enhance salt tolerance as osmolyte accumulation is often controlled by only one gene [102].

4.5. Possible Role of Osmolytes in Ion Regulation

Although the published literature has contradicting reports on the role of osmotic adjustment via osmolytes for maintaining turgor, recent patch-clamp studies suggest that osmolytes may have a significant contribution in regulating ion transporters such as K^+ outward rectifying channels (KORs), though this requires further investigation [120]. Thus, ion regulation via osmolytes may prove to be an important aspect in conferring salt tolerance. In plants, K^+ appears to be the most abundant cation in the cytosol (100–150 mM), which may account for osmotic adjustment [121], though under stressed conditions, the electrochemical gradients may lead to the loss of K^+. In halophytes, Na^+ and Cl^- seem to play a major role in osmotic adjustment [105]. Of these, Na^+ may enter the cell passively and could be used as a cheap osmoticum for maintaining cell turgor. Since Na^+ is toxic and may cause an imbalance in the cytosolic K^+/Na^+ ratio and interferes with cell metabolism, its efficient sequestration in the vacuole is thus essential. Pumping of one mole of Na^+ against the electrochemical gradient requires only 3.5 mol of ATP compared with 30–50 mol of ATP for one mole of organic osmolyte [115]. As mentioned above, some of the sugar alcohols and polyols regulate osmotic balance by sequestering Na^+ in the vacuole or apoplast. It appears that osmotic adjustment is collectively achieved by maintaining a balance between ion regulation, synthesis and accumulation of organic solutes, as well as maintenance of K^+ in the cytosol [122].

5. Effects of Salinity on Function and Protection of Photosystems

Under saline conditions, decreases in CO_2 assimilation via the Calvin cycle accompany a decrease in photochemical electron sink, which in the presence of light impacts the functioning/efficiency of photosystems [31]. In some sensitive plants such as olives, decreases in the F_v/F_m ratios indicate the incidence of photodamage under saline conditions [35]. Likewise, increases in salinity resulted in a gradual decrease in activities of PSI and PSII in four rice cultivars [123]. However, unchanged F_v/F_m hints towards sustained PSII under saline conditions [124], such as in the Mangalamahsuri variety of rice [125]. PSII-mediated electron transport increased in low salinity followed by a decrease at high salinity in the halophyte *Bruguiera parviflora* [55]. In other instances, inhibition of de novo protein synthesis, especially of the D1 protein, indicated a lack of efficient PS II repair under saline conditions [2,126,127]. A compensation mechanism of PsbO protein induction has been observed in some studies to stabilize the PSII structure under salinity [128]. Among C_3 plants, salinity reportedly resulted in poor PSII function in glycophytes such as rice and Arabidopsis [41,129] but not in the halophyte *Arthrocnemum macrostachyum* [130,131]. Several tolerant species, including halophytes such as *Sarcocornia fruticosa* [132] and *Atriplex centralasiatica* [133], also employ the xanthophyll cycle for non-photochemical quenching that dissipates excess excitation energy of PSII in the form of heat as a 'first line of defense' [31,133], thus preventing the formation of potentially cytotoxic reactive ROS. The xanthophyll cycle enzyme violaxanthin de-epoxidase consumes NADPH, which if accumulated may cause the over-reduction of reaction centers, and thereby enhance ROS (especially superoxide) formation [134]. Hence, the timely induction of the xanthophyll cycle may protect plants under stressful conditions in multiple ways. Many halophytes are reported to exhibit reversible midday photoinhibition of PSII activity to limit excitation of the PSII reaction centers [130,134]. This mechanism also minimizes the possibility of ROS formation in salt-stressed plants under high light and is considered an important ecophysiological adaptation to salinity [31]. A decrease in the antennae size due to decreased chlorophyll content was also observed in *Arthrocnemum macrostachyum* to limit PSII excitation [132].

PSI is reportedly more stress-resistant than PSII and seems to impart salt tolerance by increasing cyclic electron flow to generate ATP while avoiding the accumulation of toxic-reducing species [135–137]. Information about PSI in halophytes is scarce. PSI reaction center subunit IV protein (PsaE) was upregulated under salinity in wild halophytic rice *Porteresia coarctata* but not in conventional sensitive rice [137]. Similarly, salinity treatment caused an increase in PSI transcripts in *M. crystallinum* [138]. Formation of ATP via cyclic electron flow around PSI helped to prevent overaccumulation of Na^+ in chloroplasts of soybean [139].

Cultured plant cell lines have also been utilized to study salt-adapted tobacco cells [140,141]. Heterotrophic tobacco cells adapted to grow at 428 mM NaCl showed elevated levels of chlorophyll and lower levels of starch along with increased CO_2 fixation, oxygen evolution and photorespiration, compared to unadapted cells [140]. This was coupled with higher levels of PS-I- and PS-II-associated proteins, including Rubisco. These cells were found to have acquired a significant level of salt-tolerant photosynthetic competence [140]. Further analysis showed that oxygen evolution and CO_2 fixation were more resistant to inhibition by NaCl in the salt-adapted cells [141].

6. Effects of Salinity on CO_2 Assimilation Enzymes

Information on the effects of salinity on chloroplast CO_2 assimilation enzymes is limited among halophytes. Generally, CO_2 assimilation reactions are considered more sensitive to salinity than photochemical reactions of photosynthesis [31]. Several studies have reported that salinity generally inhibits many enzymes of the Calvin cycle [137,142,143].

Ribulose-1,5-bisphosphate carboxylase/oxygenase (Rubisco) is the key photosynthetic enzyme that catalyzes the fixation of atmospheric CO_2 in plants during the Calvin cycle [144]. It is the most abundant protein in leaves that accounts for 30% (C_4 plants) to 50% (C_3 plants) of total soluble protein in leaves [145]. In C_3 plants, it is localized in all chloroplasts while in C_4 plants with Kranz anatomy, Rubisco is localized specifically in the bundle sheath but not mesophyll chloroplasts [65]. In single-cell C_4 species, Rubisco mRNA could be targeted to the proximal or central compartment of chloroplasts [146]. The activity of Rubisco was mostly examined by direct measurement of the enzyme activity or protein levels and measurement of its carboxylase activity (V_{cmax}) [12]. Salinity exposure causes a decrease in Rubisco activity in most plant species regardless of C_3 or C_4 type [19,146–148]. In addition, the Rubisco levels also decreased under saline conditions in both halophytes and glycophytes. For example, salinity caused an inhibition (~50%) of Rubisco activity in maize, a glycophyte, and in *Atriplex spongiosa*, a halophyte [149]. In some other instances, Rubisco activity was improved both in either low [13,20] or high salinity [150]. Rubisco activity also depends on the function of a supporting enzyme, Rubisco activase, which revitalizes the active sites of Rubisco by removing inhibitory sugar phosphates [151,152]. The enhanced activity of Rubisco activase was found in rice as well as in many halophytes, such as *S. salsa* [143] and *Thellungiella salsuginea*, under saline conditions [153]. More efficient Rubisco activation was found in *T. salsuginea* compared to *Arabidopsis thaliana* [153].

Chloroplastic fructose-1,6-bisphosphatase is considered a metabolic control point of the Calvin cycle [44,154]. In vitro salt sensitivity of this enzyme was higher in salt-sensitive rice (*Oryza sativa* cv. IR26) than its wild halophytic relative *Porteresia coarctata* [142]. However, the inhibitory effects of salinity could be reversed by preincubation of the enzyme with osmolytes (effectiveness order: polyol>sugars) [142], suggesting a lower level of in vivo inhibition of chloroplastic fructose 1,6-bisphosphatase under salinity in halophytes with higher amounts of osmolytes compared to glycophytes.

Phosphoenolpyruvate carboxylase (PEPC) is the key enzyme of C_4 photosynthetic metabolism that catalyzes the β-carboxylation of phosphoenolpyruvate to form four-carbon acid oxaloacetate in the mesophyll cells [144,155]. It is considered more sensitive to salinity than Rubisco [149]. Furthermore, PEPC isolated from the halophyte *Atriplex spongiosa* was found more salt-sensitive in the in vitro studies than the one from the glycophyte maize [149]. Contrary to these observations, an increase in PEPC activity was reported in the halophyte *Mesembryanthemum crystallinum* [156] and in the C_4 species *Bienertia sinuspersici* under salinity [157]. Increased PEPC activity helps concentrate CO_2 around Rubisco and substantially reduces the incidence of photorespiration, a major cause for growth reduction and ROS formation under environmental stresses in plants.

Pyruvate orthophosphate dikinase is the rate-limiting enzyme of the C_4 cycle that catalyzes a reversible reaction to regenerate the primary CO_2 acceptor phosphoenolpyruvate (PEP) [158]. However, its role in C_3 plants is not fully understood [159]. Pyruvate orthophosphate dikinase is found in both chloroplasts and the cytoplasm irrespective of C_3 or C_4 types [160]. In C_4 plants, it can comprise up to 10% of the total protein pool [161]. Interestingly, both isoforms are encoded by a single nuclear gene [162]. The labeling of the pyruvate orthophosphate dikinase protein was observed both in mesophyll and bundle sheath chloroplasts of kranz type C_4 plant maize, albeit with higher levels in the latter rather than the earlier-mentioned chloroplasts [163]. In single-cell C_4 species, pyruvate orthophosphate dikinase mRNA could be targeted to the peripheral or distal compartment chloroplasts [146]. Information about the impacts of salinity on the abundance and activity of this enzyme is scant. Salinity caused an increase in pyruvate orthophosphate dikinase levels in both types of chloroplasts in maize [163]. These enzymes are widely studied and are important for the biochemical reactions of photosynthesis [17,164,165]. Induction of PEP activity would also help maintain C_4 functionality under salinity stress and facilitate CO_2 assimilation for biomass buildup and reduce photorespiration, as mentioned above.

Effects on Salinity on the Gas Exchange Ecophysiology of Photosynthesis

The effects of salinity on photosynthetic synthetic gas exchange, which eventually supports CO_2 assimilation at the chloroplast level, varies not only among species but also depends on the magnitude of the salinity. For instance, the net CO_2 assimilation rate (P_N or A) and stomatal conductance (*Gs*) in sugar beet improved under low (75 mM NaCl) salinity while high (250 mM NaCl) was inhibitory [166]. An increase in P_N but not in transpiration (E) under low salinity resulted in improved water-use efficiency (WUE) in sugar beet plants [166]. Salinity stress decreased the P_N and *Gs* in wild-type wheat plants [24]. P_N and *Gs* increased transiently at 200 mM NaCl in comparison to controls and 400 mM NaCl in the halophyte *Sesuvium portulacastrum* [167]. Similarly, in many other halophyte species, such as *Arthrocnemum macrostachyum* (in up to 510 mM NaCl) [133] and *Atriplex portulacoides* (200 mM NaCl) [168], low to moderate salinity improved P_N. In contrast, salinity exposure resulted in decreased P_N and *Gs* in the halophytes *Panicum antidotale* [20] and *Aster tripolium* [169]. Hence, impacts of salinity not only vary among glycophytes but also halophyte species. In many cases, decreased *Gs* improves the WUE of plants under stress conditions as a trade-off at the expense of P_N. For instance, in *Sarcocornia fruticosa*, increased WUE accompanied a decline in P_N [170]. Similarly, many halophytes exhibit C_4 and CAM modes of photosynthetic CO_2 assimilation, which not only reduce wastage of photosynthetic energy through photorespiration but also decrease the consequent H_2O_2 (a common ROS) production at the peroxisome level [31,171].

7. Effects of Salinity on Chloroplast ROS Homeostasis

Exposure of plants to salinity results in a reduction in CO_2 assimilation rates, which in turn leads to the overreduction of PSII along with diversion of electrons to molecular oxygen, which generates reactive oxygen species (ROS), particularly singlet oxygen [31,172,173]. In photosynthesizing leaves, chloroplasts are the major site for ROS production during the daytime [174]. In C_3 plants, photorespiration resulting from the oxygenase activity of Rubisco in chloroplasts is another source of ROS generation in peroxisomes [31]. Salinity-induced stimulation of electron flow to molecular oxygen has been reported in several plant species [31,175,176]. Major ROS produced in chloroplasts include singlet oxygen ($^1O_2$), superoxide radical ($O_2^{\bullet -}$), hydrogen peroxide (H_2O_2) and hydroxyl radical ($^\bullet OH$) [31,173]. Since detection of radicle-type ROS is difficult, most studies examine H_2O_2 (non-radicle ROS) formation following salinity exposure [31]. In addition, studies on ROS formation in isolated chloroplasts, particularly of halophytes, are very limited. Wiciarz et al. [153] reported that isolated thylakoids from a halophyte *Thellungiella salsuginea* produced higher H_2O_2 levels than the model glycophyte *Arabidopsis thaliana*. However, when both plant types were exposed to salt stress, even at the low level of 100 mM NaCl, Arabidopsis produced a higher H_2O_2 than *T. salsuginea* and at a 300 mM NaCl concentration. Similarly, a substantially higher H_2O_2 level was observed in chloroplasts of wild salt-tolerant tomato *Lycopersicon pennellii* compared to chloroplasts of sensitive tomato *L. esculentum* under stress-free growth conditions. However, under NaCl stress, a decrease in H_2O_2 level was noted for wild tomato while the levels were increased in the sensitive species [171]. This indicates that halophyte species have efficient mechanisms to control the production of ROS or detoxify them compared to glycophytes, either through the dissipation of excess excitation energy to alternative electron sinks, such as the plastid terminal oxidase [29,172–180] (PTOX) or ROS-scavenging system [31,174]. Alternative electron sinks not only provide 'safety valves' for the efficient functioning of the photosynthetic machinery but also act as an 'avoidance' tool for control of ROS formation. Tightly regulated levels of ROS are now acknowledged as 'signals' for the regulation of different plant processes, including the defense/tolerance response of plants [3,99]. For instance, ROS modulate the function of some plasma membrane ion transporters, such as those regulating cytosolic Na^+ and K^+ [177,181–185]. Similarly, a ROS 'surge' in response to salinity exposure may also activate chloroplast retrograde signaling pathways [180].

In order to prevent oxidative damage due to ROS accumulation, chloroplasts possess many enzymatic and nonenzymatic antioxidants [29,138,182,185]. Key enzymatic antioxidants are superoxide dismutases (SOD), enzymes of the Foyer–Halliwell–Asada pathway (also known as the ascorbate–glutathione cycle), and glutathione peroxidase (GPX), whereas ascorbate and glutathione are common nonenzymatic antioxidants of chloroplasts (Figure 2) [138,182,186]. Antioxidants in various combinations play an important role to keep the levels of ROS in 'functionally useful' ranges for signaling various plant processes and stress responses [31]. The water–water cycle is among the key processes responsible for ROS homeostasis in chloroplasts and is essential for salinity tolerance (Figure 2) [31,187]. Ground state molecular oxygen (O_2) produced during photolysis of water in chloroplasts can accept electrons from excited photosystems, particularly the thylakoid membrane-bound primary electron acceptor of PSI to form $O_2^{\bullet-}$ through a reaction called the Mehler reaction [188]. The acceptor side of the electron transport chain in PSII may also contribute to electron leakage to O_2 to generate $O_2^{\bullet-}$. Thylakoid membrane-bound copper/zinc superoxide dismutase (Cu/Zn SOD) converts $O_2^{\bullet-}$ into H_2O_2, which is finally reduced into the water by the action of thylakoid membrane-bound ascorbate peroxidase (tAPX), thus completing the 'water–water cycle' [189]. The Foyer–Halliwell–Asada pathway (also known as the ascorbate–glutathione cycle) in chloroplasts is an extension of the water–water cycle and involves quenching of ROS in chloroplasts by consuming NADPH, which also contributes to relaxing the 'overreduction of photosystems' by providing NADP (the final electron acceptor of PSI), and thereby minimizing the chances of further ROS generation (Figure 2) [167,190]. In this pathway, the H_2O_2 generated from dismutation of $O_2^{\bullet-}$ by SOD is neutralized into water by the action of stromal ascorbate peroxidase (APX) using ascorbate (AsA) as the electron donor. Oxidized ascorbate is recycled by monodehydroascorbate reductase (MDHAR) and/or dehydroascorbate reductase (DHAR). The latter consumes glutathione (GSH), which is finally recycled by the action of glutathione reductase (GR) that uses NADPH as an electron donor [167,190,191]. Often, an upregulation of enzymes involved in antioxidant processes is reported in chloroplasts under environmental stresses, with a higher magnitude of tolerance compared to sensitive species [31,181]. For instance, salinity exposure resulted in enhanced activities of SOD, APX and MDHAR in chloroplasts of halophytic wild tomato *Lycopersicon pennellii* compared to conventional sensitive tomato *L. esculentum* [172]. In addition, thioredoxin/peroxiredoxin (Trx/Prx) and glutathione peroxidase (GPX) also reportedly quenched salinity-induced excess H_2O_2 in chloroplasts [187]. Lipophilic tocopherol can protect chloroplast thylakoid membranes from oxidative damage [189]. The $^1O_2$ produced by PSII is mainly detoxified by carotenoids and tocopherols found in the chloroplast membranes [192]. Carotenoids detoxify $^1O_2$ not only through the xanthophyll cycle (NPQ) but also by direct quenching of $^1O_2$ [193]. However, some C_4 plants, especially those with NADP-malic enzyme (NADP-ME) subtypes, lack PSII in their bundle sheath chloroplasts and hence supposedly lack $^1O_2$ production [194].

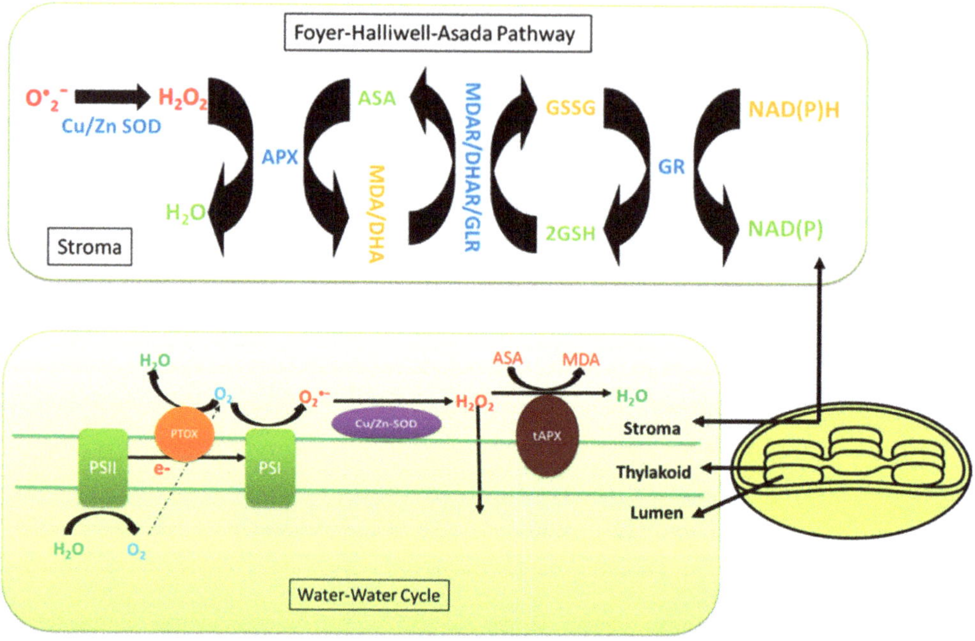

Figure 2. The Foyer–Halliwell–Asada pathway (also known as the ascorbate–glutathione cycle) and the water–water cycle are responsible to quench the superoxide radicles and hydrogen peroxide in the chloroplasts.

8. Summary

Soil salinity is one of the major challenges to the sustainable development of agriculture in different parts of the world. Salinity has detrimental effects on plant growth by imposing several constraints. For instance, salt-induced toxicity impairs the normal functioning of the organelles, such as chloroplasts—the green plastids—which house several important biochemical reactions, including photosynthesis. Chloroplast dysfunction as a result of various environmental stresses, including salinity, has been reported to have detrimental effects on plants [173]. Chloroplasts, in addition to being a site of various metabolic reactions, also act as global sensors to sense and communicate the developmental, operational and environmental changes to the nucleus.

Understanding the effect of salinity on chloroplast function and the response of various metabolic reactions to salt stress is necessary for the development of salt-tolerant crops. Little attention has been paid to how salinity affects chloroplasts and the stromal metabolic reactions. Salinity-related changes in the size, number, lamellar organization, lipid and starch accumulation, and trafficking across the chloroplast membrane are dependent on the plant species and its level of salt tolerance. Chloroplast swelling or alteration in thylakoid membranes of glycophytes may be linked with the ionic component of salinity while some halophytes are affected by the osmotic effect of high salinity (Figure 1). Most halophytes either maintain chloroplast structure or enhance grana development under salinity stress. Swelling of thylakoids and disruption of chloroplast envelopes in mesophyll cells along with intact chloroplasts in bundle sheath cells is a general C_4 response under salinity, irrespective of the subtype.

Halophytes and glycophytes have evolved different pathways to respond to salinity stress. For example, halophytes are much better adapted at maintaining a lower salt concentration in the cytoplasm compared to glycophytes. Likewise, chloroplasts in halophytes seem to have a better antioxidant system than those of glycophytes, and consequently more protected photosynthetic apparatuses under salt stress. Similarly, salinity-triggered starch deposition appears to be a damage symptom in glycophytes but a survival strategy in halophytes. The salinity-induced influx of Na^+ and Cl^- appears beneficial for halophytes but lethal for glycophytes (Figure 1). Accumulation of Na^+ or Cl^- disrupts ionic homeostasis, impairs protein synthesis and interferes with the enzymatic activities of the organelle. However, recent work suggests that the negative effects of these ions on plant health are not because of toxicity per se but are the result of interference with the absorption or metabolism of other essential ions [28]. This view stems from the evidence that K^+ influx in chloroplasts is reduced with excessive Na^+ or Cl^- accumulation. K^+ is an essential element for the plant cell and is not only required for chloroplast development but also for pH regulation, maintenance of the electron transport chain and thylakoid restacking [28,29]. Osmolyte synthesis suggests that organic solutes may help in fine adjustment along with ion transport (vacuolar compartmentation) and accumulation of cytosolic K^+ in stressed environments rather than osmotic adjustment. However, osmolytes are certainly involved in the osmoprotection of membrane transport proteins and the scavenging of ROS. Despite ion regulation and osmotic adjustment, salinity induces many changes in chloroplast functions and signaling.

Chloroplastic CO_2 fixation is generally more sensitive to salinity than the thylakoid reactions. However, CO_2 fixation in many halophytes is reportedly less prone to salinity compared to glycophytes. One major evolutionary adaptation that seems to operate in halophytes is the switching of CO_2 concentration around Rubisco under stressful environmental conditions, including salinity. The reduced photosynthetic efficiency is considered a major salt-induced constraint inhibiting plant growth, and ultimately crop productivity. However, it is not yet clear whether the decrease in photosynthesis is the cause of growth reduction or the reduction in the growth rate causes a decrease in photosynthesis under salt stress. Nevertheless, a reduced rate of photosynthesis leads to higher production of ROS and also triggers the activity of ROS-scavenging enzymes. The higher activity of the ROS-detoxifying enzymes maintains a level of these species in a functionally useful range required for cell signaling. These enzymatic systems are naturally present in plants. Although differences in the activity of these enzymes have been reported in different genotypes, it is believed to be associated with responses such as stomatal closures, reduction in the CO_2 fixation rates and an increase in photorespiration under stressful conditions [188,194]. Tight regulation of ROS alongside many chloroplastic metabolites also function as 'putative' signals for communication between chloroplasts and the nucleus (as well as other organelles) via so-called 'retrograde signaling'. Despite information on crop and model plants, our knowledge about such signaling in halophytes is still far from full comprehension. Chloroplast functions, including photosynthesis, are integrated with other basic plant metabolic mechanisms of the plant in response to stresses, including salinity, and multiple factors work together to confer tolerance against salinity [195]. These factors include ion regulation that controls uptake and transport of salt and other ions to compartments within the plant cell, synthesis of compatible solutes, antioxidative enzymes and plant hormones and changes in photosynthesis and membranes in the cell [195]. Some of these occur within the chloroplast but are not limited to that location. These mechanisms are quite complicated, and many questions remain unanswered [195,196]. Some of these questions include how the plant senses salinity to initiate the signaling process, the precise details of how salinity stress leads to stomatal closure and growth reduction and the specific targets of ion toxicity in plant cells [196]. While advances are being made, a detailed understanding of the mechanisms behind salt tolerance is not yet clear. A comprehensive understanding of these mechanisms by employing multidisciplinary approaches is necessary for their effective incorporation into salt-sensitive crops for better crop yields under stressful environments.

Author Contributions: All authors participated in the writing of different sections, editing, and proofreading of the manuscript; M.Z.A. (Figure 1) and A.H. (Figure 2) prepared the figures. All authors have read and agreed to the published version of the manuscript.

Funding: No specific funding was received for this work.

Institutional Review Board Statement: Not applicable.

Informed Consent Statement: Not applicable.

Data Availability Statement: As a review paper all data is available in the work referenced.

Acknowledgments: We are grateful to Narendra K. Singh, Auburn University (emeritus), for helpful discussions in developing this review.

Conflicts of Interest: The authors declare no conflict of interest.

References

1. Hussain, T.M.; Chandrasekhar, T.; Hazara, J.; Sultan, Z.; Saleh, B.K.; Gopal, G.R. Recent advances in salt stress biology—A review. *Biotech. Mol. Biol. Rev.* **2008**, *3*, 8–13.
2. Allakhverdiev, S.I.; Murata, N. Salt stress inhibits photosystems II and I in cyanobacteria. *Photosynth. Res.* **2008**, *98*, 529–539. [CrossRef]
3. Akyol, T.Y.; Yilmaz, O.; Uzİlday, B.; Uzİlday, R.Ö.; Türkan, İ. Plant response to salinity: An analysis of ROS formation, signaling, and antioxidant defense. *Turk. J. Bot.* **2020**, *44*, 1–13. [CrossRef]
4. Badawi, G.H.; Yamauchi, E.Y.; Shimada, R.; Sasaki, N.; Kawano, K.; Tanaka, K.; Tanaka, K. Enhanced tolerance to salt stress and water deficit by overexpressing superoxide dismutase in tobacco (*Nicotiana tabacum*) chloroplasts. *Plant Sci.* **2004**, *166*, 919–928. [CrossRef]
5. Jing, X.; Hou, P.; Lu, Y.; Deng, S.; Li, N.; Zhao, R.; Sun, J.; Wang, Y.; Han, Y.; Lang, T. Overexpression of copper/zinc superoxide dismutase from mangrove *Kandelia candel* in tobacco enhances salinity tolerance by the reduction of reactive oxygen species in chloroplast. *Front. Plant Sci.* **2015**, *5*, 23. [CrossRef]
6. Tseng, M.J.; Liu, C.-W.; Yiu, J.-C. Enhanced tolerance to sulfur dioxide and salt stress of transgenic Chinese cabbage plants expressing both superoxide dismutase and catalase in chloroplasts. *Plant Physiol. Biochem.* **2007**, *45*, 822–833. [CrossRef] [PubMed]
7. Koussevitzky, S.; Nott, A.; Mockler, T.C.; Hong, F.; Sachetto-Martins, G.; Surpin, M.; Lim, J.; Mittler, R.; Chory, J. Signals from chloroplasts converge to regulate nuclear gene expression. *Science* **2007**, *316*, 715–719. [CrossRef]
8. Munns, R.; Tester, M. Mechanisms of salinity tolerance. *Ann. Rev. Plant Biol.* **2008**, *59*, 651–681. [CrossRef] [PubMed]
9. Yan, K.; Shao, H.; Shao, C.; Chen, P.; Zhao, S.; Brestic, M.; Chen, X. Physiological adaptive mechanisms of plants grown in saline soil and implications for sustainable saline agriculture in coastal zone. *Acta Physiol. Plant* **2013**, *35*, 2867–2878. [CrossRef]
10. Abdelhamid, M.T.; Sekara, A.; Pessarakli, M.; Alarcon, J.J.; Brestic, M.; El-Ramady, H.; Gad, N.; Mohamed, H.I.; Fares, W.M.; Heba, S.S.; et al. New approaches for improving salt stress tolerance in rice. In *Rice Research for Quality Improvement: Genomics and Genetic Engineering*; Roychoudhury, A., Ed.; Springer: Singapore, 2020. [CrossRef]
11. Hajihashemi, S.; Skalicky, M.; Brestic, M.; Pavla, V. Cross-talk between nitric oxide, hydrogen peroxide and calcium in salt-stressed *Chenopodium quinoa* Willd. At seed germination stage. *Plant Physiol. Biochem.* **2020**, *154*, 657–664. [CrossRef]
12. Miransari, M.; Smith, D. Sustainable wheat (*Triticum aestivum* L.) production in saline fields: A review. *Crit. Rev. Biotechnol.* **2019**, *39*, 999–1014. [CrossRef]
13. Rasel, M.; Tahjib-Ul-Arif, M.; Hossain, M.A.; Hassan, L.; Farzana, S.; Brestic, M. Screening of salt-tolerant rice landraces by seedling stage phenotyping and dissecting biochemical determinants of tolerance mechanism. *J. Plant Growth Regul.* **2020**, 1–16. [CrossRef]
14. Ibrahimova, U.; Kumari, P.; Yadav, S.; Rastogi, A.; Antala, M.; Suleymanova, Z.; Ziveak, M.; Tahjib-Ul-Arif, M.; Hussain, S.; Abdelhamid, M.; et al. Progress in understanding salt stress response in plants using biotechnological tools. *J. Biotechnol.* **2021**, *329*, 180–191. [CrossRef] [PubMed]
15. Flowers, T.J.; Colmer, T.D. Salinity Tolerance in Halophytes. *New Phytol.* **2008**, *179*, 945–963. [CrossRef] [PubMed]
16. Asrar, H.; Hussain, T.; Hadi, S.M.S.; Gul, B.; Nielsen, B.L.; Khan, M.A. Salinity induced changes in light harvesting and carbon assimilating complexes of *Desmostachya bipinnata* (L.) Staph. *Environ. Exp. Bot.* **2017**, *135*, 86–95. [CrossRef]
17. Bellasio, C.; Quirk, J.; Beerling, D.J. Stomatal and non-stomatal limitations in savanna trees and C4 grasses grown at low, ambient and high atmospheric CO_2. *Plant Sci.* **2018**, *274*, 181–192. [CrossRef]
18. Flexas, J.; Barbour, M.M.; Brendel, O.; Cabrera, H.M.; Carriquí, M.; Díaz-Espejo, A.; Douthe, C.; Dreyer, E.; Ferrio, J.P.; Gago, J.; et al. Mesophyll diffusion conductance to CO_2: An unappreciated central player in photosynthesis. *Plant Sci.* **2012**, *193–194*, 70–84. [CrossRef]
19. Galmés, J.; Molins, A.; Flexas, J.; Conesa, M.À. Coordination between leaf CO_2 diffusion and Rubisco properties allows maximizing photosynthetic efficiency in *Limonium* species. *Plant Cell Environ.* **2017**, *40*, 2081–2094. [CrossRef]
20. Hussain, T.; Huchzermeyer, B.; Koyro, H.-W.; Khan, M.A. Linkage between leaf development and photosynthetic response at hyperosmotic salinity in the C-4 grass *Panicum antidotale*. *Flora* **2019**, *256*, 52–60. [CrossRef]
21. Nunes-Nesi, A.; Nascimento, V.d.L.; de Oliveira Silva, F.M.; Zsögön, A.; Araújo, W.L.; Sulpice, R. Natural genetic variation for morphological and molecular determinants of plant growth and yield. *J. Exp. Bot.* **2016**, *67*, 2989–3001. [CrossRef]

22. Rasouli, F.; Kiani-Pouya, A.; Tahir, A.; Shabala, L.; Chen, Z.; Shabala, S. A comparative analysis of stomatal traits and photosynthetic responses in closely related halophytic and glycophytic species under saline conditions. *Environ. Exp. Bot.* **2021**, *181*, 104300. [CrossRef]
23. Rastogi, A.; Kovar, M.; He, X.; Zivcak, M.; Kataria, S.; Kalaji, H.M.; Skalicky, M.; Ibrahimova, U.F.; Hussain, S.; Mbarki, S.; et al. JIP-test as a tool to identify salinity toelrance in sweet sorghum genotypes. *Phostosynthetica* **2020**, *58*, 518–520. [CrossRef]
24. Zuo, Z.; Ye, F.; Wang, Z.; Li, S.; Li, H.; Guo, J.; Mao, H.; Zhu, X.; Li, X. Salt acclimation induced salt tolerance in wild-type and chlrophyll b-deficient mutant wheat. *Plant Soil Environ.* **2021**, *67*, 26–32. [CrossRef]
25. Tomeo, N.J.; Rosenthal, D.M. Variable mesophyll conductance among soybean cultivars sets a tradeoff between photosynthesis and water-use-efficiency. *Plant Physiol.* **2017**, *174*, 241–257. [CrossRef] [PubMed]
26. Munns, R.; Passioura, J.B.; Colmer, T.D.; Byrt, C.S. Osmotic adjustment and energy limitations to plant growth in saline soil. *New Phytol.* **2020**, *225*, 1091–1096. [CrossRef] [PubMed]
27. Rodrigues, N.F.; da Fonseca, G.C.; Kulcheski, F.R.; Margis, R. Salt stress affects mRNA editing in soybean chloroplasts. *Genet. Mol. Biol.* **2017**, *40*, 200–208. [CrossRef] [PubMed]
28. Zhao, C.; Zhang, H.; Song, C.; Zhu, J.-K.; Shabala, S. Mechanisms of plant responses and adaptation to soil salinity. *Innovation* **2020**, *1*, 100017. [CrossRef]
29. Bose, J.; Munns, R.; Shabala, S.; Gilliham, M.; Pogson, B.; Tyerman, S.D. Chloroplast function and ion regulation in plants growing on saline soils: Lessons from halophytes. *J. Exp. Bot.* **2017**, *68*, 3129–3143. [CrossRef]
30. Pan, T.; Liu, M.; Kreslavski, V.D.; Zharmukhamedov, S.K.; Nie, C.; Yu, M.; Kuznetsov, V.V.; Allakhverdiev, S.I.; Shabala, S. Non-stomatal limitation of photosynthesis by soil salinity. *Crit. Rev. Environ. Sci. Technol.* **2021**, *51*, 791–825. [CrossRef]
31. Gulzar, S.; Hussain, T.; Gul, B.; Hameed, A. *Photosynthetic Adaptations and Oxidative Stress Tolerance in Halophytes from Warm Subtropical Region*; Springer: Cham, Switzerland, 2020. [CrossRef]
32. Maxwell, K.; Johnson, G.N. Chlorophyll fluorescence—A practical guide. *J. Exp. Bot.* **2000**, *51*, 659–668. [CrossRef]
33. Ashraf, M.; Harris, P.J.C. Photosynthesis under stressful environments: An overview. *Photosynthetica* **2013**, *51*, 163–190. [CrossRef]
34. Koyro, H.-W.; Hussain, T.; Huchzermeyer, B.; Khan, M.A. Photosynthetic and growth responses of a perennial halophytic grass *Panicum turgidum* to increasing NaCl concentrations. *Environ. Exp. Bot.* **2013**, *91*, 22–29. [CrossRef]
35. Loreto, F.; Centritto, M.; Chartzoulakis, K. Photosynthetic limitations in olive cultivars with different sensitivity to salt stress. *Plant Cell Environ.* **2003**, *26*, 595–601. [CrossRef]
36. Cooper, G.M. *The Cell: A Molecular Approach*, 2nd ed.; Sinauer Associates: Sunderland, MA, USA, 2000.
37. Papadakis, I.E.; Giannakoula, A.; Therios, I.N.; Bosabalidis, A.M.; Moustakas, M.; Nastou, A. Mn-induced changes in leaf structure and chloroplast ultrastructure of *Citrus volkameriana* (L.) plants. *J. Plant Physiol.* **2007**, *164*, 100–103. [CrossRef]
38. Blumenthal-Goldschmidt, S.; Poljakoff-Mayber, A. Effect of substrate salinity on growth and submicroscopic structure of leaf cells of *A. halimus* L. *Aust. J. Bot.* **1968**, *16*, 469–478. [CrossRef]
39. Hall, J.; Barr, R.; Al-Abbas, A.; Crane, F. The ultrastructure of chloroplasts in mineral-deficient maize leaves. *Plant Physiol.* **1972**, *50*, 404–409. [CrossRef]
40. Kelley, D.B. Salinity Effects on Growth and Fine Structure of *Atriplex halimus* L. Master's Thesis, Texas Tech University, Lubbock, TX, USA, 1974. Available online: http://hdl.handle.net/2346/20673 (accessed on 3 August 2021).
41. Štefanić, P.P.; Koffler, T.; Adler, G.; Bar-Zvi, D. Chloroplasts of salt-grown *Arabidopsis* seedlings are impaired in structure, genome copy number and transcript levels. *PLoS ONE* **2013**, *8*, e82548. [CrossRef]
42. Salama, S.; Trivedi, S.; Busheva, M.; Arafa, A.; Garab, G.; Erdei, L. Effects of NaCl salinity on growth, cation accumulation, chloroplast structure and function in wheat cultivars differing in salt tolerance. *J. Plant Physiol.* **1994**, *144*, 241–247. [CrossRef]
43. Preiss, J. Starch, sucrose biosynthesis and partition of carbon in plants are regulated by orthophosphate and triose-phosphates. *Trends Biochem. Sci.* **1984**, *9*, 24–27. [CrossRef]
44. Szabo-Nagy, A.; Galiba, G.; Erdei, L. Induction of soluble phosphatases under ionic and non-ionic osmotic stresses in wheat. *J. Plant Physiol.* **1992**, *140*, 629–633. [CrossRef]
45. Shu, S.; Guo, S.R.; Sun, J.; Yuan, L.Y. Effects of salt stress on the structure and function of the photosynthetic apparatus in Cucumis sativus and its protection by exogenous putrescine. *Physiol. Plant.* **2012**, *146*, 285–296. [CrossRef] [PubMed]
46. Chang, L.; Guo, A.; Jin, X.; Yang, Q.; Wang, D.; Sun, Y.; Huang, Q.; Wang, L.; Peng, C.; Wang, X. The beta subunit of glyceraldehyde 3-phosphate dehydrogenase is an important factor for maintaining photosynthesis and plant development under salt stress—Based on an integrative analysis of the structural, physiological and proteomic changes in chloroplasts in Thellungiella halophila. *Plant Sci.* **2015**, *236*, 223–238.
47. Delfine, S.; Alvino, A.; Zacchini, M.; Loreto, F. Consequences of salt stress on conductance to CO_2 diffusion, Rubisco characteristics and anatomy of spinach leaves. *Funct. Plant Biol.* **1998**, *25*, 395–402. [CrossRef]
48. Mitsuya, S.; Takeoka, Y.; Miyake, H. Effects of sodium chloride on foliar ultrastructure of sweet potato (*Ipomoea batatas* Lam.) plantlets grown under light and dark conditions in vitro. *J. Plant Physiol.* **2000**, *157*, 661–667. [CrossRef]
49. Wang, L.; Liang, W.; Xing, J.; Tan, F.; Chen, Y.; Huang, L.; Cheng, C.L.; Chen, W. Dynamics of chloroplast proteome in salt-stressed mangrove *Kandelia candel* (L.) Druce. *J. Proteome Res.* **2013**, *12*, 5124–5136. [CrossRef]
50. Yamane, K.; Kawasaki, M.; Taniguchi, M.; Miyake, H. Correlation between chloroplast ultrastructure and chlorophyll fluorescence characteristics in the leaves of rice (*Oryza sativa* L.) grown under salinity. *Plant Prod. Sci.* **2008**, *11*, 139–145. [CrossRef]

51. Oi, T.; Enomoto, S.; Nakao, T.; Arai, S.; Yamane, K.; Taniguchi, M. Three-dimensional ultrastructural change of chloroplasts in rice mesophyll cells responding to salt stress. *Ann. Bot.* **2020**, *125*, 833–840. [CrossRef]
52. Gao, H.-J.; Yang, H.-Y.; Bai, J.-P.; Liang, X.-Y.; Lou, Y.; Zhang, J.-L.; Wang, D.; Zhang, J.-L.; Niu, S.-Q.; Chen, Y. Ultrastructural and physiological responses of potato (*Solanum tuberosum* L.) plantlets to gradient saline stress. *Front. Plant Sci.* **2015**, *5*, 787. [CrossRef]
53. Hasan, R.; Kawasaki, M.; Taniguchi, M.; Miyake, H. Salinity stress induces granal development in bundle sheath chloroplasts of maize, an NADP-malic enzyme-type C4 plant. *Plant Prod. Sci.* **2006**, *9*, 256–265. [CrossRef]
54. Goussi, R.; Manaa, A.; Derbali, W.; Cantamessa, S.; Abdelly, C.; Barbato, R. Comparative analysis of salt stress, duration and intensity, on the chloroplast ultrastructure and photosynthetic apparatus in *Thellungiella salsuginea*. *J. Photochem. Photobiol. B Biol.* **2018**, *183*, 275–287. [CrossRef] [PubMed]
55. Parida, A.K.; Das, A.B.; Mittra, B. Effects of nacl stress on the structure, pigment complex composition, and photosynthetic activity of mangrove *Bruguiera parviflora* chloroplasts. *Photosynthet* **2003**, *41*, 191. [CrossRef]
56. Yoshimura, Y.; Kubota, F.; Ueno, O. Structural and biochemical bases of photorespiration in C4 plants: Quantification of organelles and glycine decarboxylase. *Planta* **2004**, *220*, 307–317. [CrossRef]
57. Omoto, E.; Kawasaki, M.; Taniguchi, M.; Miyake, H. Salinity induces granal development in bundle sheath chloroplasts of NADP-malic enzyme type C4 plants. *Plant Prod. Sci.* **2009**, *12*, 199–207. [CrossRef]
58. Omoto, E.; Taniguchi, M.; Miyake, H. Effects of salinity stress on the structure of bundle sheath and mesophyll chloroplasts in NAD-malic enzyme and PCK type C4 plants. *Plant Prod. Sci.* **2010**, *13*, 169–176. [CrossRef]
59. Kubínová, Z.; Janáček, J.; Lhotáková, Z.; Kubínová, L.; Albrechtová, J. Unbiased estimation of chloroplast number in mesophyll cells: Advantage of a genuine three-dimensional approach. *J. Exp. Bot.* **2013**, *65*, 609–620. [CrossRef] [PubMed]
60. Winter, H.; Robinson, D.G.; Heldt, H.W. Subcellular volumes and metabolite concentrations in barley leaves. *Planta* **1993**, *191*, 180–190. [CrossRef]
61. Marschner, H.; Possingham, J.V. Effect of K^+ and Na^+ on growth of leaf discs of sugar beet and spinach. *Z. Pflanzenphysiol.* **1975**, *75*, 6–16. [CrossRef]
62. Aldesuquy, H.; Baka, Z.; Mickky, B. Kinetin and spermine mediated induction of salt tolerance in wheat plants: Leaf area, photosynthesis and chloroplast ultrastructure of flag leaf at ear emergence. *Egypt. J. Basic Appl. Sci.* **2014**, *1*, 77–87. [CrossRef]
63. Flowers, T.J.; Munns, R.; Colmer, T.D. Sodium chloride toxicity and the cellular basis of salt tolerance in halophytes. *Ann. Bot.* **2014**, *115*, 419–431. [CrossRef] [PubMed]
64. Cosentino, C.; Fischer-Schliebs, E.; Bertl, A.; Thiel, G.; Homann, U. Na^+/H^+ antiporters are differentially regulated in response to NaCl stress in leaves and roots of *Mesembryanthemum crystallinum*. *New Phytol.* **2010**, *186*, 669–680. [CrossRef]
65. Robinson, S.; Downton, W. Potassium, sodium and chloride ion concentrations in leaves and isolated chloroplasts of the halophyte *Suaeda australis* R. Br. *Aust. J. Plant Physiol.* **1985**, *12*, 471–479. [CrossRef]
66. Robinson, S.P.; Downton, W.J.S. Potassium, sodium, and chloride content of isolated intact chloroplasts in relation to ionic compartmentation in leaves. *Arch. Biochem. Biophys.* **1984**, *228*, 197–206. [CrossRef]
67. Müller, M.; Kunz, H.-H.; Schroeder, J.I.; Kemp, G.; Young, H.S.; Neuhaus, H.E. Decreased capacity for sodium export out of Arabidopsis chloroplasts impairs salt tolerance, photosynthesis and plant performance. *Plant J.* **2014**, *78*, 646–658. [CrossRef] [PubMed]
68. Yoshida, Y.; Miyagishima, S.-y.; Kuroiwa, H.; Kuroiwa, T. The plastid-dividing machinery: Formation, constriction and fission. *Curr. Opin. Plant Biol.* **2012**, *15*, 714–721. [CrossRef] [PubMed]
69. Jarvis, P.; López-Juez, E. Biogenesis and homeostasis of chloroplasts and other plastids. *Nat. Rev. Mol. Cell Biol.* **2013**, *14*, 787–802. [CrossRef] [PubMed]
70. Osteryoung, K.W.; Pyke, K.A. Division and Dynamic Morphology of Plastids. *Annu. Rev. Plant Biol.* **2014**, *65*, 443–472. [CrossRef]
71. Soll, J. Protein import into chloroplasts. *Curr. Opin. Plant Biol.* **2002**, *5*, 529–535. [CrossRef]
72. Ling, Q.; Jarvis, P. Regulation of chloroplast protein import by the ubiquitin E3 ligase SP1 is important for stress tolerance in plants. *Curr. Biol.* **2015**, *25*, 2527–2534. [CrossRef] [PubMed]
73. Pottosin, I.; Dobrovinskaya, O. Ion channels in native chloroplast membranes: Challenges and potential for direct patch-clamp studies. *Front. Physiol.* **2015**, *6*, 396. [CrossRef]
74. Beebo, A.; Mathai, J.C.; Schoefs, B.; Spetea, C. Assessment of the requirement for aquaporins in the thylakoid membrane of plant chloroplasts to sustain photosynthetic water oxidation. *FEBS Lett.* **2013**, *587*, 2083–2089. [CrossRef]
75. Pottosin, I.; Shabala, S. Transport across chloroplast membranes: Optimizing photosynthesis for adverse environmental conditions. *Mol. Plant* **2016**, *9*, 356–370. [CrossRef]
76. Tanz, S.K.; Castleden, I.; Hooper, C.M.; Vacher, M.; Small, I.; Millar, H.A. SUBA3: A database for integrating experimentation and prediction to define the SUB cellular location of proteins in *Arabidopsis*. *Nucl. Acids Res.* **2012**, *41*, D1185–D1191. [CrossRef]
77. Hooper, C.M.; Tanz, S.K.; Castleden, I.R.; Vacher, M.A.; Small, I.D.; Millar, A.H. SUBAcon: A consensus algorithm for unifying the subcellular localization data of the *Arabidopsis* proteome. *Bioinformatics* **2014**, *30*, 3356–3364. [CrossRef]
78. Finazzi, G.; Petroutsos, D.; Tomizioli, M.; Flori, S.; Sautron, E.; Villanova, V.; Rolland, N.; Seigneurin-Berny, D. Ions channels/transporters and chloroplast regulation. *Cell Calcium* **2015**, *58*, 86–97. [CrossRef] [PubMed]
79. Slabu, C.; Zörb, C.; Steffens, D.; Schubert, S. Is salt stress of faba bean (*Vicia faba*) caused by Na^+ or Cl^- toxicity? *J. Plant Nutr. Soil Sci.* **2009**, *172*, 644–651. [CrossRef]

80. Chen, X.-Q.; Yu, B.-J. Ionic effects of Na^+ and Cl^- on photosynthesis in *Glycine max* seedlings under isoosmotic salt stress. *J. Plant Physiol. Mol. Biol.* **2007**, *33*, 294–300.
81. Subbarao, G.; Ito, O.; Berry, W.; Wheeler, R. Sodium—A functional plant nutrient. *Crit. Rev. Plant Sci.* **2003**, *22*, 391–416. [CrossRef]
82. Furumoto, T.; Yamaguchi, T.; Ohshima-Ichie, Y.; Nakamura, M.; Tsuchida-Iwata, Y.; Shimamura, M.; Ohnishi, J.; Hata, S.; Gowik, U.; Westhoff, P. A plastidial sodium-dependent pyruvate transporter. *Nature* **2011**, *476*, 472–475. [CrossRef]
83. Zhao, Y.; Ai, X.; Wang, M.; Xiao, L.; Xia, G. A putative pyruvate transporter TaBASS2 positively regulates salinity tolerance in wheat via modulation of ABI4 expression. *BMC Plant Biol.* **2016**, *16*, 1–12. [CrossRef]
84. Miyaji, T.; Kuromori, T.; Takeuchi, Y.; Yamaji, N.; Yokosho, K.; Shimazawa, A.; Sugimoto, E.; Omote, H.; Ma, J.F.; Shinozaki, K. AtPHT4;4 is a chloroplast-localized ascorbate transporter in *Arabidopsis*. *Nat. Commun.* **2015**, *6*, 1–11. [CrossRef]
85. Guo, B.; Irigoyen, S.; Fowler, T.B.; Versaw, W.K. Differential expression and phylogenetic analysis suggest specialization of plastid-localized members of the PHT4 phosphate transporter family for photosynthetic and heterotrophic tissues. *Plant Signal. Behav.* **2008**, *3*, 784–790. [CrossRef]
86. Percey, W.J.; McMinn, A.; Bose, J.; Breadmore, M.C.; Guijt, R.M.; Shabala, S. Salinity effects on chloroplast PSII performance in glycophytes and halophytes. *Funct. Plant Biol.* **2016**, *43*, 1003–1015. [CrossRef]
87. Pou, A.; Jeanguenin, L.; Milhiet, T.; Batoko, H.; Chaumont, F.; Hachez, C. Salinity-mediated transcriptional and post-translational regulation of the *Arabidopsis* aquaporin PIP2;7. *Plant Mol. Biol.* **2016**, *92*, 731–744. [CrossRef]
88. Byrt, C.S.; Zhao, M.; Kourghi, M.; Bose, J.; Henderson, S.W.; Qiu, J.; Gilliham, M.; Schultz, C.; Schwarz, M.; Ramesh, S.A. Non-selective cation channel activity of aquaporin AtPIP2; 1 regulated by Ca^{2+} and pH. *Plant Cell Environ.* **2017**, *40*, 802–815. [CrossRef]
89. Wilson, M.E.; Basu, M.R.; Bhaskara, G.B.; Verslues, P.E.; Haswell, E.S. Plastid osmotic stress activates cellular stress responses in *Arabidopsis*. *Plant Physiol.* **2014**, *165*, 119–128. [CrossRef] [PubMed]
90. Ottow, E.A.; Polle, A.; Brosche, M.; Kangasjärvi, J.; Dibrov, P.; Zörb, C.; Teichmann, T. Molecular characterization of PeNhaD1: The first member of the NhaD Na^+/H^+ antiporter family of plant origin. *Plant Mol. Biol.* **2005**, *58*, 75–88. [CrossRef] [PubMed]
91. Kunz, H.-H.; Gierth, M.; Herdean, A.; Satoh-Cruz, M.; Kramer, D.M.; Spetea, C.; Schroeder, J.I. Plastidial transporters KEA1, -2, and -3 are essential for chloroplast osmoregulation, integrity, and pH regulation in *Arabidopsis*. *Proc. Natl. Acad. Sci. USA* **2014**, *111*, 7480–7485. [CrossRef] [PubMed]
92. Armbruster, U.; Carrillo, L.R.; Venema, K.; Pavlovic, L.; Schmidtmann, E.; Kornfeld, A.; Jahns, P.; Berry, J.A.; Kramer, D.M.; Jonikas, M.C. Ion antiport accelerates photosynthetic acclimation in fluctuating light environments. *Nat. Commun.* **2014**, *5*, 1–8. [CrossRef] [PubMed]
93. Herdean, A.; Teardo, E.; Nilsson, A.K.; Pfeil, B.E.; Johansson, O.N.; Ünnep, R.; Nagy, G.; Zsiros, O.; Dana, S.; Solymosi, K. A voltage-dependent chloride channel fine-tunes photosynthesis in plants. *Nat. Commun.* **2016**, *7*, 11654. [CrossRef] [PubMed]
94. Munns, R. Why measure osmotic adjustment? *Austral. J. Plant Physiol.* **1988**, *15*, 717–726. [CrossRef]
95. Shabala, S.; Shabala, L. Ion transport and osmotic adjustment in plants and bacteria. *BioMol. Concepts* **2011**, *2*, 407–419. [CrossRef]
96. Chen, T.H.; Murata, N. Enhancement of tolerance of abiotic stress by metabolic engineering of betaines and other compatible solutes. *Curr. Opin. Plant Biol.* **2002**, *5*, 250–257. [CrossRef]
97. Delauney, A.J.; Verma, D.P.S. Proline biosynthesis and osmoregulation in plants. *Plant J.* **1993**, *4*, 215–223. [CrossRef]
98. Rivero, R.M.; Shulaev, V.; Blumwald, E. Cytokinin-dependent photorespiration and the protection of photosynthesis during water deficit. *Plant Physiol.* **2009**, *150*, 1530–1540. [CrossRef]
99. Shen, B.; Jensen, R.G.; Bohnert, H.J. Increased resistance to oxidative stress in transgenic plants by targeting mannitol biosynthesis to chloroplasts. *Plant Physiol.* **1997**, *113*, 1177–1183. [CrossRef] [PubMed]
100. Shen, B.; Jensen, R.G.; Bohnert, H.J. Mannitol protects against oxidation by hydroxyl radicals. *Plant Physiol.* **1997**, *115*, 527–532. [CrossRef] [PubMed]
101. Liang, C.; Zhang, X.Y.; Luo, Y.; Wang, G.P.; Zou, Q.; Wang, W. Overaccumulation of glycine betaine alleviates the negative effects of salt stress in wheat. *Russ. J. Plant Physiol.* **2009**, *56*, 370–376. [CrossRef]
102. Tian, F.; Wang, W.; Lianga, W.C.; Wanga, X.; Wanga, G.; Wanga, W. Overaccumulation of glycine betaine makes the function of the thylakoid membrane better in wheat under salt stress. *Crop J.* **2017**, *5*, 73–82. [CrossRef]
103. Park, E.J.; Jekni'c, Z.; Pino, M.T.; Murata, N.; Chen, T.H.H. Glycinebetaine accumulation is more effective in chloroplasts than in the cytosol for protecting transgenic tomato plants against abiotic stress. *Plant Cell Environ.* **2007**, *30*, 994–1005. [CrossRef]
104. Kurepin, L.V.; Ivanov, A.G.; Zaman, M.; Pharis, R.P.; Hurry, V.; Hüner, N.P. Interaction of glycine betaine and plant hormones: Protection of the photosynthetic apparatus during abiotic stress. In *Photosynthesis: Structures, Mechanisms, and Applications*; Hou, H.J.M., Najafpour, M.M., Moore, G.F., Allakhverdiev, S.I., Eds.; Springer: Berlin, Germany, 2017; pp. 185–202.
105. Li, H.W.; Zang, B.S.; Deng, X.W.; Wang, X.P. Overexpression of the trehalose-6-phosphate synthase gene OsTPS1 enhances abiotic stress tolerance in rice. *Planta* **2011**, *234*, 1007–1018. [CrossRef]
106. Kanayama, Y.; Watanabe, M.; Moriguchi, R.; Deguchi, M.; Kanahama, K.; Yamaki, S. Effects of Low Temperature and Abscisic Acid on the Expression of the Sorbitol-6-phosphate Dehydrogenase Gene in Apple Leaves. *J. Jpn. Soc. Hortic. Sci.* **2006**, *75*, 20–25. [CrossRef]
107. Sleator, R.D.; Hill, C. Bacterial osmoadaptation: The role of osmolytes in bacterial stress and virulence. *FEMS Microbiol. Rev.* **2002**, *26*, 49–71. [CrossRef] [PubMed]

108. Kunte, H.J. Osmoregulation in bacteria: Compatible solute accumulation and osmosensing. *Environ. Chem.* **2006**, *3*, 94–99. [CrossRef]
109. Colmer, T.D.; Flowers, T.J.; Munns, R. Use of wild relatives to improve salt tolerance in wheat. *J. Exp. Bot.* **2006**, *57*, 1059–1078. [CrossRef] [PubMed]
110. Ruffino, A.M.C.; Rosa, M.; Hilal, M.; Gonzalez, J.A.; Prado, F.E. The role of cotyledon metabolism in the establishment of quinoa (*Chenopodium quinoa*) seedlings growing under salinity. *Plant Soil.* **2010**, *326*, 213–224. [CrossRef]
111. Kohl, K.I. The effect of NaCl on growth, dry matter allocation and ion uptake in salt marsh and inland populations of Armeria maritima. *New Phytol.* **1997**, *135*, 213–225. [CrossRef]
112. Khan, M.A.; Ungar, I.A.; Showalter, A.M.; Dewald, H.D. NaCl-induced accumulation of glycinebetaine in four subtropical halophytes from Pakistan. *Physiol. Plant.* **1998**, *102*, 487–492. [CrossRef]
113. Oren, A. Bioenergetic aspects of halophilism. *Microbiol. Mol. Biol. Rev.* **1999**, *63*, 334–348. [CrossRef] [PubMed]
114. Raven, J.A. Regulation of pH and generation of osmolarity in vascular plants: A cost-benefit analysis in relation to efficiency of use of energy, nitrogen and water. *New Phytol.* **1985**, *101*, 25–77. [CrossRef]
115. Lutts, S.; Majerus, V.; Kinet, J.M. NaCl effects on proline metabolism in rice (*Oryza sativa*) seedlings. *Physiol. Plant.* **1999**, *105*, 450–458. [CrossRef]
116. Bonner, C.A.; Williams, D.S.; Aldrich, H.C.; Jensen, R.A. Antagonism by L-glutamine of toxicity and growth inhibition caused by other amino acids in suspension cultures of *Nicotiana silvestris. Plant Sci.* **1996**, *113*, 43–58. [CrossRef]
117. Hare, P.D.; Cress, W.A.; van Staden, J. Disruptive effects of exogenous proline on chloroplast and mitochondrial ultrastructure in *Arabidopsis* leaves. *S. Afr. J. Bot.* **2002**, *68*, 393–396. [CrossRef]
118. Borgo, L.; Marur, C.J.; Vieira, L.G.E. Effects of high proline accumulation on chloroplast and mitochondrial ultrastructure and on osmotic adjustment in tobacco plants. *Acta Sci.* **2015**, *37*, 191–199. [CrossRef]
119. Chatterjee, J.; Patra, B.; Mukherjee, R.; Basak, P.; Mukherjee, S.; Ray, S.; Bhattacharyya, S.; Maitra, S.; Ghosh Dastidar, K.; Ghosh, S. Cloning, characterization and expression of a chloroplastic fructose-1,6-bisphosphatase from *Porteresia coarctata* conferring salt-tolerance in transgenic tobacco. *Plant Cell Tissue Organ Cult.* **2013**, *114*, 395–409. [CrossRef]
120. Cuin, T.A.; Shabal, A.S. Exogenously supplied compatible solutes rapidly ameliorate NaCl-induced potassium efflux from barley roots. *Plant Cell Physiol.* **2005**, *46*, 1924–1933. [CrossRef]
121. Leigh, R.A. Potassium homeostasis and membrane transport. *J. Plant Nutr. Soil Sci. Pflanz. Bodenkd* **2001**, *164*, 193–198. [CrossRef]
122. Glenn, E.P.; Brown, J.J.; Blumwald, E. Salt tolerance and crop potential of halophytes. *Crit. Rev. Plant Sci.* **1999**, *18*, 227–255. [CrossRef]
123. Tiwari, B.S.; Bose, A.; Ghosh, B. Photosynthesis in rice under a salt stress. *Photosynthetica* **1998**, *34*, 303–306. [CrossRef]
124. Belkhodja, R.; Morales, F.; Abadía, A.; Medrano, H.; Abadía, J. Effects of salinity on chlorophyll fluorescence and photosynthesis of barley *(Hordeum vulgare* L.) grown under a triple-line-source sprinkler system in the field. *Photosynthetica* **1999**, *36*, 375–387. [CrossRef]
125. Faseela, P.; Sinisha, A.K.; Brestic, M.; Puthur, J.T. Chlorophyll a fluorescence parameters as indicators of a particular abiotic stress in rice. *Photosynthetica* **2020**, *58*, 293–300. [CrossRef]
126. Allakhverdiev, S.I.; Nishiyama, Y.; Miyairi, S.; Yamamoto, H.; Inagaki, N.; Kanesaki, Y.; Murata, N. Salt stress inhibits the repair of photodamaged photosystem II by suppressing the transcription and translation of *psbA*Genes in synechocystis. *Plant Physiol.* **2002**, *130*, 1443–1453. [CrossRef]
127. Takahashi, S.; Murata, N. How do environmental stresses accelerate photoinhibition? *Trends Plant Sci.* **2008**, *13*, 178–182. [CrossRef]
128. Sengupta, S.; Majumder, A.L. Insight into the salt tolerance factors of a wild halophytic rice, *Porteresia coarctata*: A physiological and proteomic approach. *Planta* **2009**, *229*, 911–929. [CrossRef]
129. Rahman, S.; Matsumuro, T.; Miyake, H.; Takeoka, Y. Salinity-induced ultrastructural alterations in leaf cells of rice (*Oryza sativa* L.). *Plant Prod. Sci.* **2000**, *3*, 422–429. [CrossRef]
130. Redondo-Gómez, S.; Mateos-Naranjo, E.; Figueroa, M.E.; Davy, A.J. Salt stimulation of growth and photosynthesis in an extreme halophyte, *Arthrocnemum macrostachyum. Plant Biol.* **2010**, *12*, 79–87. [CrossRef] [PubMed]
131. Trotta, A.; Redondo-Gómez, S.; Pagliano, C.; Clemente, M.E.F.; Rascio, N.; La Rocca, N.; Antonacci, A.; Andreucci, F.; Barbato, R. Chloroplast ultrastructure and thylakoid polypeptide composition are affected by different salt concentrations in the halophytic plant *Arthrocnemum macrostachyum. J. Plant Physiol.* **2012**, *169*, 111–116. [CrossRef]
132. Redondo-Gómez, S.; Wharmby, C.; Castillo, J.M.; Mateos-Naranjo, E.; Luque, C.J.; De Cires, A.; Luque, T.; Davy, A.J.; Enrique Figueroa, M. Growth and photosynthetic responses to salinity in an extreme halophyte, *Sarcocornia fruticosa. Physiol. Plant.* **2006**, *128*, 116–124. [CrossRef]
133. Qiu, N.; Lu, Q.; Lu, C. Photosynthesis, photosystem II efficiency and the xanthophyll cycle in the salt-adapted halophyte *Atriplex centralasiatica. New Phytol.* **2003**, *159*, 479–486. [CrossRef] [PubMed]
134. Niyogi, K.K. Photoprotection revisited: Genetic and molecular approaches. *Annu. Rev. Plant Physiol. Plant Mol. Biol.* **1999**, *50*, 333–359. [CrossRef] [PubMed]
135. Munekage, Y.; Hashimoto, M.; Miyake, C.; Tomizawa, K.-I.; Endo, T.; Tasaka, M.; Shikanai, T. Cyclic electron flow around photosystem I is essential for photosynthesis. *Nature* **2004**, *429*, 579–582. [CrossRef]

136. Johnson, G.N. Physiology of PSI cyclic electron transport in higher plants. *Biochim. Biophys. Acta (BBA) Bioenerg.* **2011**, *1807*, 384–389. [CrossRef] [PubMed]
137. Niewiadomska, E.; Wiciarz, M. Adaptations of chloroplastic metabolism in halophytic plants. In *Progress in Botany*; Springer: Berlin/Heidelberg, Germany, 2015; pp. 177–193.
138. Niewiadomska, E.; Bilger, W.; Gruca, M.; Mulisch, M.; Miszalski, Z.; Krupinska, K. CAM-related changes in chloroplastic metabolism of *Mesembryanthemum crystallinum* L. *Planta* **2011**, *233*, 275–285. [CrossRef] [PubMed]
139. He, Y.; Fu, J.; Yu, C.; Wang, X.; Jiang, Q.; Hong, J.; Lu, K.; Xue, G.; Yan, C.; James, A.; et al. Increasing cyclic electron flow is related to Na^+ sequestration into vacuoles for salt tolerance in soybean. *J. Exp. Bot.* **2015**, *66*, 6877–6889. [CrossRef] [PubMed]
140. Locy, R.D.; Chang, C.C.; Nielsen, B.L.; Singh, N.K. Photosynthesis in salt-adapted heterotrophic tobacco cells and regenerated plants. *Plant Physiol.* **1996**, *110*, 321–328. [CrossRef] [PubMed]
141. Chang, C.C.; Locy, R.D.; Smeda, R.; Sahi, S.V.; Singh, N.K. Photoautotrophic tobacco cells adapted to grow at high salinity. *Plant Cell Rep.* **1997**, *16*, 495–502. [CrossRef]
142. Ghosh, S.; Bagchi, S.; Lahiri Majumder, A. Chloroplast fructose-1,6-bisphosphatase from *Oryza* differs in salt tolerance property from the *Porteresia* enzyme and is protected by osmolytes. *Plant Sci.* **2001**, *160*, 1171–1181. [CrossRef]
143. Li, W.; Zhang, C.; Lu, Q.; Wen, X.; Lu, C. The combined effect of salt stress and heat shock on proteome profiling in *Suaeda salsa*. *J. Plant Physiol.* **2011**, *168*, 1743–1752. [CrossRef]
144. Taiz, L.; Zeiger, E.; Møller, I.M.; Murphy, A. *Plant Physiology and Development*; Sinauer Associates Incorporated: Sunderland, MA, USA, 2015.
145. Feller, U.; Anders, I.; Mae, T. Rubiscolytics: Fate of Rubisco after its enzymatic function in a cell is terminated. *J. Exp. Bot.* **2007**, *59*, 1615–1624. [CrossRef] [PubMed]
146. Sharpe, R.M.; Offermann, S. One decade after the discovery of single-cell C4 species in terrestrial plants: What did we learn about the minimal requirements of C4 photosynthesis? *Photosynth. Res.* **2014**, *119*, 169–180. [CrossRef]
147. Ziska, L.H.; Seemann, J.R.; DeJong, T.M. Salinity Induced Limitations on Photosynthesis in *Prunus salicina*, a Deciduous Tree Species 1. *Plant Physiol.* **1990**, *93*, 864–870. [CrossRef]
148. El-Shihaby, O.A.; Younis, M.E.; El-Bastawisy, Z.M.; Nemat Alla, M.M. Effect of kinetin on photosynthetic activity and carbohydrate content in waterlogged or seawater-treated *Vigna sinensis* and *Zea mays* plants. *Plant Biosyst. Int. J. Deal. All Asp. Plant Biol.* **2002**, *136*, 277–290. [CrossRef]
149. Osmond, C.B.; Greenway, H. Salt responses of carboxylation enzymes from species differing in salt tolerance. *Plant Physiol.* **1972**, *49*, 260–263. [CrossRef] [PubMed]
150. Gong, D.H.; Wang, G.Z.; Si, W.T.; Zhou, Y.; Liu, Z.; Jia, J. Effects of Salt Stress on Photosynthetic Pigments and Activity of Ribulose-1,5-bisphosphate Carboxylase/Oxygenase in *Kalidium foliatum*. *Russ. J. Plant Physiol.* **2018**, *65*, 98–103. [CrossRef]
151. Portis, A.R., Jr. Rubisco activase. *Biochim. Biophys. Acta* **1990**, *1015*, 15–28. [CrossRef]
152. Portis, A.R. Rubisco activase—Rubisco's catalytic chaperone. *Photosynth. Res.* **2003**, *75*, 11–27. [CrossRef]
153. Wiciarz, M.; Gubernator, B.; Kruk, J.; Niewiadomska, E. Enhanced chloroplastic generation of H_2O_2 in stress-resistant *Thellungiella salsuginea* in comparison to *Arabidopsis thaliana*. *Physiol. Plant.* **2015**, *153*, 467–476. [CrossRef]
154. Chueca, A.; Sahrawy, M.; Pagano, E.A.; López Gorgé, J. Chloroplast fructose-1,6-bisphosphatase: Structure and function. *Photosynth. Res.* **2002**, *74*, 235–249. [CrossRef]
155. Kanai, R.; Edwards, G.E. The biochemistry of C4 photosynthesis. *C4 Plant Biol.* **1999**, *49*, 87.
156. Yen, H.E.; Zhang, D.; Lin, J.-H.; Edwards, G.E.; Ku, M.S.B. Salt-induced changes in protein composition in light-grown callus of *Mesembryanthemum crystallinum*. *Physiol. Plant.* **1997**, *101*, 526–532. [CrossRef]
157. Leisner, C.P.; Cousins, A.B.; Offermann, S.; Okita, T.W.; Edwards, G.E. The effects of salinity on photosynthesis and growth of the single-cell C4 species *Bienertia sinuspersici* (Chenopodiaceae). *Photosynth. Res.* **2010**, *106*, 201–214. [CrossRef]
158. Parsley, K.; Hibberd, J.M. The Arabidopsis PPDK gene is transcribed from two promoters to produce differentially expressed transcripts responsible for cytosolic and plastidic proteins. *Plant Mol. Biol.* **2006**, *62*, 339–349. [CrossRef] [PubMed]
159. Burnell, J.N.; Chastain, C.J. Cloning and expression of maize-leaf pyruvate, Pi dikinase regulatory protein gene. *Biochem. Biophys. Res. Commun.* **2006**, *345*, 675–680. [CrossRef]
160. Chastain, C.J.; Heck, J.W.; Colquhoun, T.A.; Voge, D.G.; Gu, X.-Y. Posttranslational regulation of pyruvate, orthophosphate dikinase in developing rice (*Oryza sativa*) seeds. *Planta* **2006**, *224*, 924. [CrossRef]
161. Chastain, C.J.; Fries, J.P.; Vogel, J.A.; Randklev, C.L.; Vossen, A.P.; Dittmer, S.K.; Watkins, E.E.; Fiedler, L.J.; Wacker, S.A.; Meinhover, K.C.; et al. Pyruvate, orthophosphate dikinase in leaves and chloroplasts of C3 plants undergoes light-/dark-induced reversible phosphorylation. *Plant Physiol.* **2002**, *128*, 1368–1378. [CrossRef] [PubMed]
162. Ohta, S.; Ishida, Y.; Usami, S. Expression of cold-tolerant pyruvate, orthophosphate dikinase cDNA, and heterotetramer formation in transgenic maize plants. *Transgenic Res.* **2004**, *13*, 475–485. [CrossRef]
163. Omoto, E.; Taniguchi, M.; Miyake, H. Adaptation responses in C4 photosynthesis of maize under salinity. *J. Plant Physiol.* **2012**, *169*, 469–477. [CrossRef]
164. Chen, T.; Kahlen, K.; Stutzel, H. Disentangling the contributions of osmotic and ionic effects of salinity on stomatal, mesophyll, biochemical and light limitations to photosynthesis. *Plant Cell Environ.* **2015**, *38*, 1528–1542. [CrossRef] [PubMed]
165. Grassi, G.; Magnani, F. Stomatal, mesophyll conductance and biochemical limitations to photosynthesis as affected by drought and leaf ontogeny in ash and oak trees. *Plant Cell Environ.* **2005**, *28*, 834–849. [CrossRef]

166. Tahjib-Ul-Arif, M.; Sohag, A.A.M.; Afrin, S.; Bashar, K.K.; Afrin, T.; Mahamud, S.U.; Polash, M.A.S.; Hossain, M.T.; Sohel, M.A.T.; Brestic, M.; et al. Differential response of sugar beet to long-term mild to severe salinity in a soil-pot culture. *Agriculture* **2019**, *9*, 223. [CrossRef]
167. Rabhi, M.; Castagna, A.; Remorini, D.; Scattino, C.; Smaoui, A.; Ranieri, A.; Abdelly, C. Photosynthetic responses to salinity in two obligate halophytes: *Sesuvium portulacastrum* and *Tecticornia indica*. *S. Afr. J. Bot.* **2012**, *79*, 39–47. [CrossRef]
168. Benzarti, M.; Ben Rejeb, K.; Debez, A.; Messedi, D.; Abdelly, C. Photosynthetic activity and leaf antioxidative responses of *Atriplex portulacoides* subjected to extreme salinity. *Acta Physiol. Plant.* **2012**, *34*, 1679–1688. [CrossRef]
169. Geissler, N.; Hussin, S.; Koyro, H.W. Elevated atmospheric CO_2 concentration ameliorates effects of NaCl salinity on photosynthesis and leaf structure of *Aster tripolium* L. *J. Exp. Bot.* **2009**, *60*, 137–151. [CrossRef] [PubMed]
170. Duarte, B.; Santos, D.; Marques, J.C.; Caçador, I. Ecophysiological adaptations of two halophytes to salt stress: Photosynthesis, PS II photochemistry and anti-oxidant feedback—Implications for resilience in climate change. *Plant Physiol. Biochem.* **2013**, *67*, 178–188. [CrossRef]
171. Ben Hamed, K.; Dabbous, A.; Souid, A.; Abdelly, C. Antioxidant Molecules and Enzymes and Their Relevance to the Salt Adaptation of Halophytes. In *Handbook of Halophytes*; Grigore, M.N., Ed.; Springer: Cham, Switzerland, 2020. [CrossRef]
172. Ozgur, R.; Uzilday, B.; Sekmen, A.H.; Turkan, I. Reactive oxygen species regulation and antioxidant defence in halophytes. *Funct. Plant Biol.* **2013**, *40*, 832–847. [CrossRef]
173. Bose, J.; Rodrigo-Moreno, A.; Shabala, S. ROS homeostasis in halophytes in the context of salinity stress tolerance. *J. Exp. Bot.* **2013**, *65*, 1241–1257. [CrossRef] [PubMed]
174. Wrzaczek, M.; Brosché, M.; Kangasjärvi, J. ROS signaling loops—Production, perception, regulation. *Curr. Opin. Plant Biol.* **2013**, *16*, 575–582. [CrossRef]
175. Critchley, C. Stimulation of photosynthetic electron transport in a salt-tolerant plant by high chloride concentrations. *Nature* **1982**, *298*, 483–485. [CrossRef]
176. Stepien, P.; Johnson, G.N. Contrasting responses of photosynthesis to salt stress in the glycophyte Arabidopsis and the halophyte Thellungiella: Role of the plastid terminal oxidase as an alternative electron sink. *Plant Physiol.* **2008**, *149*, 1154–1165. [CrossRef] [PubMed]
177. Mittova, V.; Tal, M.; Volokita, M.; Guy, M. Salt stress induces up-regulation of an efficient chloroplast antioxidant system in the salt-tolerant wild tomato species *Lycopersicon pennellii* but not in the cultivated species. *Physiol. Plant.* **2002**, *115*, 393–400. [CrossRef] [PubMed]
178. Ahmad, N.; Michoux, F.; Nixon, P.J. Investigating the production of foreign membrane proteins in tobacco chloroplasts: Expression of an algal plastid terminal oxidase. *PLoS ONE* **2012**, *7*, e41722. [CrossRef]
179. Ahmad, N.; Khan, M.O.; Islam, E.; Wei, Z.-Y.; McAusland, L.; Lawson, T.; Johnson, G.N.; Nixon, P.J. Contrasting responses to stress displayed by tobacco overexpressing an algal plastid terminal oxidase in the chloroplast. *Front. Plant Sci.* **2020**, *11*, 501. [CrossRef]
180. Uzilday, B.; Ozgur, R.; Sekmen, A.H.; Yildiztugay, E.; Turkan, I. Changes in the alternative electron sinks and antioxidant defence in chloroplasts of the extreme halophyte *Eutrema parvulum* (*Thellungiella parvula*) under salinity. *Ann. Bot.* **2014**, *115*, 449–463. [CrossRef]
181. Jithesh, M.; Prashanth, S.; Sivaprakash, K.; Parida, A.K. Antioxidative response mechanisms in halophytes: Their role in stress defence. *J. Genet.* **2006**, *85*, 237. [CrossRef]
182. Shabala, S.; Bose, J.; Hedrich, R. Salt bladders: Do they matter? *Trends Plant Sci.* **2014**, *19*, 687–691. [CrossRef]
183. Shabala, S.; Wu, H.; Bose, J. Salt stress sensing and early signalling events in plant roots: Current knowledge and hypothesis. *Plant Sci.* **2015**, *241*, 109–119. [CrossRef] [PubMed]
184. Chan, K.X.; Phua, S.Y.; Crisp, P.; McQuinn, R.; Pogson, B.J. Learning the languages of the chloroplast: Retrograde signaling and beyond. *Annu. Rev. Plant Biol.* **2016**, *67*, 25–53. [CrossRef] [PubMed]
185. Mittler, R. Oxidative stress, antioxidants and stress tolerance. *Trends Plant Sci.* **2002**, *7*, 405–410. [CrossRef]
186. Sharma, P.; Jha, A.B.; Dubey, R.S.; Pessarakli, M. Reactive oxygen species, oxidative damage, and antioxidative defense mechanism in plants under stressful conditions. *J. Bot.* **2012**, *2012*, 217037. [CrossRef]
187. Suo, J.; Zhao, Q.; David, L.; Chen, S.; Dai, S. Salinity response in chloroplasts: Insights from gene characterization. *Int. J. Mol. Sci.* **2017**, *18*, 1011. [CrossRef] [PubMed]
188. Asada, K. Production and scavenging of reactive oxygen species in chloroplasts and their functions. *Plant Physiol.* **2006**, *141*, 391–396. [CrossRef]
189. Edreva, A. Generation and scavenging of reactive oxygen species in chloroplasts: A submolecular approach. *Agric. Ecosyst. Environ.* **2005**, *106*, 119–133. [CrossRef]
190. Noctor, G.; Foyer, C.H. Ascorbate and glutathione: Keeping active oxygen under control. *Annu. Rev. Plant Physiol. Plant Mol. Biol.* **1998**, *49*, 249–279. [CrossRef] [PubMed]
191. Miller, G.; Suzuki, N.; Ciftci-Yilmaz, S.; Mittler, R. Reactive oxygen species homeostasis and signalling during drought and salinity stresses. *Plant Cell Environ.* **2010**, *33*, 453–467. [CrossRef] [PubMed]
192. Triantaphylidès, C.; Havaux, M. Singlet oxygen in plants: Production, detoxification and signaling. *Trends Plant Sci.* **2009**, *14*, 219–228. [CrossRef]
193. Khorobrykh, S.; Havurinne, V.; Mattila, H.; Tyystjärvi, E. Oxygen and ROS in Photosynthesis. *Plants* **2020**, *9*, 91. [CrossRef]

194. Turkan, I.; Uzilday, B.; Dietz, K.J.; Bräutigam, A.; Ozgur, R. Reactive oxygen species and redox regulation in mesophyll and bundle sheath cells of C_4 plants. *J. Exp. Bot.* **2018**, *69*, 3321–3331. [CrossRef]
195. Parida, A.K.; Das, A.B. Salt tolerance and salinity effects on plants: A review. *Ecotox. Environ. Saf.* **2005**, *60*, 324–349. [CrossRef]
196. Isayenkov, S.V.; Maathuis, J.M. Plant salinity stress: Many unanswered questions remain. *Front. Plant Sci.* **2019**, *10*, 80. [CrossRef] [PubMed]

cells

Review

Updates on the Role of ABSCISIC ACID INSENSITIVE 5 (ABI5) and ABSCISIC ACID-RESPONSIVE ELEMENT BINDING FACTORs (ABFs) in ABA Signaling in Different Developmental Stages in Plants

Anna Collin, **Agata Daszkowska-Golec** * and **Iwona Szarejko**

Institute of Biology, Biotechnology and Environmental Protection, Faculty of Natural Sciences, University of Silesia in Katowice, ul. Jagiellońska 28, 40-032 Katowice, Poland; anna.skubacz@us.edu.pl (A.C.); iwona.szarejko@us.edu.pl (I.S.)

* Correspondence: agata.daszkowska@us.edu.pl; Tel.: +48-32-2009-360

Abstract: The core abscisic acid (ABA) signaling pathway consists of receptors, phosphatases, kinases and transcription factors, among them ABA INSENSITIVE 5 (ABI5) and ABRE BINDING FACTORs/ABRE-BINDING PROTEINs (ABFs/AREBs), which belong to the BASIC LEUCINE ZIPPER (bZIP) family and control expression of stress-responsive genes. ABI5 is mostly active in seeds and prevents germination and post-germinative growth under unfavorable conditions. The activity of ABI5 is controlled at transcriptional and protein levels, depending on numerous regulators, including components of other phytohormonal pathways. ABFs/AREBs act redundantly in regulating genes that control physiological processes in response to stress during vegetative growth. In this review, we focus on recent reports regarding ABI5 and ABFs/AREBs functions during abiotic stress responses, which seem to be partially overlapping and not restricted to one developmental stage in Arabidopsis and other species. Moreover, we point out that ABI5 and ABFs/AREBs play a crucial role in the core ABA pathway's feedback regulation. In this review, we also discuss increased stress tolerance of transgenic plants overexpressing genes encoding ABA-dependent bZIPs. Taken together, we show that ABI5 and ABFs/AREBs are crucial ABA-dependent transcription factors regulating processes essential for plant adaptation to stress at different developmental stages.

Keywords: ABI5; ABF; AREB; abiotic stress response; abscisic acid; phytohormone crosstalk

Citation: Collin, A.; Daszkowska-Golec, A.; Szarejko, I. Updates on the Role of ABSCISIC ACID INSENSITIVE 5 (ABI5) and ABSCISIC ACID-RESPONSIVE ELEMENT BINDING FACTORs (ABFs) in ABA Signaling in Different Developmental Stages in Plants. *Cells* **2021**, *10*, 1996. https://doi.org/10.3390/cells10081996

Academic Editor: Suleyman Allakhverdiev

Received: 7 July 2021
Accepted: 3 August 2021
Published: 5 August 2021

1. Insight into the Core ABA Signaling and ABA-Dependent bZIPs Function

Each year abiotic stresses, including drought and salinity, reduce crop yield, causing economic problems and a severe threat to food safety. Thus, it is crucial to understand plant adaptation mechanisms to unfavorable environmental conditions with aim to develop stress-tolerant cultivars [1,2]. Abscisic acid (ABA) is a significant phytohormone regulating plant responses to various abiotic stresses [3,4]. In response to stress, ABA is synthesized in plant cells triggering activation of the ABA signaling pathway. The core ABA signaling consists of ABA receptors (PYRABACTIN RESISTANCE PROTEINS/PYR-LIKE PROTEINS/REGULATORY COMPONENTS OF ABA RECEPTOR; PYR/PYL/RCAR), phosphatases (PHOSPHATASE 2Cs; PP2Cs), kinases (SNF1-RELATED PROTEIN KINASE 2; SnRK2s), and transcription factors belonging to large BASIC LEUCINE ZIPPER (bZIP) family. After perceiving the primary stress signal, ABA forms a complex with PYR/PYL/RCAR receptors and PP2C phosphatases. It prevents PP2Cs from dephosphorylating SnRK2s and releases a SnRK2s phosphorylation activity. SnRK2 kinases phosphorylate and activate bZIP transcription factors, such as ABA INSENSITIVE 5 (ABI5) and ABRE BINDING FACTORs/ABRE-BINDING PROTEINs (ABFs/AREBs) [5–7]. However, it has to be underlined that ABA-dependent bZIPs are also phosphorylated by other kinases [8–10]. Recently, it was found that RAF-LIKE KINASE 10 (RAF10) and CALCIUM-DEPENDENT PROTEIN

KINASE 6 (CPK6) interact with ABI5 and ABFs/AREBs and phosphorylate them [11,12]. Moreover, RIBOSOMAL S6 KINASE2 (S6K2), kinase active in TARGET OF RAPAMYCIN (TOR) signaling, was also shown to bind with ABI5 what in turn stimulates ABA response and drought tolerance [13]. Phosphorylated ABI5 and ABFs/AREBs recognize ABA RESPONSIVE ELEMENTs (ABRE cis-elements) containing (C/T)ACGTGGC motif and G-box coupling elements (GCEs) with ACGT/C core sequence, present in the promoters of stress-responsive genes. These bZIP factors activate or repress their expression and trigger plant adaptation to stress [5,14,15].

In our previous review [10], we described the function of AtABI5 and its homologs in the acquisition of stress tolerance in plants. However, recently numerous new data emerged about ABI5 and ABFs/AREBs function under stress. This review presents new evidence on ABI5 and ABFs/AREBs role in plant adaptation to stress and describes a tight control of their activity by other stress regulators. We also focus on ABI5 and ABFs/AREBs actions in the feedback regulation of ABA pathway and the possibility of utilizing ABA-dependent bZIPs in developing stress-tolerant cultivars.

2. New Evidence of ABI5 Regulatory Role during Seed Germination

In *Arabidopsis thaliana*, ABA-induced activation of ABI5 inhibits germination under unfavorable environmental conditions. ABI5 is responsible for the regulation of expression of stress-responsive genes, e.g., *EARLY METHIONINE-LABELED 1* (*EM1*) and *EM6* encoding LATE EMBRYOGENESIS ABUNDANT (LEA) proteins [10,16,17]. Recently, ABI5 was shown to repress the expression of *PHOSPHATE1* (*PHO1*), a gene involved in phosphate (Pi) transfer from cotyledons to radicles which promotes germination. Therefore, one of the mechanisms of ABI5-dependent germination inhibition is repression of Pi transfer [18]. Furthermore, ABI5 can directly activate *CATALASE 1* (*CAT1*), encoding a catalase responsible for scavenging H_2O_2, the main reactive oxygen species (ROS). It indicates that ABI5 is also involved in maintaining ROS homeostasis during seed germination [19] (Figure 1).

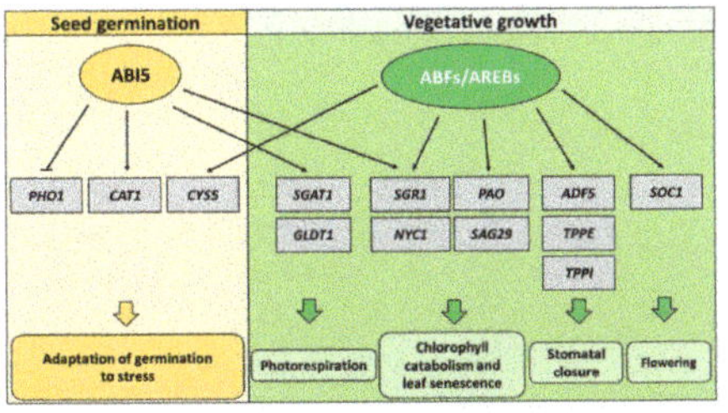

Figure 1. Partially overlapping function of ABA INSENSITIVE 5 (ABI5) and ABRE BINDING FACTORs/ABRE-BINDING PROTEINs (ABFs/AREBs). ABI5 regulates seed germination accordingly to surrounding environmental conditions. ABI5 promotes *CATALASE 1* (*CAT1*) and represses *PHOSPHATE1* (*PHO1*), responsible for reactive oxygen species (ROS) scavenging and phosphate transfer, respectively, during seed germination, while ABFs/AREBs ensure stress adaptation in vegetative tissues. Genes associated with chlorophyll catabolism (*STAY-GREEN 1—SGR1*, *NON-YELLOW COLORING 1—NYC1*), leaf senescence (*PHEOPHORBIDE A OXYGENASE—PAO*, *SENESCENCE-ASSOCIATED GENE 29—SAG29*), stomatal closure (*ACTIN-DEPOLYMERIZING FACTOR 5—ADF5*, *TREHALOSE-6-PHOSPHATE PHOSPHATASE E—TPPE*, *TPPI*) and flowering time (*SUPPRESSOR OF OVEREXPRESSION OF CO 1—SOC1*) are the direct targets of ABFs/AREBs. ABI5 is also able to promote expression of chlorophyll catabolism (*SGR1*, *NYC1*) and photorespiration (*SERINE:GLYOXYLATE AMINOTRANSFERASE 1—SGAT1*, *GDC T-PROTEIN—GLDT1*) genes in vegetative tissues, whereas ABFs/AREBs activate expression of *CYSTEINE PROTEINASE INHIBITOR 5* (*CYS5*) in seeds.

Multiple transcription factors regulate *ABI5* expression during seed germination in Arabidopsis [10,20–22]. Recently, the new regulator of *ABI5* in seeds was identified. AGAMOUS-LIKE 21 (AGL21) belongs to MCM1/AGAMOUS/DEFICIENS/SRF (MADS) box group of transcription factors. It was shown that AGL21 activates the expression of *ABI5* in ABA-treated seeds in Arabidopsis [23].

In Arabidopsis, ABI5 is also regulated at the protein level. It interacts with numerous proteins, modulating its stability or activity as a transcription factor [10,24–27]. Recently, XPO1-INTERACTING WD40 PROTEIN 1 (XIW1), a member of WD40-repeat protein (WD40) family, was found to be crucial for ABI5 stability during seed germination. XIW1 shuttles between nucleus and cytoplasm dependently on environmental conditions. ABA promotes nuclear localization of XIW1. Colocalization of XIW1 and ABI5 in the nucleus leads to their physical interaction and, thus, protects ABI5 from proteasomal degradation, which enables ABA-mediated responses in seeds [28] (Figure 2). Furthermore, the interaction between ABI5 and autophagy cargo receptor, NEIGHBOUR OF BREAST CANCER 1 (NBR1), may enhance ABI5 stability and ABA response at the germination stage [29,30]. Recently, Yang et al. [31] showed that ABI5 activity is regulated by circadian clock regulators, PSEUDO-RESPONSE REGULATOR5 (PRR5) and PRR7. Both proteins interact physically with ABI5 to stimulate ABA response in seeds and to inhibit germination according to day/night cycle (Figure 2). Moreover, Pan et al. [32] discovered that ABA-induced members of the newly identified VQ motif-containing protein (VQ) family, VQ18 and VQ26, interact physically with ABI5 and block its function (Figure 2). VQ18 and VQ26 were proposed to be a part of the fine-tuning mechanism in the frame of ABA signaling during seed germination. Furthermore, ABI5 BINDING PROTEIN 2 (AFP2) was found to interact with ABI5 and inhibits its transactivation of *SOMNUS* (*SOM*), encoding a negative regulator of GA accumulation and seed germination under high temperature [33] (Figure 2). It is noteworthy that another AFP protein, AFP1, was shown to promote ABI5 degradation [25].

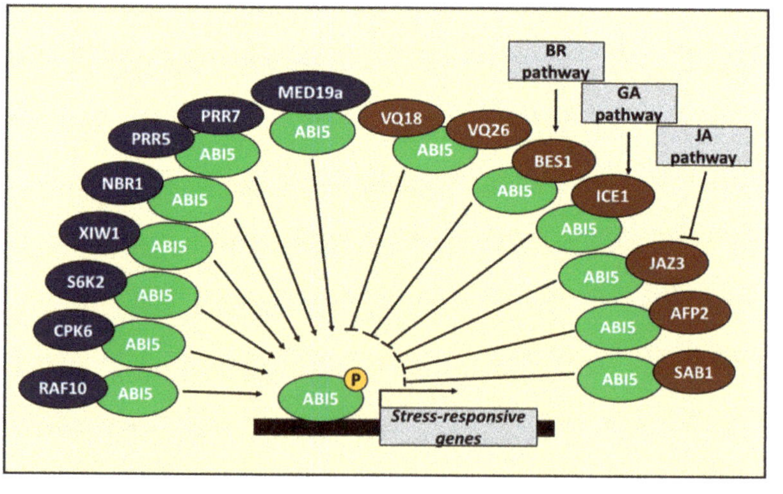

Figure 2. Positive and negative regulators of ABA INSENSITIVE 5 (ABI5) protein stability and/or function. ABI5 stability and activity is under regulation of multiple proteins. ABI5 stability and function is promoted by interaction with kinases CALCIUM-DEPENDENT PROTEIN KINASE 6 (CPK6), RAF-LIKE KINASE 10 (RAF10) and RIBOSOMAL S6 KINASE2 (S6K2), shuttle protein XPO1-INTERACTING WD40 PROTEIN 1 (XIW1), circadian clock regulators PSEUDO-RESPONSE REGULATOR 5 (PRR5) and PRR7 and mediator subunit MEDIATOR COMPLEX SUBUNIT 19a (MED19a). ABI5 is also stabilized by interaction with NEIGHBOUR OF BREAST CANCER 1 (NBR1). On the other side, ABI5 stability and/or function is negatively affected by VQ18, VQ26, BRASSINOSTEROID INSENSITIVE 1 (BRI1)-EMS-SUPPRESSOR 1 (BES1), INDUCER OF CBF EXPRESSION1 (ICE1), JASMONATE-ZIM DOMAIN PROTEIN 3 (JAZ3), ABI5 BINDING PROTEIN 2 (AFP2) and SENSITIVE TO ABA 1 (SAB1). BES1, ICE1 and JAZ3 are also involved in brassinosteroid (BR), gibberellic acid (GA) and jasmonic acid (JA) signaling, respectively. P—phosphate group.

3. Involvement of ABI5 in Phytohormonal Crosstalk at the Seed Germination Stage

ABI5 can act as a hub in phytohormonal crosstalk in Arabidopsis [10,27,34–37]. The last reports confirm this hypothesis and indicate that at the seed germination stage ABI5 plays an essential role in brassinosteroid (BR), gibberellic acid (GA), cytokinin (CK) and jasmonic acid (JA) signaling through interaction with components of these pathways [38–41]. Zhao et al. [38] showed the interaction between ABI5 and BRASSINOSTEROID INSENSITIVE 1 (BRI1)-EMS-SUPPRESSOR 1 (BES1), a BR-dependent transcription factor and negative regulator of ABA signaling. BES1 inhibits ABI5 activity and thus promotes seed germination (Figure 2). Interestingly, binding of BES1 to ABI5 prevents the ABI5 interaction with ABI3, the enhancer of ABI5 transactivation function [38]. Similarly, INDUCER OF CBF EXPRESSION1 (ICE1) transcription factor binds and represses ABI5 transactivation function in seeds. This interaction is promoted by GA, but inhibited by DELLA proteins, negative components of GA signaling, which also bind to ICE1 and therefore restore ABA signaling and ABI5 function [39]. Furthermore, CK-dependent regulators ARABIDOPSIS RESPONSE REGULATOR 4 (ARR4), ARR5 and ARR6 inhibit *ABI5* expression at germination stage [42], whereas other components of CK signaling, ARABIDOPSIS HISTIDINE KINASE 4 (AHK4), ARABIDOPSIS HISTIDINE PHOSPHOTRANSFER PROTEIN 2 (AHP2), AHP3, AHP5 and ARR12 trigger ABI5 protein degradation during cotyledon greening [36]. Moreover, JASMONATE-ZIM DOMAIN PROTEIN 3 (JAZ3), a negative regulator of JA response, interacts with ABI5 and reduces its activity as transcription factor in germinating seeds (Figure 2). However, ABA treatment promotes JAZ3 degradation and JA biosynthesis in the ABI5-dependent way [40]. Recently, it was shown that JAZ proteins repress *ABI3* and *ABI5* expression, while JA stimulates ABA response through regulation of *ABI3* and *ABI5* during seed germination inhibition [41].

4. The Function and Regulation of ABI5 during Seedling Development

ABI5 plays the essential role in repression of seedling growth under osmotic, salt and cold stress in Arabidopsis [10,43,44]. It was recently found that ABI5 may also be essential for seedling growth tolerance to aluminum (Al) through expression regulation of genes related to cell wall modification and osmoregulation [45]. New evidence also emerged about *ABI5* transcriptional regulation in Arabidopsis seedlings. It was shown that NAM/ATAF1/2/CUC2 (NAC) transcription factor, ANAC060, represses *ABI5* activity in seedlings in response to glucose. Interesting, it can be a part of negative feedback regulation between ABI5 and another ABA-dependent and glucose-related transcription factor, ABI4, which promotes *ANAC060* expression [46]. Previously, B-BOX DOMAIN PROTEIN 21 (BBX21) was shown to inhibit the *ABI5* expression under light [47]. Kang et al. [48] revealed that BBX21 represses *ABI5* by recruiting its promoter chromatin modifier, HYPERSENSITIVE TO RED AND BLUE 2/PICKLE (HRB2/PKL). Decreased *ABI5* expression results in maintenance of stomatal aperture at the level that is necessary for gas exchange, however, the precise mechanism is unknown. Therefore, these data indirectly indicate that ABI5 is responsible for the regulation of stomata movement in seedlings [48]. In the dark, *ABI5* expression is activated by PHYTOCHROME-INTERACTING FACTORs (PIF): PIF1, PIF3, PIF4 and PIF5 that are the negative regulators of photomorphogenesis [49]. Interestingly, previously ABI5 was shown to interact with PIF1 to strengthen its function as a transcription factor [50]. WRKY18, WRKY40 and WRKY60 are negative regulators of *ABI5* transcription in seedlings [10,51]. Recently, WRKY40 was found to recruit a histone 3 lysine 4 (H3K4) demethylase JUMONJI DOMAIN-CONTAINING PROTEIN 17 (JMJ17) to chromatin of *ABI5* gene. JMJ17 removes marks of transcriptionally active chromatin (H3K4me3) from *ABI5* and thus also inhibits *ABI5* expression at epigenetic level in seedlings under non-stressed conditions [52].

Other proteins also modulate ABI5 activity during seedling development in Arabidopsis. Similar to seeds, ABI5 stability is regulated by XIW1 and NBR1 at the seedling stage [28,29]. Furthermore, MEDIATOR COMPLEX SUBUNIT 19a (MED19a) physically interacts with ABI5 during ABA-dependent inhibition of root growth and cotyledon greening (Figure 2). It was shown that MED19a strengthens ABI5 binding to the promoters of *EM1* and *EM6* [53]. It has to be underlined that ABI5 is negatively regulated by another MEDIATOR subunit, MED25 [10,35]. The next identified ABI5 interactor, SENSITIVE TO ABA 1 (SAB1), in many ways inhibits ABI5 activity during early seedling growth. SAB1 belongs to REGULATOR OF CHROMATIN CONDENSATION 1 (RCC1) family and it binds to ABI5 serine at 145 position, which serves as a target of SnRK2s-mediated phosphorylation. The SAB1 binding causes reduction of ABI5 phosphorylation status and leads to its degradation (Figure 2). SAB1 also binds to *ABI5* promoter and inhibits ABI5 to auto-activate its expression. Moreover, SAB1 increases the level of histone H3K27me2, the epigenetic mark of expression repression, in the *ABI5* promoter [54]. Together, all these findings show that ABI5 activity undergoes complex and strict regulation, which shows that ABI5 plays an essential role in ABA signaling during early developmental stages. It has to be pointed out that ABI5 can interact with other transcription factors to regulate their activity. ABI5 is involved in anthocyanin accumulation in seedlings [44]. Recently, it was shown that in apple (*Malus domestica*), homolog of AtABI5, MdABI5, binds with basic helix-loop-helix 3 (MdbHLH3) which in turn enhances expression of its target genes, *DIHYDROFLAVONOL 4-REDUCTASE* (*MdDFR*) and *UDP FLAVONOID GLUCOSYL TRANSFERASE* (*MdUF3GT*), involved in anthocyanin biosynthesis. Moreover, MdABI5 strengthens interaction between MdbHLH3 and MdMYB1, another transcription factor involved in anthocyanin biosynthesis. MdABI5 can also promote directly expression of *MdbHLH3* [55].

5. Redundant Function of ABFs/AREBs in the Regulation of Plant Stress Responses in Vegetative Tissues

ABFs/AREBs: ABF1, ABF2/AREB1, ABF4/AREB2, and ABF3 regulate plant response to abiotic stresses, such as drought, salt, heat, oxidative stress, and cold, mostly in vegetative tissues of Arabidopsis [56–59]. ABFs/AREBs ensure the adaptation to unfavorable environmental conditions via promoting expression of *LEA* genes, such as *RESPONSIVE TO DESSCICATION 29B* (*RD29B*), *RESPONSIVE TO ABA 18* (*RAB18*), *COLD-RESPONSIVE 6.6* (*COR6.6/KIN2*), and regulatory genes, such as *DEHYDRATION-RESPONSIVE ELEMENT BINDING PROTEIN 2A* (*DREB2A*) [14,57,60]. However, ABFs/AREBs role in ABA-mediated adaptation to stress is highly redundant. Quadruple mutant *abf2/areb1 abf4/areb2 abf3 abf1* showed a significantly lower survival rate after drought treatment than single *abf/areb* mutants. Similar observations were made for the primary root growth of *abf/areb* mutants in the presence of ABA [14,61]. Moreover, ABFs/AREBs can interact with each other and function together to regulate expression of target genes [8,14]. On the other side, ABF2/AREB1, ABF4/AREB2, and ABF3 regulate the expression of *RD29B* or *RAB18* in a slightly different way, which indicates the partially independent role of each *ABF/AREB* during adaptation to stress [14].

ABFs/AREBs regulate redundantly also stomatal closure in Arabidopsis. *ACTIN-DEPOLYMERIZING FACTOR 5* (*ADF5*) encodes a protein responsible for actin cytoskeleton remodeling during stomatal closure in response to ABA. ABF1, ABF2/AREB1, ABF4/AREB2 and ABF3 bind to *ADF5* promoter, activate its expression, promote stomatal closure and ensure adaptation to drought [15]. It was recently shown that ABFs/AREBs-mediated stomatal closure depends on accumulation of disaccharide trehalose in seedlings. ABF1, ABF2/AREB1 and ABF4/AREB2 directly bind to the promoter of *TREHALOSE-6-PHOSPHATE PHOSPHATASE I* (*TPPI*) and promote its expression. *TPPI* gene encodes the trehalose biosynthesis enzyme [62]. Additionally, ABF2/AREB1 directly activates another gene from *TPP* family, *TPPE* [63] (Figure 1). Strikingly, *TPPI* is involved in promoting primary root growth, whereas *TPPE* takes part in ABA-dependent inhibition of root growth [62,63]. Homologs of AtABF/AREB in other dicot species are also involved in stomata regulation. In cotton (*Gossypium hirsutum*), *GhABF2D* promotes stomatal closure and thus abiotic stress tolerance [64]. Moreover, carrot (*Daucus carota*) DcABF3 activates expression of genes involved in stomata development, *SPEECHLESS* (*SPCH*), *FAMA* (*FMA*) and *MUTE*, which in turn increases number of stomata [65].

In vegetative tissues of Arabidopsis, ABFs/AREBs are regulated at transcriptional, post-transcriptional and protein levels. Expression of *ABFs/AREBs* is under control of other stress-responsive transcription factors, NACs. NAC016 and NAC-LIKE, ACTIVATED BY AP3/PI (NAP) were shown to repress *ABF2/AREB1* under drought [66]. Additionally, NAP negatively regulates also salt stress response by inhibiting *ABF2/AREB1* [67]. Furthermore, *ABF3* expression can be downregulated at the post-transcriptional level by miR399f to release ABA-mediated growth arrest under stress [68]. ABFs/AREBs activity is also modulated by NACs at the protein level. Previously, ABF2/AREB1 and ABF4/AREB2 were found to bind with ANAC096 and cooperatively regulate the expression of stress-responsive genes [69]. Interestingly, the NAC072 interacts with ABF3 to enhance expression of *RD29A* and to decrease *RD29B* activity. Therefore, NAC072 exerts a differential type of action on ABF3 function in ABA signaling [70]. Recently, the activity of ABF2/AREB1 was shown to be likely enhanced by interaction with GA-related DELLA proteins. This could promote stomatal closure and drought tolerance. It was evidenced that the ABF2/AREB1 was also involved in ABA-GA crosstalk to ensure the efficient response to drought stress [71].

6. The Partially Overlapping Function of ABI5 and ABFs/AREBs in Seeds, Seedlings, and Vegetative Tissues

Although ABI5 is considered as the main ABA-dependent bZIP factor regulating stress responses in seeds and seedlings, ABFs/AREBs also seem to be active during Arabidopsis' early developmental stages. The function of ABFs/AREBs during germination and early seedling development was already indicated by Kim et al. [72], Finkelstein et al. [73] and Sharma et al. [59]. They observed faster germination of *abf1*, *abf4/areb2* and *abf3* mutants under optimal growth conditions and better germination rate of *abf4/areb2* and *abf3* under ABA treatment [59,72]. Moreover, ABF3 acts redundantly with ABI5 in seeds and during post-germinative growth under abiotic stress [73]. It was also shown that ABF3 directly promotes *ABI5* expression in salt-treated seedlings [74]. Furthermore, ABF3 together with ABF1 regulate seed germination and post-germinative growth under heat stress. They directly activate expression of *CYSTEINE PROTEINASE INHIBITOR 5* (*CYS5*), a gene encoding an inhibitor of cysteine protease, leading to thermotolerant germination and growth of primary root [75] (Figure 1). *ABF1*, *ABF3* and *ABF4/AREB2* also regulate seed germination under salt and osmotic stress, downstream of *DE-ETIOLATED 1* (*DET1*), a negative regulator involved in light signaling pathway [76].

On the other side, ABI5 is active in vegetative tissues of Arabidopsis. Recently, You et al. [77] showed that ABI5 takes part in the adaptation of plant growth to a low CO_2 level [78]. It was noted that ABI5 binds to the promoters of *SERINE:GLYOXYLATE AMINOTRANSFERASE 1* (*SGAT1*) and *GDC T-PROTEIN* (*GLDT1*) genes encoding enzymes associated with photorespiration process, and activates their expression [77] (Figure 1). Moreover, ABI5 is important for regulation of plant juvenile-to-adult transition. In response to ABA, MYB33, the main target of miR159, directly activates *ABI5* transcription. Next, ABI5 influences positively on *MIR156* expression, what in turn delays vegetative development under abiotic stress [79]. Furthermore, function of ABI5 and ABFs/AREBs can be overlapping in vegetative tissues. It has to be underlined that ABI5 can form heterodimers with ABFs/AREBs, which is the evidence of their synergistic role in ABA responses [8]. ABI5 was shown to inhibit photosynthesis and promote chlorophyll catabolism and leaf senescence [10,80,81] (Figure 1). However, ABFs/AREBs are also involved in regulating ABA-induced chlorophyll catabolism and leaf senescence in a similar way. ABF2/AREB1, ABF4/AREB2 and ABF3 directly activate expression of genes associated with chlorophyll catabolism: *STAY-GREEN 1/NON-YELLOWING 1* (*SGR1/NYE1*), *PHEOPHORBIDE A OXYGENASE* (*PAO*), *NON-YELLOW COLORING 1* (*NYC1*), and leaf senescence, *SENESCENCE-ASSOCIATED GENE 29* (*SAG29*) [82] (Figure 1). Besides Arabidopsis, ABI5 function was also observed during inhibition of chloroplast-related processes in potato (*Solanum tuberosum*). Similar to AtABI5, StABI5 negatively impacts photosynthesis and promotes chlorophyll catabolism and leaf senescence via positive regulation of *CHLOROPLAST VESICULATION* (*StCV*) and *StNYC1*, encoding proteins involved in chlorophyll degradation [83]. Moreover, in apple MdABI5 binds directly to promoters of *MdNYE1* and *MdNYC1* to promote chlorophyll catabolism and leaf senescence. MdABI5 transcriptional activity during regulation of leaf senescence is enhanced by its physical interaction with MdWRKY40 and MdbZIP44, but weakened by binding with MdBBX22 [84]. On the other side, dicot AtABF/AREB homologs, tomato (*Solanum lycopersicum*) SlAREB1 and sweet-potato (*Ipomoea batatas*) IbABF4 positively regulate photosynthesis efficiency in response to stress [85,86].

In vegetative tissues ABI5 can be also involved in response to biotic stress. *ABI5* expression is under epigenetic control of HOOKLESS1 (HLS1), which takes part in plant response to pathogens. It might imply that ABI5 is related to plant defense reaction against biotic stress [87]. Recently, it was shown that in tobacco (*Nicotiana benthamiana*), homolog of AtABI5, NbABI5 is able to directly repress the gene responsible for chloroplast electron transfer chain, *FERREDOXIN 1* (*NbFD1*), in rice stripe virus (RSV)-infected tobacco plants. *NbFD1* is also involved in callose deposition in plasmodesmata, which in turn protects plant against viral infection. Therefore, RSV-mediated activation of *NbABI5* can inhibit plant defense mechanism to virus [88]. It should be also underlined that tomato SlAREB1 participates in regulation of biotic stress-responsive genes such as *PATHOGENESIS-RELATED GENE 5* (*PR5*) or *CHITINASE3* (*CHI3*) [85]. Together, all these observations indicate that function of *ABI5* and *ABFs/AREBs* is not only restricted to seed germination and vegetative tissues, respectively. Furthermore, *ABI5* can function synergistically with *ABFs/AREBs* during ABA-dependent responses.

7. The Role of ABI5 and ABFs/AREBs in Flowering Regulation and Fruit Ripening

In Arabidopsis, ABFs/AREBs and ABI5 take part in ABA-dependent regulation of flowering timing. ABF4/AREB2 and ABF3 were shown to regulate flowering time and mediate the drought-escape mechanism. Both ABFs/AREBs interact with NUCLEAR FACTOR Y, SUBUNITs C3/4/9 (NF-YC3/4/9) in the promoter of *SUPPRESSOR OF OVER-EXPRESSION OF CO 1* (*SOC1*), to promote its expression. *SOC1* encodes a MADS box transcription factor, an important regulator of flowering timing. ABA-induced expression of *SOC1* accelerates flowering under drought [89] (Figure 1). Moreover, Xiong et al. [90] observed that *ABI5* undergoes regulation at the post-transcriptional level during flowering transition. In the presence of ABA, splicing regulator, U2AF65b, increases efficiency of intron splicing from *ABI5* pre-mRNA, which promotes abundance of *ABI5* mature transcripts. This type of the *ABI5* expression control was observed in a shoot apex during ABA-mediated regulation of flowering transition [90].

Recently, some evidence emerged that ABI5 and ABFs/AREBs are also involved in fruit ripening and have an effect on fruit quality [91]. In Japanese plum (*Prunus salicina*), PsABI5 can directly activate the expression of ethylene biosynthesis gene, *ACC SYNTHASE1* (*PsACS1*) and thus, it takes part in fruit maturation [92]. It was also found that expression of *MdABI5* in apple is directly activated by KNOTTED1-LIKE HOMEOBOX 19 (MdKNOX19), ABA-responsive transcription factor involved in inhibition of fruit and seed development [93]. Additionally, apple MdAREB2 promotes accumulation of soluble sugars through regulation of expression of sugar transporter (*SUCROSE TRANSPORTER 2, MdSUT2* and *TONOPLAST MONOSACCHARIDE TRANSPORTER1, MdTMT1*) and α-/β-amylase (*MdAMY1, MdAMY3, MdBAM1* and *MdBAM3*) genes, which in turn affects fruit quality [94]. Moreover, MaABI5, a banana homolog of AtABI5, is associated with ABA-induced cold tolerance of banana (*Musa acuminata*) fruits [95].

8. Function of ABI5 and ABFs/AREBs Homologs in Monocots

In monocots, a function of ABI5 and ABFs/AREBs homologs is usually not restricted to the specific developmental stage as it is observed in Arabidopsis. Monocot homologs of AtABI5 can be active in seeds and vegetative tissues during ABA-dependent responses to abiotic stress [10,96–99]. It was observed that rice (*Oryza sativa*) OsABI5, OsABF1, OsABF2/ABI5-Like1 (OsABL1)/OsbZIP46, OsABF4/OsbZIP72, OsTRAB1, wheat

(*Triticum aestivum*) TaABL1, ABRE BINDING PROTEIN 1 (TaABP1), TaAREB3, wABI5 and maize (*Zea mays*) ZmABP9, ZmABI5 are crucial components of abiotic stress responses in vegetative tissues because of ABA-dependent regulation of stress-responsive genes [97–108]. Moreover, barley (*Hordeum vulgare*) HvABI5, rice OsABF2/OsABL1/OsbZIP46, OsABF4/OsbZIP72 and maize ZmABP9 regulate ABA-dependent responses at seed germination stage [96,103,105–107]. Furthermore, another monocot ABI5 and ABFs/AREBs homologs, barley HvABF1, HvABF2, sorghum (*Sorghum bicolor*) SbABI5 and wheat TaABF1, are important for ABA-GA crosstalk during seed germination [109–112] (Table 1).

Table 1. Function of *ABI5* and *ABFs/AREBs* homologs in monocots.

Name	Source	GenBank ID/Ensembl Plants ID	Function	References
HvABF1	*Hordeum vulgare*	DQ786408/ HORVU3Hr1G084360	Inhibition of GA-induced expression of *Amy32b* in aleurone cells	[110]
HvABF2	*Hordeum vulgare*	DQ786409/ HORVU7Hr1G035500	Inhibition of GA-induced expression of *Amy32b* in aleurone cells	[110]
HvABI5	*Hordeum vulgare*	AY150676/ HORVU5Hr1G068230	ABA-dependent activation of *HVA1* and *HVA22* in aleurone cells	[96]
		AY150676/ HORVU5Hr1G068230	Direct activation of *HvCAT2* in dormant seeds	[113]
		HQ456390/ HORVU5Hr1G068230	ABA-dependent regulation of drought response including stomatal closure, flavonoid biosynthesis, photosynthesis inhibition and activation of stress-responsive and ABA pathway genes, regulation of seed germination under ABA	[114]
OsABF1	*Oryza sativa*	GQ904238/ Os01t0867300	Positive regulation of drought and salt stress responses through activation of stress-responsive genes	[104]
OsABF2/ OsABL1/ OsbZIP46	*Oryza sativa*	GU552783, XM_015785510/ Os06t0211200	Positive regulation of drought, salt and oxidative stress responses, ABA-dependent regulation of stress-responsive genes including *WRKYs*, participation in auxin responses, regulation of seed germination under ABA	[105,106,115]
OsABF4/ OsbZIP72	*Oryza sativa*	AK065873, XM_015757064/ Os09g0456200	Positive regulation of drought response, regulation of seed germination under ABA, activation of chlorophyll catabolism genes: *OsSGR1* and *OsNYC1*	[103,116]
OsABI5	*Oryza sativa*	EF199631/ Os01t0859300	Negative regulation of drought and salt stress responses, involved in expression regulation of stress-responsive genes and in pollen maturation	[97]
			Interaction with OsKEAP1 at seed germination stage	[117]
OsTRAB1	*Oryza sativa*	AB023288/ Os08t0472000	ABA-dependent regulation of stress-responsive genes	[102]
SbABI5	*Sorghum bicolor*	XM_002454559/ SORBI_3004G309600	Activation of GA catabolism gene *SbGA2ox3* in embryos and promotion of seed dormancy	[111,118]
TaABF1	*Triticum aestivum*	AF519804/ TraesCS3A02G371800	Inhibition of GA-induced expression of *Amy32b* in aleurone cells via repression of *GAMyb* expression	[109,112]
TaABP1	*Triticum aestivum*	HQ166718/ TraesCS3B02G404300	Positive regulation of drought response	[108]

Table 1. *Cont.*

Name	Source	GenBank ID/Ensembl Plants ID	Function	References
TaABI5	*Triticum aestivum*	AB238932/ TraesCS3D02G364900	Interaction with TaJAZ1, a negative regulator of JA signaling, at seed germination stage	[40]
			Regulation of seed dormancy and germination under ABA, activation of *LEA* expression	[119]
TaABL1	*Triticum aestivum*	BJ267580/ TraesCS6A02G333600	Positive regulation of drought, salt, freezing and oxidative stresses responses through promotion of chlorophyll accumulation, stomatal closure and stress-responsive genes expression, regulation of seed germination under ABA	[100]
TaAREB3	*Triticum aestivum*	-	Positive regulation of drought response through activation of stress-responsive genes	[101]
wABI5	*Triticum aestivum*	AB193553/ TraesCS5A02G237200	Positive regulation of drought, salt and freezing stresses responses through activation of stress-responsive genes	[98]
ZmABP9	*Zea mays*	GU237073/ Zm00001eb147240	Positive regulation of drought, salt, freezing and oxidative stresses responses through promotion of photosynthesis, stomatal closure, ROS scavenging and stress-responsive genes expression; regulation of seed germination under ABA; direct activation of *ZmCAT1*	[107]
ZmABI5	*Zea mays*	EU968937/ Zm00001d018178	Negative regulation of drought, salt, heat and cold stresses responses through promotion of chlorophyll catabolism and inhibition of detoxifying enzymes activity and proline accumulation, regulation of stress-responsive genes	[99]
			Activation of raffinose biosynthesis gene, *ZmGOLS2*, in seeds	[120]

Recently, a set of new data has emerged regarding monocot AtABI5 and AtABFs/AREBs homologs. Ishibashi et al. [113] found that HvABI5, similarly to AtABI5, directly activates expression of *HvCAT2* and thus reduces H_2O_2 level in embryos and promotes seed dormancy. HvABI5 was also described as an ABA-dependent regulator of drought response in barley vegetative tissues via participation in stomatal closure, photosynthesis inhibition and flavonoid accumulation. It has to be underlined that HvABI5 enables drought adaptation by activation of stress-responsive genes and the induction of genes encoding core ABA pathway components [114]. Furthermore, Zhang et al. [120] found that *ZmABI5* is active in seeds and the encoded transcription factor directly promotes *GALACTINOL SYNTHASE2* (*ZmGOLS2*) gene expression associated with raffinose biosynthesis, which is essential for seed viability. Therefore, *HvABI5* and *ZmABI5* expression regulate ABA-dependent responses during seed germination and further developmental stages. However, in wheat, *TaABI5* was shown to be active only in seeds. It promotes *LEA* expression and maintains seeds in dormancy stage. Transgenic Arabidopsis lines overexpressing *TaABI5* were ABA-hypersensitive during seed germination [119,121]. Furthermore, as observed in Arabidopsis, TaJAZ1 is able to interact with TaABI5 and thus it inhibits ABA-dependent TaABI5 activity and promotes seed germination [40]. Similar activity of TaABI5 and AtABI5 indicates that TaABI5 can be a functional ortholog of AtABI5 in wheat. Piao et al. [116] found that OsABF4/OsbZIP72 is involved in chlorophyll catabolism and leaf senescence, such as it was observed for AtABI5 and AtABFs/AREBs. They showed that OsABF4/OsbZIP72 binds to promoters of genes responsible for chlorophyll catabolism, *OsSGR1* and *OsNYC1*, and activates their expression. Moreover, OsABI5 was shown to interact with KELCH-LIKE ECH-ASSOCIATED PROTEIN 1 (OsKEAP1) which in turn promotes OsABI5 proteasomal degradation in germinating seeds under non-stressed conditions [117]. To summarize, all these findings indicate that function of monocot ABA-dependent bZIPs under abiotic stress can be diversified. It might be more or less similar to their homologs in Arabidopsis (Table 1). However, it should be noted that monocot homologs of ABI5 and ABF/AREBs can be active at different developmental stages, whereas in dicot species their function is generally observed during specific timing. On the other side, some evidence has emerged about the role of dicot ABI5 and ABF/AREBs in ABA signaling during later developmental stages e.g., flowering, fruit development and during plant response to biotic stress. These findings have not yet been confirmed for ABI5 and ABF/AREBs in monocot species. It should be also pointed out that dicot ABI5 and ABF/AREBs positively regulate response to abiotic stress, whereas monocot ABI5 and ABF/AREB homologs can function as positive or negative regulators of stress response. Moreover, AtABI5 plays the important role in crosstalk between ABA and other phytohormonal pathways at germination stage. This function of ABI5 has not yet been fully revealed in monocot species. Therefore, further studies are needed to reveal their proper function under stress at different developmental stages.

9. Role of ABI5 and ABFs/AREBs in Feedback Regulation of Core ABA Pathway

Recently, growing evidence indicated the crucial role of ABI5 and ABFs/AREBs in the feedback regulation of the core ABA signaling and ABA biosynthesis. In Arabidopsis, Wang et al. [122] found that *ABI1* and *ABI2*, genes encoding group A PP2C phosphatases, negative regulators of ABA signaling, are directly activated by ABI5, ABF1, ABF2/AREB1, ABF4/AREB2 and ABF3 in the presence of ABA (Figure 3). Moreover, they noticed that *ABF2/AREB1* is a target gene of ABF1 and ABF4/AREB2. It shows that ABA-dependent bZIPs are part of the negative feedback loop in the ABA signaling. However, they can also strengthen their expression. Recently, ABI5 was described as a direct activator of genes encoding ABA receptors, *PYL11*, and *PYL12*, during seed germination under ABA (Figure 3). Furthermore, the pattern of eight other *PYL* genes' expression is disturbed in *abi5* mutant [123]. Thus, ABI5 is also crucial for the reinforcement of ABA signaling. Moreover, the ABA pathway's feedback regulation depends on intermediate regulators such as *FYVE DOMAIN PROTEIN REQUIRED FOR ENDOSOMAL SORTING 1* (*FREE1/FYVE1*). *FYVE1* encodes a protein localized in the peripheral membrane of late endosomal compartments and is involved in protein sorting. The expression of *FYVE1* is directly induced by ABF4/AREB2 [124] (Figure 3). However, FYVE1 protein interacts with all PYR/PYL/RCAR receptors and stimulates their degradation [124,125]. Moreover, SnRK2 kinases, SnRK2.2 and SnRK2.3, which mostly phosphorylate and activate ABI5 and ABF/AREBs, are also able to phosphorylate FYVE1 promoting its nuclear localization in response to ABA. In the nucleus, FYVE1 binds with ABI5 and ABF4/AREB2 and reduces their transactivation function [126]. Thus, ABF4/AREB2 is also involved in negative modulation of ABA-mediated stress responses through *FYVE1* regulation.

Similar interactions were also observed in other species. The overexpression of *StABI5* induces expression of *PYR/PYL/RCAR* and *SnRK2* genes in potato [83]. Furthermore, it was shown in barley that expression of *HvSnRK2.1*, *HvPP2C4* and key ABA biosynthesis gene, *CAROTENOID CLEAVAGE DIOXYGENASE 1* (*HvNCED1*), is activated in *hvabi5* mutant, which presumably exhibits enhanced activity of HvABI5 protein. *HvSnRK2.1*, *HvPP2C4*, and *HvNCED1* are putative ABI5 direct target genes because of ABRE elements in their promoters [114]. It is noteworthy that rice ABA-related bZIP transcription factor, OsbZIP23, can promote directly the expression of *OsPP2C49* and *OsNCED4* [127]. Taken together, ABI5 and ABFs/AREBs act in the feedback regulation of ABA pathway, which is important for the fine-tuning of ABA-dependent responses, according to the surrounding environmental conditions.

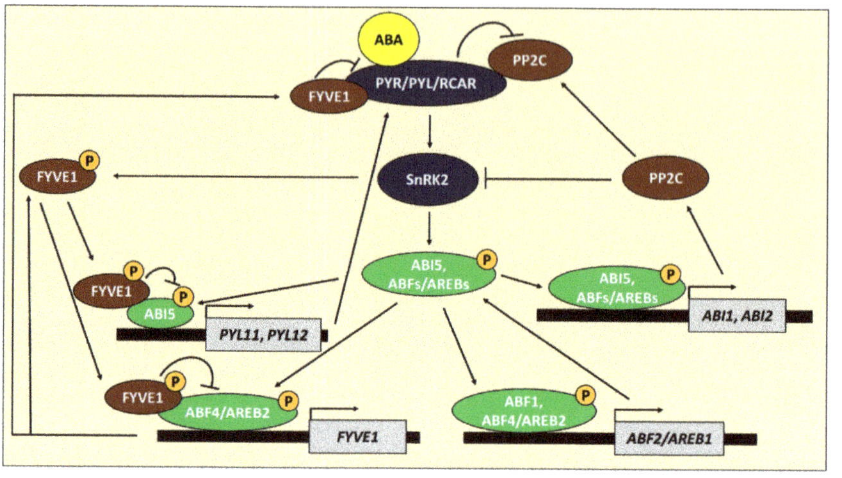

Figure 3. ABA INSENSITIVE 5 (ABI5) and ABRE BINDING FACTORs/ABRE-BINDING PROTEINs (ABFs/AREBs) function in feedback regulation of abscisic acid (ABA) pathway in Arabidopsis. In the presence of ABA, ABA receptors PYRABACTIN RESISTANCE PROTEINS/PYR-LIKE PROTEINS/REGULATORY COMPONENTS OF ABA RECEPTORs (PYR/PYL/RCARs) bind and inhibit phosphatases PHOSPHATASE 2Cs (PP2Cs), what in turn activates kinases, SNF1-RELATED PROTEIN KINASE 2s (SnRK2s). SnRK2s phosphorylate and activate ABI5 and ABFs/AREBs. Phosphatases PP2Cs, when not bound with ABA receptors PYR/PYL/RCAR, inhibit SnRK2s activity and thus ABI5 and ABFs/AREBs function. ABI5 is able to promote expression of *PYL11* and *PYL12* genes, which in turn strengthen ABA perception through PYR/PYL/RCAR receptors. ABF1 and ABF4/AREB2 trigger *ABF2/AREB1* expression. On the other side, ABI5 and ABFs/AREBs activate expression of *ABI1* and *ABI2*, genes encoding PP2Cs, which inhibits activity of SnRK2. Furthermore, ABF4/AREB2 promotes *FYVE DOMAIN PROTEIN REQUIRED FOR ENDOSOMAL SORTING 1* (*FYVE1*) expression. FYVE1 positively regulates degradation of PYR/PYL/RCARs and diminishes ABF4/AREB2 and ABI5 activity. Inhibition of ABF4/AREB2 and ABI5 activity by FYVE1 is possible after its phosphorylation carried out by SnRK2s. P—phosphate group.

10. Stress Tolerance of Transgenic Plants Overexpressing ABA-Dependent bZIPs

In Arabidopsis, transgenic lines overexpressing *ABFs/AREBs* showed increased tolerance to multiple abiotic stresses [57,72,128,129]. Moreover, rice, soybean (*Glycine max*) and cotton overexpressing Arabidopsis *ABF3, ABF2/AREB1* and *ABI5*, respectively, exhibited stress-tolerant phenotypes [130–132]. Therefore, ectopic expression of *ABFs/AREBs, ABI5* and their homologues can serve as a biotechnological tool for developing stress-tolerant cultivars. It was recently shown that potato overexpressing *ABF4/AREB2* was more tolerant to drought and salt stress. Plants with constitutive *ABF4/AREB2* expression accumulated more proline and exhibited decreased stomatal conductance and transpiration rate. Furthermore, tuber yield of transgenic lines was better than the wild-type, both, in the presence of optimal growth conditions and under stress [133]. In cotton, the overexpression of *AtABF3* and its cotton homolog, *GhABF2D*, resulted in drought tolerance related to the faster stomatal closure and reduced transpiration. However, transgenic lines also showed photosynthesis inhibition under drought [64]. Ectopic expression of *ABF3* also ensured increased tolerance of alfalfa (*Medicago sativa*) to drought, salt and oxidative stress. Drought tolerance of *ABF3* overexpression lines was caused by the reduction in transpiration rate, decreased ROS accumulation and higher chlorophyll content. On the other side, leaves of transgenic lines showed reduced size under optimal growth conditions [134]. The observed growth retardation could arise from constitutive expression of *ABFs/AREBs*. However, the application of stress-inducible or tissue-specific promoters can help to avoid this problem [64,72,129]. Na and Metzger [135] obtained transgenic plants of tomato and tobacco

(*Nicotiana tabacum*) overexpressing *ABF4/AREB2* only in guard-cells of stomata. Transgenic lines showed reduced transpiration and ensured drought tolerance. Utilization of a guard-cell specific promoter reduced negative effect of *ABF4/AREB2* overexpression on plant development (Table 2).

Table 2. Abiotic stress tolerance of transgenic plants overexpressing *ABFs/AREBs* and their homologs, generated in recent years.

Gene	Source	Target Species	Effect	Reference
ABF3	*Arabidopsis thaliana*	*Medicago sativa*	Tolerance to drought, salt and oxidative stresses. Reduced transpiration rate, ROS content and higher chlorophyll content under stress. Smaller leaf area under optimal growth conditions.	[134]
ABF3	*Arabidopsis thaliana*	*Gossypium hirsutum*	Tolerance to drought. Reduced transpiration and photosynthetic rates. Slower growth and smaller leaves under optimal growth conditions.	[64]
ABF4/AREB2	*Arabidopsis thaliana*	*Solanum tuberosum*	Tolerance to drought and salt stress. Lower transpiration rate and higher proline content under stress. Improved tuber yield under optimal growth conditions and under stress.	[133]
ABF4/AREB2	*Arabidopsis thaliana*	*Nicotiana tabacum* and *Solanum lycopersicum*	Tolerance to drought. Reduced transpiration under stress. Mild growth reduction under optimal growth conditions.	[135]
BnaABF2	*Brassica napus*	*Arabidopsis thaliana*	Tolerance to drought and salt. Reduced stomatal aperture and water loss under stress. Induced expression of *LEA* genes under stress.	[136]
GhABF2D	*Gossypium hirsutum*	*Gossypium hirsutum*	Tolerance to drought. Reduced transpiration and photosynthetic rates. Slower growth under optimal growth conditions.	[64]
IbABF4	*Ipomoea batatas*	*Arabidopsis thaliana/ Ipomoea batatas*	Tolerance to drought, salt and oxidative stresses. Higher photosynthetic efficiency, endogenous ABA content and lower ROS content under stress. Induced expression of *LEA* genes under stress. Better seed germination under salt and osmotic stress in Arabidopsis.	[86]
TaAREB3	*Triticum aestivum*	*Arabidopsis thaliana*	Tolerance to drought and freezing stresses. Lower ion leakage under stress. Induced expression of *LEA* genes under stress.	[101]
VvABF2	*Vitis vinifera*	*Arabidopsis thaliana*	Tolerance to osmotic stress. Higher activity of detoxifying enzymes activity and better ROS scavenging under stress. Induced expression of *LEA* genes under stress.	[137]
ZmABP9	*Zea mays*	*Gossypium hirsutum*	Tolerance to drought, salt and oxidative stresses. Higher chlorophyll, proline and soluble sugars content, higher activity of detoxifying enzymes, reduced stomatal aperture and ROS accumulation under stress, induced expression of stress-responsive genes. Better seed germination under salt and osmotic stress.	[138]

Transgenic plants overexpressing homologs of *ABI5* and *ABFs/AREBs* from other species than Arabidopsis also showed better performance under stress [98,100,103,107,108,115]. Overexpression of wheat *TaAREB3* in Arabidopsis induced *LEA* expression and conferred freezing and drought tolerance [101]. Transfer of rapeseed (*Brassica napus*) gene, *BnaABF2*, to Arabidopsis ensured better tolerance to drought and salt due to the smaller stomatal aperture and induced expression of *LEAs*: *RD29B*, *RAB18* and *KIN2* [136]. Overexpression of maize *ZmABP9* in cotton resulted in tolerance to drought, salt and oxidative stresses. Transgenic lines showed a higher chlorophyll, proline and soluble sugars content, increased detoxifying enzymes activity, decreased ROS level and reduced stomatal aperture in response to stress. They also exhibited increased expression of stress-responsive genes under stress. Furthermore, seeds of transgenic lines germinated better under salt and osmotic stress [138]. Expression of *VvABF2* from grapevine (*Vitis vinifera*) in Arabidopsis caused increased activity of detoxifying enzymes, better ability to scavenge ROS and higher *LEA* expression. Together, it conferred tolerance to osmotic stress of transgenic lines [137]. Moreover, overexpression of sweet potato *IbABF4* in Arabidopsis and in sweet potato ensured tolerance to drought, salt and oxidative stresses, and lower ROS level, elevated ABA content, better performance of photosynthesis, and higher expression of *LEA* genes associated with stress. Additionally, Arabidopsis transgenic lines showed better germination rate under salt and osmotic stress [86] (Table 2).

Together, these data show that ABA-dependent bZIPs are a promising tool for developing cultivars with enhanced tolerance to abiotic stresses. However, the activity of ABA-dependent bZIPs can depend on a target species, type of applied stress and analyzed developmental stage. Moreover, overexpression of *ABFs/AREBs* and their homologs can also cause undesirable growth retardation under optimal conditions. Thus, laborious preliminary studies are needed before utilizing ABA-dependent bZIPs for obtaining stress-tolerant crops.

11. Concluding Remarks

ABA-dependent regulation of plant response to abiotic stress involves the activity of many different components. ABA-dependent bZIPs, ABI5 and ABFs/AREBs are a group of transcription factors that trigger plant adaptation to unfavorable stress conditions. Their activity is strictly controlled by multiple regulators at transcriptional and protein level to ensure the accurate response to surrounding environmental conditions. In Arabidopsis, ABI5 and ABFs/AREBs are mostly active in seeds and in vegetative tissues, respectively. However, very often their function is overlapping and together, in response to ABA, they regulate many processes including seed germination, chlorophyll catabolism and flowering time. In monocots, ABI5 and ABFs/AREBs homologs are involved in regulation of stress responses, however, their activity is usually observed during different developmental stages. Function of ABI5 and ABFs/AREBs is also crucial for the feedback regulation of core ABA pathway, which results in promotion or repression of ABA-dependent plant response to the stress. Finally, *ABI5*, *ABFs/AREBs* and their homologs can serve as candidates for developing transgenic plants with increased tolerance to abiotic stress. However, more studies are necessary to understand the precise function of numerous ABA-dependent bZIPs in regulation of plant responses to multiple abiotic stresses at different developmental stages.

Author Contributions: A.C. wrote the manuscript and prepared figures. A.D.-G. and I.S. contributed to the writing of manuscript and revised it critically for important intellectual content. All authors have read and agreed to the published version of the manuscript.

Funding: This work was supported by the National Science Centre, Poland, project PRELUDIUM 2017/25/N/NZ9/01941 "The role of HvABI5 transcription factor in spring barley (*Hordeum vulgare* L.) response to drought stress".

Institutional Review Board Statement: Not applicable.

Informed Consent Statement: Not applicable.

Data Availability Statement: Not applicable.

Conflicts of Interest: The authors declare no conflict of interest.

References

1. Mahmood, T.; Khalid, S.; Abdullah, M.; Ahmed, Z.; Shah, M.K.N.; Ghafoor, A.; Du, X. Insights into Drought Stress Signaling in Plants and the Molecular Genetic Basis of Cotton Drought Tolerance. *Cells* **2019**, *9*, 105. [CrossRef] [PubMed]
2. Blum, A. Towards a conceptual ABA ideotype in plant breeding for water limited environments. *Funct. Plant Biol.* **2015**, *42*, 502–513. [CrossRef]
3. Sah, S.K.; Reddy, K.R.; Li, J. Abscisic acid and abiotic stress tolerance in crop plants. *Front. Plant Sci.* **2016**, *7*, 571. [CrossRef] [PubMed]
4. Martignago, D.; Rico-Medina, A.; Blasco-Escámez, D.; Fontanet-Manzaneque, J.B.; Caño-Delgado, A.I. Drought Resistance by Engineering Plant Tissue-Specific Responses. *Front. Plant Sci.* **2020**, *10*, 1676. [CrossRef] [PubMed]
5. Banerjee, A.; Roychoudhury, A. Abscisic-acid-dependent basic leucine zipper (bZIP) transcription factors in plant abiotic stress. *Protoplasma* **2017**, *254*, 3–16. [CrossRef]
6. Dejonghe, W.; Okamoto, M.; Cutler, S.R. Small molecule probes of ABA biosynthesis and signaling. *Plant Cell Physiol.* **2018**, *59*, 1490–1499. [CrossRef]
7. Yoshida, T.; Christmann, A.; Yamaguchi-Shinozaki, K.; Grill, E.; Fernie, A.R. Revisiting the Basal Role of ABA—Roles Outside of Stress. *Trends Plant Sci.* **2019**, *24*, 625–635. [CrossRef]
8. Lynch, T.; Erickson, B.J.; Finkelstein, R.R. Direct interactions of ABA-insensitive(ABI)-clade protein phosphatase(PP)2Cs with calcium-dependent protein kinases and ABA response element-binding bZIPs may contribute to turning off ABA response. *Plant Mol. Biol.* **2012**, *80*, 647–658. [CrossRef] [PubMed]
9. Zhou, X.; Hao, H.; Zhang, Y.; Bai, Y.; Zhu, W.; Qin, Y.; Yuan, F.; Zhao, F.; Wang, M.; Hu, J.; et al. SOS2-LIKE PROTEIN KINASE5, an SNF1-RELATED PROTEIN KINASE3-Type protein kinase, is important for abscisic acid responses in arabidopsis through phosphorylation of ABSCISIC ACID-INSENSITIVE5. *Plant Physiol.* **2015**, *168*, 659–676. [CrossRef]
10. Skubacz, A.; Daszkowska-Golec, A.; Szarejko, I. The role and regulation of ABI5 (ABA-insensitive 5) in plant development, abiotic stress responses and phytohormone crosstalk. *Front. Plant Sci.* **2016**, *7*, 1884. [CrossRef] [PubMed]
11. Nguyen, Q.T.C.; Lee, S.J.; Choi, S.W.; Na, Y.J.; Song, M.R.; Hoang, Q.T.N.; Sim, S.Y.; Kim, M.S.; Kim, J.I.; Soh, M.S.; et al. Arabidopsis Raf-Like Kinase Raf10 Is a Regulatory Component of Core ABA Signaling. *Mol. Cells* **2019**, *42*, 646–660. [CrossRef] [PubMed]
12. Zhang, H.; Liu, D.; Yang, B.; Liu, W.Z.; Mu, B.; Song, H.; Chen, B.; Li, Y.; Ren, D.; Deng, H.; et al. Arabidopsis CPK6 positively regulates ABA signaling and drought tolerance through phosphorylating ABA-responsive element-binding factors. *J. Exp. Bot.* **2020**, *71*, 188–203. [CrossRef]
13. Li, L.; Zhu, T.; Song, Y.; Feng, L.; Farag, E.A.H.; Ren, M. ABSCISIC ACID INSENSITIVE5 Interacts With RIBOSOMAL S6 KINASE2 to Mediate ABA Responses During Seedling Growth in Arabidopsis. *Front. Plant Sci.* **2021**, *11*, 598654. [CrossRef]
14. Yoshida, T.; Fujita, Y.; Sayama, H.; Kidokoro, S.; Maruyama, K.; Mizoi, J.; Shinozaki, K.; Yamaguchi-Shinozaki, K. AREB1, AREB2, and ABF3 are master transcription factors that cooperatively regulate ABRE-dependent ABA signaling involved in drought stress tolerance and require ABA for full activation. *Plant J.* **2010**, *61*, 672–685. [CrossRef] [PubMed]
15. Qian, D.; Zhang, Z.; He, J.; Zhang, P.; Ou, X.; Li, T.; Niu, L.; Nan, Q.; Niu, Y.; He, W.; et al. Arabidopsis ADF5 promotes stomatal closure by regulating actin cytoskeleton remodeling in response to ABA and drought stress. *J. Exp. Bot.* **2019**, *70*, 435–446. [CrossRef] [PubMed]
16. Finkelstein, R.R.; Lynch, T.J. The Arabidopsis abscisic acid response gene ABI5 encodes a basic leucine zipper transcription factor. *Plant Cell* **2000**, *12*, 599–609. [CrossRef]
17. Carles, C.; Bies-Etheve, N.; Aspart, L.; Léon-Kloosterziel, K.M.; Koornneef, M.; Echeverria, M.; Delseny, M. Regulation of Arabidopsis thaliana Em genes: Role of ABI5. *Plant J.* **2002**, *30*, 373–383. [CrossRef] [PubMed]
18. Huang, Y.; Sun, M.M.; Ye, Q.; Wu, X.Q.; Wu, W.H.; Chen, Y.F. Abscisic acid modulates seed germination via ABA INSENSITIVE5-mediated PHOSPHATE1. *Plant Physiol.* **2017**, *175*, 1661–1668. [CrossRef]
19. Bi, C.; Ma, Y.; Wu, Z.; Yu, Y.T.; Liang, S.; Lu, K.; Wang, X.F. Arabidopsis ABI5 plays a role in regulating ROS homeostasis by activating CATALASE 1 transcription in seed germination. *Plant Mol. Biol.* **2017**, *94*, 197–213. [CrossRef] [PubMed]
20. Chen, C.; Wu, C.; Miao, J.; Lei, Y.; Zhao, D.; Sun, D.; Yang, G.; Huang, J.; Zheng, C. Arabidopsis SAG protein containing the MDN1 domain participates in seed germination and seedling development by negatively regulating ABI3 and ABI5. *J. Exp. Bot.* **2014**, *65*, 35–45. [CrossRef]
21. Feng, C.Z.; Chen, Y.; Wang, C.; Kong, Y.H.; Wu, W.H.; Chen, Y.F. Arabidopsis RAV1 transcription factor, phosphorylated by SnRK2 kinases, regulates the expression of ABI3, ABI4, and ABI5 during seed germination and early seedling development. *Plant J.* **2014**, *80*, 654–668. [CrossRef]
22. Dekkers, B.J.W.; He, H.; Hanson, J.; Willems, L.A.J.; Jamar, D.C.L.; Cueff, G.; Rajjou, L.; Hilhorst, H.W.M.; Bentsink, L. The Arabidopsis Delay of Germination 1 gene affects Abscisic Acid Insensitive 5 (ABI5) expression and genetically interacts with ABI3 during Arabidopsis seed development. *Plant J.* **2016**, *85*, 451–465. [CrossRef]
23. Yu, L.H.; Wu, J.; Zhang, Z.S.; Miao, Z.Q.; Zhao, P.X.; Wang, Z.; Xiang, C. Bin Arabidopsis MADS-Box Transcription Factor AGL21 Acts as Environmental Surveillance of Seed Germination by Regulating ABI5 Expression. *Mol. Plant* **2017**, *10*, 834–845. [CrossRef]
24. Nakamura, S.; Lynch, T.J.; Finkelstein, R.R. Physical interactions between ABA response loci of Arabidopsis. *Plant J.* **2001**, *26*, 627–635. [CrossRef]
25. Lopez-Molina, L.; Mongrand, S.; Kinoshita, N.; Chua, N.H. AFP is a novel negative regulator of ABA signaling that promotes ABI5 protein degradation. *Genes Dev.* **2003**, *17*, 410–418. [CrossRef]

26. Stone, S.L.; Williams, L.A.; Farmer, L.M.; Vierstra, R.D.; Callis, J. KEEP ON GOING, a RING E3 ligase essential for Arabidopsis growth and development, is involved in abscisic acid signaling. *Plant Cell* **2006**, *18*, 3415–3428. [CrossRef] [PubMed]
27. Hu, Y.; Yu, D. BRASSINOSTEROID INSENSITIVE2 Interacts with ABSCISIC ACID INSENSITIVE5 to Mediate the Antagonism of Brassinosteroids to Abscisic Acid during Seed Germination in Arabidopsis. *Plant Cell* **2014**, *26*, 4394–4408. [CrossRef] [PubMed]
28. Xu, X.; Wan, W.; Jiang, G.; Xi, Y.; Huang, H.; Cai, J.; Chang, Y.; Duan, C.G.; Mangrauthia, S.K.; Peng, X.; et al. Nucleocytoplasmic Trafficking of the Arabidopsis WD40 Repeat Protein XIW1 Regulates ABI5 Stability and Abscisic Acid Responses. *Mol. Plant* **2019**, *12*, 1598–1611. [CrossRef] [PubMed]
29. Tarnowski, L.; Rodriguez, M.C.; Brzywczy, J.; Piecho-Kabacik, M.; Krčkova, Z.; Martinec, J.; Wawrzynska, A.; Sirko, A. A selective autophagy cargo receptor NBR1 modulates abscisic acid signalling in Arabidopsis thaliana. *Sci. Rep.* **2020**, *10*, 7778. [CrossRef] [PubMed]
30. Zhang, Y.; Chen, Z. Broad and Complex Roles of NBR1-Mediated Selective Autophagy in Plant Stress Responses. *Cells* **2020**, *9*, 2562. [CrossRef]
31. Yang, M.; Han, X.; Yang, J.; Jiang, Y.; Hu, Y. The Arabidopsis circadian clock protein PRR5 interacts with and stimulates ABI5 to modulate abscisic acid signaling during seed germination. *Plant Cell* **2021**, koab168. [CrossRef] [PubMed]
32. Pan, J.; Wang, H.; Hu, Y.; Yu, D. Arabidopsis vq18 and vq26 proteins interact with abi5 transcription factor to negatively modulate aba response during seed germination. *Plant J.* **2018**, *95*, 529–544. [CrossRef]
33. Chang, G.; Wang, C.; Kong, X.; Chen, Q.; Yang, Y.; Hu, X. AFP2 as the novel regulator breaks high-temperature-induced seeds secondary dormancy through ABI5 and SOM in Arabidopsis thaliana. *Biochem. Biophys. Res. Commun.* **2018**, *501*, 232–238. [CrossRef]
34. Piskurewicz, U.; Jikumaru, Y.; Kinoshita, N.; Nambara, E.; Kamiya, Y.; Lopez-Molina, L. The gibberellic acid signaling repressor RGL2 inhibits Arabidopsis seed germination by stimulating abscisic acid synthesis and ABI5 activity. *Plant Cell* **2008**, *20*, 2729–2745. [CrossRef]
35. Chen, R.; Jiang, H.; Li, L.; Zhai, Q.; Qi, L.; Zhou, W.; Liu, X.; Li, H.; Zheng, W.; Sun, J.; et al. The Arabidopsis Mediator Subunit MED25 Differentially Regulates Jasmonate and Abscisic Acid Signaling through Interacting with the MYC2 and ABI5 Transcription Factors. *Plant Cell* **2012**, *24*, 2898–2916. [CrossRef]
36. Guan, C.; Wang, X.; Feng, J.; Hong, S.; Liang, Y.; Ren, B.; Zuo, J. Cytokinin antagonizes abscisic acid-mediated inhibition of cotyledon greening by promoting the degradation of ABSCISIC ACID INSENSITIVE5 protein in Arabidopsis. *Plant Physiol.* **2014**, *164*, 1515–1526. [CrossRef] [PubMed]
37. Yuan, T.-T.; Xu, H.-H.; Zhang, K.-X.; Guo, T.-T.; Lu, Y.-T. Glucose inhibits root meristem growth via ABA INSENSITIVE 5, which represses PIN1 accumulation and auxin activity in Arabidopsis. *Plant Cell Environ.* **2014**, *37*, 1338–1350. [CrossRef] [PubMed]
38. Zhao, X.; Dou, L.; Gong, Z.; Wang, X.; Mao, T. BES1 hinders ABSCISIC ACID INSENSITIVE5 and promotes seed germination in Arabidopsis. *New Phytol.* **2019**, *221*, 908–918. [CrossRef]
39. Hu, Y.; Han, X.; Yang, M.; Zhang, M.; Pan, J.; Yu, D. The transcription factor INDUCER OF CBF EXPRESSION1 interacts with ABSCISIC ACID INSENSITIVE5 and DELLA proteins to fine-tune abscisic acid signaling during seed germination in arabidopsis. *Plant Cell* **2019**, *31*, 1520–1538. [CrossRef] [PubMed]
40. Ju, L.; Jing, Y.; Shi, P.; Liu, J.; Chen, J.; Yan, J.; Chu, J.; Chen, K.M.; Sun, J. JAZ proteins modulate seed germination through interaction with ABI5 in bread wheat and Arabidopsis. *New Phytol.* **2019**, *223*, 246–260. [CrossRef] [PubMed]
41. Pan, J.; Hu, Y.; Wang, H.; Guo, Q.; Chen, Y.; Howe, G.A.; Yu, D. Molecular mechanism underlying the synergetic effect of jasmonate on abscisic acid signaling during seed germination in arabidopsis. *Plant Cell* **2020**, *32*, 3846–3865. [CrossRef] [PubMed]
42. Wang, Y.; Li, L.; Ye, T.; Zhao, S.; Liu, Z.; Feng, Y.Q.; Wu, Y. Cytokinin antagonizes ABA suppression to seed germination of Arabidopsis by downregulating ABI5 expression. *Plant J.* **2011**, *68*, 249–261. [CrossRef]
43. Lopez-Molina, L.; Mongrand, S.; Chua, N.H. A postgermination developmental arrest checkpoint is mediated by abscisic acid and requires the ABI5 transcription factor in Arabidopsis. *Proc. Natl. Acad. Sci. USA* **2001**, *98*, 4782–4787. [CrossRef] [PubMed]
44. Brocard, I.M.; Lynch, T.J.; Finkelstein, R.R. Regulation and Role of the Arabidopsis Abscisic Acid-Insensitive 5 Gene in Abscisic Acid, Sugar, and Stress Response. *Plant Physiol.* **2002**, *129*, 1533–1543. [CrossRef] [PubMed]
45. Fan, W.; Xu, J.M.; Wu, P.; Yang, Z.X.; Lou, H.Q.; Chen, W.W.; Jin, J.F.; Zheng, S.J.; Yang, J.L. Alleviation by abscisic acid of Al toxicity in rice bean is not associated with citrate efflux but depends on ABI5-mediated signal transduction pathways. *J. Integr. Plant Biol.* **2019**, *61*, 140–154. [CrossRef] [PubMed]
46. Yu, B.; Wang, Y.; Zhou, H.; Li, P.; Liu, C.; Chen, S.; Peng, Y.; Zhang, Y.; Teng, S. Genome-wide binding analysis reveals that ANAC060 directly represses sugar-induced transcription of ABI5 in Arabidopsis. *Plant J.* **2020**, *103*, 965–979. [CrossRef]
47. Xu, D.; Li, J.; Gangappa, S.N.; Hettiarachchi, C.; Lin, F. Convergence of Light and ABA Signaling on the ABI5 Promoter. *PLoS Genet.* **2014**, *10*, e1004197. [CrossRef]
48. Kang, X.; Xu, G.; Lee, B.; Chen, C.; Zhang, H.; Kuang, R.; Ni, M. HRB2 and BBX21 interaction modulates Arabidopsis ABI5 locus and stomatal aperture. *Plant Cell Environ.* **2018**, *41*, 1912–1925. [CrossRef]
49. Qi, L.; Liu, S.; Li, C.; Fu, J.; Jing, Y.; Cheng, J.; Li, H.; Zhang, D.; Wang, X.; Dong, X.; et al. PHYTOCHROME-INTERACTING FACTORS Interact with the ABA Receptors PYL8 and PYL9 to Orchestrate ABA Signaling in Darkness. *Mol. Plant* **2020**, *13*, 414–430. [CrossRef]
50. Kim, J.; Kang, H.; Park, J.; Kim, W.; Yoo, J.; Lee, N.; Kim, J.; Yoon, T.Y.; Choi, G. PIF1-interacting transcription factors and their binding sequence elements determine the in vivo targeting sites of PIF1. *Plant Cell* **2016**, *28*, 1388–1405. [CrossRef]

51. Liu, Z.Q.; Yan, L.; Wu, Z.; Mei, C.; Lu, K.; Yu, Y.T.; Liang, S.; Zhang, X.F.; Wang, X.F.; Zhang, D. Cooperation of three WRKY-domain transcription factors WRKY18, WRKY40, and WRKY60 in repressing two ABA-responsive genes ABI4 and ABI5 in Arabidopsis. *J. Exp. Bot.* **2012**, *63*, 6371–6392. [CrossRef]
52. Wang, T.J.; Huang, S.; Zhang, A.; Guo, P.; Liu, Y.; Xu, C.; Cong, W.; Liu, B.; Xu, Z.Y. JMJ17–WRKY40 and HY5–ABI5 modules regulate the expression of ABA-responsive genes in Arabidopsis. *New Phytol.* **2021**, *230*, 567–584. [CrossRef]
53. Li, X.; Yang, R.; Gong, Y.; Chen, H. The Arabidopsis Mediator Complex Subunit MED19a is Involved in ABI5-mediated ABA Responses. *J. Plant Biol.* **2018**, *61*, 97–110. [CrossRef]
54. Ji, H.; Wang, S.; Cheng, C.; Li, R.; Wang, Z.; Jenkins, G.I.; Kong, F.; Li, X. The RCC1 family protein SAB1 negatively regulates ABI5 through multidimensional mechanisms during postgermination in Arabidopsis. *New Phytol.* **2019**, *222*, 907–922. [CrossRef]
55. An, J.P.; Zhang, X.W.; Liu, Y.J.; Wang, X.F.; You, C.X.; Hao, Y.J. ABI5 regulates ABA-induced anthocyanin biosynthesis by modulating the MYB1-bHLH3 complex in apple. *J. Exp. Bot.* **2021**, *72*, 1460–1472. [CrossRef] [PubMed]
56. Choi, H.I.; Hong, J.H.; Ha, J.O.; Kang, J.Y.; Kim, S.Y. ABFs, a family of ABA-responsive element binding factors. *J. Biol. Chem.* **2000**, *275*, 1723–1730. [CrossRef] [PubMed]
57. Fujita, Y.; Fujita, M.; Satoh, R.; Maruyama, K.; Parvez, M.M.; Seki, M.; Hiratsu, K.; Ohme-Takagi, M.; Shinozaki, K.; Yamaguchi-Shinozaki, K. AREB1 is a transcription activator of novel ABRE-dependent ABA signaling that enhances drought stress tolerance in Arabidopsis. *Plant Cell* **2005**, *17*, 3470–3488. [CrossRef] [PubMed]
58. Abdeen, A.; Schnell, J.; Miki, B. Transcriptome analysis reveals absence of unintended effects in drought-tolerant transgenic plants overexpressing the transcription factor ABF3. *BMC Genom.* **2010**, *11*, 69. [CrossRef]
59. Sharma, P.D.; Singh, N.; Ahuja, P.S.; Reddy, T.V. Abscisic acid response element binding factor 1 is required for establishment of Arabidopsis seedlings during winter. *Mol. Biol. Rep.* **2011**, *38*, 5147–5159. [CrossRef] [PubMed]
60. Kim, J.S.; Mizoi, J.; Yoshida, T.; Fujita, Y.; Nakajima, J.; Ohori, T.; Todaka, D.; Nakashima, K.; Hirayama, T.; Shinozaki, K.; et al. An ABRE promoter sequence is involved in osmotic stress-responsive expression of the DREB2A gene, which encodes a transcription factor regulating drought-inducible genes in Arabidopsis. *Plant Cell Physiol.* **2011**, *52*, 2136–2146. [CrossRef]
61. Yoshida, T.; Fujita, Y.; Maruyama, K.; Mogami, J.; Todaka, D.; Shinozaki, K.; Yamaguchi-Shinozaki, K. Four Arabidopsis AREB/ABF transcription factors function predominantly in gene expression downstream of SnRK2 kinases in abscisic acid signalling in response to osmotic stress. *Plant Cell Environ.* **2015**, *38*, 35–49. [CrossRef]
62. Lin, Q.; Wang, S.; Dao, Y.; Wang, J.; Wang, K. Arabidopsis thaliana trehalose-6-phosphate phosphatase gene TPPI enhances drought tolerance by regulating stomatal apertures. *J. Exp. Bot.* **2020**, *71*, 4285–4297. [CrossRef]
63. Wang, W.; Chen, Q.; Xu, S.; Liu, W.C.; Zhu, X.; Song, C.P. Trehalose-6-phosphate phosphatase E modulates ABA-controlled root growth and stomatal movement in Arabidopsis. *J. Integr. Plant Biol.* **2020**, *62*, 1518–1534. [CrossRef]
64. Kerr, T.C.C.; Abdel-Mageed, H.; Aleman, L.; Lee, J.; Payton, P.; Cryer, D.; Allen, R.D. Ectopic expression of two AREB/ABF orthologs increases drought tolerance in cotton (*Gossypium hirsutum*). *Plant Cell Environ.* **2018**, *41*, 898–907. [CrossRef] [PubMed]
65. Wang, Y.H.; Que, F.; Li, T.; Zhang, R.R.; Khadr, A.; Xu, Z.S.; Tian, Y.S.; Xiong, A.S. DcABF3, an ABF transcription factor from carrot, alters stomatal density and reduces ABA sensitivity in transgenic Arabidopsis. *Plant Sci.* **2021**, *302*, 110699. [CrossRef] [PubMed]
66. Sakuraba, Y.; Kim, Y.S.; Han, S.H.; Lee, B.D.; Paek, N.C. The arabidopsis transcription factor NAC016 promotes drought stress responses by repressing AREB1 transcription through a trifurcate feed-forward regulatory loop involving NAP. *Plant Cell* **2015**, *27*, 1771–1787. [CrossRef] [PubMed]
67. Seok, H.Y.; Woo, D.H.; Nguyen, L.V.; Tran, H.T.; Tarte, V.N.; Mehdi, S.M.M.; Lee, S.Y.; Moon, Y.H. Arabidopsis AtNAP functions as a negative regulator via repression of AREB1 in salt stress response. *Planta* **2017**, *245*, 329–341. [CrossRef]
68. Baek, D.; Chun, H.J.; Kang, S.; Shin, G.; Park, S.J.; Hong, H.; Kim, C.; Kim, D.H.; Lee, S.Y.; Kim, M.C.; et al. A role for arabidopsis miR399f in salt, drought, and ABA signaling. *Mol. Cells* **2016**, *39*, 111–118. [CrossRef]
69. Xu, Z.Y.; Kim, S.Y.; Hyeon, D.Y.; Kim, D.H.; Dong, T.; Park, Y.; Jin, J.B.; Joo, S.H.; Kim, S.K.; Hong, J.C.; et al. The Arabidopsis NAC transcription factor ANAC096 cooperates with bZIP-type transcription factors in dehydration and osmotic stress responses. *Plant Cell* **2013**, *25*, 4708–4724. [CrossRef]
70. Li, X.; Li, X.; Li, M.; Yan, Y.; Liu, X.; Li, L. Dual function of NAC072 in ABF3-mediated ABA-responsive gene regulation in Arabidopsis. *Front. Plant Sci.* **2016**, *7*, 1075. [CrossRef]
71. Wang, Z.; Liu, L.; Cheng, C.; Ren, Z.; Xu, S.; Li, X. GAI functions in the plant response to dehydration stress in Arabidopsis thaliana. *Int. J. Mol. Sci.* **2020**, *21*, 819. [CrossRef] [PubMed]
72. Kim, S.; Kang, J.Y.; Cho, D.I.; Park, J.H.; Soo, Y.K. ABF2, an ABRE-binding bZIP factor, is an essential component of glucose signaling and its overexpression affects multiple stress tolerance. *Plant J.* **2004**, *40*, 75–87. [CrossRef] [PubMed]
73. Finkelstein, R.; Gampala, S.S.L.; Lynch, T.J.; Thomas, T.L.; Rock, C.D. Redundant and distinct functions of the ABA response loci ABA-insensitive(ABI)5 and ABRE-binding factor (ABF)3. *Plant Mol. Biol.* **2005**, *59*, 253–267. [CrossRef] [PubMed]
74. Chang, H.C.; Tsai, M.C.; Wu, S.S.; Chang, I.F. Regulation of ABI5 expression by ABF3 during salt stress responses in Arabidopsis thaliana. *Bot. Stud.* **2019**, *60*, 16. [CrossRef]
75. Song, C.; Kim, T.; Chung, W.S.; Lim, C.O. The Arabidopsis phytocystatin AtCYS5 enhances seed germination and seedling growth under heat stress conditions. *Mol. Cells* **2017**, *40*, 577–586. [CrossRef]
76. Fernando, V.C.D.; Al Khateeb, W.; Belmonte, M.F.; Schroeder, D.F. Role of Arabidopsis ABF1/3/4 during det1 germination in salt and osmotic stress conditions. *Plant Mol. Biol.* **2018**, *97*, 149–163. [CrossRef] [PubMed]

77. You, L.; Zhang, J.; Li, L.; Xiao, C.; Feng, X.; Chen, S.; Guo, L.; Hu, H. Involvement of abscisic acid, ABI5, and PPC2 in plant acclimation to low CO2. *J. Exp. Bot.* **2020**, *71*, 4093–4108. [CrossRef] [PubMed]
78. Sunil, B.; Saini, D.; Bapatla, R.B.; Aswani, V.; Raghavendra, A.S. Photorespiration is complemented by cyclic electron flow and the alternative oxidase pathway to optimize photosynthesis and protect against abiotic stress. *Photosynth. Res.* **2019**, *139*, 67–79. [CrossRef]
79. Guo, C.; Jiang, Y.; Shi, M.; Wu, X.; Wu, G. ABI5 acts downstream of miR159 to delay vegetative phase change in Arabidopsis. *New Phytol.* **2021**, *231*, 339–350. [CrossRef]
80. Sakuraba, Y.; Jeong, J.; Kang, M.Y.; Kim, J.; Paek, N.C.; Choi, G. Phytochrome-interacting transcription factors PIF4 and PIF5 induce leaf senescence in Arabidopsis. *Nat. Commun.* **2014**, *5*, 4636. [CrossRef]
81. Su, M.; Huang, G.; Zhang, Q.; Wang, X.; Li, C.; Tao, Y.; Zhang, S.; Lai, J.; Yang, C.; Wang, Y. The LEA protein, ABR, is regulated by ABI5 and involved in dark-induced leaf senescence in Arabidopsis thaliana. *Plant Sci.* **2016**, *247*, 93–103. [CrossRef]
82. Gao, S.; Gao, J.; Zhu, X.; Song, Y.; Li, Z.; Ren, G.; Zhou, X.; Kuai, B. ABF2, ABF3, and ABF4 Promote ABA-Mediated Chlorophyll Degradation and Leaf Senescence by Transcriptional Activation of Chlorophyll Catabolic Genes and Senescence-Associated Genes in Arabidopsis. *Mol. Plant* **2016**, *9*, 1272–1285. [CrossRef] [PubMed]
83. Zhu, T.; Li, L.; Feng, L.; Ren, M. StABI5 involved in the regulation of chloroplast development and photosynthesis in potato. *Int. J. Mol. Sci.* **2020**, *21*, 1068. [CrossRef] [PubMed]
84. An, J.P.; Zhang, X.W.; Liu, Y.J.; Zhang, J.C.; Wang, X.F.; You, C.X.; Hao, Y.J. MdABI5 works with its interaction partners to regulate abscisic acid-mediated leaf senescence in apple. *Plant J.* **2021**, *105*, 1566–1581. [CrossRef] [PubMed]
85. Orellana, S.; Yañez, M.; Espinoza, A.; Verdugo, I.; González, E.; Ruiz-Lara, S.; Casaretto, J.A. The transcription factor SlAREB1 confers drought, salt stress tolerance and regulates biotic and abiotic stress-related genes in tomato. *Plant Cell Environ.* **2010**, *33*, 2191–2208. [CrossRef]
86. Wang, W.; Qiu, X.; Yang, Y.; Kim, H.S.; Jia, X.; Yu, H.; Kwak, S.S. Sweetpotato BZIP transcription factor IBABF4 confers tolerance to multiple abiotic stresses. *Front. Plant Sci.* **2019**, *10*, 630. [CrossRef]
87. Liao, C.J.; Lai, Z.; Lee, S.; Yun, D.J.; Mengiste, T. Arabidopsis HOOKLESS1 regulates responses to pathogens and abscisic acid through interaction with MED18 and acetylation of WRKY33 and ABI5 chromatin. *Plant Cell* **2016**, *28*, 1662–1681. [CrossRef]
88. Cui, W.; Wang, S.; Han, K.; Zheng, E.; Ji, M.; Chen, B.; Wang, X.; Chen, J.; Yan, F. Ferredoxin 1 is downregulated by the accumulation of abscisic acid in an ABI5-dependent manner to facilitate rice stripe virus infection in Nicotiana benthamiana and rice. *Plant J.* **2021**. [CrossRef]
89. Hwang, K.; Susila, H.; Nasim, Z.; Jung, J.Y.; Ahn, J.H. Arabidopsis ABF3 and ABF4 Transcription Factors Act with the NF-YC Complex to Regulate SOC1 Expression and Mediate Drought-Accelerated Flowering. *Mol. Plant* **2019**, *12*, 489–505. [CrossRef]
90. Xiong, F.; Ren, J.J.; Yu, Q.; Wang, Y.Y.; Lu, C.C.; Kong, L.J.; Otegui, M.S.; Wang, X.L. AtU2AF65b functions in abscisic acid mediated flowering via regulating the precursor messenger RNA splicing of ABI5 and FLC in Arabidopsis. *New Phytol.* **2019**, *223*, 277–292. [CrossRef]
91. Li, S.; Chen, K.; Grierson, D. Molecular and Hormonal Mechanisms Regulating Fleshy Fruit Ripening. *Cells* **2021**, *10*, 1136. [CrossRef] [PubMed]
92. Sadka, A.; Qin, Q.; Feng, J.; Farcuh, M.; Shlizerman, L.; Zhang, Y.; Toubiana, D.; Blumwald, E. Ethylene response of plum ACC synthase 1 (ACS1) promoter is mediated through the binding site of abscisic acid insensitive 5 (ABI5). *Plants* **2019**, *8*, 117. [CrossRef] [PubMed]
93. Jia, P.; Xing, L.; Zhang, C.; Zhang, D.; Ma, J.; Zhao, C.; Han, M.; Ren, X.; An, N. MdKNOX19, a class II knotted-like transcription factor of apple, plays roles in ABA signalling/sensitivity by targeting ABI5 during organ development. *Plant Sci.* **2021**, *302*, 110701. [CrossRef]
94. Ma, Q.J.; Sun, M.H.; Lu, J.; Liu, Y.J.; Hu, D.G.; Hao, Y.J. Transcription factor AREB2 is involved in soluble sugar accumulation by activating sugar transporter and amylase genes1. *Plant Physiol.* **2017**, *174*, 2348–2362. [CrossRef]
95. Chen, J.; Li, Y.; Li, F.; Wu, Q.; Jiang, Y.; Yuan, D. Banana MaABI5 is involved in ABA-induced cold tolerance through interaction with a RING E3 ubiquitin ligase, MaC3HC4-1. *Sci. Hortic.* **2018**, *237*, 239–246. [CrossRef]
96. Casaretto, J.; Ho, T.D. The transcription factors HvABI5 and HvVP1 are required for the abscisic acid induction of gene expression in barley aleurone cells. *Plant Cell* **2003**, *15*, 271–284. [CrossRef]
97. Zou, M.; Guan, Y.; Ren, H.; Zhang, F.; Chen, F. A bZIP transcription factor, OsABI5, is involved in rice fertility and stress tolerance. *Plant Mol. Biol.* **2008**, *66*, 675–683. [CrossRef]
98. Kobayashi, F.; Maeta, E.; Terashima, A.; Takumi, S. Positive role of a wheat HvABI5 ortholog in abiotic stress response of seedlings. *Physiol. Plant.* **2008**, *134*, 74–86. [CrossRef]
99. Yan, F.; Deng, W.; Wang, X.; Yang, C.; Li, Z. Maize (Zea mays L.) homologue of ABA-insensitive (ABI) 5 gene plays a negative regulatory role in abiotic stresses response. *Plant Growth Regul.* **2012**, *68*, 383–393. [CrossRef]
100. Xu, D.B.; Gao, S.Q.; Ma, Y.Z.; Xu, Z.S.; Zhao, C.P.; Tang, Y.M.; Li, X.Y.; Li, L.C.; Chen, Y.F.; Chen, M. ABI-like transcription factor gene TaABL1 from wheat improves multiple abiotic stress tolerances in transgenic plants. *Funct. Integr. Genomics* **2014**, *14*, 717–730. [CrossRef]
101. Wang, J.; Li, Q.; Mao, X.; Li, A.; Jing, R. Wheat transcription factor TaAREB3 participates in drought and freezing tolerances in Arabidopsis. *Int. J. Biol. Sci.* **2016**, *12*, 257–269. [CrossRef] [PubMed]
102. Kagaya, Y.; Hobo, T.; Murata, M.; Ban, A.; Hattori, T. Abscisic acid-induced transcription is mediated by phosphorylation of an abscisic acid response element binding factor, TRAB1. *Plant Cell* **2002**, *14*, 3177–3189. [CrossRef]

103. Lu, G.; Gao, C.; Zheng, X.; Han, B. Identification of OsbZIP72 as a positive regulator of ABA response and drought tolerance in rice. *Planta* **2009**, *229*, 605–615. [CrossRef]
104. Hossain, M.A.; Lee, Y.; Cho, J.I.; Ahn, C.H.; Lee, S.K.; Jeon, J.S.; Kang, H.; Lee, C.H.; An, G.; Park, P.B. The bZIP transcription factor OsABF1 is an ABA responsive element binding factor that enhances abiotic stress signaling in rice. *Plant Mol. Biol.* **2010**, *72*, 557–566. [CrossRef]
105. Hossain, M.A.; Cho, J.I.; Han, M.; Ahn, C.H.; Jeon, J.S.; An, G.; Park, P.B. The ABRE-binding bZIP transcription factor OsABF2 is a positive regulator of abiotic stress and ABA signaling in rice. *J. Plant Physiol.* **2010**, *167*, 1512–1520. [CrossRef]
106. Yang, X.; Yang, Y.N.; Xue, L.J.; Zou, M.J.; Liu, J.Y.; Chen, F.; Xue, H.W. Rice ABI5-like1 regulates abscisic acid and auxin responses by affecting the expression of ABRE-containing genes. *Plant Physiol.* **2011**, *156*, 1397–1409. [CrossRef] [PubMed]
107. Zhang, X.; Wang, L.; Meng, H.; Wen, H.; Fan, Y.; Zhao, J. Maize ABP9 enhances tolerance to multiple stresses in transgenic Arabidopsis by modulating ABA signaling and cellular levels of reactive oxygen species. *Plant Mol. Biol.* **2011**, *75*, 365–378. [CrossRef] [PubMed]
108. Cao, X.Y.; Chen, M.; Xu, Z.S.; Chen, Y.F.; Li, L.C.; Yu, Y.H.; Liu, Y.N.; Ma, Y.Z. Isolation and Functional Analysis of the bZIP Transcription Factor Gene TaABP1 from a Chinese Wheat Landrace. *J. Integr. Agric.* **2012**, *11*, 1580–1591. [CrossRef]
109. Johnson, R.R.; Shin, M.; Shen, J.Q. The wheat PKABA1-interacting factor TaABF1 mediates both abscisic acid-suppressed and abscisic acid-induced gene expression in bombarded aleurone cells. *Plant Mol. Biol.* **2008**, *68*, 93–103. [CrossRef] [PubMed]
110. Schoonheim, P.J.; Costa Pereira, D.D.; De Boer, A.H. Dual role for 14-3-3 proteins and ABF transcription factors in gibberellic acid and abscisic acid signalling in barley (Hordeum vulgare) aleurone cells. *Plant Cell Environ.* **2009**, *32*, 439–447. [CrossRef]
111. Cantoro, R.; Crocco, C.D.; Benech-Arnold, R.L.; Rodríguez, M.V. In vitro binding of Sorghum bicolor transcription factors ABI4 and ABI5 to a conserved region of a GA 2-OXIDASE promoter: Possible role of this interaction in the expression of seed dormancy. *J. Exp. Bot.* **2013**, *64*, 5721–5735. [CrossRef] [PubMed]
112. Harris, L.J.; Martinez, S.A.; Keyser, B.R.; Dyer, W.E.; Johnson, R.R. Functional analysis of TaABF1 during abscisic acid and gibberellin signalling in aleurone cells of cereal grains. *Seed Sci. Res.* **2013**, *23*, 89–98. [CrossRef]
113. Ishibashi, Y.; Aoki, N.; Kasa, S.; Sakamoto, M.; Kai, K.; Tomokiyo, R.; Watabe, G.; Yuasa, T.; Iwaya-Inoue, M. The interrelationship between abscisic acid and reactive oxygen species plays a key role in barley seed dormancy and germination. *Front. Plant Sci.* **2017**, *8*, 275. [CrossRef]
114. Collin, A.; Daszkowska-Golec, A.; Kurowska, M.; Szarejko, I. Barley ABI5 (Abscisic Acid INSENSITIVE 5) Is Involved in Abscisic Acid-Dependent Drought Response. *Front. Plant Sci.* **2020**, *11*, 1138. [CrossRef] [PubMed]
115. Tang, N.; Zhang, H.; Li, X.; Xiao, J.; Xiong, L. Constitutive activation of transcription factor OsbZIP46 improves drought tolerance in rice. *Plant Physiol.* **2012**, *158*, 1755–1768. [CrossRef] [PubMed]
116. Piao, W.; Kim, S.H.; Lee, B.D.; An, G.; Sakuraba, Y.; Paek, N.C. Rice transcription factor OsMYB102 delays leaf senescence by down-regulating abscisic acid accumulation and signaling. *J. Exp. Bot.* **2019**, *70*, 2699–2715. [CrossRef]
117. Liu, Y.H.; Jiang, M.; Li, R.Q.; Huang, J.Z.; Shu, Q.Y. Oskeap1 interacts with osabi5 and its downregulation increases the transcription of osabi5 and the aba response genes in germinating rice seeds. *Plants* **2021**, *10*, 527. [CrossRef]
118. Rodríguez, M.V.; Mendiondo, G.M.; Maskin, L.; Gudesblat, G.E.; Iusem, N.D.; Benech-Arnold, R.L. Expression of ABA signalling genes and ABI5 protein levels in imbibed Sorghum bicolor caryopses with contrasting dormancy and at different developmental stages. *Ann. Bot.* **2009**, *104*, 975–985. [CrossRef]
119. Utsugi, S.; Ashikawa, I.; Nakamura, S.; Shibasaka, M. TaABI5, a wheat homolog of Arabidopsis thaliana ABA insensitive 5, controls seed germination. *J. Plant Res.* **2020**, *133*, 245–256. [CrossRef]
120. Zhang, Y.; Sun, Q.; Zhang, C.; Hao, G.; Wang, C.; Dirk, L.M.A.; Downie, A.B.; Zhao, T. Maize VIVIPAROUS1 Interacts with ABA INSENSITIVE5 to Regulate GALACTINOL SYNTHASE2 Expression Controlling Seed Raffinose Accumulation. *J. Agric. Food Chem.* **2019**, *67*, 4214–4223. [CrossRef] [PubMed]
121. Zhou, K.; Yang, J.; Wang, Z.X.; Wang, J.R. Sequence analysis and expression profiles of TaABI5, a pre-harvest sprouting resistance gene in wheat. *Genes Genom.* **2017**, *39*, 161–171. [CrossRef]
122. Wang, X.; Guo, C.; Peng, J.; Li, C.; Wan, F.; Zhang, S.; Zhou, Y.; Yan, Y.; Qi, L.; Sun, K.; et al. ABRE-BINDING FACTORS play a role in the feedback regulation of ABA signaling by mediating rapid ABA induction of ABA co-receptor genes. *New Phytol.* **2019**, *221*, 341–355. [CrossRef]
123. Zhao, H.; Nie, K.; Zhou, H.; Yan, X.; Zhan, Q.; Zheng, Y.; Song, C.P. ABI5 modulates seed germination via feedback regulation of the expression of the PYR/PYL/RCAR ABA receptor genes. *New Phytol.* **2020**, *228*, 596–608. [CrossRef] [PubMed]
124. Pan, W.; Zheng, P.; Zhang, C.; Wang, W.; Li, Y.; Fan, T.; Liu, Y.; Cao, S. The effect of ABRE BINDING FACTOR 4-mediated FYVE1 on salt stress tolerance in Arabidopsis. *Plant Sci.* **2020**, *296*, 110489. [CrossRef]
125. Belda-Palazon, B.; Rodriguez, L.; Fernandez, M.A.; Castillo, M.C.; Anderson, E.M.; Gao, C.; Gonzalez-Guzman, M.; Peirats-Llobet, M.; Zhao, Q.; De Winne, N.; et al. FYVE1/FREE1 interacts with the PYL4 ABA receptor and mediates its delivery to the vacuolar degradation pathway. *Plant Cell* **2016**, *28*, 2291–2311. [CrossRef]
126. Li, H.; Li, Y.; Zhao, Q.; Li, T.; Wei, J.; Li, B.; Shen, W.; Yang, C.; Zeng, Y.; Rodriguez, P.L.; et al. The plant ESCRT component FREE1 shuttles to the nucleus to attenuate abscisic acid signalling. *Nat. Plants* **2019**, *5*, 512–524. [CrossRef] [PubMed]
127. Zong, W.; Tang, N.; Yang, J.; Peng, L.; Ma, S.; Xu, Y.; Li, G.; Xiong, L. Feedback regulation of ABA signaling and biosynthesis by a bZIP transcription factor targets drought-resistance-related genes. *Plant Physiol.* **2016**, *171*, 2810–2825. [CrossRef] [PubMed]

128. Kang, J.Y.; Choi, H.I.; Im, M.Y.; Soo, Y.K. Arabidopsis basic leucine zipper proteins that mediate stress-responsive abscisic acid signaling. *Plant Cell* **2002**, *14*, 343–357. [CrossRef] [PubMed]
129. Kim, J.B.; Kang, J.Y.; Soo, Y.K. Over-expression of a transcription factor regulating ABA-responsive gene expression confers multiple stress tolerance. *Plant Biotechnol. J.* **2004**, *2*, 459–466. [CrossRef]
130. Oh, S.J.; Sang, I.S.; Youn, S.K.; Jang, H.J.; Soo, Y.K.; Kim, M.; Kim, Y.K.; Baek, H.N.; Kim, J.K. Arabidopsis CBF3/DREB1A and ABF3 in transgenic rice increased tolerance to abiotic stress without stunting growth. *Plant Physiol.* **2005**, *138*, 341–351. [CrossRef] [PubMed]
131. Barbosa, E.G.G.; Leite, J.P.; Marin, S.R.R.; Marinho, J.P.; de Fátima Corrêa Carvalho, J.; Fuganti-Pagliarini, R.; Farias, J.R.B.; Neumaier, N.; Marcelino-Guimarães, F.C.; de Oliveira, M.C.N.; et al. Overexpression of the ABA-Dependent AREB1 Transcription Factor from Arabidopsis thaliana Improves Soybean Tolerance to Water Deficit. *Plant Mol. Biol. Rep.* **2013**, *31*, 719–730. [CrossRef]
132. Mittal, A.; Gampala, S.S.L.; Ritchie, G.L.; Payton, P.; Burke, J.J.; Rock, C.D. Related to ABA-Insensitive3(ABI3)/Viviparous1 and AtABI5 transcription factor coexpression in cotton enhances drought stress adaptation. *Plant Biotechnol. J.* **2014**, *12*, 578–589. [CrossRef] [PubMed]
133. Muñiz García, M.N.; Cortelezzi, J.I.; Fumagalli, M.; Capiati, D.A. Expression of the Arabidopsis ABF4 gene in potato increases tuber yield, improves tuber quality and enhances salt and drought tolerance. *Plant Mol. Biol.* **2018**, *98*, 137–152. [CrossRef]
134. Wang, Z.; Su, G.; Li, M.; Ke, Q.; Kim, S.Y.; Li, H.; Huang, J.; Xu, B.; Deng, X.P.; Kwak, S.S. Overexpressing Arabidopsis ABF3 increases tolerance to multiple abiotic stresses and reduces leaf size in alfalfa. *Plant Physiol. Biochem.* **2016**, *109*, 199–208. [CrossRef]
135. Na, J.K.; Metzger, J.D. Guard-cell-specific expression of Arabidopsis ABF4 improves drought tolerance of tomato and tobacco. *Mol. Breed.* **2017**, *37*, 154. [CrossRef]
136. Zhao, B.Y.; Hu, Y.F.; Li, J.J.; Yao, X.; Liu, K.-D. BnaABF2, a bZIP transcription factor from rapeseed (*Brassica napus* L.), enhances drought and salt tolerance in transgenic Arabidopsis. *Bot. Stud.* **2016**, *57*, 12. [CrossRef]
137. Liu, J.; Chu, J.; Ma, C.; Jiang, Y.; Ma, Y.; Xiong, J.; Cheng, Z.M. Overexpression of an ABA-dependent grapevine bZIP transcription factor, VvABF2, enhances osmotic stress in Arabidopsis. *Plant Cell Rep.* **2019**, *38*, 587–596. [CrossRef] [PubMed]
138. Wang, C.; Lu, G.; Hao, Y.; Guo, H.; Guo, Y.; Zhao, J.; Cheng, H. ABP9, a maize bZIP transcription factor, enhances tolerance to salt and drought in transgenic cotton. *Planta* **2017**, *246*, 453–469. [CrossRef]

cells

Review

Pentose Phosphate Pathway Reactions in Photosynthesizing Cells

Thomas D. Sharkey

MSU-DOE Plant Research Laboratory, Plant Resilience Institute, Department of Biochemistry and Molecular Biology, Michigan State University, East Lansing, MI 48823, USA; tsharkey@msu.edu

Abstract: The pentose phosphate pathway (PPP) is divided into an oxidative branch that makes pentose phosphates and a non-oxidative branch that consumes pentose phosphates, though the non-oxidative branch is considered reversible. A modified version of the non-oxidative branch is a critical component of the Calvin–Benson cycle that converts CO_2 into sugar. The reaction sequence in the Calvin–Benson cycle is from triose phosphates to pentose phosphates, the opposite of the typical direction of the non-oxidative PPP. The photosynthetic direction is favored by replacing the transaldolase step of the normal non-oxidative PPP with a second aldolase reaction plus sedoheptulose-1,7-bisphosphatase. This can be considered an anabolic version of the non-oxidative PPP and is found in a few situations other than photosynthesis. In addition to the strong association of the non-oxidative PPP with photosynthesis metabolism, there is recent evidence that the oxidative PPP reactions are also important in photosynthesizing cells. These reactions can form a shunt around the non-oxidative PPP section of the Calvin–Benson cycle, consuming three ATP per glucose 6-phosphate consumed. A constitutive operation of this shunt occurs in the cytosol and gives rise to an unusual labeling pattern of photosynthetic metabolites while an inducible shunt in the stroma may occur in response to stress.

Keywords: 13C; 14C; aldol; Calvin-Benson cycle; light respiration; photosynthesis; isotope labeling

Citation: Sharkey, T.D. Pentose Phosphate Pathway Reactions in Photosynthesizing Cells. *Cells* **2021**, *10*, 1547. https://doi.org/10.3390/cells10061547

Academic Editor: Suleyman Allakhverdiev

Received: 30 May 2021
Accepted: 11 June 2021
Published: 18 June 2021

1. Introduction

The PPP consists of two reaction sequences, often referred to as the oxidative and non-oxidative branches [1] (Figure 1). In the oxidative branch, glucose 6-phosphate (G6P) is converted to ribulose 5-phosphate (Ru5P) and CO_2, with the reduction of two molecules of $NADP^+$ to NADPH. The NADPH is needed for anabolic reactions, especially synthesis of lipids, and for ROS scavenging systems. The Ru5P and other pentose phosphates are used for nucleic acid synthesis, among other things. In the non-oxidative branch of the PPP, three molecules of Ru5P are converted to two molecules of fructose 6-phosphate (F6P) and one molecule of glyceraldehyde 3-phosphate (GAP) [1]. The pathway can also supply erythrose 4-phosphate (E4P) for synthesis of phenylpropanoids, including several amino acids. The PPPs are important in nearly all biological organisms. The PPPs, like glycolysis, are considered to be evolutionarily ancient, perhaps preceding the first living organisms [2].

Photosynthetic carbon metabolism, specifically the Calvin–Benson cycle of CO_2 uptake and reduction, is considered a variant of the non-oxidative branch of the PPP. It is not uncommon for photosynthetic carbon metabolism to be referred to as the "reductive PPP." However, the reduction step of the Calvin–Benson cycle is not related to pentose phosphate metabolism but rather to gluconeogenesis (the reversal of glycolysis) [3,4]. The Calvin–Benson cycle has more gluconeogenic enzymes and reactions than PPP reactions (Table 1, Figure 2).

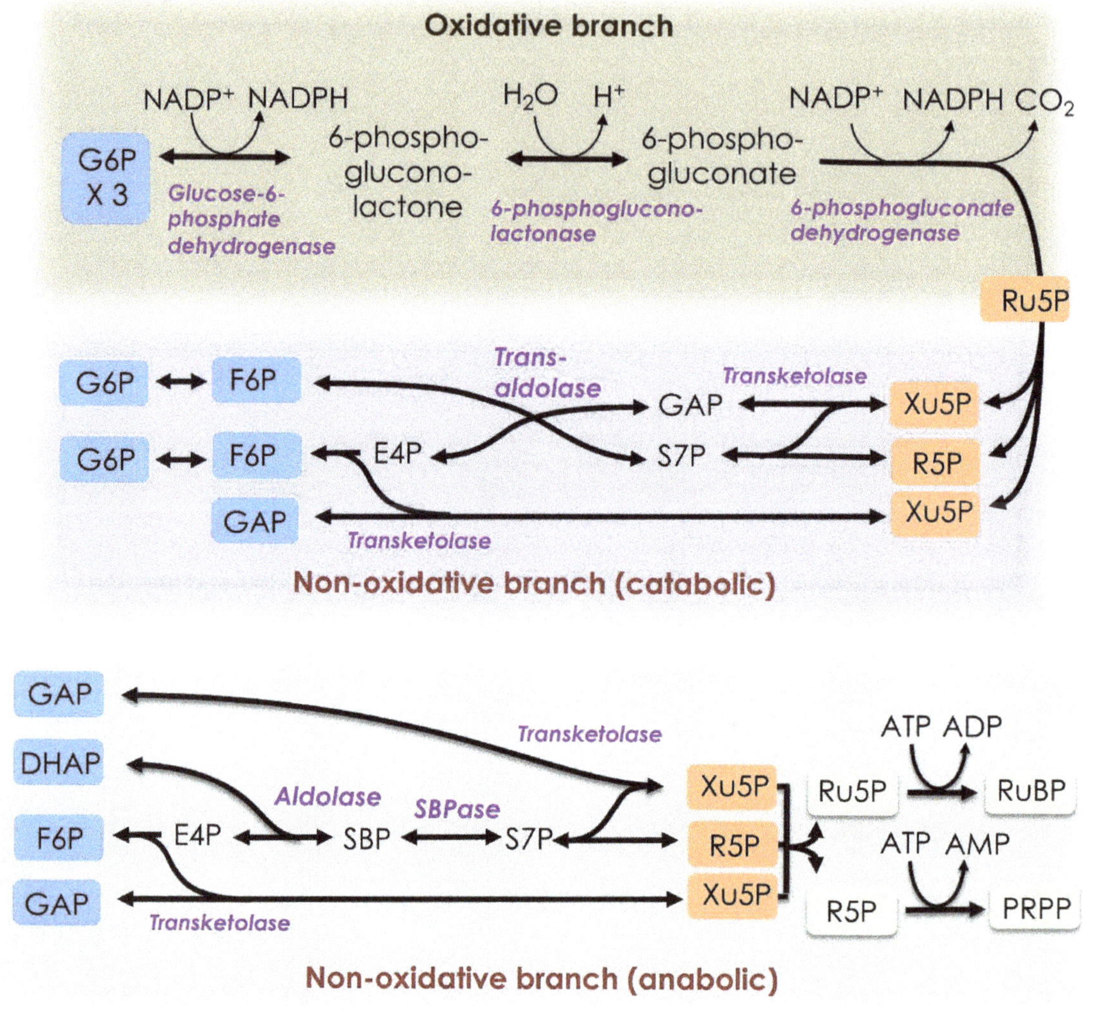

Figure 1. Pentose phosphate pathways. The oxidative branch (top) and catabolic non-oxidative branch (middle) can be shown as one pathway releasing one CO_2 molecule per glucose 6-phosphate consumed. At the bottom, the catabolic non-oxidative PPP, as found in the Calvin–Benson cycle, is shown. DHAP = dihydroxyacetone phosphate, E4P = erythrose 4-phosphate, F6P = fructose 6-phosphate, G6P = glucose 6-phosphate, GAP = glyceraldehyde 3-phosphate, PRPP = 5-*O*-phosphono- ribose 1-diphosphate, R5P = ribose 5-phosphate, Ru5P = ribulose 5-phosphate, RuBP = ribulose 1,5-bisphosphate, S7P = sedoheptulose 7-phosphate, SBP = sedoheptulose 1,7-bisphosphate, SBPase = sedoheptulose 1,7-bisphosphatase, and Xu5P = xylulose 5-phosphate.

Table 1. Number of enzymes and reaction of the Calvin–Benson cycle which are gluconeogenic or pentose phosphate pathway reactions.

Pathway	Enzymes	Reactions
Gluconeogenesis	5	16
Pentose phosphate pathway	4	6

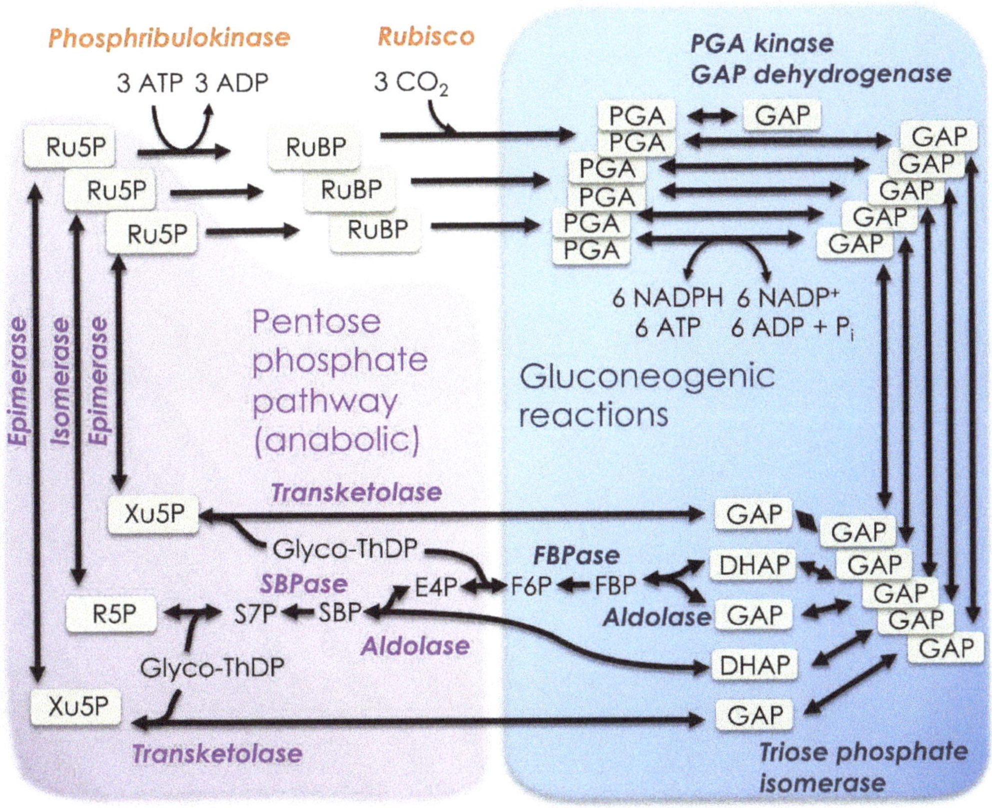

Figure 2. The Calvin–Benson cycle. Two reactions unique (or nearly so) to photosynthetic carbon metabolism link catabolic non-oxidative pentose phosphate reactions with gluconeogenic reactions to form the Calvin–Benson cycle. DHAP = dihydroxyacetone phosphate, E4P = erythrose 4-phosphate, F6P = fructose 6-phosphate, FBP = fructose 1,6-bisphosphate, FBPase = FBP phosphatase, GAP = glyceraldehyde 3-phosphate, PGA = 3-phosphoglycerate, R5P = ribose 5-phosphate, Ru5P = ribulose 5-phosphate, RuBP = ribulose 1,5-bisphosphate, S7P = sedoheptulose 7-phosphate, SBP = sedoheptulose 1,7-bisphosphate, SBPase = SBP phosphatase, ThDP = thiamine diphosphate, and Xu5P = xylulose 5-phosphate.

Using radioactive $^{14}CO_2$, it was quickly determined that PGA is the first product of CO_2 fixation by algae and plants [5]. There remained two mysteries about the path of carbon in photosynthesis that was blocking its discovery [3]. First was the nature of the reduction step and second was the nature of the "two-carbon acceptor", known to be necessary in order for 3-PGA to be the first product. Answers to both mysteries were proposed in a landmark 1954 paper [4]; the reduction step was the reversal of the oxidation step in glycolysis and the "two-carbon acceptor" was the top two carbons of RuBP, which had three different sources related to the PPP. Ribose 5-phosphate (R5P) is made by successive aldol additions of carbon to GAP (Figure 3). The carbon donor is DHAP. After the aldol addition of DHAP to GAP, two of the carbons of DHAP are transferred to thiamine diphosphate (ThDP) as a glycolaldehyde fragment, leaving one carbon added to GAP. The resulting E4P undergoes another aldol addition and removal of two of the three carbons, resulting in R5P. Each of the two glycolaldehyde fragments attached to ThDP is transferred to a molecule of GAP, resulting in two molecules of Xu5P. An isomerase converts R5P to Ru5P and an epimerase converts Xu5P to Ru5P. The Ru5P is then phosphorylated to make ribulose 1,5-bisphosphate, the CO_2 acceptor. The glycolaldehyde fragment is transferred by transketolase, which was discovered at the same time as the Calvin–Benson cycle was being worked out [6]. However, while the non-oxidative branch of the PPP was being studied for its ability to convert pentose phosphates to hexose and triose phosphates, the Calvin–Benson cycle requires flux in the opposite direction, triose phosphates must be converted to pentose phosphates.

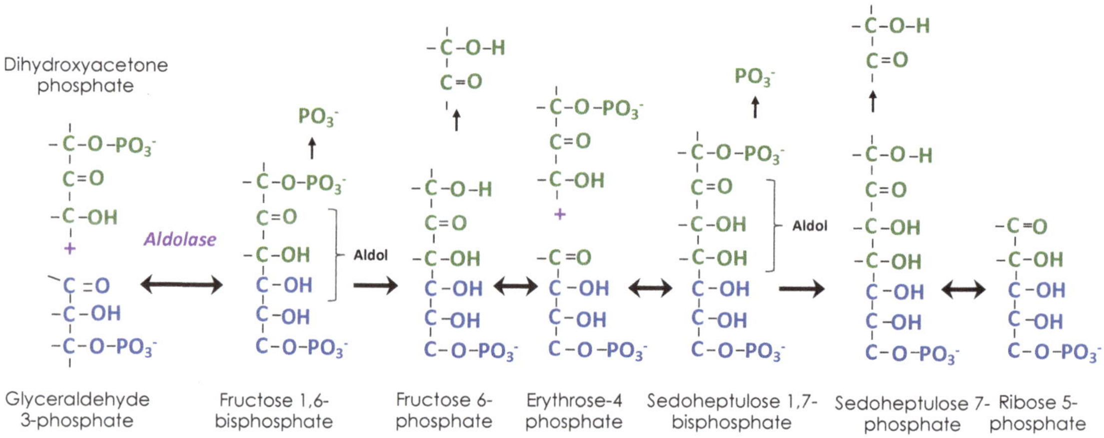

Figure 3. Building ribose 5-phosphate using aldol chemistry. Combining dihydroxyacetone phosphate and glyceraldehyde 3-phosphate results in an aldol (ALDehyde/alcohOL). Dephosphorylation is followed by removal of two carbons to make a new aldose, erythrose 4-phosphate. A second dihydroxyacetone phosphate is added followed by dephosphorylation and removal of two carbons, resulting in ribose 5-phosphate.

2. Historical Perspective

Pentose phosphate metabolism and the path of carbon in photosynthesis were under intense study around 1950. The oxidative branch of pentose phosphate metabolism was discovered first. For example, Horecker et al. [7] demonstrated the formation of ribose 5-phosphate from 6-phosphogluconate. However, it was the non-oxidative branch of pentose phosphate metabolism that was most relevant to photosynthetic carbon metabolism. In this case, the studies of Benson were especially useful to those studying pentose phosphate metabolism, especially the role of sedoheptulose phosphate, which was discovered by Benson [8,9] and quickly incorporated into proposed pathways for the non-oxidative PPP [10,11]. However, while the non-oxidative PPP involved sedoheptulose 7-phosphate synthesis by transaldolase, in the Calvin–Benson cycle, sedoheptulose 7-phosphate is made by dephosphorylation of sedoheptulose 1,7-bisphosphate. This important distinction can now be understood as optimizing the non-oxidative PPPs for either degradation of pentoses (catabolic) or synthesis of pentoses (anabolic).

3. The Catabolic and Anabolic Non-Oxidative Pentose Phosphate Pathways

The oxidative PPP makes NADPH and pentose phosphates, but if a cell needs more NADPH than pentose phosphates, the excess pentose phosphates can be converted back eventually to glucose 6-phosphate, basically a catabolic, non-oxidative PPP. However, when more pentose phosphate than NADPH is needed, the non-oxidative branch of the

PPP can operate in the opposite, anabolic direction. Just as gluconeogenesis involves most but not all of the enzymes in glycolysis, the anabolic PPP uses most, but not all, of the catabolic non-oxidative branch of the PPP. Specifically, the anabolic PPP uses SBPase instead of transaldolase. The anabolic PPP is best known from photosynthesis but it has also been found in yeast [12], where it is called the riboneogenesis pathway, and Kinetoplasta (including trypanosomes) [13]. The parasitic amoeba *Entamoeba histolytica* uses a similar pathway to make pentoses for ribonucleotides. The riboneogenesis pathway of yeast has been used to bioengineer an efficient cycle for converting F6P to acetyl phosphate [14].

4. The Calvin–Benson Cycle and the Anabolic Pentose Phosphate Pathway

The Calvin–Benson cycle combines gluconeogenic reactions with the anabolic PPP (also riboneogenesis pathway) (Figure 2). The reactions along the top in Figure 2 begin with reactions that are nearly unique to photosynthesis. The first reaction is formation of RuBP catalyzed by phosphoribulokinase. Early on, Benson [15] had recognized the potential for RuBP to be the CO_2 acceptor. Most organisms use phosphoribulokinase but some archea can use other enzymes and make 5-*O*-phosphono- ribose 1-diphosphate, convert that to ribose 1,5-bisphosphate and then finally to RuBP [16]. However, it is possible that this is more important for nucleoside degradation than for photoautotrophic CO_2 fixation [17].

The next step is catalyzed by rubisco, a fascinating enzyme that has been the subject of a great deal of research and so will not be discussed here. Rubisco introduces a carboxylic acid which must be reduced to the oxidation status of a sugar. The uncertainty about the reduction step, which was evident as late as 1952 [18], was solved by invoking gluconeogenic reactions. Further gluconeogenic reactions result in one molecule of F6P, one molecule of DHAP, and two molecules of GAP. These are the starting materials for the anabolic PPP. Among these starting materials there are four phosphate esters but at the end there are three pentose phosphates, one phosphate ester is lost. This causes the metabolism to be energetically favorable in the direction of making pentose phosphates.

4.1. The Chemistry

Building pentoses from trioses involves two types of chemistry. The first is aldol formation by adding a ketose (DHAP) to an aldehyde (first GAP, then E4P) (Figure 3). Each time three carbons are added, two are removed by transketolase and so the original GAP molecule is built up to a pentose by two one-carbon additions. When $^{14}CO_2$ is fed, carbon 1 of GAP (the aldehyde carbon) will be labeled, whereas carbon 3 of DHAP will be labeled following isomerization by triose phosphate isomerase. Carbon 3 of DHAP is the carbon that is added in the successive aldolase reactions so that the resulting R5P is labeled at carbons 1 and 2 from DHAP and carbon 3 from GAP.

The second type of chemistry is ThDP-dependent transfer of two carbons from a ketose to an aldose to make a new ketose (hence "transketolase" [6]) (Figure 4). Typically, the transketolase reactions are written with specific precursors and products. For example, two carbons are removed from F6P and donated to GAP. When two substrates are involved, the reaction can be ordered and sequential, with the two-carbon donor and acceptor both bound to the enzyme, or it can be a ping-pong mechanism, with some products leaving the enzyme before the acceptor substrate binds. Flux balance analysis is consistent with the ping-pong mechanism [19] for transketolase. The donor ketose molecule binds, transfers two carbons to ThDP, then the remaining carbons leave the enzyme. In the case of the Calvin–Benson cycle, this means that E4P or R5P leave transketolase before GAP binds to accept the glycolaldehyde fragment that is attached to ThDP.

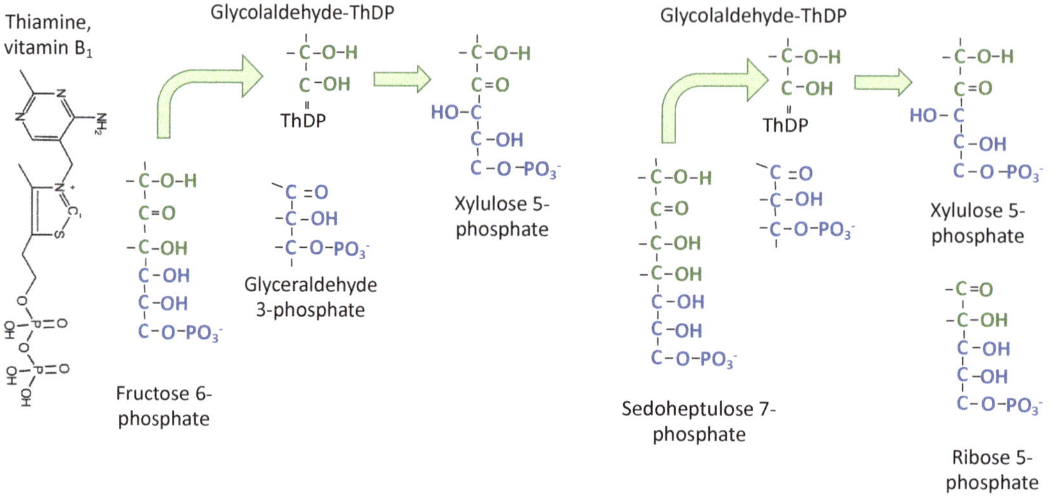

Figure 4. Transketolase transfers the two carbon fragments. Thiamine diphosphate forms a covalent bond between a carbon in the thiazole ring of thiamine diphosphate and the carbonyl carbon of fructose 6-phosphate. The bottom four carbons leave the enzyme and then the top two carbons are transferred to glyceraldehyde 3-phosphate to make xylulose 5-phosphate, thus transferring the keto group from fructose 6-phosphate to xylulose 5-phosphate.

One consequence of this is that the transketolase reaction can be nonspecific. Many ketose sugars can act as a donor and many aldose sugars can act as an acceptor [20]. When the two-carbon fragment is donated to G6P, an octulose 8-phosphate molecule can be formed. This has been reported [21–23] (Figure 5). This pathway has sometimes been called the liver version of the non-oxidative pathway [23], but the observations of octulose 8-phosphate, even in photosynthetic organisms [24], would appear to be a consequence of the flexibility of transketolase, and not necessarily indicative of significant flux through octulose 8-phosphate [1] during photosynthesis. The production of G6P from F6P in the stroma during photosynthesis is highly regulated by phosphoglucoisomerase [25]; low concentrations of G6P in the stroma might limit the formation of octulose 8-phosphate.

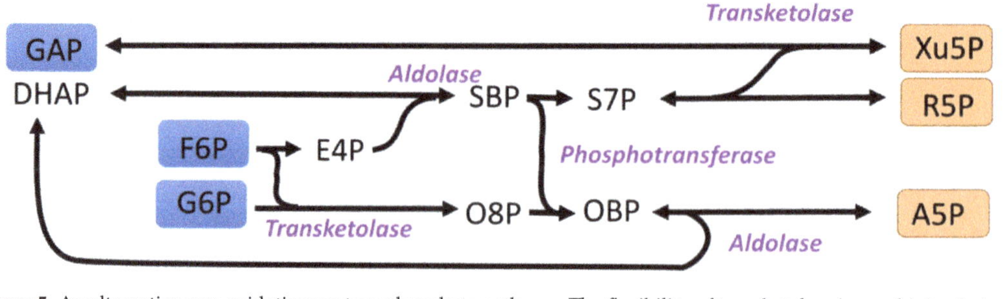

Figure 5. An alternative non-oxidative pentose phosphate pathway. The flexibility of transketolase in combining ketoses with aldoses can allow alternatives to the non-oxidative PPP. A5P = arabinose 5 phosphate, DHAP = dihydroxyacetone phosphate, E4P = erythrose 4-phosphate, F6P = fructose 6-phosphate, G6P = glucose 6-phosphate, GAP = glyceraldehyde 3-phosphate, O8P = octulose 8 phosphate, OBP = octulose 1,8-bisphosphate, R5P = ribose 5-phosphate, S7P = sedoheptulose 7-phosphate, and SBP = sedoheptulose 1,7-bisphosphate.

The major flux catalyzed by transketolase transfers a glycolaldehyde fragment bound to ThDP from F6P to GAP, resulting in Xu5P. The GAP is labeled at the 1 position, which becomes the 3 position of Xu5P. Similarly, a glycolaldehyde fragment bound to ThDP is transferred from sedoheptulose 7-phosphate to GAP, resulting in another Xu5P. The two Xu5P produced by transketolase reactions and the R5P produced by sequential addition of carbon by aldol chemistry are converted to ribulose 5-phosphate (Ru5P). Carbon 3 of both Xu5P molecules and also of the R5P are derived from carbon 3 of GAP and so carbon 3 of RuBP labels very heavily. Carbons 1 and 2 of the R5P are derived from carbon 3 of DHAP. None of these molecules would be labeled at carbons 4 and 5 (Figure 6). This gives rise to the labeling pattern used to deduce the Calvin–Benson cycle [4]. However, carbon 3 was more heavily labeled than predicted. Instead of three-fold more label in carbon 3 relative to carbon 1 or 2, it was over six-fold more labeled (Table 2). This can be explained by a lack of isotopic equilibrium between the triose phosphates [3] and it has been reported that triose phosphate isomerase is not in chemical equilibrium during photosynthesis [26]. The thermodynamics of the aldolase reaction in the direction of making FBP are more favorable if TPI is not in equilibrium [3].

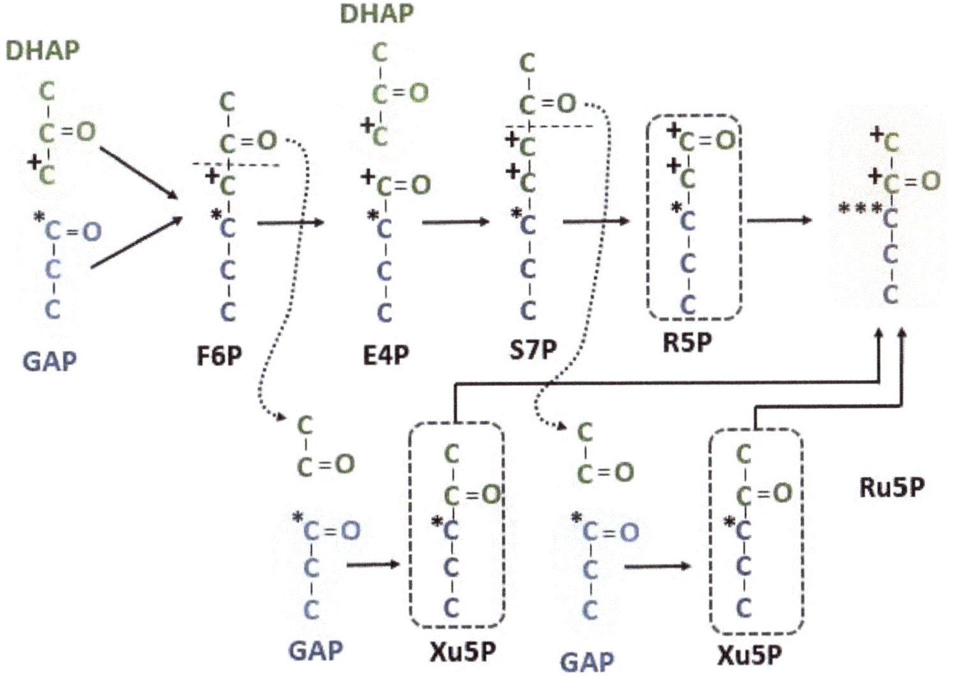

Figure 6. Intramolecular labeling pattern of the Calvin–Benson cycle. Summing the labeling resulting from three different pentose phosphate sources shows that one of the Ru5Ps will have carbon 1 and 2 labeled from carbon 3 of DHAP, carbon 3 of RuBP of all Ru5Ps will have a label from carbon 1 of GAP and carbons 4 and 5 should have no label after one occurrence of the pentose phosphate reactions of the Calvin–Benson cycle. DHAP = dihydroxyacetone phosphate, E4P = erythrose 4-phosphate, F6P = fructose 6-phosphate, GAP = glyceraldehyde 3-phosphate, R5P = ribose 5-phosphate, Ru5P = ribulose 5-phosphate, S7P = sedoheptulose 7-phosphate and Xu5P = xylulose 5-phosphate.

Table 2. Relative radioactivity observed versus predicted.

Carbon Number	Relative Radioactivity	Theoretical
1	1	1
2	0.9	1
3	6.3	3
4	0.4	0
5	0.3	0

The radioactivity in each carbon position relative to radioactivity in carbon 1 reported by Bassham et al. [4]. The theoretical values are based on the metabolism shown in Figure 4 and assume that triose phosphate isomerase results in isotopic equilibrium between glyceraldehyde 3-phosphate and dihydroxyacetone phosphate.

4.2. No Transaldolase in the Anabolic Pentose Phosphate Pathway

Dephosphorylation of SBP makes pentose phosphate production energetically favorable relative to triose phosphate production from pentose phosphates. This dephosphorylation is carried out by a specific phosphorylase (E.C. 3.1.3.37) in plants, with the abbreviation SBPase, while in yeast and other organisms the enzyme symbol is Shb17 [12]. In cyanobacteria, one of the FBPases has SBPase activity while a second form does not. In plants, FBPase does not have SBPase activity despite the similarity in sequence. In *Clostridia*, the SBPase function is carried out by pyrophosphate-dependent phosphofructokinase (PFP), potentially preserving energy. In plants, PFP activity is confined to the cytosol [27] and plastids have an active pyrophosphatase, so PFP dephosphorylation of SBP is not possible. The loss of energy upon dephosphorylation will contribute to moving metabolites from trioses to pentoses, i.e., in the anabolic or photosynthetic direction.

The presence of transaldolase in the catabolic, non-oxidative PPP allows interconversion of pentoses to hexoses and trioses without significant changes in free energy. In the case of the Calvin–Benson cycle, E4P made from F6P by transketolase could be converted to S7P by transaldolase (the other product would be GAP). In this scheme there would be no irreversible reactions between two F6P plus one GAP as inputs and three pentose phosphates as outputs. Simultaneous operation of transaldolase and aldolase/SBPase would defeat the utility of the aldolase/SBPase metabolism in providing directional driving force toward pentose phosphates.

This puts SBPase in a critical position for regulating the Calvin–Benson cycle. Modifying the expression of SBPase can have significant effects on photosynthesis rates [28–32]. Very marked increases in photosynthetic rate were observed when both aldolase and SBPase were overexpressed [32].

5. Oxidative Pentose Phosphate Pathways during Photosynthesis

The Calvin–Benson cycle is, in large measure, the anabolic, non-oxidative PPP but there is now evidence that the oxidative PPP is also important in photosynthesizing cells. The oxidative PPP in the cytosol likely operates continuously and supplies Ru5P to the chloroplast at a low but continuous rate [33]. The plastidial oxidative PPP is normally off during photosynthesis [34] but can be stimulated by H_2O_2 [35]. Because the shunts would cost 3 ATP per G6P, significant PS 1 cyclic electron flow would be required, and this could add to the myriad of processes that protect photosystems from excess energy.

It is believed that the cytosol of plant cells has most or all of the oxidative branch of the PPP but lacks critical enzymes of the non-oxidative branch [36,37]. It was thought that the oxidative PPP is not active in photosynthesizing leaves. If it were, a futile cycle, in which 3 ATP are consumed for each G6P that undergoes oxidation as shown in Scheme 1, would be created.

F6P ↔ G6P	PGI
G6P + 2 · $NADP^+$ → Ru5P + 2 · NADPH + CO_2	OPPP
Ru5P + **ATP** → RuBP + **ADP**	PRK
RuBP + CO_2 → 2 · 3PGA	Rubisco
2 · 3PGA + **2 · ATP** + 2 · NADPH ↔ 2 · TP + **2 · ADP** + **2 · P_i** + 2 · $NADP^+$	PGK & GAPDH
2TP ↔ FBP → F6P + **P_i**	Aldolase
3 · ATP → 3 · ADP + 3 · P_i	**Net**

Scheme 1. The energy-requiring or -releasing reactions when the oxidative pentose phosphate pathway bypasses the anabolic pentose phosphate pathway of the Calvin-Benson cycle. F6P = fructose 6-phosphate, G6P = glucose 6-phosphate, 3PGA = 3-phosphoglycewrate, R5P = ribose 5-phosphate, Ru5P = ribulose 5-phosphate, RuBP = ribulose 1,5-bisphosphate, and TP = triose phosphate.

There are six G6P dehydrogenases in *Arabidopsis*, of which three are located in the chloroplast [38]. It was found that G6PDH1 is very sensitive to redox status, normally having very low activity (high K_m) in a reducing environment [35,39] as a result of reduction of two cysteine residues [40]. On the other hand, there is no indication that G6PDH enzymes in the cytosol, especially Arabidopsis G6PDH 5 and 6, are regulated to reduce their activity in the light [40]. The question arises, what would prevent the oxidative PPP from operating in the cytosol during photosynthesis?

It has recently been hypothesized that the oxidative PPP *is* active during photosynthesis and it makes a shunt, bypassing the conversion of triose and hexose phosphates to pentose phosphates in the Calvin–Benson cycle [41]. There is no evidence that cytosolic G6PDH enzymes in the cytosol are turned off in the light and there is a significant pool of G6P in the cytosol in leaves in the light [42–45]. Both static [33] and dynamic [46] analyses of $^{13}CO_2$ feeding results have found that the operation of the oxidative PPP as a shunt bypassing much of the Calvin–Benson cycle is strongly supported. It is estimated to proceed at approximately 5% of the rate of net CO_2 assimilation.

6. Explanatory Power of the Cytosolic Oxidative Pentose Phosphate Pathway

6.1. Labeling Kinetics of Calvin–Benson Cycle Intermediates

Labeling of Calvin–Benson cycle intermediates provided the insights needed to deduce the path of carbon in photosynthesis. However, in the 1970s, Canvin and colleagues published results of experiments showing that, although the Calvin–Benson cycle intermediates labeled quickly at first, once they were ~80% labeled, the rate of labeling slowed significantly [47–49]. This was shown to be also true of isoprene labeling [50], which is now known to reflect labeling kinetics of the Calvin–Benson cycle intermediates [33]. The disappearance of ^{12}C in isoprene when leaves were fed with 99+% $^{13}CO_2$ showed two distinct phases when plotted on a log scale (Figure 7). Szecowka et al. [45] also pointed out the biphasic labeling pattern they found when $^{13}CO_2$ was fed to Arabidopsis rosettes and similar data were reported by Hasunuma et al. [51] and Ma et al. [52]. When the CO_2 supply was switched back to $^{12}CO_2$, the disappearance of $^{13}CO_2$ from emitted isoprene was not biphasic and appeared to have the same time constant as the fast phase of disappearance of $^{12}CO_2$ in a $^{13}CO_2$ atmosphere (Figure 7).

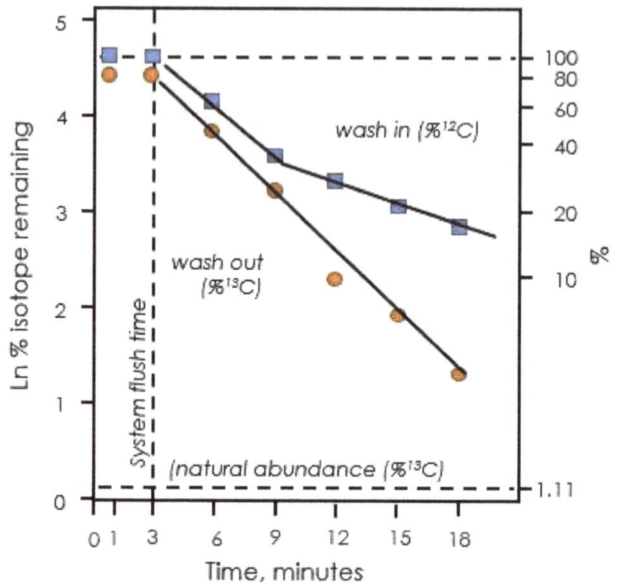

Figure 7. Exponential decay of label in isoprene when leaves are fed $^{13}CO_2$. Upon switching from $^{12}CO_2$ to $^{13}CO_2$, there are two distinct phases of exponential decay of ^{12}C remaining (blue squares). Upon switching back to $^{12}CO_2$, only one phase is observed (orange circles). Data redrawn from Delwiche and Sharkey [50].

To account for this slow phase of labeling, pools of Calvin–Benson cycle intermediates disconnected from metabolism (inactive metabolites) were hypothesized [45,52]. However, the slow-labeling component was not static, the loss of ^{12}C during feeding $^{13}CO_2$ showed a very clear two-phased exponential decay [50]. Similar two-phased exponential decay kinetics are seen in the data of [33,45]. This indicates that there are at least two processes affecting the kinetics of labeling, one fast and one slower. The time constant for the two phases of exponential decay of ^{12}C in Calvin–Benson cycle intermediates was 0.23 min^{-1} for the fast phase and 0.014 min^{-1} for the slow phase.

It is now hypothesized that the slow phase reflects a slow flow of unlabeled carbon, from glucose, fructose, or sucrose acted on by invertase back into the Calvin–Benson cycle [33]. Hexokinase or fructokinase [53] can return unphosphorylated hexoses to the hexose phosphate pool. Hexose phosphates are generally not exchanged across chloroplast membranes [54,55] but the oxidative PPP provides a path of carbon from hexose phosphates in the cytosol to the chloroplast by import through the xylulose-5-phosphate transporter (XPT) [56].

An examination of the data of Szecowka et al. [45] supports this mechanism (Figure 8). After 20 min of feeding $^{13}CO_2$, the fast-labeling phase is over but there is little label in free glucose and fructose. Intermediates of the Calvin–Benson cycle (Figure 8) as well as the chloroplastic methylerythritol 4-phosphate pathway, as evidenced by the label in isoprene [50], are labeled to the same degree, usually between 80 and 90%. When the Calvin–Benson cycle intermediates are labeled to 80%, carbon export from the Calvin–Benson cycle is removing $^{12}CO_2$ at a rate of 0.2 times the net rate of CO_2 assimilation. If the net rate of CO_2 assimilation is 10 μmol m^{-2} s^{-1}, there must be a ^{12}C input back into the Calvin–Benson cycle at a rate of 2 μmol m^{-2} s^{-1}. If the oxidative PPP supplies this from unlabeled glucose, and five carbons are supplied for each glucose molecule that follows the pathway, then the oxidative PPP would need to proceed at a rate of 2 μmol m^{-2} s^{-1} divided by 5 carbons per glucose for a G6PDH rate of 0.4 μmol m^{-2} s^{-1}. If the source for the oxidative PPP has some label, the rate would need to be proportionally more but if some of the CO_2 released in the oxidative PPP is refixed, less would be needed.

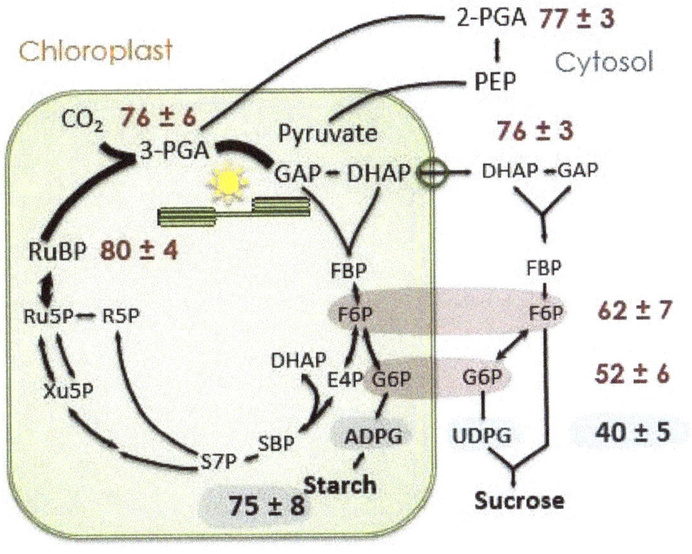

Figure 8. Label in selected photosynthesis intermediates. The proportion of carbon atoms that are ^{13}C 20 min after beginning to feed $^{13}CO_2$. Data taken from Szecowka et al. [45]). Data for F6P and G6P are for both the stromal and cytosolic pools, presumably the value for ADPG reflects the degree of label in the stromal F6P and G6P pools while UDPG (mostly) reflects the values for cytosolic F6P and G6P. ADPG = ADP-glucose, DHAP = dihydroxyacetone phosphate, E4P = erythrose 4-phosphate, F6P = fructose 6-phosphate, FBP = fructose 1,6-bisphosphate, G6P = glucose 6-phosphate, GAP = glyceraldehyde 3-phosphate, PEP = phosphoenolpyruvate, PGA = 3-phosphoglycerate, R5P = ribose 5-phosphate, Ru5P = ribulose 5-phosphate, RuBP = ribulose 1,5-bisphosphate, S7P = sedoheptulose 7-phosphate, SBP = sedoheptulose 1,7-bisphosphate, and UDPG = UDP-glucose.

6.2. Respiration in the Light Explained by the Cytosolic G6P Shunt through the Oxidative Pentose Phosphate Pathway

The widely used model of photosynthetic carbon metabolism published by Farquhar et al. [57] (now often referred to as the FvCB model for the three authors) the net rate of CO_2 assimilation, A, was given as

$$A = v_c - 0.5v_o - R_d \quad (1)$$

where v_c is the rate of carboxylation of RuBP and v_o is the rate of oxygenation (the first step of photorespiration). The last parameter was called dark respiration to account for mitochondrial respiration that might continue in the light [57]. The definition was broadened to include any process that releases CO_2 during photosynthesis and so, to keep the same symbol, was renamed day respiration [58]. However, recently this has been relabeled to R_L, for respiration in the light.

There are three methods commonly used to estimate R_L. The first is the Laisk method [59], which came into common usage after its use by Brooks and Farquhar [60]. The method uses CO_2 response curves measured at different (but low) light intensities. These curves cross over at a CO_2 level where v_c is one half of v_o so that A is equal to $-R_L$. There have been a number of refinements to the technique, but the basic premise remains and the advances in Laisk method measurements will not be considered here. A second measurement is based on extrapolation of a light response curve. When A is plotted as a function of light there is a distinct reduction in slope at about the light intensity where A switches from negative to positive (called the Kok effect). The lower slope of the light response above the compensation point is extrapolated to zero light and the difference between this number and respiration actually measured in darkness is taken to be R_L. A related technique makes use of chlorophyll fluorescence estimates of photosynthetic electron transport rates [61,62]. This improves on the Kok method for determining R_L, but again makes use of photosynthesis at low light intensities.

A method for measuring R_L using isotopes of carbon was developed by Loreto et al. [63]. This method is based on the assumption that carbon sources for R_L are not quickly labeled when $^{13}CO_2$ is fed to leaves and so R_L can be measured as an efflux of $^{12}CO_2$ into a $^{13}CO_2$ atmosphere. An advantage to this method is that it can be used at higher rates of photosynthesis. However, it is unknown whether these different methods for measuring R_L are measuring the same phenomenon given that they are made a very low light and/or CO_2 in most cases but not in the Loreto method. The cytosolic oxidative PPP and slow-labeling carbon pool can provide an explanation for the isotope method measure of R_L.

If the cytosolic G6P shunt operates as indicated above, CO_2 will be released with the same degree of label as the G6P pool in the cytosol. The rate of that release would be the rate of the cytosolic G6P shunt, estimated above to be on the order of 0.4 µmol m^{-2} s^{-1}. This is in the range of reported values for R_L [64]. The slow-labeling phase when $^{13}CO_2$ is fed to leaves likely reflects labeling of the cytosolic hexose phosphate pool. As this pool labels, the CO_2 released by the G6P shunt will become labeled and the apparent R_L measured by the isotopic method should decline. This is in fact observed in unpublished data of Alyssa Preiser (Figure 9).

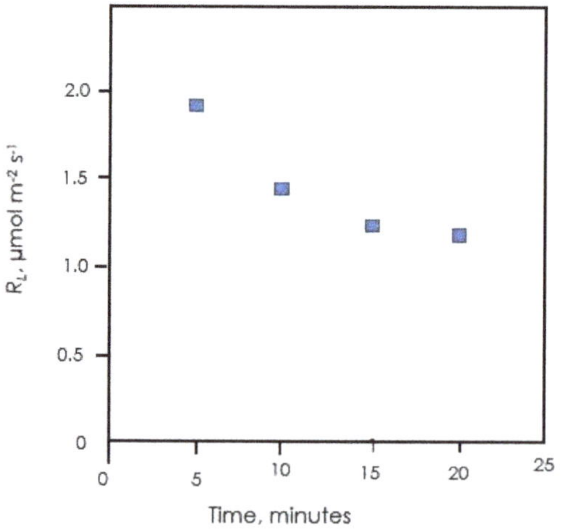

Figure 9. R_L over time after feeding $^{13}CO_2$. R_L was measured as $^{12}CO_2$ emission into a $^{13}CO_2$ atmosphere (Loreto method). Data of Alyssa Preiser.

If the cytosolic G6P shunt is responsible for R_L measured isotopically, is it also responsible for R_L measured by the Laisk method, near the CO_2 compensation point, or the Yin and Struik method, near the light compensation point? The controls on the rate of the G6P shunt are unknown. If it is controlled simply by the amount of G6P in the cytosol it will be possible to learn more about how R_L changes with conditions although there currently are few reported measurements of G6P levels in the cytosol.

7. Explanatory Power of the Stromal Oxidative Pentose Phosphate Pathway

7.1. Stimulation of Cyclic Electron Flow during Stress

The oxidative PPP in the chloroplast is inhibited in the light by reduction of two cysteine residues of a stromal G6PDH [40]. However, this inhibition, which prevents the G6P shunt futile cycle, can be overcome by high concentration of G6P or H_2O_2 [35]. Various stresses can cause H_2O_2 to accumulate in chloroplasts. Detecting flux through the chloroplastic G6P shunt when the cytosolic shunt is occurring is difficult, but one indicator is the label in 6-PG relative to that in ADP-glucose, an indicator of the degree of label in stromal G6P, and UDP glucose, an indicator of the degree of label in the cytosol, although UDPglucose label will be affected somewhat by a small pool of UDP glucose in the chloroplast, presumably related to lipid synthesis [65]. When poplar leaves were fed $^{13}CO_2$ at 30 °C, 6PG labeling was indistinguishable from UDPglucose, indicating little or no plastidial G6P shunt. However, at 40 °C, 6PG labeling was much closer to ADP glucose, indicating a significant rate of plastidial G6P shunt under high temperature stress [33].

Operation of the plastidial G6P shunt will consume ATP but not NADPH (Scheme 1), which will allow/require cyclic electron flow around photosystem I. This could help balance the ATP/NADPH demand or help protect photosystem I when too much energy is arriving there. Photosystem I can reduce oxygen to superoxide, and superoxide dismutase can convert that to H_2O_2. Thus, H_2O_2 is an indicator of too much energy at photosystem I. By stimulating the stromal G6P shunt, ATP will be consumed, stimulating cyclic electron flow, and so relieve the excess pressure at photosystem I. Stimulation of cyclic electron flow by H_2O_2 has been reported [66].

7.2. Excess CO_2 Release during Photorespiration

The stromal oxidative PPP was proposed to explain two traits found in mutants lacking hydroxypyruvate reductase, needed in photorespiration. These plants have excess CO_2 release during photosynthesis. This was calculated assuming that R_L is constant during various treatments and changes in CO_2 release result from changes in the stoichiometry of photorespiratory CO_2 release [67]. However, Li et al. [68] showed that these plants also expressed a gene for a G6P transporter (*GPT2*) that is normally silent in photosynthetic tissue. It was hypothesized that this allows metabolites to bypass stromal triose phosphate isomerase, an enzyme significantly inhibited by 2-phosphoglycolate of photorespiration [26,69,70]. This would allow G6P to flood into the chloroplast and high levels of G6P can overcome the inhibition of stromal G6PDH [35]. These plants also accumulate H_2O_2 [26] another plastidial G6PDH activating factor. Activation of the plastidial oxidative PPP would form a shunt that would release CO_2, offering an alternative explanation for the excess CO_2 release in *hdr1* mutants [26].

If a G6P shunt in the chloroplast is responsible for the excess CO_2 release seen in *hpr1* plants, then these plants should also have increased cyclic electron flow to compensate for the excess ATP consumed in the shunt. This has been observed [68].

8. Energetics

The operation of either the plastidial or cytosolic G6P shunts results in the consumption of three ATP per G6P oxidized. These ATP are consumed entirely inside the chloroplast whether the shunt is plastidial or cytosolic. Using data reported in Sharkey et al. [33], the added costs of photorespiration and G6P shunts at 30 and 40 °C are shown in Table 3. Linear photosynthetic electron transport is believed to provide 1.285 ATPs per NADPH (assuming a constitutive Q cycle and 14 protons per 3 ATP). Since CO_2 assimilation costs 1.5 ATP per NADPH there is an ATP deficit that can be made up by cyclic electron flow around photosystem I or by the Mehler reaction, also called the water-water cycle. At 30 °C, photorespiration and the G6P shunt added 0.06 to the ATP deficit on top of the 0.215 deficit for CO_2 fixation.

Table 3. Effect of G6P shunts on NADPH and ATP required.

30 °C	*A*	Photorespiration	G6P Shunts
Rate, μmol m^{-2} s^{-1}	15.8	5.0	0.79
NADPH (cumulative)	31.6	41.6	41.6
ATP (cumulative)	47.4	64.9	67.3
ATP/NADPH ratio cumulative	1.5	1.56	1.62
40 °C			
Rate, μmol m^{-2} s^{-1}	6.6	2.8	1.19
NADPH (cumulative)	13.2	18.8	18.8
ATP (cumulative)	19.8	29.6	33.2
ATP/NADPH ratio cumulative	1.5	1.57	1.77

Data from the experiment reported in Sharkey et al. [33] with the effect of adding photorespiration or photorespiration plus the G6P shunts to energy used in photosynthesis. Both the cytosolic and stromal G6P shunts increase the ATP requirement (by three for each occurrence of the G6P shunt) but do not increase the requirement for NADPH. It is estimated that linear electron flow provides 1.285 ATP per NADPH and so as the ATP/NADPH ratio goes above 1.285 cyclic electron flow, or a similar mechanism, is required.

At 40 °C, both photorespiration and the G6P shunts were stimulated. In this experiment photorespiration added 0.07 while the shunts added 0.20 to the ATP deficit relative to linear electron flow, nearly as much as the deficit for CO_2 fixation alone. At 40 °C the total ATP/NADPH ratio required for CO_2 fixation plus photorespiration plus the G6P shunts was 1.77, requiring substantial extra ATP, which could be supplied by cyclic electron flow. Stimulation of cyclic electron flow at high temperature has been reported often [71–77].

9. Conclusions

The role of the anabolic, non-oxidative PPP in photosynthesis was proposed in 1954 [4] and has stood the test of time. Now, recent data indicate that the oxidative PPP is also very important in photosynthesizing leaves. This pathway may operate constitutively in the cytosol at a rate of approximately 4% of the rate CO_2 fixation. In the plastid, this pathway appears to normally be insignificant in unstressed leaves, as had been thought, but new evidence indicates that the plastidial pathway can be stimulated during stress; H_2O_2 likely plays a role in stimulating this pathway in the light.

Funding: My photosynthesis research is supported by the Division of Chemical Sciences, Geosciences and Biosciences, Office of Basic Energy Sciences of the United States Department of Energy (Grant DE-FG02-91ER20021), my isoprene research is supported by NSF grant 2022495, and I receive partial salary support from Michigan AgBioResearch.

Conflicts of Interest: The author declares no conflict of interest.

Abbreviations

DHAP	dihydroxyacetone phosphate
E4P	erythrose 4-phosphate
F6P	fructose 6-phosphate
FBP	fructose 1,6-bisphosphate
G6P	glucose 6-phosphate
G6PDH	G6P dehydrogenase
GAP	glyceraldehyde 3-phosphate (not to be confused with G3P, glycerol 3-phosphate, which is important in lipid synthesis)
PGA	3-phosphoglycerate
6-PG	6-phosphogluconate
PPP	pentose phosphate pathway
PRPP	5-O-phosphono- ribose 1-diphosphate,
R5P	ribose 5-phosphate
Ru5P	ribulose 5-phosphate

RuBP	ribulose 1,5-bisphosphate
S7P	sedoheptulose 7-phosphate
SBP	sedoheptulose 1,7-bisphosphate
SBPase	sedoheptulose 1,7-bisphosphatase
ThDP	thiamine diphosphate
Xu5P	xylulose 5-phosphate

References

1. Stincone, A.; Prigione, A.; Cramer, T.; Wamelink, M.M.C.; Campbell, K.; Cheung, E.; Olin-Sandoval, V.; Grüning, N.-M.; Krüger, A.; Tauqeer Alam, M.; et al. The return of metabolism: Biochemistry and physiology of the pentose phosphate pathway. *Biol. Rev.* **2015**, *90*, 927–963. [CrossRef]
2. Keller, M.A.; Turchyn, A.V.; Ralser, M. Non-enzymatic glycolysis and pentose phosphate pathway-like reactions in a plausible Archean ocean. *Mol. Syst. Biol.* **2014**, *10*, 725. [CrossRef] [PubMed]
3. Sharkey, T.D. Discovery of the canonical Calvin–Benson cycle. *Photosynth. Res.* **2019**, *140*, 235–252. [CrossRef] [PubMed]
4. Bassham, J.A.; Benson, A.A.; Kay, L.D.; Harris, A.Z.; Wilson, A.T.; Calvin, M. The path of carbon in photosynthesis. XXI. The cyclic regeneration of carbon dioxide acceptor. *J. Am. Chem. Soc.* **1954**, *76*, 1760–1770. [CrossRef]
5. Benson, A.A.; Bassham, J.A.; Calvin, M.; Goodale, T.C.; Haas, V.A.; Stepka, W. The path of carbon in photosynthesis. V. Paper chromatography and radioautography of the products. *J. Am. Chem. Soc.* **1950**, *72*, 1710–1718. [CrossRef]
6. Racker, E.; Haba, G.D.L.; Leder, I.G. Thiamine pyrophosphate, a coenzyme of transketolase. *J. Am. Chem. Soc.* **1953**, *75*, 1010–1011. [CrossRef]
7. Horecker, B.L.; Smyrniotis, P.Z.; Seegmiller, J.E. The enzymatic conversion of 6-phosphogluconate to ribulose-5-phosphate and ribose-5-phosphate. *J. Biol. Chem.* **1951**, *193*, 383–396. [CrossRef]
8. Benson, A.A.; Bassham, J.A.; Calvin, M. Sedoheptulose in photosynthesis by plants. *J. Am. Chem. Soc.* **1951**, *73*, 2970. [CrossRef]
9. Benson, A.A.; Bassham, J.A.; Calvin, M.; Hall, A.G.; Hirsch, H.E.; Kawaguchi, S.; Lynch, V.; Tolbert, N.E. The path of carbon in photosynthesis XV. Ribulose and sedoheptulose. *J. Biol. Chem.* **1952**, *196*, 703–716. [CrossRef]
10. Horecker, B.L.; Smyrniotis, P.Z. The enzymatic formation of sedoheptulose phosphate from pentose phosphate. *J. Am. Chem. Soc.* **1952**, *74*, 2123. [CrossRef]
11. Horecker, B.L.; Gibbs, M.; Klenow, H.; Smyrniotis, P.Z. The mechanism of pentose phosphate conversion to hexose monophosphate: I. With a liver enzyme preparation. *J. Biol. Chem.* **1954**, *207*, 393–404. [CrossRef]
12. Clasquin, M.F.; Melamud, E.; Singer, A.; Gooding, J.R.; Xu, X.; Dong, A.; Cui, H.; Campagna, S.R.; Savchenko, A.; Yakunin, A.F.; et al. Riboneogenesis in yeast. *Cell* **2011**, *145*, 969–980. [CrossRef] [PubMed]
13. Hannaert, V.; Bringaud, F.; Opperdoes, F.R.; Michels, P.A. Evolution of energy metabolism and its compartmentation in Kinetoplastida. *Kinetoplastid Biol. Dis.* **2003**, *2*, 11. [CrossRef] [PubMed]
14. Hellgren, J.; Godina, A.; Nielsen, J.; Siewers, V. Promiscuous phosphoketolase and metabolic rewiring enables novel non-oxidative glycolysis in yeast for high-yield production of acetyl-CoA derived products. *Metab. Eng.* **2020**, *62*, 150–160. [CrossRef]
15. Benson, A.A. Identification of ribulose in $C^{14}O_2$ photosynthesis products. *J. Am. Chem. Soc.* **1951**, *73*, 2971–2972. [CrossRef]
16. Finn, M.W.; Tabita, F.R. Modified pathway to synthesize ribulose 1,5-bisphosphate in methanogenic archaea. *J. Bacteriol.* **2004**, *186*, 6360–6366. [CrossRef] [PubMed]
17. Aono, R.; Sato, T.; Imanaka, T.; Atomi, H. A pentose bisphosphate pathway for nucleoside degradation in Archaea. *Nat. Chem. Biol.* **2015**, *11*, 355–360. [CrossRef] [PubMed]
18. Calvin, M.; Massini, P. The path of carbon in photosynthesis. XX. The steady state. *Experientia* **1952**, *8*, 445–457. [CrossRef]
19. Kleijn, R.J.; van Winden, W.A.; van Gulik, W.M.; Heijnen, J.J. Revisiting the ^{13}C-label distribution of the non-oxidative branch of the pentose phosphate pathway based upon kinetic and genetic evidence. *FEBS J.* **2005**, *272*, 4970–4982. [CrossRef]
20. Ebenhöh, O.; Spelberg, S. The importance of the photosynthetic Gibbs effect in the elucidation of the Calvin–Benson–Bassham cycle. *Biochem. Soc. Trans.* **2018**, *46*, 131. [CrossRef]
21. Williams, J.F.; MacLeod, J.K. The metabolic significance of octulose phosphates in the photosynthetic carbon reduction cycle in spinach. *Photosynth. Res.* **2006**, *90*, 125–148. [CrossRef]
22. Flanigan, I.L.; MacLeod, J.K.; Williams, J.F. A re-investigation of the path of carbon in photosynthesis utilizing GC/MS methodology. Unequivocal verification of the participation of octulose phosphates in the pathway. *Photosynth. Res.* **2006**, *90*, 149–159. [CrossRef]
23. Horecker, B.L.; Paoletti, F.; Williams, J.F. Occurrence and significance of octulose phosphates in liver. *Ann. N. Y. Acad. Sci.* **1982**, *378*, 215–224. [CrossRef]
24. Williams, J.F.; Arora, K.K.; Longenecker, J.P. The pentose pathway: A random harvest: Impediments which oppose acceptance of the classical (F-type) pentose cycle for liver, some neoplasms and photosynthetic tissue. The case for the L-type pentose pathway. *Int. J. Biochem.* **1987**, *19*, 749–817. [CrossRef]
25. Preiser, A.L.; Banerjee, A.; Weise, S.E.; Renna, L.; Brandizzi, F.; Sharkey, T.D. Phosphoglucoisomerase Is an Important regulatory enzyme in partitioning carbon out of the Calvin-Benson cycle. *Front. Plant Sci.* **2020**, *11*, 1967. [CrossRef] [PubMed]

26. Li, J.; Weraduwage, S.M.; Preiser, A.L.; Tietz, S.; Weise, S.E.; Strand, D.D.; Froehlich, J.E.; Kramer, D.M.; Hu, J.; Sharkey, T.D. A cytosolic bypass and G6P shunt in plants lacking peroxisomal hydroxypyruvate reductase. *Plant Physiol.* **2019**, *180*, 783–792. [CrossRef] [PubMed]
27. Botha, F.C.; Small, J.G. Comparison of the activities and some properties of pyrophosphate and ATP dependent fructose-6-phosphate 1-phosphotransferases of *Phaseolus vulgaris* seeds. *Plant Physiol.* **1987**, *83*, 772–777. [CrossRef]
28. Harrison, E.P.; Willingham, N.M.; Lloyd, J.C.; Raines, C.A. Reduced sedoheptulose-1,7-bisphosphatase levels in transgenic tobacco lead to decreased photosynthetic capacity and altered carbohydrate accumulation. *Planta* **1998**, *204*, 27–36. [CrossRef]
29. Raines, C.A.; Lloyd, J.C.; Dyer, T.A. New insights into the structure and function of sedoheptulose-1,7-bisphosphatase; an important but neglected Calvin cycle enzyme. *J. Exp. Bot.* **1999**, *50*, 1–8.
30. Raines, C.A.; Harrison, E.P.; Ölçer, H.; Lloyd, J.C. Investigating the role of the thiol-regulated enzyme sedoheptulose-1,7-bisphosphatase in the control of photosynthesis. *Physiol. Plant.* **2000**, *110*, 303–308. [CrossRef]
31. Ölçer, H.; Lloyd, J.C.; Raines, C.A. Photosynthetic capacity is differentially affected by reductions in sedoheptulose-1,7-bisphosphatase activity during leaf development in transgenic tobacco plants. *Plant Physiol.* **2001**, *125*, 982–989. [CrossRef]
32. Simkin, A.J.; Lopez-Calcagno, P.E.; Davey, P.A.; Headland, L.R.; Lawson, T.; Timm, S.; Bauwe, H.; Raines, C.A. Simultaneous stimulation of sedoheptulose 1,7-bisphosphatase, fructose 1,6-bisphophate aldolase and the photorespiratory glycine decarboxylase-H protein increases CO (2) assimilation, vegetative biomass and seed yield in Arabidopsis. *Plant Biotechnol. J.* **2017**, *15*, 805–816. [CrossRef] [PubMed]
33. Sharkey, T.D.; Preiser, A.L.; Weraduwage, S.M.; Gog, L. Source of ^{12}C in Calvin-Benson cycle intermediates and isoprene emitted from plant leaves fed with $^{13}CO_2$. *Biochem. J.* **2020**, *477*, 3237–3252. [CrossRef]
34. Scheibe, R. Light/dark modulation: Regulation of chloroplast metabolism in a new light. *Bot. Acta* **1990**, *103*, 327–334. [CrossRef]
35. Preiser, A.L.; Fisher, N.; Banerjee, A.; Sharkey, T.D. Plastidic glucose-6-phosphate dehydrogenases are regulated to maintain activity in the light. *Biochem. J.* **2019**, *476*, 1539–1551. [CrossRef]
36. Caillau, M.; Quick, W.P. New insights into plant transaldolase. *Plant J.* **2005**, *43*, 1–16. [CrossRef]
37. Schnarrenberger, C.; Flechner, A.; Martin, W. Enzymatic evidence for a complete oxidative pentose phosphate pathway in chloroplasts and an incomplete pathway in the cytosol of spinach leaves. *Plant Physiol.* **1995**, *108*, 609–614. [CrossRef] [PubMed]
38. Wakao, S.; Benning, C. Genome-wide analysis of glucose-6-phosphate dehydrogenases in Arabidopsis. *Plant J.* **2005**, *41*, 243–256. [CrossRef] [PubMed]
39. Scheibe, R.; Geissler, A.; Fickenscher, K. Chloroplast glucose-6-phosphate dehydrogenase: K_m shift upon light modulation and reduction. *Arch. Biochem. Biophys.* **1989**, *274*, 290–297. [CrossRef]
40. Wenderoth, I.; Scheibe, R.; von Schaewen, A. Identification of the cysteine residues involved in redox modification of plant plastidic glucose-6-phosphate dehydrogenase. *J. Biol. Chem.* **1997**, *272*, 26985–26990. [CrossRef]
41. Sharkey, T.D.; Weise, S.E. The glucose 6-phosphate shunt around the Calvin-Benson cycle. *J. Exp. Bot.* **2016**, *67*, 4067–4077. [CrossRef]
42. Gerhardt, R.; Stitt, M.; Heldt, H.W. Subcellular metabolite levels in spinach leaves. Regulation of sucrose synthesis during diurnal alterations in photosynthetic partitioning. *Plant Physiol.* **1987**, *83*, 399–407. [CrossRef]
43. Sharkey, T.D.; Vassey, T.L. Low oxygen inhibition of photosynthesis is caused by inhibition of starch synthesis. *Plant Physiol.* **1989**, *90*, 385–387. [CrossRef] [PubMed]
44. Weise, S.E.; Schrader, S.M.; Kleinbeck, K.R.; Sharkey, T.D. Carbon balance and circadian regulation of hydrolytic and phosphorolytic breakdown of transitory starch. *Plant Physiol.* **2006**, *141*, 879–886. [CrossRef] [PubMed]
45. Szecowka, M.; Heise, R.; Tohge, T.; Nunes-Nesi, A.; Vosloh, D.; Huege, J.; Feil, R.; Lunn, J.; Nikoloski, Z.; Stitt, M.; et al. Metabolic fluxes in an illuminated *Arabidopsis* rosette. *Plant Cell Online* **2013**, *25*, 694–714. [CrossRef]
46. Xu, Y.; Fu, X.; Sharkey, T.D.; Shachar-Hill, Y.; Walker, B. The metabolic origins of non-photorespiratory CO_2 release during photosynthesis: A metabolic flux analysis. *Plant Physiol.* **2021**. [CrossRef] [PubMed]
47. McVetty, P.B.E.; Canvin, D.T. Inhibition of photosynthesis by low oxygen concentrations. *Can. J. Bot.* **1981**, *59*, 721–725. [CrossRef]
48. Mahon, J.D.; Fock, H.; Canvin, D.T. Changes in specific radioactivities of sunflower leaf metabolites during photosynthesis in $^{14}CO_2$ and $^{12}CO_2$ at normal and low oxygen. *Planta* **1974**, *120*, 125–134. [CrossRef]
49. Mahon, J.D.; Fock, H.; Canvin, D.T. Changes in specific radioactivity of sunflower leaf metabolites during photosynthesis in $^{14}CO_2$ and $^{14}CO_2$ at three concentrations of CO_2. *Planta* **1974**, *120*, 245–254. [CrossRef]
50. Delwiche, C.F.; Sharkey, T.D. Rapid appearance of ^{13}C in biogenic isoprene when $^{13}CO_2$ is fed to intact leaves. *Plant Cell Environ.* **1993**, *16*, 587–591. [CrossRef]
51. Hasunuma, T.; Harada, K.; Miyazawa, S.-I.; Kondo, A.; Fukusaki, E.; Miyake, C. Metabolic turnover analysis by a combination of in vivo ^{13}C-labelling from $^{13}CO_2$ and metabolic profiling with CE-MS/MS reveals rate-limiting steps of the C_3 photosynthetic pathway in *Nicotiana tabacum* leaves. *J. Exp. Bot.* **2010**, *61*, 1041–1051. [CrossRef]
52. Ma, F.; Jazmin, L.J.; Young, J.D.; Allen, D.K. Isotopically nonstationary ^{13}C flux analysis of changes in *Arabidopsis thaliana* leaf metabolism due to high light acclimation. *Proc. Natl. Acad. Sci. USA* **2014**, *111*, 16967–16972. [CrossRef]
53. Granot, D.; Kelly, G.; Stein, O.; David-Schwartz, R. Substantial roles of hexokinase and fructokinase in the effects of sugars on plant physiology and development. *J. Exp. Bot.* **2014**, *65*, 809–819. [CrossRef]

54. Weise, S.E.; Liu, T.; Childs, K.L.; Preiser, A.L.; Katulski, H.M.; Perrin-Porzondek, C.; Sharkey, T.D. Transcriptional regulation of the glucose-6-phosphate/phosphate translocator 2 is related to carbon exchange across the chloroplast envelope. *Front. Plant Sci.* **2019**, *10*, 827. [CrossRef]
55. Kammerer, B.; Fischer, K.; Hilpert, B.; Schubert, S.; Gutensohn, M.; Weber, A.; Flügge, U.I. Molecular characterization of a carbon transporter in plastids from heterotrophic tissues: The glucose 6-phosphate phosphate antiporter. *Plant Cell* **1998**, *10*, 105–117. [CrossRef]
56. Eicks, M.; Maurino, V.; Knappe, S.; Flügge, U.-I.; Fischer, K. The plastidic pentose phosphate translocator represents a link between the cytosolic and the plastidic pentose phosphate pathways in plants. *Plant Physiol.* **2002**, *128*, 512–522. [CrossRef]
57. Farquhar, G.D.; von Caemmerer, S.; Berry, J.A. A biochemical model of photosynthetic CO_2 assimilation in leaves of C_3 species. *Planta* **1980**, *149*, 78–90. [CrossRef]
58. von Caemmerer, S.; Farquhar, G.D. Some relationships between the biochemistry of photosynthesis and the gas exchange of leaves. *Planta* **1981**, *153*, 376–387. [CrossRef] [PubMed]
59. Laisk, A. *Kinetics of Photosynthesis and Photorespiration of C3 Plants (in Russian)*; Nauka: Moscow, Russia, 1977; p. 195.
60. Brooks, A.; Farquhar, G.D. Effects of temperature on the O_2/CO_2 specificity of ribulose-1,5-bisphosphate carboxylase/oxygenase and the rate of respiration in the light. Estimates from gas exchange measurements on spinach. *Planta* **1985**, *165*, 397–406. [CrossRef] [PubMed]
61. Yin, X.; Sun, Z.; Struik, P.C.; Gu, J. Evaluating a new method to estimate the rate of leaf respiration in the light by analysis of combined gas exchange and chlorophyll fluorescence measurements. *J. Exp. Bot.* **2011**, *62*, 3489–3499. [CrossRef] [PubMed]
62. Yin, X.; Struik, P.C.; Romero, P.; Harbinson, J.; Evers, J.B.; Van Der Putten, P.E.L.; Vos, J.A.N. Using combined measurements of gas exchange and chlorophyll fluorescence to estimate parameters of a biochemical C3 photosynthesis model: A critical appraisal and a new integrated approach applied to leaves in a wheat (Triticum aestivum) canopy. *Plant Cell Environ.* **2009**, *32*, 448–464. [CrossRef] [PubMed]
63. Loreto, F.; Velikova, V.; Di Marco, G. Respiration in the light measured by $^{12}CO_2$ emission in $^{13}CO_2$ atmosphere in maize leaves. *Aust. J. Plant Physiol.* **2001**, *28*, 1103–1108.
64. Tcherkez, G.; Cornic, G.; Bligny, R.; Gout, E.; Ghashghaie, J. In vivo respiratory metabolism of illuminated leaves. *Plant Physiol.* **2005**, *138*, 1596. [CrossRef] [PubMed]
65. Okazaki, Y.; Shimojima, M.; Sawada, Y.; Toyooka, K.; Narisawa, T.; Mochida, K.; Tanaka, H.; Matsuda, F.; Hirai, A.; Hirai, M.Y.; et al. A chloroplastic UDP-glucose pyrophosphorylase from *Arabidopsis* is the committed enzyme for the first step of sulfolipid biosynthesis. *Plant Cell* **2009**, *21*, 892–909. [CrossRef]
66. Strand, D.D.; Livingston, A.K.; Satoh-Cruz, M.; Froehlich, J.E.; Maurino, V.G.; Kramer, D.M. Activation of cyclic electron flow by hydrogen peroxide in vivo. *Proc. Natl. Acad. Sci. USA* **2015**, *112*, 5539–5544. [CrossRef]
67. Cousins, A.; Walker, B.; Pracharoenwattana, I.; Smith, S.; Badger, M. Peroxisomal hydroxypyruvate reductase is not essential for photorespiration in Arabidopsis but its absence causes an increase in the stoichiometry of photorespiratory CO_2 release. *Photosynth. Res.* **2011**, *108*, 91–100. [CrossRef]
68. Li, J.; Tietz, S.; Cruz, J.A.; Strand, D.D.; Xu, Y.; Chen, J.; Kramer, D.M.; Hu, J. Photometric screens identified Arabidopsis peroxisome proteins that impact photosynthesis under dynamic light conditions. *Plant J.* **2018**, *97*, 460–474. [CrossRef] [PubMed]
69. Flügel, F.; Timm, S.; Arrivault, S.; Florian, A.; Stitt, M.; Fernie, A.R.; Bauwe, H. The photorespiratory metabolite 2-phosphoglycolate regulates photosynthesis and starch accumulation in Arabidopsis. *Plant Cell* **2017**, *29*, 2537–2551. [CrossRef]
70. Anderson, L.E. Chloroplast and cytoplasmic enzymes II. Pea leaf triose phosphate isomerases. *Biochim. Biophys. Acta (BBA) Enzymol.* **1971**, *235*, 237–244. [CrossRef]
71. Sun, Y.; Geng, Q.; Du, Y.; Yang, X.; Zhai, H. Induction of cyclic electron flow around photosystem I during heat stress in grape leaves. *Plant Sci. Int. J. Exp. Plant Biol.* **2017**, *256*, 65–71. [CrossRef]
72. Agrawal, D.; Allakhverdiev, S.I.; Jajoo, A. Cyclic electron flow plays an important role in protection of spinach leaves under high temperature stress. *Russ. J. Plant Physiol.* **2016**, *63*, 210–215. [CrossRef]
73. Zhang, R.; Sharkey, T.D. Photosynthetic electron transport and proton flux under moderate heat stress. *Photosynth. Res.* **2009**, *100*, 29–43. [CrossRef]
74. Wang, P.; Duan, W.; Takabayashi, A.; Endo, T.; Shikanai, T.; Ye, J.Y.; Mi, H. Chloroplastic NAD(P)H dehydrogenase in tobacco leaves functions in alleviation of oxidative damage caused by temperature stress. *Plant Physiol.* **2006**, *141*, 465–474. [CrossRef]
75. Bukhov, N.G.; Wiese, C.; Neimanis, S.; Heber, U. Heat sensitivity of chloroplasts and leaves: Leakage of protons from thylakoids and reversible activation of cyclic electron transport. *Photosynth. Res.* **1999**, *59*, 81–93. [CrossRef]
76. Pastenes, C.; Horton, P. Effect of high temperature on photosynthesis in beans 1. Oxygen evolution and chlorophyll fluorescence. *Plant Physiol.* **1996**, *112*, 1245–1251. [CrossRef] [PubMed]
77. Havaux, M. Short-term responses of photosystem I to heat stress—Induction of a PS II-independent electron transport through PS I fed by stromal components. *Photosynth. Res.* **1996**, *47*, 85–97. [CrossRef]

MDPI
St. Alban-Anlage 66
4052 Basel
Switzerland
www.mdpi.com

Cells Editorial Office
E-mail: cells@mdpi.com
www.mdpi.com/journal/cells

www.ingramcontent.com/pod-product-compliance
Lightning Source LLC
Chambersburg PA
CBHW061826210926
51850CB00006B/816

* 9 7 8 3 0 3 6 5 9 1 4 5 2 *